Grid Generation Methods

Second Edition

Scientific Computation

For other titles published in this series, go to
www.springer.com/series/718

Vladimir D. Liseikin

Grid Generation Methods

Second Edition

Springer

Professor Vladimir D. Liseikin
Russian Academy of Sciences
Institute of Computational Technologies
Pr. Lavrentjeva 6
Novosibirsk 630090
Russia
liseikin@ict.nsc.ru, ivd@ict.nsc.ru

and

Novosibirsk State University
Pirogova St. 2
Novosibirsk 630090
Russia

ISSN 1434-8322
ISBN 978-94-007-3078-6 e-ISBN 978-90-481-2912-6
DOI 10.1007/978-90-481-2912-6
Springer Dordrecht Heidelberg London New York

Cover design: eStudio Calamar S.L.

Printed on acid-free paper

Springer is part of Springer Science+Business Media (www.springer.com)

Preface to the Second Edition

This second edition is significantly expanded by new material that discusses recent advances in grid generation technology based on the numerical solution of Beltrami and diffusion equations in control metrics. It gives a more detailed and practice-oriented description of the control metrics for providing the generation of adaptive, field-aligned, and balanced numerical grids. Some numerical algorithms are described for generating surface and domain grids. Applications of the algorithms to the generation of numerical grids with individual and balanced properties are demonstrated.

Grid generation codes represent an indispensable tool for solving field problems in nearly all areas of applied mathematics and computational physics. The use of these grid codes significantly enhances the productivity and reliability of the numerical analysis of problems with complex geometry and complicated solutions. The science of grid generation is still growing fast; new developments are continually occurring in the fields of grid methods, codes, and practical applications. Therefore there exists an evident need of students, researchers, and practitioners in applied mathematics for new books which coherently complement the existing ones with a description of new developments in grid methods, grid codes, and the concomitant areas of grid technology.

The objective of this book is to give a clear, comprehensive, and easily learned description of all essential methods of grid generation technology for two major classes of grids: structured and unstructured. These classes rely on two somewhat opposite basic concepts. The basic concept of the former class is adherence to order and organization, while the latter is prone to the absence of any restrictions.

The present monograph discusses the current state of the art in methods of grid generation and describes new directions and new techniques aimed at the enhancement of the efficiency and productivity of the grid process. The emphasis is put on mathematical formulations, explanations, and examples of various aspects of grid generation.

Special attention is paid to a review of those promising approaches and methods which have been developed recently and/or have not been sufficiently covered in other monographs. In particular, the book includes a stretching method adjusted to the numerical solution of singularly perturbed equations having large scale solution variations, e.g. those modeling high-Reynolds-number flows. A number of functionals related to conformality, orthogonality, energy, and alignment are described. The book includes differential and variational techniques for generating uniform, conformal, and harmonic coordinate transformations on hypersurfaces for the development of a comprehensive approach to the construction of both fixed and adaptive grids in the interior and on the boundary of domains in a unified manner. The monograph is also concerned with the description of all essential grid quality measures such as skewness, curvature, torsion, angle and length values, and conformality. Emphasis is given to a clear style and new angles of consideration where it is not intended to include unnecessary abstractions.

The major area of attention of this book is structured grid techniques. However, the author has also included an elementary introduction to basic unstructured approaches to grid generation. A more detailed description of unstructured grid techniques can be found in *Computational Grids: Adaptation and Solution Strategies* by Carey (1997), *Delaunay Triangulation and Meshing* by George and Borouchaki (1998), and *Mesh Generation Application to Finite Elements* by Frey and George (2008).

Since grid technology has widespread application to nearly all field problems, this monograph may have some interest for a broad range of readers, including teachers, students, researchers, and practitioners in applied mathematics, mechanics, and physics.

The first chapter gives a general introduction to the subject of grids. There are two fundamental forms of mesh: structured and unstructured. Structured grids are commonly obtained by mapping a standard grid into the physical region with a transformation from a reference computational domain. The most popular structured grids are coordinate grids. The cells of such grids are curvilinear hexahedrons, and the identification of neighboring points is done by incrementing coordinate indices. Unstructured grids are composed of cells of arbitrary shape and, therefore, require the generation of a connectivity table to allow the identification of neighbors. The chapter outlines structured, unstructured, and composite grids and delineates some basic approaches to their generation. It also includes a description of various types of grid topology and touches on certain issues of big grid codes.

Chapter 2 deals with some relations, necessary only for grid generation, connected with and derived from the metric tensors of coordinate transformations. As an example of an application of these relations, the chapter presents a technique aimed at obtaining conservation-law equations in new fixed or time-dependent coordinates. In the procedures described, the deduction of the expressions for the transformed equations is based only on the formula for differentiation of the Jacobian.

Very important issues of grid generation, connected with a description of grid quality measures in forms suitable for formulating grid techniques and efficiently analyzing the necessary mesh properties, are discussed in Chap. 3. The definitions of the grid quality measures are based on the metric tensors and on the relations between the metric elements considered in Chap. 2. Special attention is paid to the invariants of the metric tensors, which are the basic elements for the definition of many important grid quality measures. Clear algebraic and geometric interpretations of the invariants are presented.

Equations with large variations of the solution, such as those modeling high-Reynolds-number flows, are one of the most important areas of the application of adaptive grids and of demonstration of the efficiency of grid technology. The numerical analysis of such equations on special grids obtained by a stretching method has a definite advantage in comparison with the classical analytic expansion method in that it requires only a minimum knowledge of the qualitative properties of the physical solution. The fourth chapter is concerned with the description of such stretching methods aimed at the numerical analysis of equations with singularities.

The first part of Chap. 4 acquaints the reader with various types of singularity arising in solutions to equations with a small parameter affecting the higher derivatives. The solutions of these equations undergo large variations in very small zones, called boundary or interior layers. The chapter gives a concise description of the qualitative properties of solutions in boundary and interior layers and an identification of the invariants governing the location and structure of these layers. Besides the well-known exponential layers, three types of power layer which are common to bisingular problems having com plementary singularities arising from reduced equations, are described. Such equations are widespread in applications, for example, in gas dynamics. Simple examples of one- and two-dimensional problems which realize different types of boundary and interior layers are demonstrated, in particular, the exotic case where the interior layer approaches infinitely close to the boundary as the parameter tends to zero, so that the interior layer turns out to be a boundary layer of the reduced problem. This interior layer exhibits one more phenomenon: it is composed of layers of two basic types, exponential on one side of the center of the layer and power-type on the other side.

The second part of Chap. 4 describes a stretching method based on the application of special nonuniform stretching coordinates in regions of large variation of the solution. The use of stretching coordinates is extremly effective for the numerical solution of problems with boundary and interior layers. The method requires only a very basic knowledge of the qualitative properties of the physical solution in the layers. The specification of the stretching functions is given for each type of basic singularity. The functions are defined in such a way that the singularities are automatically smoothed with respect to the new stretching coordinates. The chapter ends with the description of a procedure to generate intermediate coordinate transformations which are suitable for smoothing both exponential and power layers. The grids derived with such stretching coordinates are often themselves well adapted to the expected physical features. Therefore, they make it easier to provide dynamic adaptation by taking part of the adaptive burden on themselves.

The simplest and fastest technique of grid generation is the algebraic method based on transfinite interpolation. Chapter 5 describes formulas for general unidirectional transfinite interpolations. Multidirectional interpolation is defined by Boolean summation of unidirectional interpolations. The grid lines across block interfaces can be made completely continuous by using Lagrange interpolation or to have slope continuity by using the Hermite technique. Of central importance in transfinite interpolation are the blending functions (positive univariate quantities depending only on one chosen coordinate) which provide the matching of the grid lines at the boundary and interior surfaces. Detailed relations between the blending functions and approaches to their specification are discussed in this chapter. Examples of various types of blending function are reviewed, in particular, the functions defined through the basic stretching coordinate transformations for singular layers described in Chap. 4. These transformations are dependent on a small parameter so that the resulting grid automatically adjusts to the respective physical parameter, e.g. viscosity, Reynolds number, or shell thickness, in practical applications. The chapter ends with a description of a procedure for generating triangular, tetrahedral, or prismatic grids through the method of transfinite interpolation.

Chapter 6 is concerned with grid generation techniques based on the numerical solution of systems of partial differential equations. Generation of grids from these systems of equations is largely based on the numerical solution of elliptic, hyperbolic, and parabolic equations for the coordinates of grid lines which are specified on the boundary segments. The elliptic and parabolic systems reviewed in the chapter provide grid generation within blocks with specified boundary point distributions. These systems are also used to smooth algebraic, hyperbolic, and unstructured grids. A very important role is currently played in grid codes by a system of Poisson equations defined as a sum of Laplace equations and control functions. This system was originally considered by Godunov and Prokopov and further generalized, developed, and implemented for practical applications by Thompson, Thames, Mastin, and others. The chapter describes the properties of the Poisson system and specifies expressions for the control functions required to construct nearly orthogonal coordinates at the boundaries. Hyperbolic systems are useful when an outer boundary is free of specification. The control of the grid spacing in the hyperbolic method is largely performed through the specification of volume distribution functions. Special hyperbolic and elliptic systems are presented for generating orthogonal and nearly orthogonal coordinate lines, in particular, those proposed by Ryskin and Leal. The chapter also reviews some parabolic and high-order systems for the generation of structured grids.

Effective adaptation is one of the most important requirements put on grid technology. The basic aim of adaptation is to increase the accuracy and productivity of the numerical solution of partial differential equations through a redistribution of the grid points and refinement of the grid cells. Chapter 7 describes some basic techniques of dynamic adaptation. The chapter starts with the equidistribution method, first suggested in difference form by Boor and further applied and extended by Dwyer, Kee, Sanders, Yanenko, Liseikin, Danaev, and others. In this method, the lengths of the cell edges are defined through a weight function modeling some measure of the solution error. An interesting fact about the uniform convergence of the numerical solution of some singularly perturbed equations on a uniform grid is noted and explained. The chapter also describes adaptation in the elliptic method, performed by the control functions. Features and effects of the control functions are discussed and the specification of the control functions used in practical applications is presented. Approaches to the generation of moving grids for the numerical solution of nonstationary problems are also reviewed. The most important feature of a structured grid is the Jacobian of the coordinate transformation from which the grid is derived. A method based on the specification of the values of the Jacobian to keep it positive, developed by Liao, is presented.

Chapter 8 reviews the developments of variational methods applied to grid generation. Variational grid generation relies on functionals related to grid quality: smoothness, orthogonality, regularity, aspect ratio, adaptivity, etc. By the minimization of a combination of these functionals, a user can define a compromise grid with the desired properties. The chapter discusses the variational approach of error minimization introduced by Morrison and further developed by Babuŝka, Tihonov, Yanenko, Liseikin, and others. Functionals for generating uniform, conformal, quasiconformal, orthogonal, and adaptive grids, suggested by Brackbill, Saltzman, Winslow, Godunov, Prokopov, Yanenko, Liseikin, Liao, Steinberg, Knupp, Roache, and others are also presented. A variational approach using functionals dependent on invariants of the metric tensor is also considered. The chapter discusses a new variational approach for generating harmonic maps through the minimization of energy functionals, which was suggested by Dvinsky. Several versions of the functionals from which harmonic maps can be derived are identified.

Methods developed for the generation of grids on curves and surfaces are discussed in Chap. 9. The chapter describes the development and application of hyperbolic, elliptic, and variational techniques for the generation of grids on parametrically defined curves and surfaces. The differential approaches are based on the Beltrami equations proposed by Warsi and Thomas, while the variational methods rely on functionals of surface grid quality measures. The chapter includes also a description of the approach to constructing conformal mappings on surfaces developed by Khamaysen and Mastin.

Chapter 10 is devoted to the author variant of the implementation of an idea of Eiseman for generating adaptive grids by projecting quasiuniform grids from monitor hypersurfaces. The monitor hypersurface is formed as a surface of the values of some vector function over the physical geometry. The vector function can be a solution to the problem of interest, a combination of its components or derivatives, or any other variable quantity that suitably monitors the way that the behavior of the solution influences the efficiency of the calculations. For the purpose of commonality a general approach based on differential and variational methods for the generation of quasiuniform grids on arbitrary hypersurfaces is considered. The variational method of generating quasiuniform grids, developed by the author, is grounded on the minimization of a generalized functional of grid smoothness on hypersurfaces, which was introduced for domains by Brackbill and Saltzman. The chapter also includes an expansion of the method by introducing more general control metrics in the physical geometry. The control metrics provide efficient and straightforwardly defined conditions for various types of grid adaptation, particularly grid clustering according to given function values and/or gradients, grid alignment with given vector fields, and combinations thereof. Using this approach, one can generate both adaptive and fixed grids in a unified manner, in arbitrary domains and on their boundaries. This allows code designers to merge the two tasks of surface grid generation and volume grid generation into one task while developing a comprehensive grid generation code.

The subject of unstructured grid generation is discussed in Chap. 11. Unstructured grids may be composed of cells of arbitrary shape, but they are generally composed of triangles and tetrahedrons. Tetrahedral grid methods described in the chapter include Delaunay procedures and the advancing-front method. The Delaunay approach connects neighboring points (of some previously defined set of nodes in the region) to form tetrahedral cells in such a way that the sphere through the vertices of any tetrahedron does not contain any other points. In the advancing-front method, the grid is generated by building cells one at a time, marching from the boundary into the volume by successively connecting new points to points on the front until all the unmeshed space is filled with grid cells.

The book ends with a list of references.

Acknowledgement. The author is greatly thankful to G. Liseikin who prepared the text of the manuscript in LaTeX code. Thanks go as well to G. Lukas for correcting the author's English.

The author is very grateful for the helpful suggestions in the comments for Chaps. 1, 3, and 11 made by E. Ivanov, a leading expert in up-to-date codes, grid quality measures, and methods for unstructured grid generation. The author is also greatly obliged to the researchers who responded to his requests and sent him their papers, namely, T.J. Baker, D.A. Field, E. Ivanov, P. Knupp, G. Liao, M.S. Shephard, N.P. Weatherill, and P.P. Zegeling.

The work over the book was supported in part by an Integrated Grant of the Siberian Branch of the Russian Academy of Sciences (2009–2011): Award No 94, and by the Russian Foundation for Basic Research (RFBR): Award 09-01-12023. Specifically, the efforts related to computing figures of grids made by A. Kharitonchick, A. Kofanov, Yu. Likhanova, and I. Vaseva, whom the author thanks very much, were remunerated by payments from these grants.

Novosibirsk *Vladimir D. Liseikin*

Contents

1

General Considerations

1.1 Introduction

An important element of the numerical solution of partial differential equations by finite-element or finite-difference methods on general regions is a grid which represents the physical domain in a discrete form. In fact, the grid is a preprocessing tool or a foundation on which physical, continuous quantities are described by discrete functions and on which the differential equations are approximated by algebraic relations for discrete values that are then numerically analyzed by the application of computational codes. The grid technique also has the capacity, based on an appropriate distribution of the grid points, to enhance the computational efficiency of the numerical solution of complex problems.

The efficiency of a numerical study of a boundary value problem is estimated from the accuracy of the computed solution and from the cost and time of the computation.

The accuracy of the numerical solution in the physical domain depends on both the error of the solution at the grid points and the error of interpolation. Commonly, the error of the numerical computation at the grid points arises from several distinct sources. First, mathematical models do not represent physical phenomena with absolute accuracy. Second, an error arises at the stage of the numerical approximation of the mathematical model. Third, the error is influenced by the size and shape of the grid cells. Fourth, an error is contributed by the computation of the discrete physical quantities satisfying the equations of the numerical approximation. And fifth, an error in the solution is caused by the inaccuracy of the process of interpolation of the discrete solution. Of course, the accurate evaluation of the errors due to there sources remains a formidable task. It is apparent, however, that the quantitative and qualitative properties of the grid play a significant role in controlling the influence of the third and fifth sources of the error in the numerical analysis of physical problems.

V.D. Liseikin, *Grid Generation Methods*, Scientific Computation,
DOI 10.1007/978-90-481-2912-6_1, © Springer Science+Business Media B.V. 2010

Another important characteristic of a numerical algorithm that influences its efficiency is the cost of the operation of obtaining the solution. From this point of view, the process of generating a sophisticated grid may increase the computational costs of the numerical solution and encumber the computer tools with the requirement of additional memory. On the other hand, there may be a significant profit in accuracy which allows one to use a smaller number of grid points. Any estimation of the contributions of these opposing factors can help in choosing an appropriate grid. In any case, since grid generation is an important component of numerical modeling, research in this field is aimed at creating techniques which are not too costly but which give a significant improvement in the accuracy of the solution. The utilization of these techniques provides one with the real opportunities to enhance the efficiency of the numerical solution of complex problems. Thus grid generation helps to satisfy the constant demand for enhancement of the efficiency of the numerical analysis of practical problems.

The first efforts aimed at the development of grid techniques were undertaken in the 1960s. Now, a significant number of advanced methods have been created: algebraic, elliptic, hyperbolic, parabolic, variational, Delaunay, advancing-front, etc. The development of these methods has reached a stage where calculations in fairly complicated domains and on surfaces that arise while analyzing multidimensional problems are possible. Because of its successful development, the field of numerical grid generation has already formed a separate mathematical discipline with its own methodology, approaches, and technology.

At the end of the 1980s there started a new stage in the development of grid generation technique. It is characterized by the creation of comprehensive, multipurpose, three-dimensional grid generation codes which are aimed at providing a uniform environment for the construction of grids in arbitrary multidimensional geometries.

The current chapter presents a framework for the subject of grid generation. It outlines the most general concepts and techniques, which will be expounded in the following chapters in more detail.

1.2 General Concepts Related to Grids

There are two general notions of a grid in an n-dimensional bounded domain or on a surface. One of these considers the grid as a set of algorithmically specified points of the domain or the surface. The points are called the grid nodes. The second considers the grid as an algorithmically described collection of standard n-dimensional volumes covering the necessary area of the domain or surface. The standard volumes are referred to as the grid cells. The cells are bounded curvilinear volumes, whose boundaries are divided into a few segments which are $(n-1)$-dimensional cells. Therefore they can be formulated successively from one dimension to higher dimensions. The boundary points of the one-dimensional cells are called the cell vertices. These vertices are the grid nodes. Thus the grid nodes are consistent with the grid cells in that they coincide with the cell vertices.

This section discusses some general concepts related to grids and grid cells.

1.2.1 Grid Cells

For cells in an n-dimensional domain or surface, there are commonly used n-dimensional volumes of simple standard shapes (see Fig. 1.1 for $n = 1, 2, 3$).

In one dimension the cell is a closed line or segment, whose boundary is composed of two points referred to as the cell vertices.

A general two-dimensional cell is a two-dimensional simply connected domain, whose boundary is divided into a finite number of one-dimensional cells referred to as the edges of the cell. Commonly, the cells of two-dimensional domains or surfaces are constructed in the form of triangles or quadrilaterals. The boundary of a triangular cell is composed of three segments, while the boundary of a quadrilateral is represented by four segments. These segments are the one-dimensional grid cells.

By a general three-dimensional cell there is meant a simply connected three-dimensional polyhedron whose boundary is partitioned into a finite number of two-dimensional cells called its faces. In practical applications, three-dimensional cells typically have the shape of tetrahedrons or hexahedrons. The boundary of a tetrahedral cell is composed of four triangular cells, while a hexahedron is bounded by six quadrilaterals. Thus a hexahedral cell has six faces, twelve edges, and eight vertices. Some applications also use volumes in the form of prisms as three-dimensional cells. A prism has two triangular and three quadrilateral faces, nine edges, and six vertices.

Commonly, the edges and the faces of the cells are linear. Linear triangles and tetrahedrons are also referred to as two-dimensional simplexes and three-dimensional simplexes, respectively. The notion of the simplex can be formulated for arbitrary dimensions. Namely, by an n-dimensional simplex there is meant a domain of n-dimensional space whose nodes are defined by the equation

$$\boldsymbol{x} = \sum_{i=1}^{N+1} \alpha^i \boldsymbol{x}_i,$$

where $\boldsymbol{x}_i$, $i = 1, \dots, N+1$, are some specified points which are the vertices of the simplex, and α^i, $i = 1, \dots, N+1$, are real numbers satisfying the relations

$$\sum_{i=1}^{N+1} \alpha^i = 1, \quad \alpha^i \geq 0.$$

In this respect a one-dimensional linear cell is the one-dimensional simplex. The boundary of an n-dimensional simplex is composed of $n + 1$ $(n - 1)$-dimensional simplexes.

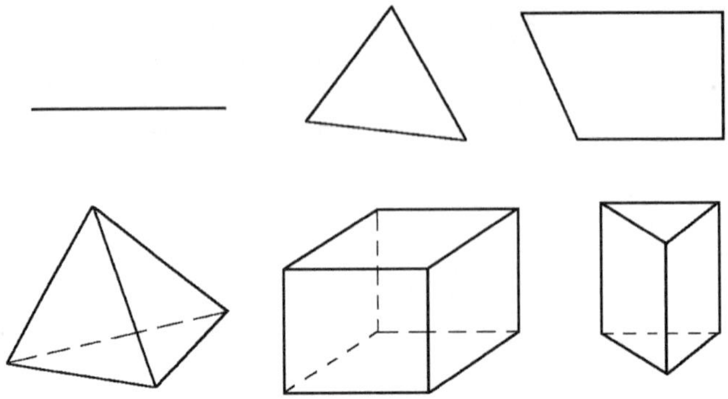

Fig. 1.1. Typical grid cells

The selection of the shapes shown in Fig. 1.1 to represent the standard cells is justified, first, by their geometrical simplicity and, second, because the existing procedures for the numerical simulation of physical problems are largely based on approximations of partial differential equations using these elemental volumes. The specific choice of cell shape depends on the geometry and physics of the particular problem and on the method of solution. In particular, tetrahedrons (triangles in two dimensions) are well suited for finite-element methods, while hexahedrons are commonly used for finite-difference techniques.

Some applications consider curvilinear cells as well. These grid cells are obtained by deformation of ordinary linear segments, triangles, tetrahedrons, squares, cubes, and prisms.

The major advantage of hexahedral cells (quadrilaterals in two dimensions) is that their faces (or edges) may be aligned with the coordinate surfaces (or curves). In contrast, no coordinates can be aligned with tetrahedral meshes. However, strictly hexahedral meshes may be ineffective near boundaries with sharp corners.

Prismatic cells are generally placed near boundary surfaces which have previously been triangulated. The surface triangular cells serve as faces of prisms, which are grown out from these triangles. Prismatic cells are efficient for treating boundary layers, since they can be constructed with a high aspect ratio in order to resolve the layers, but without small angles, as would be the case for tetrahedral cells.

Triangular cells are the simplest two-dimensional elements and can be produced from quadrilateral cells by constructing interior edges. Analogously, tetrahedral cells are the simplest three-dimensional elements and can be derived from hexahedrons and prisms by constructing interior faces. The strength of triangular and tetrahedral cells is in their applicability to virtually any type of domain configuration. The drawback is that the integration of the physical equations becomes a few times more expensive with these cells in comparison with quadrilateral or hexahedral cells.

The vertices of the cells define grid points which approximate the physical domain. Alternatively, the grid points in the domain may have been generated previously by some other process. In this case the construction of the grid cells requires special techniques.

1.2.2 Requirements Imposed on Grids

The grid should discretize the physical domain or surface in such a manner that the computation of the physical quantities is carried out as efficiently as desired. The accuracy, which is one of the components of the efficiency of the computation, is influenced by a number of grid factors, such as grid size, grid topology, cell shape and size, and consistency of the grid with the geometry and with the solution. A very general consideration of these grid factors is given in this subsection.

1.2.2.1 Grid Size and Cell Size

The grid size is indicated by the number of grid points, while the cell size implies the maximum value of the lengths of the cell edges. Grid generation requires techniques which possess the intrinsic ability to increase the number of grid nodes. At the same time the edge lengths of the resulting cells should be reduced in such a manner that they approach zero as the number of nodes tends to infinity.

An instructive example of a grid on the interval $[0,1]$ which does not satisfy the requirement of unlimited reduction of the cell sizes when the number of the nodes is increased is a grid generated by a rule in which the steps are in a geometrical progression:

$$\frac{h_{i+1}}{h_i} = a, \quad a > 0,\ a \neq 1, \tag{1.1}$$

where $h_i = x_{i+1} - x_i$, $i = 0, \dots, N-1$, are the steps of the grid nodes x_i, $i = 0, \dots, N$, with $x_0 = 0$, $x_N = 1$. The grid nodes x_i satisfying (1.1) are computed for arbitrary N by the formula

$$x_i = \frac{a-1}{a^{N-1} - a} \sum_{j=1}^{i} a^j, \quad i = 1, \dots, N,$$

and consequently we obtain

$$h_i = \frac{a^{i+1}(a-1)}{a^{N-1} - a}, \quad i = 0, \dots, N-1.$$

Therefore

$$\lim_{N \to \infty} h_0 = 1 - a \qquad \text{if } 0 < a < 1,$$

$$\lim_{N \to \infty} h_{N-1} = \frac{a-1}{a} \quad \text{if } a > 1,$$

i.e. the left-hand boundary cell of this grid, if $a < 1$, or the right-hand boundary cell, if $a > 1$, does not approach zero even though the number of grid points tends to infinity.

Small cells are necessary to obtain more accurate solutions and to investigate phenomena associated with the physical quantities on small scales, such as transition layers and turbulence. Also, the opportunity to increase the number of grid points and to reduce the size of the cells enables one to study the convergence rate of a numerical code and to improve the accuracy of the solution by multigrid approaches.

1.2.2.2 Grid Organization

There also is a requirement on grids to have some organization of their nodes and cells, which is aimed at facilitating the procedures for formulating and solving the algebraic equations substituted for the differential equations. This organization should identify neighboring points and cells. The grid organization is especially important for that class of finite-difference methods whose procedures for obtaining the algebraic equations consist of substituting differences for derivatives. To a lesser degree, this organization is needed for finite-volume methods because of their inherent compatibility with irregular meshes.

1.2.2.3 Cell and Grid Deformation

The cell deformation characteristics can be formulated as some measures of the departure of the cell from a standard, least deformed one. Such standard triangular and tetrahedral cells are those with edges of equal lengths. The least distorted quadrilaterals and hexahedrons are squares and cubes, respectively. The standard prism is evidently the prism with standard linear faces. Cells with low deformity are preferable from the point of view of simplicity and uniformity of the construction of the algebraic equations approximating the differential equations.

Typically, cell deformation is characterized through the aspect ratio, the angles between the cell edges, and the volume (area in two dimensions) of the cell.

The major requirement for the grid cells is that they must not be folded or degenerate at any points or lines, as demonstrated in Fig. 1.2. Unfolded cells are obtained from standard cells by a one-to-one deformation. Commonly, the value of any grid generation method is judged by its ability to yield unfolded grids in regions with complex geometry.

The grid deformity is also characterized by the rate of the change of the geometrical features of contiguous cells. Grids whose neighboring cells do not change abruptly are referred to as smooth grids.

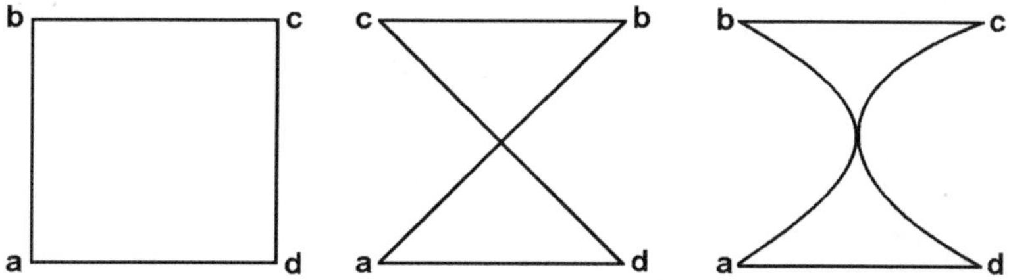

Fig. 1.2. Normal (*left*) and badly deformed (*center, right*) quadrilateral cells

1.2.2.4 Consistency with Geometry

The accuracy of the numerical solution of a partial differential equation and of the interpolation of a discrete function is considerably influenced by the degree of compatibility of the mesh with the geometry of the physical domain. First of all, the grid nodes must adequately approximate the original geometry, that is, the distance between any point of the domain and the nearest grid node must not be too large. Moreover, this distance must approach zero when the grid size tends to infinity. This requirement of adequate geometry approximation by the grid nodes is indispensible for the accurate computation and interpolation of the solution over the whole region.

The second requirement for consistency of the grid with the geometry is concerned with the approximation of the boundary of the physical domain by the grid, i.e. there is to be a sufficient number of nodes which can be considered as the boundary ones, so that a set of edges (in two dimensions) and cell faces (in three dimensions) formed by these nodes models efficiently the boundary. In this case, the boundary conditions may be applied more easily and accurately. If these points lie on the boundary of the domain, then the grid is referred to as a boundary-fitting or boundary-conforming grid.

1.2.2.5 Consistency with Solution

It is evident that distribution of the grid points and the form of the grid cells should be dependent on the features of the physical solution. In particular, it is better to generate the cells in the shape of hexahedrons or prisms in boundary layers. Often, the grid points are aligned with some preferred directions, e.g. streamlines. Furthermore, a nonuniform variation of the solution requires clustering of the grid point in regions of high gradients, so that these areas of the domain have finer resolution. Local grid clustering is needed because the uniform refinement of the entire domain may be very costly for multidimensional computations. It is especially true for problems whose solutions have localized regions of very rapid variation (layers). Without grid clustering in the layers, some important features of the solution can be missed, and the accuracy of the solution can be degraded. Problems with boundary and interior layers occur in many areas of application, for example in fluid dynamics, combustion, solidification, solid mechanics and wave propagation. Typically the

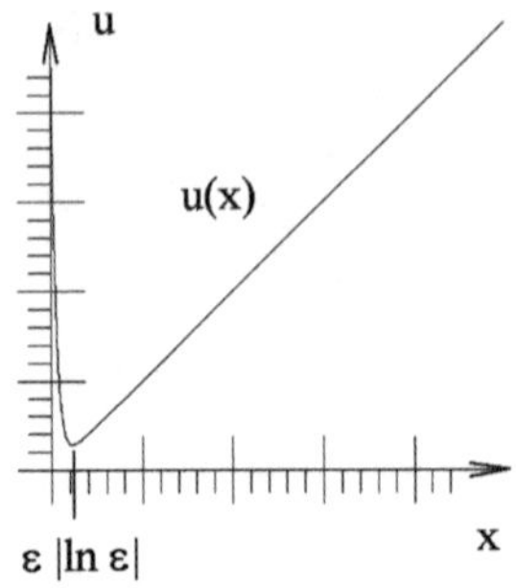

Fig. 1.3. Boundary layer function for $\epsilon = 10^{-2}$

locations where the high resolution is needed are not known beforehand but are found in the process of computation. Consequently, a suitable mesh, tracking the necessary features of the physical quantities, as the solution evolves, is required.

A local grid refinement is accomplished in two ways: (a) by moving a fixed number of grid nodes, with clustering of them in zones where this is necessary, and coarsening outside of these zones, and (b) by inserting new points in the zones of the domain where they are needed. Local grid refinement in zones of large variation of the solution commonly results in the following improvements:

(1) the solution at the grid points is obtained more accurately;
(2) the solution is interpolated over the whole region more precisely;
(3) oscillations of the solution are eliminated;
(4) larger time steps can be taken in the process of computing solutions of time-dependent problems.

The typical pattern of a solution with large local variation is illustrated by the following univariate monotonic function

$$u(x) = \exp(-x/\epsilon) + x, \quad 0 \leq x \leq 1,$$

with a positive parameter ϵ. This function is a solution to the two-point boundary value problem

$$\begin{aligned} &\epsilon u'' + u' = 1, \quad 1 > x > 0, \\ &u(0) = 1, \qquad u(1) = 1 + \exp(-1/\epsilon). \end{aligned}$$

When the parameter ϵ is very small, then $u(x)$ has a boundary layer of rapid variation (Fig. 1.3). Namely, in the interval $[0, \epsilon|\ln \epsilon|]$ the function $u(x)$ changes from 1 to $\epsilon + \epsilon|\ln \epsilon|$. For example, if $\epsilon = 10^{-5}$, then $\epsilon|\ln \epsilon| = 5 \times 10^{-5} \ln 10 < 2 \times 10^{-4}$. In this small interval the variation of the function $u(x)$ is more than 1–2×10^{-4}.

Let the number of uniform grid points required for the accurate approximation of $u(x)$ on the boundary layer be N_0. Then the number of uniform grid points on $[0, 1]$ with the same step as in the boundary layer will be

$$N = N_0/\epsilon\,|\ln\epsilon| \geq 2 \times 10^4 N_0.$$

However, in order to approximate $u(x)$ with the same accuracy in the interval $[\epsilon|\ln\epsilon|, 1]$ there is no necessity to use more than N_0 points of the uniform grid, since $u(x)$ is monotonic and changes with nearly the same variation in this interval as it does in the boundary layer. Thus, instead of $2 \times 10^4 N_0$, we can restrict the number of grid nodes on the interval $[0, 1]$ to $2N_0$ in order to obtain the same accuracy of interpolation. This spectacular reduction in the number of grid points is obtained at the expense of using a finer grid in the boundary layer only.

This example clearly demonstrates that local grid refinement for problems where the solution quantities have narrow zones in which the dominant length scales are very small is more promising than the uniform refinement of the entire region, since a significant reduction in the total number of grid nodes and consequently in the solution time can be attained. Local refinement becomes indispensable for complex geometries in three dimensions, since otherwise the cost of grid generation may be even higher than the cost of the numerical solution of a physical problem on the grid.

1.2.2.6 Compatibility with Numerical Methods

The locations of the zones of local refinement are also dependent on the numerical approximation to the physical equations. In particular, the areas of high solution error require more refined grid cells. However, the error is estimated through the derivatives of the solution and the size of the grid cells. Thus, ultimately, the grid point locations are to be defined in accordance with the derivatives of the solution.

In general, numerical methods for solving partial differential equations can be divided into two classes: methods based on direct approximations of the derivatives in the differential equation and methods that approximate the solution of the continuum differential equation by linear combinations of trial functions. Finite-difference methods belong to the first class. This difference in methods has a direct impact on the construction of the numerical grid. For the finite-difference methods it is desirable to locate the grid points along directions of constant coordinates in the physical region in order to provide a natural approximation of the derivatives: on the other hand, the methods in the second class that approximate the solution with trial functions do not impose such a restriction on the grid, since the approximate derivatives are obtained after substitution of the approximate solution.

1.3 Grid Classes

There are two fundamental classes of grid popular in the numerical solution of boundary value problems in multidimensional regions: structured and unstructured. These classes differ in the way in which the mesh points are locally organized. In the most general sense, this means that if the local organization of the grid points and the form of the grid cells do not depend on their position but are defined by a general rule, the mesh is considered as structured. When the connection of the neighboring grid nodes varies from point to point, the mesh is called unstructured. As a result, in the structured case the connectivity of the grid is implicitly taken into account, while the connectivity of unstructured grids must be explicitly described by an appropriate data structure procedure.

The two fundamental classes of mesh give rise to three additional subdivisions of grid types: block-structured, overset, and hybrid. These kinds of mesh possess to some extent the features of both structured and unstructured grids, thus occupying an intermediate position between the purely structured and unstructured grids.

1.3.1 Structured Grids Generated by Mapping Approach

The most popular and efficient structured grids are those whose generation relies on a mapping concept. According to this concept the nodes and cells of the grid in an n-dimensional region $X^n \subset R^n$ are defined by mapping the nodes and cells of a reference (generally uniform) grid in some standard n-dimensional domain Ξ^n with a certain transformation

$$\boldsymbol{x}(\boldsymbol{\xi}) : \Xi^n \to X^n, \quad \boldsymbol{\xi} = (\xi^1, \dots, \xi^n), \ \boldsymbol{x} = (x^1, \dots, x^n), \tag{1.2}$$

from Ξ^n onto X^n. The domain Ξ^n is referred to as the logical or computational domain.

The mapping concept was borrowed from examples of grids generated for geometries that are described by analytic coordinate transformations. In particular, two-dimensional transformations have often been defined by analytic functions of a complex variable and by direct shearing. This is the case, for example, for the polar coordinate system for circular regions

$$\boldsymbol{x}(\boldsymbol{\xi}) = \exp(\xi^1)(\cos\xi^2, \sin\xi^2), \quad r_0 \le \xi^1 \le r_1, \ 0 \le \xi^2 \le 2\pi.$$

As an illustrative example of a three-dimensional transformation, the following scaled cylindrical transformation may be considered:

$$\boldsymbol{x}(\boldsymbol{\xi}) : \Xi^3 \to X^3, \quad \boldsymbol{\xi} = (\xi^1, \xi^2, \xi^3), \ 0 \le \xi^i \le 1, \ i = 1, 2, 3,$$

described by

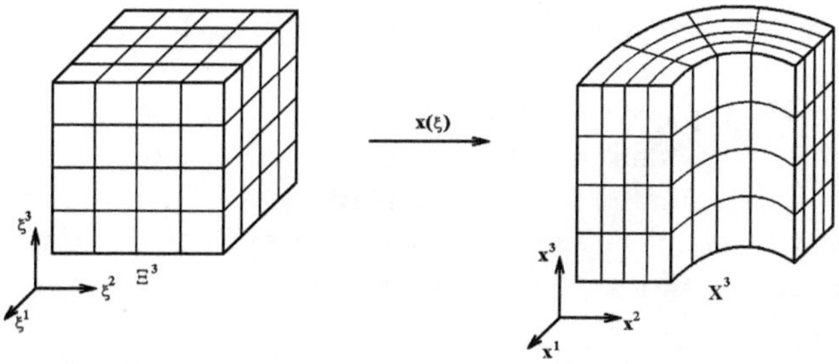

Fig. 1.4. Cylindrical structured grid

$$\begin{aligned} x^1(\boldsymbol{\xi}) &= r\cos\theta, \\ x^2(\boldsymbol{\xi}) &= r\sin\theta, \\ x^3(\boldsymbol{\xi}) &= H\xi^3, \end{aligned} \tag{1.3}$$

where

$$r = r_0 + (r_1 - r_0)\xi^1, \qquad \theta = \theta_0 + (\theta_1 - \theta_0)\xi^2, \qquad H > 0,$$

with

$$0 < r_0 < r_1, \qquad 0 \le \theta_0 < \theta_1 \le 2\pi.$$

If $\theta_1 = 2\pi$ then this function transforms the unit three-dimensional cube into a space bounded by two cylinders of radii r_0 and r_1 and by the two planes $x^3 = 0$ and $x^3 = H$. The reference uniform grid in Ξ^3 is defined by the nodes

$$\boldsymbol{\xi}_{ijk} = (ih, jh, kh), \qquad 0 \le i, j, k \le N, \ h = 1/N,$$

where i, j, k and N are positive integers. The cells of this grid are the three-dimensional cubes bounded by the coordinate planes $\xi^1_i = ih$, $\xi^2_j = jh$, and $\xi^3_k = kh$. Corresponding, the structured grid in the domain X^3 is determined by the nodes

$$\boldsymbol{x}_{ijk} = \boldsymbol{x}(\boldsymbol{\xi}_{ijk}), \quad 0 \le i, j, k \le N.$$

The cells of the grid in X^3 are the curvilinear hexahedrons bounded by the curvilinear coordinate surfaces derived from the parametrization $\boldsymbol{x}(\boldsymbol{\xi})$ (Fig. 1.4).

1.3.1.1 Realization of Grid Requirements

The notion of using a transformation to generate a mesh is very helpful. The idea is to choose a computational domain Ξ^n with a simpler geometry than that of the physical domain X^n and then to find a transformation $\boldsymbol{x}(\boldsymbol{\xi})$ between these domains which eliminates the need for a nonuniform mesh when approximating the physical quantities. That is, if the computational area and the transformation are well chosen, the transformed boundary value problem can be accurately represented by a small number of equally spaced mesh points. Emphasis is placed on a small number of points, because any transformed problem (provided only that the transformation is nonsingular) may be accurately approximated with a sufficiently fine, uniform mesh. In practice, there will be a trade-off between the difficulty of finding the transformation and the number of uniformly spaced points required to find the solution to a given accuracy.

The idea of using mappings to generate grids is extremely appropriate for finding the conditions that the grid must satisfy for obtaining accurate solutions of partial differential equations in the physical domain X^n, because these conditions can be readily defined in terms of the transformations. For example, the grid requirements described in Sect. 1.2.2 are readily formulated through the transformation concept.

Since a solution which is a linear function is computed accurately at the grid points and is approximated accurately over the whole region, an attractive possible method for generating structured grids is to find a transformation $\boldsymbol{x}(\boldsymbol{\xi})$ such that the solution is linear in Ξ^n. Though in practice this requirement for the transformation is not attained even theoretically (except in the case of strongly monotonic univariate functions), it is useful in the sense of an ideal that the developers of structured grid generation techniques should bear in mind. One modification of this requirement which can be practically realized consists of the requirement of a local linearity of the solution in Ξ^n.

The requirements imposed on the grid and the cell size are realized by the construction of a uniform grid in Ξ^n and a smooth function $\boldsymbol{x}(\boldsymbol{\xi})$. The grid cells are not folded if $\boldsymbol{x}(\boldsymbol{\xi})$ is a one-to-one mapping. Consistency with the geometry is satisfied with a transformation $\boldsymbol{x}(\boldsymbol{\xi})$ that maps the boundary of Ξ^n onto the boundary of X^n. Grid concentration in zones of large variation of a function $\boldsymbol{u}(\boldsymbol{x})$ is accomplished with a mapping $\boldsymbol{x}(\boldsymbol{\xi})$ which provides variations of the function $\boldsymbol{u}[\boldsymbol{x}(\boldsymbol{\xi})]$ in the domain Ξ^n that are not large.

1.3.1.2 Coordinate Grids

Among structured grids, coordinate grids in which the nodes and cell faces are defined by the intersection of lines and surfaces of a coordinate system in X^n are very popular in finite-difference methods. The range of values of this system defines a computation region Ξ^n in which the cells of the uniform grid are rectangular n-dimensional parallelepipeds, and the coordinate values define the function $\boldsymbol{x}(\boldsymbol{\xi}) : \Xi^n \to X^n$.

The simplest of such grids are the Cartesian grids obtained by the intersection of the Cartesian coordinates in X^n. The cells of these grids are rectangular parallelepipeds (rectangles in two dimensions). The use of Cartesian coordinates avoids the need to transform the physical equations. However, the nodes of the Cartesian grid do not coincide with the curvilinear boundary, which leads to difficulties in implementing the boundary conditions with second-order accuracy.

1.3.1.3 Boundary-Conforming Grids

An important subdivision of structured grids is the boundary-fitted or boundary-conforming grids. These grids are obtained from one-to-one transformations $\boldsymbol{x}(\boldsymbol{\xi})$ which map the boundary of the domain Ξ^n onto the boundary of X^n.

The most popular of these, for finite-difference methods, have become the coordinate boundary-fitted grids whose points are formed by intersection of the coordinate lines, while the boundary of X^n is composed of a finite number of coordinate surfaces (lines in two dimensions) $\xi^i = \xi^i_0$. Consequently, in this case the computation region Ξ^n is a rectangular domain, the boundaries of which are determined by $(n-1)$-dimensional coordinate planes in R^n, and the uniform grid in Ξ^n is the Cartesian grid. Thus the physical region is represented as a deformation of a rectangular domain and the generated grid as a deformed lattice (Fig. 1.5).

These grids give a good approximation to the boundary of the region and are therefore suitable for the numerical solution of problems with boundary singularities, such as those with boundary layers in which the solution depends very much on the accuracy of the approximation of the boundary conditions.

The requirements imposed on boundary-conforming grids are naturally satisfied with the coordinate transformations $\boldsymbol{x}(\boldsymbol{\xi})$.

The algorithm for the organization of the nodes of boundary-fitted coordinate grids consists of the trivial identification of neighboring points by incrementing the coordinate indices, while the cells are curvilinear hexahedrons. This kind of grid is very suitable for algorithms with parallelization.

Its design makes it easy to increase or change the number of nodes as required for multigrid methods or in order to estimate the convergence rate and error, and to improve the accuracy of numerical methods for solving boundary value problems.

With boundary-conforming grids there is no necessity to interpolate the boundary conditions of the problem, and the boundary values of the region

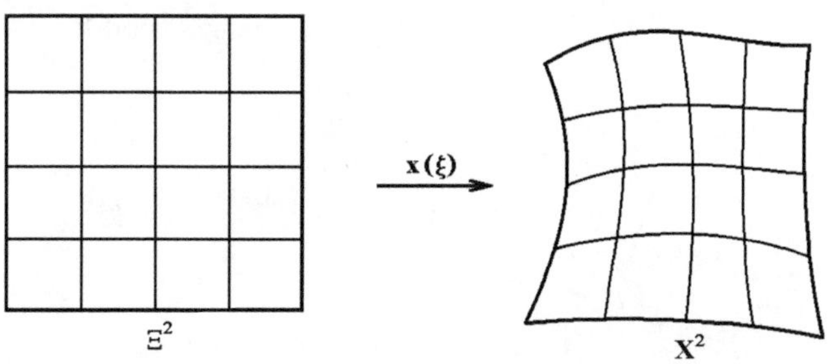

Fig. 1.5. Boundary-conforming quadrilateral grid

can be considered as input data to the algorithm, so automatic codes for grid generation can be designed for a wide class of regions and problems.

In the case of unsteady problems the most direct way to set up a moving grid is to do it via a coordinate transformation. These grids do not require a complicated data structure, since they are obtained from uniform grids in simple fixed domains such as rectangular ones, where the grid data structure remains intact.

1.3.1.4 Shape of Computational Domains

The idea of the structured approach is to transform a complex physical domain X^n to a simpler domain Ξ^n with the help of the parametrization $\boldsymbol{x}(\boldsymbol{\xi})$. The region Ξ^n in (1.2), which is called the computational or logical region, can be either rectangular or of a different matching qualitatively the geometry of the physical domain; in particular, shape it can be triangular for $n = 2$ (Fig. 1.6) or tetrahedral for $n = 3$. Using such parametrizations, a numerical solution of a partial differential equation in a physical region of arbitrary shape can be carried out in a standard computational domain, and codes can be developed that require only changes in the input.

The cells of the uniform grid can be rectangular or of a different shape. Schematic illustrations of two-dimensional triangular and quadrilateral grids are presented in Figs. 1.6 and 1.7, respectively. Note that regions in the form of curvilinear triangles, such as that shown in Fig. 1.6, are more suitable for

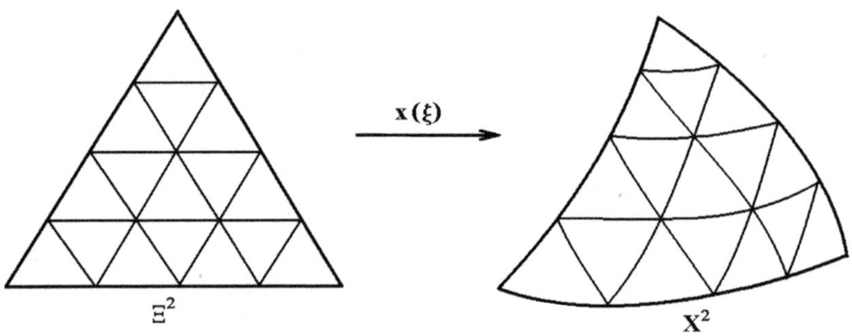

Fig. 1.6. Boundary-conforming triangular grid

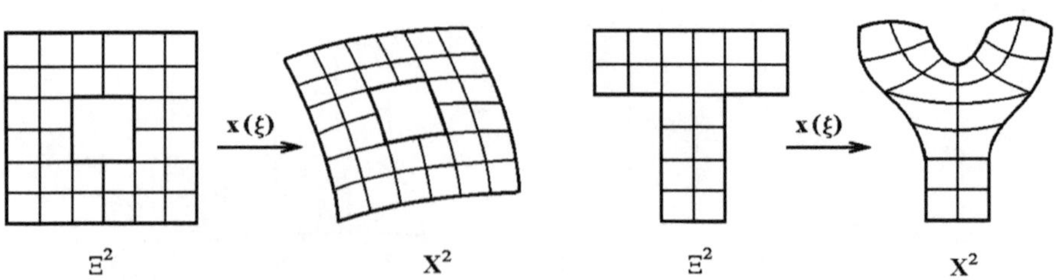

Fig. 1.7. Computational domains adjusted to the physical domains

gridding in the structured approach by triangular cells than by quadrilateral ones. One approach for such gridding is described in Sect. 5.6.

1.3.2 Unstructured Grids

Many field problems of interest involve very complex geometries that are not easily amenable to the framework of the pure structured grid concept. Structured grids may lack the required flexibility and robustness for handling domains with complicated boundaries, or the grid cells may become too skewed and twisted, thus prohibiting an efficient numerical solution. An unstructured grid concept is considered as one of the appropriate solutions to the problem of producing grids in regions with complex shapes.

Unstructured grids have irregularly distributed nodes and their cells are not obliged to have any one standard shape. Besides this, the connectivity of neighboring grid cells is not subject to any restrictions; in particular, the cells can overlap or enclose one another. Thus, unstructured grids provide the most flexible tool for the discrete description of a geometry.

These grids are suitable for the discretization of domains with a complicated shape, such as regions around aircraft surfaces or turbomachinery blade rows. They also allow one to apply a natural approach to local adaptation, by either insertion or removal of nodes. Cell refinement in an unstructured system can be accomplished locally by dividing the cells in the appropriate zones into a few smaller cells. Unstructured grids also allow excessive resolution to be removed by deleting grid cells locally over regions in which the solution does not vary appreciably. In practice, the overall time required to generate unstructured grids in complex geometries is much shorter than for structured or block structured grids.

However, the use of unstructured grids complicates the numerical algorithm because of the inherent data management problem, which demands a special program to number and order the nodes, edges, faces, and cells of the grid, and extra memory is required to store information about the connections between the cells of the mesh. One further disadvantage of unstructured grids that causes excessive computational work is associated with increased numbers of cells, cell faces, and edges in comparison with those for hexahedral meshes. For example, a tetrahedral mesh of N points has roughly $6N$ cells, $12N$ faces, and $7N$ edges, while a mesh of hexahedra has roughly N cells, $3N$ faces, and $3N$ edges. Furthermore, moving boundaries or moving internal surfaces of physical domains are difficult to handle with unstructured grids. Besides this, linearized difference scheme operators on unstructured grids are not usually band matrices, which makes it more difficult to use implicit schemes. As a result, the numerical algorithms based on an unstructured grid topology are the most costly in terms of operations per time step and memory per grid point.

Originally, unstructured grids were mainly used in the theory of elasticity and plasticity, and in numerical algorithms based on finite-element methods.

However, the field of application of unstructured grids has now expanded considerably and includes computational fluid dynamics. Some important aspects of the construction of unstructured grids are considered in Chap. 11.

1.3.3 Block-Structured Grids

In the commonly applied block strategy, the region is divided without holes or overlaps into a few contiguous subdomains, which may be considered as the cells of a coarse, generally unstructured grid. And then a separate structured grid is generated in each block. The union of these local grids constitutes a mesh referred to as a block-structured or multi-block grid. Grids of this kind can thus be considered as locally structured at the level of an individual block, but globally unstructured when viewed as a collection of blocks. Thus a common idea in the block-structured grid technique is the use of different structured grids, or coordinate systems, in different regions, allowing the most appropriate grid configuration to be used in each region.

Block-structured grids are considerably more flexible in handling complex geometries than structured grids. Since these grids retain the simple regular connectivity pattern of a structured mesh on a local level, these block-structured grids maintain, in nearly the same manner as structured grids, compatibility with efficient finite-difference or finite-volume algorithms used to solve partial differential equations. However, the generation of block-structured grids may take a fair amount of user interaction and, therefore, requires the implementation of an automation technique to lay out the block topology.

The main reasons for using multi-block grids rather than single-block grids are that

(1) the geometry of the region is complicated, having a multiply connected boundary, cuts, narrow protuberances, cavities, etc.;
(2) the physical problem is heterogeneous relative to some of the physical quantities, so that different mathematical models are required in different zones of the domain to adequately describe the physical phenomena;
(3) the solution of the problem behaves non-uniformly: zones of smooth and rapid variation of different scales may exist.

The blocks of locally structured grids in a three-dimensional region are commonly homeomorphic to a three-dimensional cube, thus having the shape of a curvilinear hexahedron. However, some domains can be more effectively partitioned with the use of cylindrical blocks as well. Cylindrical blocks are commonly applied to the numerical solution of problems in regions with holes and to the calculation of flows past aircraft or aircraft components (wings, fuselages, etc.). For many problems it is easier to take into account the geometry of the region and the structure of the solution by using cylindrical blocks. Also, the total number of blocks and sections might be smaller than when using only blocks homeomorphic to a cube.

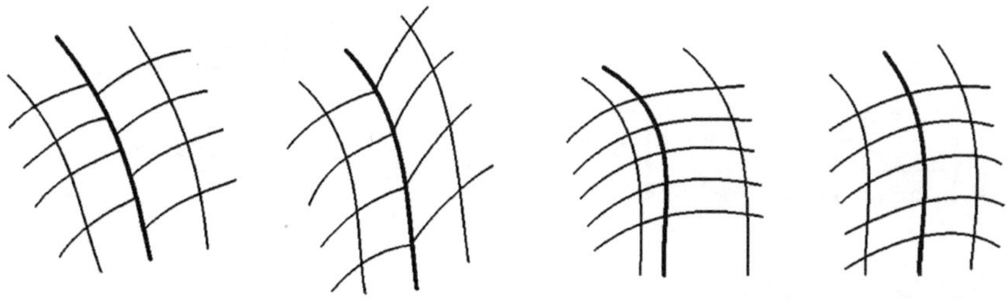

Fig. 1.8. Types of interface between contiguous blocks (**a** discontinuous; **b**, **c** nonsmooth; **d** smooth)

1.3.3.1 Communication of Adjacent Coordinate Lines

The requirement of mutual positioning or "communication" of adjacent grid blocks can also have a considerable influence on the construction of locally structured grids and on the efficiency of the numerical calculations. The coordinate lines defining the grid nodes of two adjacent blocks need not have points in common, and can join smoothly or nonsmoothly (Fig. 1.8). If all adjacent grid blocks join smoothly, interpolation is not required. If the coordinate lines do not join, then during the calculation the solution values at the nodes of one block must be transferred to those of the adjacent block in the neighborhood of their intersection. This is done by interpolation or (in mechanics) using conservation laws.

The types of interaction between adjacent grid blocks are selected on the basis of the features of the physical quantities in the region of their intersection. If the gradient of the physical solution is not high in the vicinity of a boundary between two adjacent blocks and interpolation can, therefore, be performed with high accuracy, the coordinate lines do not need to join. This greatly simplifies the algorithm for constructing the grid in a block. If there are high gradients of the solution near the intersection of two blocks, a smooth matching is usually performed between the coordinate lines of the two blocks. This kind of conformity poses a serious difficulty for structured mesh generation methods. Currently the problem is overcome by an algebraic technique using Hermitian interpolation, or by elliptic methods, involving a choice of control functions. A combination of Laplace and Poisson equations, yielding equations of fourth or even sixth order, is also used for this purpose.

1.3.3.2 Topology of the Grid

The correct choice of the topology in a block, depending on the geometry of the computational region and the choice of the transformation of the region into the block, has a considerable influence on the quality of the grid. There are two ways of specifying the computational region for a block:

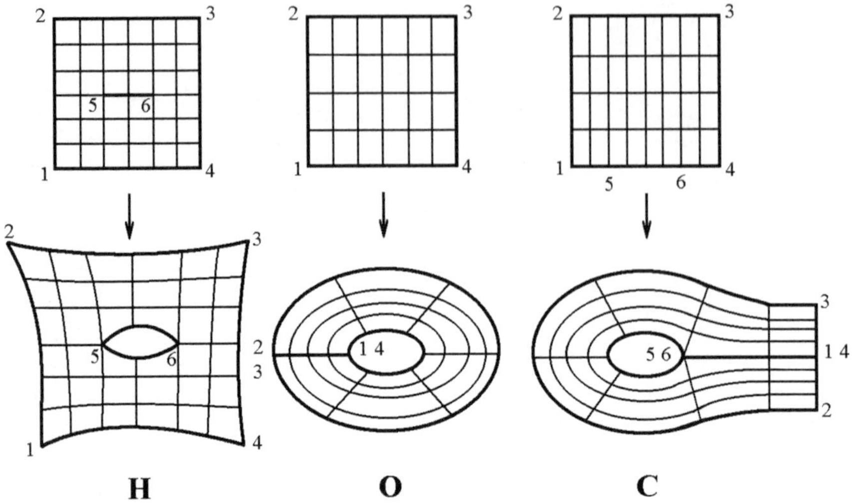

Fig. 1.9. Patterns of grid topology

(1) as a complicated polyhedron which maintains the schematic form of the block subdomain (Fig. 1.7);
(2) simply as a solid cube or a cube with cuts (Fig. 1.9).

With the first approach, the problem of constructing the coordinate transformation $\boldsymbol{x}(\boldsymbol{\xi})$ is simplified, and this method is often used to generate a single-blocked grid in a complicated domain. The second approach relies on a simplified geometry of the computational domain but requires sophisticated methods to derive suitable transformations $\boldsymbol{x}(\boldsymbol{\xi})$.

In a block which is homeomorphic to a cylinder with thick walls, the grid topology is determined by the topology of the two-dimensional grids in the transverse sections. In applications, for sections of this kind, which are annular planes or surfaces with a hole, wide use is made of three basic grid topologies: H, O and C (see Fig. 1.9).

In H-type grids, the computational region is a square with an interior cut which is opened by the construction of the coordinate transformation and mapped onto an interior boundary of the region X^2. The outer boundary of the square is mapped onto the exterior of X^2. The interior boundary has two points with singularities where one coordinate line splits. H-type grids are used, for instance, when calculating the flow past thin bodies (aircraft wings, turbine blades, etc.).

In O-type grids, the computational region is a solid square. In this case the system of coordinates is obtained by bending the square, sticking two opposite sides together and then deforming. The stuck sides determine the cut, called the fictive edge, in the block. An example of O-type grid is the nodes and cells of a polar system of coordinates. The O-type grid can be constructed without singularities when the boundary of the region is smooth. Grids of this kind are used when calculating the flow past bulky aircraft components (fuselages, gondolas, etc.) and, in combination with H-type grids, for multilayered block structures.

The computational region is also a solid square in a C-type grid, but the mapping onto X^2 involves the identification of some segments of one of its sides and then deforming it. In the C-type grid, the coordinate lines of one family leave the outer boundary, circle the inner boundary and return again to the outer boundary. There is one point on the inner boundary which has the same type of singularity as in the H-type grid. The C-type grids are commonly used in regions with holes and long protuberances.

The O and C-type techniques in fact introduce artificial interior cuts in multiply connected regions to generate single block-structured grids. The cuts are used to join the disconnected components of the domain boundary in order to reduce their number. Theoretically, this operation can allow one to generate a single coordinate transformation in a multiply connected domain.

The choice of the grid topology in a block depends on the structure of the solution, the geometry of the domain, and, in the case of continuous or smooth grid-line communication, on the topology of the grid in the adjacent block as well. For complicated domains, such as those near aircraft surfaces or turbines with a large number of blades, it is difficult to choose the grid topology of the blocks, because each component of the system (wing, fuselage, etc.) has its own natural type of grid topology, but these topologies are usually incompatible with each other.

1.3.3.3 Conditions Imposed on Grids in Blocks

A grid in a block must satisfy the conditions which are required to obtain an acceptable solution. In any specific case, these conditions are determined by features of the computer, the methods of grid generation available, the topology and conditions of interaction of the blocks, the numerical algorithms, and the type of data to be obtained.

One of the main requirements imposed on the grid is its adaptation to the solution. Multidimensional computations are likely to be very costly without the application of adaptive grid techniques. The basic aim of adaptation is to enhance the efficiency of numerical algorithms for solving physical problems by a special nonuniform distribution of grid nodes. The appropriate adaptive displacement of the nodes, depending on the physical solution, can increase the accuracy and rate of convergence and reduce oscillations and the interpolation error.

In addition to adaptation, the construction of locally structured grids often requires the coordinate lines to cross the boundary of the domain or the surface in an orthogonal or nearly orthogonal fashion. The orthogonality at the boundary can greatly simplify the specification of boundary conditions. Also, a more accurate representation of algebraic models of turbulence, the equations of a boundary layer, and parabolic Navier–Stokes equations is possible in this case. If for grids of O and C-type the coordinate lines are orthogonal to the boundary of each block and its interior cuts, the global block-structured grid will be smooth. It is also desirable for the coordinate lines to be orthogonal or nearly orthogonal inside the blocks. This will improve the convergence of the difference algorithms, and the equations, if written in orthogonal variables, will have a simpler form.

For unsteady gas-dynamics problems, some coordinates in the entire domain or on the boundary are required to have Lagrange or nearly Lagrange properties. With Lagrangian coordinates the computational region remains fixed in time and simpler expressions for the equations can be obtained in this case.

It is also important that the grid cells do not collapse, the changes in the steps are not too abrupt, the lengths of the cell sides are not very different, and the cells are finer in any domain of high gradient, large error, or slow convergence. Requirements of this kind are taken into account by introducing quantitative and qualitative characteristics of the grid, both with the help of coordinate transformations and by using the sizes of cell edges, faces, angles, and volumes. The characteristics used include the deviation from orthogonality, the Lagrange properties, the values of the transformation Jacobian or cell volume, and the smoothness and adaptivity of the transformation. For cell faces, the deviation from a parallelogram, rectangle, or square, as well as the ratio of the area of the face to its perimeter, is also used.

1.3.4 Overset Grids

Block-structured grids require the partition of the domain into blocks that are restricted so as to abut each other. Overset grids are exempt from this restriction. With the overset concept the blocks are allowed to overlap, which significantly simplifies the problem of the selection of the blocks covering the physical region. In fact, each block may be a subdomain which is associated only with a single geometry or physical feature. The global grid is obtained as an assembly of structured grids which are generated separately in each block. These structured grids are overset on each other, with data communicated by interpolation in overlapping areas of the blocks (Fig. 1.10).

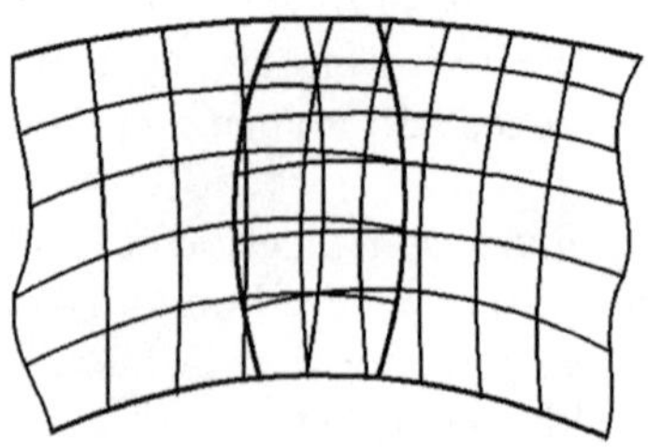

Fig. 1.10. Fragment of an overset grid

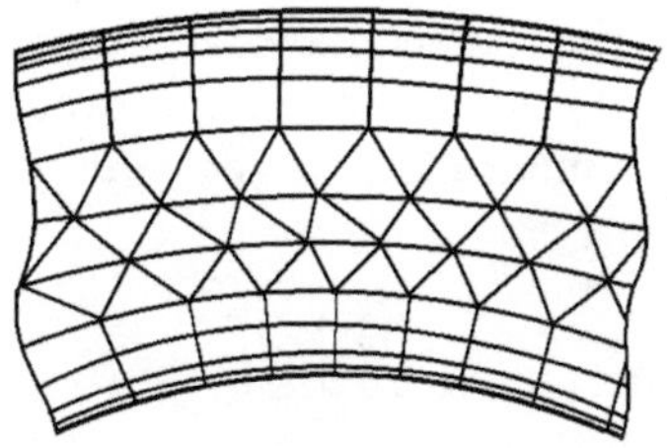

Fig. 1.11. Fragment of a hybrid grid

1.3.5 Hybrid Grids

Hybrid numerical grids are meshes which are obtained by combining both structured and unstructured grids. These meshes are widely used for the numerical analysis of boundary value problems in regions with a complex geometry and with a solution of complicated structure. They are formed by joining structured and unstructured grids on different parts of the region or surface. Commonly, a structured grid is generated about each chosen boundary segment. These structured grids are required not to overlap. The remainder of the domain is filled with the cells of an unstructured grid (Fig. 1.11). This construction is widely applied for the numerical solution of problems with boundary layers.

1.4 Approaches to Grid Generation

The unique aspect of grid generation on general domains is that grid generation has a high degree of freedom, i.e. mesh techniques are not obliged to have any specified formulation, so any foundation may be suitable for this purpose if the grid generated is acceptable.

The chief practical difficulty facing grid generation techniques is that of formulating satisfactory techniques which can realize the user's requirements. Grid generation techniques should develop methods that can help in handling problems with multiple variables, each varying over many orders of magnitude. These methods should be capable of generating grids which are locally compressed by large factors when compared with uniform grids.

The methods should incorporate specific control tools, with simple and clear relationships between these control tools and characteristics of the grid such as mesh spacing, skewness, smoothness, and aspect ratio, in order to provide a reliable way to influence the efficiency of the computation. And finally, the methods should be computationally efficient and easy to code.

A number of techniques for grid generation have been developed. Every method has its strengths and its weaknesses. Therefore, there is also the question of how to choose the most efficient method for the solution of any specific problem, taking into account the geometrical complexity, the computing cost for generating the grid, the grid structure, and other factors.

The goal of the development of these methods is to provide effective and acceptable grid generation systems.

1.4.1 Methods for Structured Grids

The most efficient structured grids are boundary-conforming grids. The generation of these grids can be performed by a number of approaches and techniques. Many of these methods are specifically oriented to the generation of grids for the finite-difference method.

A boundary-fitted coordinate grid in the region X^n is commonly generated first on the boundary of X^n and then successively extended from the boundary to the interior of X^n. This process is analogous to the interpolation of a function from a boundary or to the solution of a differential boundary value problem. On this basis there have been developed three basic groups of methods of grid generation:

(1) algebraic methods, which use various forms of interpolation or special functions;
(2) differential methods, based mainly on the solution of elliptic, parabolic, and hyperbolic equations in a selected transformed region;
(3) variational methods, based on optimization of grid quality properties.

1.4.1.1 Algebraic Methods

In the algebraic approach the interior points of the grid are commonly computed through formulas of transfinite interpolation.

Algebraic methods are simple; they enable the grid to be generated rapidly and the spacing and slope of the coordinate lines to be controlled by the blending coefficients in the transfinite interpolation formulas. However, in regions of complicated shape the coordinate surfaces obtained by algebraic methods can become degenerate or the cells can overlap or cross the boundary. Moreover, they basically preserve the features of the boundary surfaces, in particular, discontinuities. Algebraic approaches are commonly used to generate grids in regions with smooth boundaries that are not highly deformed, or as an initial approximation in order to start the iterative process of an elliptic grid solver.

1.4.1.2 Differential Methods

For regions with arbitrary boundaries, differential methods based on the solution of elliptic and parabolic equations are commonly used. The interior coordinate lines derived through these methods are always smooth, being a solution of these equations, and thus discontinuties on the boundary surface do not extend into the region. The use of parabolic and elliptic systems enables orthogonal and clustering coordinate lines to be constructed, while, in many cases, the maximum principle, which is typical for these systems, ensures that the coordinate transformations are nondegenerate. Elliptic equations are also used to smooth algebraic or unstructured grids.

In practice, hyperbolic equations are simpler then nonlinear elliptic ones and enable marching methods to be used and an orthogonal system of coordinates to be constructed, while grid adaptation can be performed using the coefficients of the equations. However, methods based on the solution of hyperbolic equations are not always mathematically correct and they are not applicable to regions in which the complete boundary surface is strictly defined. Therefore hyperbolic methods are mainly used for simple regions which have several lateral faces for which no special nodal distribution is required. Hyperbolic generation is particularly well suited for use with the overset grid approach. The marching procedure for the solution of hyperbolic equations allows one to decompose only the boundary geometry in such a way that neighboring boundary grids overlap. Volume grids will overlap naturally if sufficient overlap is provided on the boundary. In practice, a separate coordinate grid around each subdomain can be generated by this approach.

1.4.1.3 Variational Methods

Variational methods are widely used to generate grids which are required to satisfy more than one condition, such as nondegeneracy, smoothness, uniformity, near-orthogonality, or adaptivity, which cannot be realized simultaneously with algebraic or differential techniques. Variational methods take into account the conditions imposed on the grid by constructing special functionals defined on a set of smooth or discrete transformations. A compromise grid, with properties close to those required, is obtained with the optimum transformation for a combination of these functionals.

At present, variational techniques are not widely applied to practical grid generation, mainly because their formulation does not always lead to a well-posed mathematical problem. However, the variational approach has been cited repeatedly as the most promising method for the development of future grid generation techniques, owing to its underlying, latent, powerful potential.

1.4.2 Methods for Unstructured Grids

Unstructured grids can be obtained with cells of arbitrary shape, but are generally composed of tetrahedrons (triangles in two dimensions). There are, fundamentally, three approaches to the generation of unstructured grids: octree methods, Delaunay procedures, and advancing-front techniques.

1.4.2.1 Octree Approach

In the octree approach the region is first covered by a regular Cartesian grid of cubic cells (squares in two dimensions). Then the cubes containing segments of the domain surface are recursively subdivided into eight cubes (four squares in two dimensions) until the desired resolution is reached. The cells intersecting the body surfaces are formed into irregular polygonal cells. The grid generated by this octree approach is not considered as the final one, but serves to simplify the geometry of the final grid, which is commonly composed of tetrahedral (or triangular) cells built from the polygonal cells and the remaining cubes.

The main drawback of the octree approach is the inability to match a prescribed boundary surface grid, so the grid on the surface is not constructed beforehand as desired but is derived from the irregular volume cells that intersect the surface. Another drawback of the grid is its rapid variation in cell size near the boundary. In addition, since each surface cell is generated by the intersection of a hexahedron with the boundary there arise problems in controlling the variation of the surface cell size and shape.

1.4.2.2 Delaunay Approach

The Delaunay approach connects neighboring points (of some previously specified set of nodes in the region) to form tetrahedral cells in such a way that the circumsphere through the four vertices of a tetrahedral cell does not contain any other point. The points can be generated in two ways; they can be defined at the start by some technique or they can be inserted within the tetrahedra as they are created, starting with very coarse elements connecting boundary points and continuing until the element size criteria are satisfied. In the latter case a new Delaunay triangulation is constructed at every step using usually Watson's and Rebay's incremental algorithms.

The major drawback of the Delaunay approach is that it requires the insertion of additional boundary nodes, since the boundary cells may not become the boundary segments of the Delaunay volume cells. Either the Delaunay criterion must be mitigated near the boundaries or boundary points must be added as necessary to avert breakthrough of the boundary.

1.4.2.3 Advancing-Front Techniques

In these techniques the grid is generated by building cells progressively one at a time and marching from the boundary into the volume by successively connecting new points to points on the front until all previously unmeshed space is filled with grid cells. Some provision must be made to keep the marching front from intersecting.

To find a suitable vertices for the new cells is very difficult task in this approach, since significant searches must be made to adjust the new cells to the existing elements. Commonly, the marching directions for the advancing front must take into account the surface normals and also the adjacent surface points. A particular difficulty of this method occurs in the closing stage of the procedure, when the front folds over itself and the final vertices of the empty space are replaced by tetrahedra. Serious attention must also be paid to the marching step size, depending on the size of the front faces as well as the shape of the unfilled domain that is left.

Unstructured grids, after they have been completed, are generally smoothed by a Laplacian-type or other smoother to enhance their qualitative properties.

A major drawback remaining for unstructured techniques is the increased computational cost of the numerical solution of partial differential equations in comparison with solution on structured grids.

1.5 Big Codes

A "big grid generation code" is an effective system for generating structured and unstructured grids, as well as hybrid and overset combinations, in general regions. Such systems also are referred to as "comprehensive grid generation codes".

The development of such codes is a considerable problem in its own right. The present comprehensive grid generation codes developed for the solution of multidimensional problems have to incorporate combinations of block-structured, hybrid, and overset grid methods and are still rather cumbersome, rely on interactive tools, and take too many man-hours to generate a complicated grid. Efforts to increase the efficiency and productivity of these codes are mainly being conducted in two interconnected research areas.

The first, the "array area", is concerned with the automation of those routine processes of grid generation which presently require interactive tools and a great deal of human time and effort. Some of these are:

(1) the decomposition of a domain into a set of contiguous or overlapping blocks consistent with the distinctive features of the domain geometry, the singularities of the physical medium and the sought-for solution, and the computer architecture;
(2) numbering the set of blocks, their faces, and their edges with a connectivity hierarchy and determining the order in which the grids are constructed in the blocks and their boundaries;
(3) choosing the grid topology and the requirements placed on the qualitative and quantitative characteristics of the internal and boundary grids and on their communication between the blocks;
(4) selecting appropriate methods to satisfy the requirements put on the grid in accordance with a particular geometry and solution;
(5) assessment and enhancement of grid quality.

The second, more traditional, "methods area" deals with developing new, more reliable, and more elaborate methods for generating, adapting, and smoothing grids in domains in a unified manner, irrespective of the geometry of the domain or surface and of the qualitative and quantitative characteristics the grids should possess, so that these methods, when incorporated in the comprehensive codes, should ease the bottlenecks of the array area, in particular, by enabling a considerable reduction of the number of blocks required.

There are many demands that are made on the codes. The code must be efficient, expandable, portable, and configurable. It should incorporate state-of-the-art techniques for generating grids. Besides this, the code should include pre- and post-processing tools in order to start from prescribed data of the geometry and end with the final generation of the grid in the proper format for use with the specified partial differential codes. The code should have the ability to be updated by the addition of new features and the removal of obsolete ones.

The overall purpose of the development of these comprehensive grid generation codes is to create a system which enables one to generate grids in a "black box" mode without or with only a slight human interaction. Currently, however, the user has to take active role and be fully occupied in the grid generation process. The user has to make conclusions about qualitative properties of the grid and undertake corrective measures when necessary. The present codes include significant measures to increase the productivity of such human activity, namely, graphical interactive systems and user-friendly interfaces. Efforts to eliminate the "human component" of the codes are directed towards developing new techniques, in particular, new grid generation methods and automated block decomposition techniques.

1.5.1 Interactive Systems

An important element of the current comprehensive grid generation codes is an interactive system which includes extensive graphical tools to display all elements of the grid generation process and graphical feedback to monitor progress in grid efficiency and to verify, as well as correct, errors and faults easily. All existing comprehensive codes possess well-developed interactive systems which are used, in particular, to define grid boundaries and surface normals on block faces; to generate multiblock topologies and domain decompositions; to specify connectivity data, grid density, and spatial distribution in the normal direction at the boundaries; and to provide attraction to chosen points or lines. The graphical systems of the codes provide a display of data and domain and surface elements with different colors and markers; a representation of surface grids and their boundaries by specific colors; a visual representation of the qualitative and quantitative properties of the grid in terms of cell skewness, aspect ratio, surface and volume Jacobian checks, estimates of truncation errors, and measures of grid continuity across blocks; and views of surface and block grids at various levels coarseness.

These capabilities allow any portion of a multiple-block grid to be displayed in a manner that is quickly discernible to the user. All functions of the generation process are invoked through interactive screens and menus. In an interactive environment the user can continually examine and correct the surfaces and grids as they are developed.

1.5.2 New Techniques

The present codes are designed in a modular fashion to facilitate both the addition of new techniques in a straightforward manner as they become available and the removal of obsolete ones. Although most of the methods included in the codes have provided appropriate results for specific applications, they lack the desired generality, flexibility, efficiency, automation, and robustness. Efforts to create new techniques are directed towards the automation of domain decomposition, interactive and automated generation of block connectivity, the development of new, more effective methods for generating grids within the blocks, interactive local adaptive adjustment of the control functions in elliptic equations, the optimal specification and modification of the distribution functions in hyperbolic and advancing-front grid generators, interactive local quality enhancement of the grid, and interactive and generally applicable interpolation techniques to transfer data between the separate component grids.

1.5.2.1 Domain Decomposition

To perform domain zoning well, some expertise is required: the user must have experience with composite zonal grid methods, familiarity with the grid generation capabilities available, knowledge of the behavior of the zonal technique to be used, knowledge of the physical behavior, some expectation of the important physical features of the problem to be solved, and criteria for evaluating the zonings. To perform zoning quickly, the user must have both expertise and interactive, graphical, easy-to-use tools.

However, even with the interactive techniques available, the generation of the block structure is the most difficult and time-consuming task in the grid generation process. Therefore any automation of domain decomposition is greatly desirable.

The first attempts to overcome the problem of domain decomposition were presented in the 1980s. The proposed approaches laid a foundation for an automated approach to 3-D domain decomposition which relies mainly on observation of how experts perform the task and on a knowledge-based programming approach, typically described by means of examples. The user first represents all components of the domain schematically as rectangular sets of blocks and then the codes develop a schematic block structure.

1.5.2.2 New Methods

Recent results in the field of grid generation methods have largely been related to the application of harmonic function theory to adaptive grid generation. The suggestion to use harmonic functions for generating adaptive grids was made by Dvinsky (1991). Adaptive grids can be generated by mapping the reference grid into the domain with a coordinate transformation which is inverse to a harmonic vector function (in terms of Riemannian manifolds). Adaptation is performed by a specified adaptive metric in the domain which converts it into a Riemannian manifold. Each harmonic function minimizes some functional of the total energy, and hence it can be found by the numerical solution either of a variational problem or of a boundary value problem for a system of Euler–Lagrange equations.

One version of the harmonic approach, proposed by Liseikin (1991a, 2004), uses a method of generating smooth hypersurface grids. Specifically, the adaptive grid with node clustering in the zones of large values of a vector function is obtained as a projection of a quasiuniform grid from a monitor surface generated as a surface of the function values over the physical space. The vector function can be the physical solution or a combination of its components or derivatives, or it can be any other quantity that suitably monitors the behavior of the solution. A generalization of this approach for generating grids with another properties was performed by Liseikin (2004, 2007). The method developed allows the designer to merge the two tasks of surface grid generation and volume grid generation into one task while developing a comprehensive grid generation code. It also eases the array bottlenecks of the codes by allowing a decrease of the number of blocks required for the decomposition of a complicated region.

1.6 Comments

Detailed descriptions of the most popular structured methods and their theoretical and logical justifications and numerical implementations were given in the monographs by Thompson, Warsi, and Mastin (1985), Knupp and Steinberg (1993), and Liseikin (1999). Particular issues concerned with the generation of one-dimensional moving grids for gas-dynamics problems, the stretching technique for the numerical solution of singularly perturbed equations, and nonstationary grid techniques were considered in the books by Alalykin et al. (1970), Liseikin and Petrenko (1989), Liseikin (2001a), and Zegeling (1993), respectively.

A considerable number of general structured grid generation methods were reviewed in surveys by Thompson, Warsi, and Mastin (1982), Thompson (1984a, 1996), Eiseman (1985), Liseikin (1991b), Thompson and Weatherill (1993), and in the handbook of grid generation edited by Thompson, Soni, and Weatherill (1999).

Adaptive structured grid methods were first surveyed by Anderson (1983) and Thompson (1984b, 1985). Then a series of surveys on general adaptive methods was presented by Eiseman (1987), Hawken, Gottlieb, and Hansen (1991), Liseikin (1996b), and Baker (1997). Adaptive techniques for moving grids were described by Hedstrom and Rodrigue (1982) and Zegeling (1993).

A description of the types of mesh topology and the singular points of the grids around wing-body shapes was carried out by Eriksson (1982).

Methods for unstructured grids were reviewed by Thacker (1980), Ho-Le (1988), Shephard et al. (1988a), Baker (1995, 1997), Field (1995), Carey (1997), George and Borouchaki (1998), Krugljakova et al. (1998), and Owen (1998). An exhaustive survey of both structured and unstructured techniques has been given by Thompson and Weatherill (1993), and Thompson (1996).

The multiblock strategy for generating grids around complicated shapes was originally proposed by Lee et al. (1980); however, the idea of using different coordinates in different subregions of the domain can be traced back to Thoman and Szewezyk (1969). The overset grid approach was introduced by Atta and Vadyak (1982), Berger and Oliger (1983), Benek, Steger, and Dougherty (1983), Miki and Takagi (1984), and Benek, Buning, and Steger (1985). The first attempts to overcome the problem of domain decomposition were discussed by Andrews (1988), Georgala and Shaw (1989), Allwright (1989), and Vogel (1990). The concept of blocks with a continuous alignment of grid lines across adjacent block boundaries was described by Weatherill and Forsey (1984) and Thompson (1987b). Thomas (1982) and Eriksson (1983) applied the concept of continuous line slope, while a discontinuity in slope was discussed by Rubbert and Lee (1982). A shape recognition technique based on an analysis of a physical domain and an interactive construction of a computational domain with a similar geometry was proposed by Takahashi and Shimizu (1991) and extended by Chiba et al. (1998). The embedding technique was considered by Albone and Joyce (1990) and Albone (1992). Some of the first applications of block-structured grids to the numerical solution of three-dimensional fluid-flow problems in realistic configurations were demonstrated by Rizk and Ben-Shmuel (1985), Sorenson (1986), Atta, Birchelbaw, and Hall (1987), and Belk and Whitfield (1987).

The first comprehensive grid codes were described by Holcomb (1987), Thompson (1987a), Thomas, Bache, and Blumenthal (1990), Widhopf et al. (1990), and Steinbrenner, Chawner, and Fouts (1990). These codes have stimulated the development of better ones, reviewed by Thompson (1996). This paper also describes the current domain decomposition techniques developed by Shaw and Weatherill (1992), Stewart (1992), Dannenhoffer (1995), Wulf and Akrag (1995), Schonfeld, Weinerfelt, and Jenssen (1995), and Kim and Eberhardt (1995).

Since the time that these pioneering codes appeared many codes more advanced and sophisticated have been developed by researchers. Exhaustive surveys of the most popular recent codes can be found at the meshing pages http://www.andrew.cmu.edu/user/sowen/softsurv.html and http://www-users.informatik.rwth-aachen.de/~roberts/software.html. In particular, very useful descriptions of mesh generation codes and codes with parallel mesh generation techniques are given in the books of Frey and George (2008) and Ivanov (2008), respectively. An informal survey of software vendors, research labs, and educational institutions that develop grid generation codes was presented by Owen (1998).

2

Coordinate Transformations

2.1 Introduction

Partial differential equations in the physical domain X^n can be solved on a structured numerical grid obtained by mapping a reference grid in the logical region Ξ^n into X^n with a coordinate transformation $\boldsymbol{x}(\boldsymbol{\xi}) : \Xi^n \to X^n$. The structured grid concept also gives an alternative way to obtain a numerical solution to a partial differential equation, by solving the transformed equation with respect to the new independent variables ξ^i on the reference grid in the logical domain Ξ^n. Some notions and relations concerning the coordinate transformations yielding structured grids are discussed in this chapter. These notions and relations are used to represent some conservation-law equations in the new logical coordinates in a convenient form. The relations presented will be used in Chap. 3 to formulate various grid properties.

Conservation-law equations in curvilinear coordinates are typically deduced from the equations in Cartesian coordinates through the classical formulas of tensor calculus, by procedures which include the substitution of tensor derivatives for ordinary derivatives. The formulation and evaluation of the tensor derivatives is rather difficult, and they retain some elements of mystery. However, these derivatives are based on specific transformations of tensors, modeling in the equations some dependent variables, e.g. the components of a fluid velocity vector, which after the transformation have a clear interpretation in terms of the contravariant components of the vector. With this concept, the conservation-law equations are readily written out in this chapter without application of the tensor derivatives, but utilizing instend only some specific transformations of the dependent variables, ordinary derivatives, and one basic identity of coordinate transformations derived from the formula for differentiation of the Jacobian.

For generality, the transformations of the coordinates are mainly considered for arbitrary n-dimensional domains, though in practical applications the dimension n equals 1, 2, 3, or 4 for time-dependent transformations of three-dimensional domains. We also apply chiefly a standard vector notation for

V.D. Liseikin, *Grid Generation Methods*, Scientific Computation,
DOI 10.1007/978-90-481-2912-6_2,

the coordinates, as variables with indices. Sometimes, however, particularly in figures, the ordinary designation for three-dimensional coordinates, namely x, y, z for the physical coordinates and ξ, η, ζ for the logical ones, is used to simplify the presentation.

2.2 General Notions and Relations

This section presents some basic relations between Cartesian and curvilinear coordinates.

2.2.1 Jacobi Matrix

Let

$$\boldsymbol{x}(\boldsymbol{\xi}) : \Xi^n \to X^n, \quad \boldsymbol{\xi} = (\xi^1, \dots, \xi^n), \ \boldsymbol{x} = (x^1, \dots, x^n),$$

be a smooth invertible coordinate transformation of the physical region $X^n \subset R^n$ from the parametric domain $\Xi^n \subset R^n$. If Ξ^n is a standard logical domain, then, in accordance with Chap. 1, this coordinate transformation can be used to generate a structured grid in X^n. Here and later R^n presents the Euclidean space with the Cartesian basis $\boldsymbol{e}_1, \dots, \boldsymbol{e}_n$, which represents an orthogonal system of vectors, i.e.

$$\boldsymbol{e}_i \cdot \boldsymbol{e}_j = \begin{cases} 1 & \text{if } i = j, \\ 0 & \text{if } i \neq j. \end{cases}$$

Thus we have

$$\boldsymbol{x} = x^1 \boldsymbol{e}_1 + \cdots + x^n \boldsymbol{e}_n,$$
$$\boldsymbol{\xi} = \xi^1 \boldsymbol{e}_1 + \cdots + \xi^n \boldsymbol{e}_n.$$

The values x^i, $i = 1, \dots, n$, are called the Cartesian coordinates of the vector $\boldsymbol{x}$. The coordinate transformation $\boldsymbol{x}(\boldsymbol{\xi})$ defines, in the domain X^n, new coordinates $\xi^1, \dots, \xi^n$, which are called the curvilinear coordinates. The matrix

$$\jmath = \left(\frac{\partial x^i}{\partial \xi^j} \right), \quad i, j = 1, \dots, n,$$

is referred to as the Jacobi matrix, and its Jacobian is designated by J:

$$J = \det \left(\frac{\partial x^i}{\partial \xi^j} \right), \quad i, j = 1, \dots, n. \tag{2.1}$$

The inverse transformation to the coordinate mapping $\boldsymbol{x}(\boldsymbol{\xi})$ is denoted by

$$\boldsymbol{\xi}(\boldsymbol{x}) : X^n \to \Xi^n.$$

This transformation can be considered analogously as a mapping introducing a curvilinear coordinate system $x^1, \dots, x^n$ in the domain $\Xi^n \subset R^n$. It is obvious that the inverse to the matrix $\jmath$ is

$$\jmath^{-1} = \left(\frac{\partial \xi^i}{\partial x^j}\right), \quad i, j = 1, \dots, n,$$

and consequently

$$\det\left(\frac{\partial \xi^i}{\partial x^j}\right) = \frac{1}{J}, \quad i, j = 1, \dots, n. \tag{2.2}$$

In the case of two-dimensional space the elements of the matrices $(\partial x^i / \partial \xi^j)$ and $(\partial \xi^i / \partial x^j)$ are connected by

$$\begin{aligned} \frac{\partial \xi^i}{\partial x^j} &= (-1)^{i+j} \frac{\partial x^{3-j}}{\partial \xi^{3-i}} \Big/ J, \\ \frac{\partial x^i}{\partial \xi^j} &= (-1)^{i+j} J \frac{\partial \xi^{3-j}}{\partial x^{3-i}}, \quad i, j = 1, 2. \end{aligned} \tag{2.3}$$

Similar relations between the elements of the corresponding three-dimensional matrices have the form

$$\begin{aligned} \frac{\partial \xi^i}{\partial x^j} &= \frac{1}{J}\left(\frac{\partial x^{j+1}}{\partial \xi^{i+1}} \frac{\partial x^{j+2}}{\partial \xi^{i+2}} - \frac{\partial x^{j+1}}{\partial \xi^{i+2}} \frac{\partial x^{j+2}}{\partial \xi^{i+1}}\right), \\ \frac{\partial x^i}{\partial \xi^j} &= J\left(\frac{\partial \xi^{j+1}}{\partial x^{i+1}} \frac{\partial \xi^{j+2}}{\partial x^{i+2}} - \frac{\partial \xi^{j+1}}{\partial x^{i+2}} \frac{\partial \xi^{j+2}}{\partial x^{i+1}}\right), \quad i, j = 1, 2, 3, \end{aligned} \tag{2.4}$$

where for each superscript or subscript index, say l, $l + 3$ is equivalent to l. With this condition the sequence of indices $(l, l + 1, l + 2)$ is the result of a cyclic permutation of $(1, 2, 3)$ and vice versa; the indices of a cyclic sequence (i, j, k) satisfy the relation $j = i + 1, \; k = i + 2$.

2.2.2 Tangential Vectors

The value of the function $\boldsymbol{x}(\boldsymbol{\xi}) = [x^1(\boldsymbol{\xi}), \dots, x^n(\boldsymbol{\xi})]$ in the Cartesian basis $(\boldsymbol{e}_1, \dots, \boldsymbol{e}_n)$, i.e.

$$\boldsymbol{x}(\boldsymbol{\xi}) = x^1(\boldsymbol{\xi})\boldsymbol{e}_1 + \cdots + x^n(\boldsymbol{\xi})\boldsymbol{e}_n,$$

is a position vector for every $\boldsymbol{\xi} \in \Xi^n$. This vector-valued function $\boldsymbol{x}(\boldsymbol{\xi})$ generates the nodes, edges, faces, etc. of the cells of the coordinate grid in the domain X^n. Each edge of the cell corresponds to a coordinate line ξ^i for some i and is defined by the vector

$$\Delta_i \boldsymbol{x} = \boldsymbol{x}(\boldsymbol{\xi} + h\boldsymbol{e}^i) - \boldsymbol{x}(\boldsymbol{\xi}),$$

where h is the step size of the uniform grid in the ξ^i direction in the logical domain Ξ^n. We have

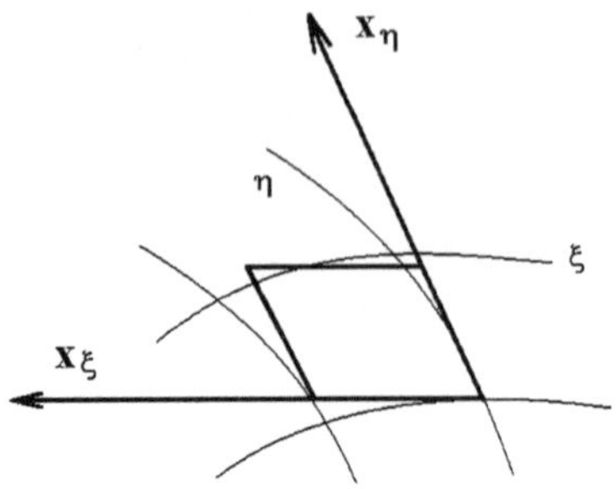

Fig. 2.1. Grid cell and contracted parallelogram

$$\Delta_i \boldsymbol{x} = h\boldsymbol{x}_{\xi^i} + \boldsymbol{t},$$

where

$$\boldsymbol{x}_{\xi^i} = \left(\frac{\partial x^1}{\partial \xi^i}, \dots, \frac{\partial x^n}{\partial \xi^i}\right)$$

is the vector tangential to the coordinate curve ξ^i, and $\boldsymbol{t}$ is a residual vector whose length does not exceed the following quantity:

$$\frac{1}{2}\max|\boldsymbol{x}_{\xi^i\xi^i}|h^2.$$

Thus the cells in the domain X^n whose edges are formed by the vectors $h\boldsymbol{x}_{\xi^i}$, $i = 1, \dots, n$, are approximately the same as those obtained by mapping the uniform coordinate cells in the computational domain Ξ^n with the transformation $\boldsymbol{x}(\boldsymbol{\xi})$. Consequently, the uniformly contracted parallelepiped spanned by the tangential vectors $\boldsymbol{x}_{\xi^i}$, $i = 1, \dots, n$, represents to a high order of accuracy with respect to h the cell of the coordinate grid at the corresponding point in X^n (see Fig. 2.1). In particular, for the length l_i of the ith grid edge we have

$$l_i = h|\boldsymbol{x}_{\xi^i}| + O(h^2).$$

The volume V_h (area in two dimensions) of the cell is expressed as follows:

$$V_h = h^n V + O(h^{n+1}),$$

where V is the volume of the n-dimensional parallelepiped determined by the tangential vectors $\boldsymbol{x}_{\xi^i}$, $i = 1, \dots, n$.

The tangential vectors $\boldsymbol{x}_{\xi^i}$, $i = 1, \dots, n$, are called the base covariant vectors since they comprise a vector basis. The sequence $\boldsymbol{x}_{\xi^1}, \dots, \boldsymbol{x}_{\xi^n}$ of the tangential vectors has a right-handed orientation if the Jacobian of the transformation $\boldsymbol{x}(\boldsymbol{\xi})$ is positive. Otherwise, the base vectors $\boldsymbol{x}_{\xi^i}$ have a left-handed orientation.

The operation of the dot product on these vectors produces elements of the covariant metric tensor. These elements generate the coefficients that appear in the transformed governing equations that model the conservation-law equations of mechanics. Besides this, the metric elements play a primary role in studying and formulating various geometric characteristics of the grid cells.

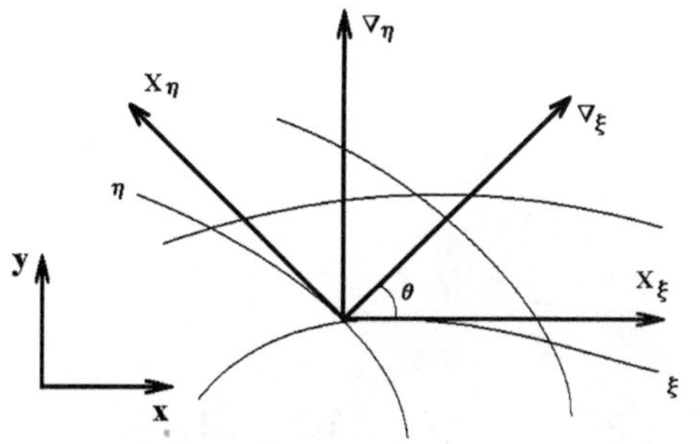

Fig. 2.2. Disposition of the base tangential and normal vectors in two dimensions

2.2.3 Normal Vectors

For a fixed i the vector

$$\left(\frac{\partial \xi^i}{\partial x^1}, \dots, \frac{\partial \xi^i}{\partial x^n}\right),$$

which is the gradient of $\xi^i(\boldsymbol{x})$ with respect to the Cartesian coordinates $x^1, \dots, x^n$, is denoted by $\boldsymbol{\nabla}\xi^i$. The set of the vectors $\boldsymbol{\nabla}\xi^i$, $i = 1, \dots, n$, is called the set of base contravariant vectors.

Similarly, as the tangential vectors relate to the coordinate curves, the contravariant vectors $\boldsymbol{\nabla}\xi^i$, $i = 1, \dots, n$, are connected with their respective coordinate surfaces (curves in two dimensions). A coordinate surface is defined by the equation $\xi^i = \xi^i_0$; i.e. along the surface all of the coordinates $\xi^1, \dots, \xi^n$ except ξ^i are allowed to vary. For all of the tangent vectors $\boldsymbol{x}_{\xi^j}$ to the coordinate lines on the surface $\xi^i = \xi^i_0$ we have the obvious identity

$$\boldsymbol{x}_{\xi^j} \cdot \boldsymbol{\nabla}\xi^i = 0, \quad i \neq j,$$

and thus the vector $\boldsymbol{\nabla}\xi^i$ is a normal to the coordinate surface $\xi^i = \xi^i_0$. Therefore the vectors $\boldsymbol{\nabla}\xi^i$, $i = 1, \dots, n$, are also called the normal base vectors.

Since

$$\boldsymbol{x}_{\xi^i} \cdot \boldsymbol{\nabla}\xi^i = 1$$

for each fixed $i = 1, \dots, n$, the vectors $\boldsymbol{x}_{\xi^i}$ and $\boldsymbol{\nabla}\xi^i$ intersect each other at an angle θ which is less than $\pi/2$. Now, taking into account the orthogonality of the vector $\boldsymbol{\nabla}\xi^i$ to the surface $\xi^i = \xi^i_0$, we find that these two vectors $\boldsymbol{x}_{\xi^i}$ and $\boldsymbol{\nabla}\xi^i$ are directed to the same side of the coordinate surface (curve in two dimensions). An illustration of this fact in two dimensions is given in Fig. 2.2.

The length of any normal base vector $\boldsymbol{\nabla}\xi^i$ is linked to the distance d_i between the corresponding opposite boundary segments (joined by the vector $\boldsymbol{x}_{\xi^i}$) of the n-dimensional parallelepiped formed by the base tangential vectors, namely,

$$d_i = 1/|\boldsymbol{\nabla}\xi^i|, \quad |\boldsymbol{\nabla}\xi^i| = \sqrt{\boldsymbol{\nabla}\xi^i \cdot \boldsymbol{\nabla}\xi^i}.$$

To prove this relation we recall that the vector $\boldsymbol{\nabla}\xi^i$ is a normal to all of the vectors $\boldsymbol{x}_{\xi^j}$, $j \neq i$, and therefore to the boundary segments formed by these

$n-1$ vectors. Hence, the unit normal vector $\boldsymbol{n}_i$ to these segments is expressed by

$$\boldsymbol{n}_i = \boldsymbol{\nabla}\xi^i/|\boldsymbol{\nabla}\xi^i|.$$

Now, taking into account that

$$d_i = \boldsymbol{x}_{\xi^i} \cdot \boldsymbol{n}_i,$$

we readily obtain

$$d_i = \boldsymbol{x}_{\xi^i} \cdot \boldsymbol{\nabla}\xi^i/|\boldsymbol{\nabla}\xi^i| = 1/|\boldsymbol{\nabla}\xi^i|.$$

Let l_i denote the distance between a grid point on the coordinate surface $\xi^i = c$ and the nearest point on the neighboring coordinate surface $\xi^i = c+h$; then

$$l_i = hd_i + O(h^2) = h/|\boldsymbol{\nabla}\xi^i| + O(h^2).$$

This equation shows that the inverse length of the normal vector $\boldsymbol{\nabla}\xi^i$ multiplied by h represents with high accuracy the distance between the corresponding faces of the coordinate cells in the domain X^n.

Note that the volume of the parallelepiped spanned by the tangential vectors equals J, so we find from (2.2) that the volume of the n-dimensional parallelepiped defined by the normal vectors $\boldsymbol{\nabla}\xi^i$, $i = 1, \ldots, n$, is equal to $1/J$. Thus both the base normal vectors $\boldsymbol{\nabla}\xi^i$ and the base tangential vectors $\boldsymbol{x}_{\xi^i}$ have the same right-handed or left-handed orientation.

If the coordinate system $\xi^1, \ldots, \xi^n$ is orthogonal, i.e.

$$\boldsymbol{x}_{\xi^i} \cdot \boldsymbol{x}_{\xi^j} = P(\boldsymbol{x})\delta^i_j, \quad P(\boldsymbol{x}) > 0, \ i, j = 1, \ldots, n,$$

then for each fixed $i = 1, \ldots, n$ the vector $\boldsymbol{\nabla}\xi^i$ is parallel to $\boldsymbol{x}_{\xi^i}$. Here and later, δ^i_j is the Kronecker symbol, i.e.

$$\delta^i_j = 0 \quad \text{if } i \neq j, \qquad \delta^i_j = 1 \quad \text{if } i = j.$$

2.2.4 Representation of Vectors Through the Base Vectors

If there are n independent base vectors $\boldsymbol{a}_1, \ldots, \boldsymbol{a}_n$ of the Euclidean space R^n then any vector $\boldsymbol{b}$ with components $b^1, \ldots, b^n$ in the Cartesian basis $\boldsymbol{e}_1, \ldots, \boldsymbol{e}_n$ is represented through the vectors $\boldsymbol{a}^i$, $i = 1, \ldots, n$, by

$$\boldsymbol{b} = a^{ij}(\boldsymbol{b} \cdot \boldsymbol{a}_j)\boldsymbol{a}_i, \quad i, j = 1, \ldots, n, \tag{2.5}$$

where a^{ij} are the elements of the matrix (a^{ij}) which is the inverse of the tensor (a_{ij}), $a_{ij} = \boldsymbol{a}_i \cdot \boldsymbol{a}_j$, $i, j = 1, \ldots, n$. It is assumed in (2.5) and later that a summation is carried out over repeated indices unless otherwise noted.

The components of the vector $\boldsymbol{b}$ in the natural basis of the tangential vectors $\boldsymbol{x}_{\xi^i}$, $i = 1, \ldots, n$, are called contravariant. Let them be denoted by $\overline{b}^i$, $i = 1, \ldots, n$. Thus

$$\boldsymbol{b} = \overline{b}^1 \boldsymbol{x}_{\xi^1} + \cdots + \overline{b}^n \boldsymbol{x}_{\xi^n}.$$

Assuming in (2.5) $\boldsymbol{a}_i = \boldsymbol{x}_{\xi^i}$, $i = 1, \ldots, n$, we obtain

$$\overline{b}^i = a^{mj} \left(b^k \frac{\partial x^k}{\partial \xi^j} \right) \frac{\partial x^i}{\partial \xi^m}, \quad i, j, k, m = 1, \ldots, n, \tag{2.6}$$

where $b^1, \ldots, b^n$ are the components of the vector $\boldsymbol{b}$ in the Cartesian basis $\boldsymbol{e}_1, \ldots, \boldsymbol{e}_n$. Since

$$a_{ij} = \frac{\partial x^k}{\partial \xi^i} \frac{\partial x^k}{\partial \xi^j}, \quad i, j, k = 1, \ldots, n,$$

we have

$$a^{ij} = \frac{\partial \xi^i}{\partial x^k} \frac{\partial \xi^j}{\partial x^k}, \quad k = 1, \ldots, n.$$

Therefore, from (2.6),

$$\overline{b}^i = b^j \frac{\partial \xi^i}{\partial x^j}, \quad i, j = 1, \ldots, n, \tag{2.7}$$

or, using the dot product notation

$$\overline{b}^i = \boldsymbol{b} \cdot \boldsymbol{\nabla} \xi^i, \quad i = 1, \ldots, n. \tag{2.8}$$

Thus, in this case (2.5) has the form

$$\boldsymbol{b} = (\boldsymbol{b} \cdot \boldsymbol{\nabla} \xi^i) \boldsymbol{x}_{\xi^i}, \quad i = 1, \ldots, n. \tag{2.9}$$

For example, the normal base vector $\boldsymbol{\nabla} \xi^i$ is expanded through the base tangential vectors $\boldsymbol{x}_{\xi^j}$, $j = 1, \ldots, n$, by the following formula:

$$\boldsymbol{\nabla} \xi^i = \frac{\partial \xi^i}{\partial x^j} \frac{\partial \xi^k}{\partial x^j} \boldsymbol{x}_{\xi^k}, \quad i, j, k, = 1, \ldots, n. \tag{2.10}$$

Analogously, a component $\overline{b}_i$ of the vector $\boldsymbol{b}$ in the basis $\boldsymbol{\nabla} \xi^i$, $i = 1, \ldots, n$, is expressed by the formula

$$\overline{b}_i = b^j \frac{\partial x^j}{\partial \xi^i} = \boldsymbol{b} \cdot \boldsymbol{x}_{\xi^i}, \quad i = 1, \ldots, n, \tag{2.11}$$

and consequently

$$\boldsymbol{b} = \overline{b}_i \boldsymbol{\nabla} \xi^i = (\boldsymbol{b} \cdot \boldsymbol{x}_{\xi^i}) \boldsymbol{\nabla} \xi^i, \quad i = 1, \ldots, n. \tag{2.12}$$

These components $\overline{b}_i$, $i = 1, \ldots, n$, of the vector $\boldsymbol{b}$ are called covariant. In particular, the base tangential vector $\boldsymbol{x}_{\xi^i}$ is expressed through the base normal vectors $\boldsymbol{\nabla} \xi^j$, $j = 1, \ldots, n$, as follows:

$$\boldsymbol{x}_{\xi^i} = \frac{\partial x^j}{\partial \xi^i} \frac{\partial x^j}{\partial \xi^k} \boldsymbol{\nabla} \xi^k, \quad i, j, k = 1, \ldots, n. \tag{2.13}$$

2.2.5 Metric Tensors

Many grid generation algorithms, in particular those based on the calculus of variations, are typically formulated in terms of fundamental features of coordinate transformations and the corresponding mesh cells. These features are compactly described with the use of the metric notation, which is discussed in this subsection.

2.2.5.1 Covariant Metric Tensor

The matrix

$$(g_{ij}), \quad i,j=1,\dots,n,$$

whose elements g_{ij} are the dot products of the pairs of the basic tangential vectors $\boldsymbol{x}_{\xi^i}$,

$$g_{ij}=\boldsymbol{x}_{\xi^i}\cdot\boldsymbol{x}_{\xi^j}=\frac{\partial x^k}{\partial\xi^i}\frac{\partial x^k}{\partial\xi^j}, \quad i,j,k=1,\dots,n, \tag{2.14}$$

is called a covariant metric tensor of the domain X^n in the coordinates $\xi^1,\dots,\xi^n$. Geometrically, each diagonal element g_{ii} of the matrix (g_{ij}) is the length of the tangent vector $\boldsymbol{x}_{\xi^i}$ squared:

$$g_{ii}=|\boldsymbol{x}_{\xi^i}|^2, \quad i=1,\dots,n.$$

Also,

$$g_{ij}=|\boldsymbol{x}_{\xi^i}||\boldsymbol{x}_{\xi^j}|\cos\theta=\sqrt{g_{ii}}\sqrt{g_{jj}}\cos\theta, \tag{2.15}$$

where θ is the angle between the tangent vectors $\boldsymbol{x}_{\xi^i}$ and $\boldsymbol{x}_{\xi^j}$. In these expressions for g_{ii} and g_{ij} the subscripts ii and jj are fixed, i.e. here the summation over the repeated indices is not carried out.

The matrix (g_{ij}) is called the metric tensor because it defines distance measurements with respect to the coordinates $\xi^1,\dots,\xi^n$:

$$\mathrm{d}s=\sqrt{g_{ij}\mathrm{d}\xi^i\mathrm{d}\xi^j}, \quad i,j=1,\dots,n.$$

Thus the length s of the curve in X^n prescribed by the parametrization

$$\boldsymbol{x}[\boldsymbol{\xi}(t)]:[a,b]\to X^n$$

is computed by the formula

$$s=\int_a^b\sqrt{g_{ij}\frac{\mathrm{d}\xi^i}{\mathrm{d}t}\frac{\mathrm{d}\xi^j}{\mathrm{d}t}}\mathrm{d}t.$$

We designate by g the Jacobian of the covariant matrix (g_{ij}). It is evident that

$$(g_{ij})=\jmath\jmath^T,$$

and hence

$$J^2 = g.$$

The covariant metric tensor is a symmetric matrix, i.e. $g_{ij} = g_{ji}$. If a coordinate system at a point $\boldsymbol{\xi}$ is orthogonal then the tensor (g_{ij}) has a simple diagonal form at this point. Note that these advantageous properties are in general not possessed by the Jacobi matrix $(\partial x^i/\partial \xi^i)$ from which the covariant metric tensor (g_{ij}) is defined.

2.2.5.2 Contravariant Metric Tensor

The contravariant metric tensor of the domain X^n in the coordinates $\xi^1, \dots, \xi^n$ is the matrix

$$(g^{ij}), \quad i,j = 1, \dots, n,$$

inverse to (g_{ij}), i.e.

$$g_{ij}g^{jk} = \delta_i^k, \quad i,j,k = 1, \dots, n. \tag{2.16}$$

Therefore

$$\det(g^{ij}) = \frac{1}{g}.$$

It is easily shown that (2.16) is satisfied if and only if

$$g^{ij} = \nabla \xi^i \cdot \nabla \xi^j = \frac{\partial \xi^i}{\partial x^k} \frac{\partial \xi^j}{\partial x^k}, \quad i,j,k = 1, \dots, n. \tag{2.17}$$

Thus, each diagonal element g^{ii} (where i is fixed) of the matrix (g^{ij}) is the square of the length of the vector $\nabla \xi^i$:

$$g^{ii} = |\nabla \xi^i|^2. \tag{2.18}$$

Geometric Interpretation

Now we discuss the geometric meaning of a fixed diagonal element g^{ii}, say g^{11}, of the matrix (g^{ij}). Let us consider a three-dimensional coordinate transformation $\boldsymbol{x}(\boldsymbol{\xi}) : \Xi^3 \to X^3$. Its tangential vectors $\boldsymbol{x}_{\xi^1}$, $\boldsymbol{x}_{\xi^2}$, $\boldsymbol{x}_{\xi^3}$ represent geometrically the edges of the parallelepiped formed by these vectors. For the distance d_1 between the opposite faces of the parallelepiped which are defined by the vectors $\boldsymbol{x}_{\xi^2}$ and $\boldsymbol{x}_{\xi^3}$, we have

$$d_1 = \boldsymbol{x}_{\xi^1} \cdot \boldsymbol{n}_1,$$

where $\boldsymbol{n}_1$ is the unit normal to the plane spanned by the vectors $\boldsymbol{x}_{\xi^2}$ and $\boldsymbol{x}_{\xi^3}$. It is clear, that

$$\nabla_{\xi^1} \cdot \boldsymbol{x}_{\xi^j} = 0, \quad j = 2, 3,$$

and hence the unit normal $\boldsymbol{n}_1$ is parallel to the normal base vector ∇_{ξ^1}. Thus we obtain

$$\boldsymbol{n}_1 = \boldsymbol{\nabla}_{\xi^1}/|\boldsymbol{\nabla}_{\xi^1}| = \boldsymbol{\nabla}_{\xi^1}/\sqrt{g^{11}}.$$

Therefore

$$d_1 = \boldsymbol{\nabla}_{\xi^1} \cdot \boldsymbol{\nabla}_{\xi^1}/\sqrt{g^{11}} = 1/\sqrt{g^{11}},$$

and consequently

$$g^{11} = 1/(d_1)^2.$$

Analogous relations are valid for g^{22} and g^{33}, i.e. in three dimensions the diagonal element g^{ii} for a fixed i means the inverse square of the distance d_i between those faces of the parallelepiped which are connected by the vector $\boldsymbol{x}_{\xi^i}$. In two-dimensional space the element g^{ii} (where i is fixed) is the inverse square of the distance between the edges of the parallelogram defined by the tangential vectors $\boldsymbol{x}_{\xi^1}$ and $\boldsymbol{x}_{\xi^2}$.

The same interpretation of g^{ii} is valid for general multidimensional coordinate transformations:

$$g^{ii} = 1/(d_i)^2, \quad i = 1, \ldots, n, \tag{2.19}$$

where the index i is fixed, and d_i is the distance between those faces of the n-dimensional parallelepiped which are linked by the tangential vector $\boldsymbol{x}_{\xi^i}$.

2.2.5.3 Relations Between Covariant and Contravariant Elements

Now, in analogy with (2.3) and (2.4), we write out very convenient formulas for natural relations between the contravariant elements g^{ij} and the covariant ones g_{ij} in two and three dimensions.

For $n = 2$,

$$\begin{aligned} g^{ij} &= (-1)^{i+j} \frac{g_{3-i\ 3-j}}{g}, \\ g_{ij} &= (-1)^{i+j} g g^{3-i\ 3-j}, \quad i, j = 1, 2, \end{aligned} \tag{2.20}$$

where the indices i, j on the right-hand side of the relations (2.20) are fixed, i.e. summation over the repeated indices is not carried out here. For $n = 3$ we have

$$\begin{aligned} g^{ij} &= \frac{1}{g}(g_{i+1\ j+1}\, g_{i+2\ j+2} - g_{i+1\ j+2}\, g_{i+2\ j+1}), \\ g_{ij} &= g(g^{i+1\ j+1}\, g^{i+2\ j+2} - g^{i+1\ j+2}\, g^{i+2\ j+1}), \quad i, j = 1, 2, 3, \end{aligned} \tag{2.21}$$

with the convention that any index, say l, is identified with $l \pm 3$, so, for instance, $g_{45} = g_{12}$.

We also note that, in accordance with the expressions (2.14), (2.17) for g_{ij} and g^{ij}, respectively, the relations (2.10) and (2.13) between the basic vectors $\boldsymbol{x}_{\xi^i}$ and $\boldsymbol{\nabla}\xi^j$ can be written in the form

$$\begin{aligned} \boldsymbol{x}_{\xi^i} &= g_{ik} \boldsymbol{\nabla}\xi^k, \\ \boldsymbol{\nabla}\xi^i &= g^{ik} \boldsymbol{x}_{\xi^k}, \quad i, k = 1, \ldots, n. \end{aligned} \tag{2.22}$$

So the first derivatives $\partial x^i/\partial \xi^j$ and $\partial \xi^k/\partial x^m$ of the transformations $\boldsymbol{x}(\boldsymbol{\xi})$ and $\boldsymbol{\xi}(\boldsymbol{x})$, respectively, are connected through the metric elements:

$$\begin{aligned} \frac{\partial x^i}{\partial \xi^j} &= g_{mj}\frac{\partial \xi^m}{\partial x^i}, \\ \frac{\partial \xi^i}{\partial x^j} &= g^{mi}\frac{\partial x^j}{\partial \xi^m}, \qquad i,j,m = 1,\dots,n. \end{aligned} \tag{2.23}$$

2.2.6 Cross Product

In addition to the dot product there is another important operation on three-dimensional vectors. This is the cross product, $\times$, which for any two vectors $\boldsymbol{a} = (a^1, a^2, a^3)$, $\boldsymbol{b} = (b^1, b^2, b^3)$ is expressed as the determinant of a matrix:

$$\boldsymbol{a} \times \boldsymbol{b} = \det \begin{pmatrix} \boldsymbol{e}_1 & \boldsymbol{e}_2 & \boldsymbol{e}_3 \\ a^1 & a^2 & a^3 \\ b^1 & b^2 & b^3 \end{pmatrix}, \tag{2.24}$$

where $(\boldsymbol{e}_1, \boldsymbol{e}_2, \boldsymbol{e}_3)$ is the Cartesian vector basis of the Euclidean space R^3. Thus

$$\boldsymbol{a} \times \boldsymbol{b} = (a^2b^3 - a^3b^2,\ a^3b^1 - a^1b^3,\ a^1b^2 - a^2b^1),$$

or, with the previously mentioned convention in three dimensions of the identification of any index j with $j \pm 3$,

$$\boldsymbol{a} \times \boldsymbol{b} = (a^{i+1}b^{i+2} - a^{i+2}b^{i+1})\boldsymbol{e}_i, \quad i = 1,2,3. \tag{2.25}$$

We will now state some facts connected with the cross product operation.

2.2.6.1 Geometric Meaning

We can readily see that $\boldsymbol{a} \times \boldsymbol{b} = \boldsymbol{0}$ if the vectors $\boldsymbol{a}$ and $\boldsymbol{b}$ are parallel. Also, from (2.25) we find that $\boldsymbol{a} \cdot (\boldsymbol{a} \times \boldsymbol{b}) = 0$ and $\boldsymbol{b} \cdot (\boldsymbol{a} \times \boldsymbol{b}) = 0$, i.e. the vector $\boldsymbol{a} \times \boldsymbol{b}$ is orthogonal to each of the vectors $\boldsymbol{a}$ and $\boldsymbol{b}$. Thus, if these vectors are not parallel then

$$\boldsymbol{a} \times \boldsymbol{b} = \alpha|\boldsymbol{a} \times \boldsymbol{b}|\boldsymbol{n}, \tag{2.26}$$

where $\alpha = 1$ or $\alpha = -1$ and $\boldsymbol{n}$ is a unit normal vector to the plane determined by the vectors $\boldsymbol{a}$ and $\boldsymbol{b}$.

Now we show that the length of the vector $\boldsymbol{a} \times \boldsymbol{b}$ equals the area of the parallelogram formed by the vectors $\boldsymbol{a}$ and $\boldsymbol{b}$, i.e.

$$|\boldsymbol{a} \times \boldsymbol{b}| = |\boldsymbol{a}||\boldsymbol{b}|\sin\theta, \tag{2.27}$$

where θ is the angle between the two vectors $\boldsymbol{a}$ and $\boldsymbol{b}$. To prove (2.27) we first note that

$$|\boldsymbol{a}|^2|\boldsymbol{b}|^2 \sin^2\theta = |\boldsymbol{a}|^2|\boldsymbol{b}|^2(1-\cos^2\theta) = |\boldsymbol{a}|^2|\boldsymbol{b}|^2 - (\boldsymbol{a}\cdot\boldsymbol{b})^2.$$

We have, further,

$$\begin{aligned}|\boldsymbol{a}|^2|\boldsymbol{b}|^2 - |\boldsymbol{a}\cdot\boldsymbol{b}|^2 &= \left(\sum_{i=1}^{3} a^i a^i\right)\left(\sum_{j=1}^{3} b^j b^j\right) - \left(\sum_{k=1}^{3} a^k b^k\right)^2 \\ &= \sum_{k=1}^{3}[(a^l)^2(b^m)^2 + (a^m)^2(b^l)^2 - 2a^l b^l a^m b^m] \\ &= \sum_{k=1}^{3}(a^l b^m - a^m b^l)^2,\end{aligned}$$

where (k,l,m) are cyclic, i.e. $l = k+1$, $m = k+2$ with the convention that $j+3$ is equivalent to j for any index j. According to (2.25) the quantity $a^l b^m - a^m b^l$ for the cyclic sequence (k,l,m) is the kth component of the vector $\boldsymbol{a}\times\boldsymbol{b}$, so we find that

$$|\boldsymbol{a}||\boldsymbol{b}|\sin^2\theta = |\boldsymbol{a}|^2|\boldsymbol{b}|^2 - |\boldsymbol{a}\cdot\boldsymbol{b}|^2 = |\boldsymbol{a}\times\boldsymbol{b}|^2, \tag{2.28}$$

what proves (2.27). Thus we obtain the result that if the vectors $\boldsymbol{a}$ and $\boldsymbol{b}$ are not parallel then the vector $\boldsymbol{a}\times\boldsymbol{b}$ is orthogonal to the parallelogram formed by these vectors and its length equals the area of the parallelogram. Therefore the three vectors $\boldsymbol{a}$, $\boldsymbol{b}$ and $\boldsymbol{a}\times\boldsymbol{b}$ are independent in this case and represent a base vector system in the three-dimensional space R^3. Moreover, the vectors $\boldsymbol{a}$, $\boldsymbol{b}$ and $\boldsymbol{a}\times\boldsymbol{b}$ form a right-handed triad since $\boldsymbol{a}\times\boldsymbol{b}\neq\boldsymbol{0}$, and consequently the Jacobian of the matrix determined by $\boldsymbol{a}$, $\boldsymbol{b}$, and $\boldsymbol{a}\times\boldsymbol{b}$ is positive; it equals

$$\boldsymbol{a}\times\boldsymbol{b}\cdot\boldsymbol{a}\times\boldsymbol{b} = (\boldsymbol{a}\times\boldsymbol{b})^2.$$

2.2.6.2 Relation to Volumes

Let $\boldsymbol{c} = (c^1, c^2, c^3)$ be one more vector. The volume V of the parallelepiped whose edges are the vectors $\boldsymbol{a}$, $\boldsymbol{b}$ and $\boldsymbol{c}$ equals the area of the parallelogram formed by the vectors $\boldsymbol{a}$ and $\boldsymbol{b}$ multiplied by the modulas of the dot product of the vector $\boldsymbol{c}$ and the unit normal $\boldsymbol{n}$ to the parallelogram. Thus

$$V = |\boldsymbol{a}\times\boldsymbol{b}||\boldsymbol{n}\cdot\boldsymbol{c}|$$

and from (2.26) we obtain

$$V = |(\boldsymbol{a}\times\boldsymbol{b})\cdot\boldsymbol{c}|. \tag{2.29}$$

Taking into account (2.25), we obtain

$$(\boldsymbol{a}\times\boldsymbol{b})\cdot\boldsymbol{c} = c^1(a^2b^3 - a^3b^2) + c^2(a^3b^1 - a^1b^3) + c^3(a^1b^2 - a^2b^1).$$

The right-hand side of this equation is the Jacobian of the matrix whose rows are formed by the vectors $\boldsymbol{a}$, $\boldsymbol{b}$, and $\boldsymbol{c}$, i.e.

$$(\boldsymbol{a} \times \boldsymbol{b}) \cdot \boldsymbol{c} = \det \begin{pmatrix} a^1 & a^2 & a^3 \\ b^1 & b^2 & b^3 \\ c^1 & c^2 & c^3 \end{pmatrix}. \tag{2.30}$$

From this equation we readily obtain

$$(\boldsymbol{a} \times \boldsymbol{b}) \cdot \boldsymbol{c} = \boldsymbol{a} \cdot (\boldsymbol{b} \times \boldsymbol{c}) = (\boldsymbol{c} \times \boldsymbol{a}) \cdot \boldsymbol{b}.$$

Thus the volume of the parallelepiped determined by the vectors $\boldsymbol{a}$, $\boldsymbol{b}$, and $\boldsymbol{c}$ equals the Jacobian of the matrix formed by the components of these vectors. In particular, we obtain from (2.1) that the Jacobian of a three-dimensional coordinate transformation $\boldsymbol{x}(\boldsymbol{\xi})$ is expressed as follows:

$$J = \boldsymbol{x}_{\xi^1} \cdot (\boldsymbol{x}_{\xi^2} \times \boldsymbol{x}_{\xi^3}). \tag{2.31}$$

2.2.6.3 Relation to Base Vectors

Applying the operation of the cross product to two base tangential vectors $\boldsymbol{x}_{\xi^l}$ and $\boldsymbol{x}_{\xi^m}$, we find that the vector $\boldsymbol{x}_{\xi^l} \times \boldsymbol{x}_{\xi^m}$ is a normal to the coordinate surface $\xi^i = \xi_0^i$ with (i, l, m) cyclic. The base normal vector $\nabla \xi^i$ is also orthogonal to the surface and therefore it is a scalar multiple of $\boldsymbol{x}_{\xi^l} \times \boldsymbol{x}_{\xi^m}$, i.e.

$$\nabla \xi^i = c(\boldsymbol{x}_{\xi^l} \times \boldsymbol{x}_{\xi^m}).$$

Multiplying this equation for a fixed i by $\boldsymbol{x}_{\xi^i}$, using the operation of the dot product, we obtain, using (2.31),

$$1 = cJ,$$

and therefore

$$\nabla \xi^i = \frac{1}{J}(\boldsymbol{x}_{\xi^l} \times \boldsymbol{x}_{\xi^m}). \tag{2.32}$$

Thus the elements of the three-dimensional contravariant metric tensor (g^{ij}) are computed through the tangential vectors $\boldsymbol{x}_{\xi^i}$ by the formula

$$g^{ij} = \frac{1}{g}(\boldsymbol{x}_{\xi^{i+1}} \times \boldsymbol{x}_{\xi^{i+2}}) \cdot (\boldsymbol{x}_{\xi^{j+1}} \times \boldsymbol{x}_{\xi^{j+2}}), \quad i, j = 1, 2, 3.$$

Analogously, every base vector $\boldsymbol{x}_{\xi^i}$, $i = 1, 2, 3$, is expressed by the tensor product of the vectors $\nabla \xi^j$, $j = 1, 2, 3$:

$$\boldsymbol{x}_{\xi^i} = J(\nabla \xi^l \times \nabla \xi^k), \quad i = 1, 2, 3, \tag{2.33}$$

where $l = i + 1$, $k = i + 2$, and m is equivalent to $m + 3$ for any index m. Accordingly, we have

$$g_{ij} = g(\nabla\xi^{i+1} \times \nabla\xi^{i+2}) \cdot (\nabla\xi^{j+1} \times \nabla\xi^{j+2}), \quad i, j = 1, 2, 3.$$

Using the relations (2.32) and (2.33) in (2.31), we also obtain

$$\frac{1}{J} = \nabla\xi^1 \cdot \nabla\xi^2 \times \nabla\xi^3. \tag{2.34}$$

Thus the volume of the parallelepiped formed by the base normal vectors $\nabla\xi^1$, $\nabla\xi^2$, and $\nabla\xi^3$ is the modulus of the inverse of the Jacobian J of the transformation $\boldsymbol{x}(\boldsymbol{\xi})$.

2.3 Relations Concerning Second Derivatives

The elements of the covariant and contravariant metric tensors are defined by the dot products of the base tangential and normal vectors, respectively. These elements are suitable for describing the internal features of the cells such as the lengths of the edges, the areas of the faces, their volumes, and the angles between the edges and the faces. However, as they are derived from the first derivatives of the coordinate transformation $\boldsymbol{x}(\boldsymbol{\xi})$, the direct use of the metric elements is not sufficient for the description of the dynamic features of the grid (e.g. curvature), which reflect changes between adjacent cells. This is because the formulation of these grid features relies not only on the first derivatives but also on the second derivatives of $\boldsymbol{x}(\boldsymbol{\xi})$. Therefore there is a need to study relations connected with the second derivatives of the coordinate parametrizations.

This section presents some notations and formulas which are concerned with the second derivatives of the components of the coordinate transformations. These notations and relations will be used to describe the curvature and eccentricity of the coordinate lines and to formulate some equations of mechanics in new independent variables.

2.3.1 Christoffel Symbols

The edge of a grid cell in the ξ^i direction can be represented with high accuracy by the base vector $\boldsymbol{x}_{\xi^i}$ contracted by the factor h, which represents the step size of a uniform grid in Ξ^n. Therefore the local change of the edge in the ξ^j direction is characterized by the derivative of $\boldsymbol{x}_{\xi^i}$ with respect to ξ^j, i.e. by $\boldsymbol{x}_{\xi^i\xi^j}$.

Since the second derivatives may be used to formulate quantitative measures of the grid, we describe these vectors $\boldsymbol{x}_{\xi^i\xi^j}$ through the base tangential and normal vectors using certain three-index quantities known as Christoffel symbols. The Christoffel symbols are commonly used in formulating measures of the mutual interaction of the cells and in formulas for differential equations.

Let us denote by Γ^k_{ij} the kth contravariant component of the vector $\boldsymbol{x}_{\xi^i\xi^j}$ in the base tangential vectors $\boldsymbol{x}_{\xi^k}$, $k = 1, \ldots, n$. The superscript k in this designation relates to the base vector $\boldsymbol{x}_{\xi^k}$ and the subscript ij corresponds to the mixed derivative with respect to ξ^i and ξ^j. Thus

$$\boldsymbol{x}_{\xi^i\xi^j} = \Gamma^k_{ij}\boldsymbol{x}_{\xi^k}, \quad i,j,k = 1,\dots,n, \tag{2.35}$$

and consequently

$$\frac{\partial^2 x^p}{\partial\xi^j\partial\xi^k} = \Gamma^m_{kj}\frac{\partial x^p}{\partial\xi^m}, \quad j,k,m,p = 1,\dots,n. \tag{2.36}$$

In accordance with (2.7), we have

$$\Gamma^i_{kj} = \frac{\partial^2 x^l}{\partial\xi^k\partial\xi^j}\frac{\partial\xi^i}{\partial x^l}, \quad i,j,k,l = 1,\dots,n, \tag{2.37}$$

or in vector form,

$$\Gamma^i_{kj} = \boldsymbol{x}_{\xi^k\xi^j}\cdot\boldsymbol{\nabla}\xi^i. \tag{2.38}$$

Equation (2.37) is also obtained by multiplying (2.36) by $\partial\xi^i/\partial x^p$ and summing over p.

The quantities Γ^i_{kj} are called the space Christoffel symbols of the second kind and the expression (2.35) is a form of the Gauss relation representing the second derivatives of the position vector $\boldsymbol{x}(\boldsymbol{\xi})$ through the tangential vectors $\boldsymbol{x}_{\xi^i}$.

Analogously, the components of the second derivatives of the position vector $\boldsymbol{x}(\boldsymbol{\xi})$ expanded in the base normal vectors $\boldsymbol{\nabla}\xi^i$, $i = 1,\dots,n$, are referred to as the space Christoffel symbols of the first kind. The mth component of the vector $\boldsymbol{x}_{\xi^k\xi^j}$ in the base vectors $\boldsymbol{\nabla}\xi^i$, $i = 1,\dots,n$, is denoted by $[kj,m]$. Thus, according to (2.11),

$$[kj,m] = \boldsymbol{x}_{\xi^k\xi^j}\cdot\boldsymbol{x}_{\xi^m} = \frac{\partial^2 x^l}{\partial\xi^k\partial\xi^j}\frac{\partial x^l}{\partial\xi^m}, \quad j,k,l,m = 1,\dots,n, \tag{2.39}$$

and consequently

$$\boldsymbol{x}_{\xi^k\xi^j} = [kj,m]\boldsymbol{\nabla}\xi^m. \tag{2.40}$$

So, in analogy with (2.36), we obtain

$$\frac{\partial^2 x^l}{\partial\xi^j\partial\xi^k} = [kj,m]\frac{\partial\xi^m}{\partial x^i}, \quad i,j,k,m = 1,\dots,n. \tag{2.41}$$

Multiplying (2.39) by g^{im} and summing over m we find that the space Christoffel symbols of the first and second kind are connected by the following relation:

$$\Gamma^i_{kj} = g^{im}[kj,m], \quad i,j,k,m = 1,\dots,n. \tag{2.42}$$

Conversely, from (2.37),

$$[kj,m] = g_{ml}\Gamma^l_{kj}, \quad j,k,l,m = 1,\dots,n. \tag{2.43}$$

The space Christoffel symbols of the first kind $[kj, m]$ can be expressed through the first derivatives of the covariant elements g_{ij} of the metric tensor (g_{ij}) by the following readily verified formula:

$$[kj, m] = \frac{1}{2}\left(\frac{\partial g_{jm}}{\partial \xi^k} + \frac{\partial g_{km}}{\partial \xi^j} - \frac{\partial g_{kj}}{\partial \xi^m}\right), \quad i, j, k, m = 1, \dots, n. \tag{2.44}$$

Thus, taking into account (2.42), we see that the space Christoffel symbols of the second kind Γ^i_{kj} can be written in terms of metric elements and their first derivatives. In particular, in the case of an orthogonal coordinate system ξ^i, we obtain from (2.42), (2.44)

$$\Gamma^i_{kj} = \frac{1}{g} g^{ii}\left(\frac{\partial g_{ii}}{\partial \xi^k} + \frac{\partial g_{ii}}{\partial \xi^j} - \frac{\partial g_{kj}}{\partial \xi^i}\right).$$

Here the index i is fixed, i.e. the summation over i is not carried out.

2.3.2 Differentiation of the Jacobian

Of critical importance in obtaining compact conservation-law equations with coefficients derived from the metric elements in new curvilinear coordinates $\xi^1, \dots, \xi^n$ is the formula for differentiation of the Jacobian

$$\frac{\partial J}{\partial \xi^k} \equiv J \frac{\partial^2 x^i}{\partial \xi^k \partial \xi^m} \frac{\partial \xi^m}{\partial x^i} \equiv J \frac{\partial}{\partial x^i}\left(\frac{\partial x^i}{\partial \xi^k}\right) \equiv J \mathrm{div}_x \frac{\partial \boldsymbol{x}}{\partial \xi^k}, \quad i, k, m = 1, \dots, n. \tag{2.45}$$

In accordance with (2.37), this identity can also be expressed through the space Christoffel symbols of the second kind Γ^i_{kj} by

$$\frac{\partial J}{\partial \xi^k} = J \Gamma^i_{ik}, \quad i, k = 1, \dots, n,$$

with the summation convention over the repeated index i.

In order to prove the identity (2.45) we note that in the case of an arbitrary matrix (a_{ij}) the first derivative of its Jacobian with respect to ξ^k is obtained by the process of differentiating the first row (the others are left unchanged), then performing the same operation on the second row, and so on with all of the rows of the matrix. The summation of the Jacobians of the matrices derived in such a manner gives the first derivative of the Jacobian of the original matrix (a_{ij}). Thus

$$\frac{\partial}{\partial \xi^k} \det(a_{ij}) = \frac{\partial a_{im}}{\partial \xi^k} G^{im}, \quad i, j, k, m = 1, \dots, n, \tag{2.46}$$

where G^{im} is the cofactor of the element a_{im}. For the Jacobi matrix $(\partial x^i / \partial \xi^j)$ of the coordinate transformation $\boldsymbol{x}(\boldsymbol{\xi})$ we have

$$G^{im} = J \frac{\partial \xi^m}{\partial x^i}, \quad i, j = 1, \dots, n.$$

Therefore, applying (2.46) to the Jacobi matrix, we obtain (2.45).

2.3.3 Basic Identity

The identity (2.45) implies the extremely important relation

$$\frac{\partial}{\partial \xi^j}\left(J\frac{\partial \xi^j}{\partial x^i}\right) \equiv 0, \quad i,j=1,\dots,n, \tag{2.47}$$

which leads to specific forms of new dependent variables for conservation-law equations. To prove (2.47) we first note that

$$\frac{\partial^2 \xi^p}{\partial x^k \partial x^j}\frac{\partial x^l}{\partial \xi^p} = -\frac{\partial^2 x^l}{\partial \xi^p \partial \xi^m}\frac{\partial \xi^m}{\partial x^k}\frac{\partial \xi^p}{\partial x^j}.$$

Multiplying this equation by $\partial \xi^i/\partial x^l$ and summing over l, we obtain a formula representing the second derivative $\partial^2 \xi^i/\partial x^k \partial x^m$ of the functions $\xi^i(\boldsymbol{x})$ through the second derivatives $\partial^2 x^m/\partial \xi^l \partial \xi^p$ of the functions $x^m(\boldsymbol{\xi})$, $m = 1,\dots,n$:

$$\frac{\partial^2 \xi^i}{\partial x^k \partial x^m} = -\frac{\partial^2 x^p}{\partial \xi^l \partial \xi^j}\frac{\partial \xi^j}{\partial x^k}\frac{\partial \xi^l}{\partial x^m}\frac{\partial \xi^i}{\partial x^p}, \quad i,j,k,l,m,p=1,\dots,n. \tag{2.48}$$

Now, using this relation and the formula (2.45) for differentiation of the Jacobian in the identity

$$\frac{\partial}{\partial \xi^j}\left(J\frac{\partial \xi^j}{\partial x^i}\right) = \frac{\partial J}{\partial \xi^j}\frac{\partial \xi^j}{\partial x^i} + J\frac{\partial^2 \xi^j}{\partial x^i \partial x^k}\frac{\partial x^k}{\partial \xi^j},$$

we obtain

$$\begin{aligned}\frac{\partial}{\partial \xi^j}\left(J\frac{\partial \xi^j}{\partial x^i}\right) &= J\frac{\partial^2 x^k}{\partial \xi^p \partial \xi^j}\frac{\partial \xi^p}{\partial x^k}\frac{\partial \xi^j}{\partial x^i} - J\frac{\partial^2 x^p}{\partial \xi^l \partial \xi^m}\frac{\partial \xi^m}{\partial x^i}\frac{\partial \xi^l}{\partial x^k}\frac{\partial \xi^j}{\partial x^p}\frac{\partial x^k}{\partial \xi^j}\\ &= J\frac{\partial^2 x^k}{\partial \xi^p \partial \xi^j}\frac{\partial \xi^p}{\partial x^k}\frac{\partial \xi^j}{\partial x^i} - J\frac{\partial^2 x^p}{\partial \xi^l \partial \xi^m}\frac{\partial \xi^l}{\partial x^p}\frac{\partial \xi^m}{\partial x^i} = 0,\\ &i,j,k,l,m,p=1,\dots,n,\end{aligned}$$

i.e. (2.47) has been proved.

The identity (2.47) is obvious when $n = 1$ or $n = 2$. For example, for $n = 2$ we have from (2.3)

$$J\frac{\partial \xi^j}{\partial x^i} = (-1)^{i+j}\frac{\partial x^{3-i}}{\partial \xi^{3-j}}, \quad i,j=1,2,$$

with fixed indices i and j, and therefore

$$\frac{\partial}{\partial \xi^j}\left(J\frac{\partial \xi^j}{\partial x^i}\right) = (-1)^{i+1}\left(\frac{\partial}{\partial \xi^1}\frac{\partial x^{3-i}}{\partial \xi^2} - \frac{\partial}{\partial \xi^2}\frac{\partial x^{3-i}}{\partial \xi^1}\right) = 0, \quad i,j=1,2.$$

An inference from (2.47) for $n = 3$ follows from the differentiation of the cross product of the base tangential vectors $\boldsymbol{r}_{\xi^i}$, $i = 1,2,3$. Taking into account (2.25), we readily obtain the following formula for the differentiation of the cross product of two three-dimensional vector-valued functions $\boldsymbol{a}$ and $\boldsymbol{b}$:

$$\frac{\partial}{\partial \xi^i}(\boldsymbol{a} \times \boldsymbol{b}) = \frac{\partial}{\partial \xi^i}\boldsymbol{a} \times \boldsymbol{b} + \boldsymbol{a} \times \frac{\partial}{\partial \xi^i}\boldsymbol{b}, \quad i = 1, 2, 3.$$

With this formula we obtain

$$\sum_{i=1}^{3} \frac{\partial}{\partial \xi^i}(\boldsymbol{x}_{\xi^j} \times \boldsymbol{x}_{\xi^k}) = \sum_{i=1}^{3} \boldsymbol{x}_{\xi^j \xi^i} \times \boldsymbol{x}_{\xi^k} + \sum_{i=1}^{3} \boldsymbol{x}_{\xi^j} \times \boldsymbol{x}_{\xi^k \xi^i}, \tag{2.49}$$

where the indices $(i,\ j,\ k)$ are cyclic, i.e. $j = i + 1,\ k = i + 2$, m is equivalent to $m + 3$. For the last summation of the above formula, we obtain

$$\sum_{i=1}^{3} \boldsymbol{x}_{\xi^j} \times \boldsymbol{x}_{\xi^k \xi^i} = \sum_{i=1}^{3} \boldsymbol{x}_{\xi^k} \times \boldsymbol{x}_{\xi^i \xi^j}.$$

Therefore, from (2.49),

$$\sum_{i=1}^{3} \frac{\partial}{\partial \xi^i}(\boldsymbol{x}_{\xi^j} \times \boldsymbol{x}_{\xi^k}) = 0,$$

since

$$\boldsymbol{x}_{\xi^i} \times \boldsymbol{x}_{\xi^j \xi^k} = -\boldsymbol{x}_{\xi^j \xi^k} \times \boldsymbol{x}_{\xi^i}$$

and (2.32) implies (2.47) for $n = 3$.

The identity (2.47) can help one to obtain conservative or compact forms of some differential expressions and equations in the curvilinear coordinates $\xi^1, \dots, \xi^n$. For example, for the first derivative of a function $f(\boldsymbol{x})$ with respect to x^i we obtain, using (2.47),

$$\frac{\partial f}{\partial x^i} = \frac{1}{J}\frac{\partial}{\partial \xi^j}\left(J\frac{\partial \xi^j}{\partial x^i} f\right), \quad j = 1, \dots, n. \tag{2.50}$$

For the Laplacian

$$\nabla^2 f = \frac{\partial}{\partial x^j}\frac{\partial f}{\partial x^j}, \quad j = 1, \dots, n \tag{2.51}$$

we have, substituting the quantity $\partial f/\partial x^i$ for f in (2.50),

$$\begin{aligned} \nabla^2 f &= \frac{1}{J}\frac{\partial}{\partial \xi^j}\left(J\frac{\partial \xi^j}{\partial x^i}\frac{\partial f}{\partial x^i}\right) = \frac{1}{J}\frac{\partial}{\partial \xi^j}\left(J\frac{\partial \xi^j}{\partial x^i}\frac{\partial \xi^m}{\partial x^i}\frac{\partial f}{\partial \xi^m}\right) \\ &= \frac{1}{J}\frac{\partial}{\partial \xi^j}\left(J g^{mj}\frac{\partial f}{\partial \xi^m}\right), \quad i, j, m = 1, \dots, n. \end{aligned} \tag{2.52}$$

Therefore the Poisson equation

$$\nabla^2 f = P \tag{2.53}$$

has the form

$$\frac{1}{J}\frac{\partial}{\partial \xi^j}\left(Jg^{mj}\frac{\partial f}{\partial \xi^m}\right) = P \tag{2.54}$$

with respect to the independent variables $\xi^1, \ldots, \xi^n$.

2.4 Conservation Laws

This section utilizes the relations described in Sects. 2.2 and 2.3, in particular the identity (2.47), in order to describe some conservation-law equations of mechanics in divergent or compact form in new independent curvilinear coordinates $\xi^1, \ldots, \xi^n$. For this purpose the dependent physical variables are also transformed to new dependent variables using some specific formulas. The essential advantage of the equations described here is that their coefficients are derived from the elements of the covariant metric tensor (g_{ij}).

2.4.1 Scalar Conservation Laws

Let $\boldsymbol{A}$ be an n-dimensional vector with components A^i, $i = 1, \ldots, n$, in the Cartesian coordinates $x^1, \ldots, x^n$. The operator

$$\mathrm{div}_x \boldsymbol{A} = \frac{\partial A^i}{\partial x^i}, \quad i = 1, \ldots, n, \tag{2.55}$$

is commonly used in mechanics for the representation of scalar conservation laws, commonly in the form

$$\mathrm{div}_x \boldsymbol{A} = F.$$

Using (2.47) we easily obtain

$$\mathrm{div}_x \boldsymbol{A} = \frac{1}{J}\frac{\partial}{\partial \xi^j}(J\overline{A}^j) = F, \quad j = 1, \ldots, n, \tag{2.56}$$

where $\overline{A}^j$ is the jth contravariant component of the vector $\boldsymbol{A}$ in the coordinates ξ^i, $i = 1, \ldots, n$, i.e. in accordance with (2.7):

$$\overline{A}^j = A^i \frac{\partial \xi^j}{\partial x^i}, \quad i, j = 1, \ldots, n. \tag{2.57}$$

Therefore a divergent form of the conservation-law equation represented by (2.55) is obtained in the new coordinates when the dependent variables A^i are replaced by new dependent variables $\overline{A}^i$ defined by the rule (2.57). Some examples of scalar conservation-law equations are given below.

2.4.1.1 Mass Conservation Law

As an example of the application of (2.56), we consider the equation of conservation of mass for steady gas flow

$$\frac{\partial \rho u^i}{\partial x^i} = 0, \quad i = 1, \dots, n, \tag{2.58}$$

where ρ is the gas density, and u^i is the ith component of the flow velocity vector $\boldsymbol{u}$ in the Cartesian coordinates $x^1, \dots, x^n$. With the substitution $A^i = \rho u^i$, (2.58) is transformed to the following divergent form with respect to the new dependent variables ρ and $\overline{u}^i$ in the coordinates $\xi^1, \dots, \xi^n$:

$$\frac{\partial}{\partial \xi^i}(J\rho\overline{u}^i) = 0, \quad i = 1, \dots, n. \tag{2.59}$$

Here $\overline{u}^i$ is the ith contravariant component of the flow velocity vector $\boldsymbol{u}$ in the basis $\boldsymbol{x}_{\xi^i}$, $i = 1, \dots, n$, i.e.

$$\overline{u}^i = u^j \frac{\partial \xi^i}{\partial x^j}, \quad i, j = 1, \dots, n. \tag{2.60}$$

2.4.1.2 Convection–Diffusion Equation

Another example is the conservation equation for the steady convection–diffusion of a transport variable ϕ, which can be expressed as

$$-\frac{\partial}{\partial x^i}\left(\epsilon \frac{\partial \phi}{\partial x^i}\right) + \frac{\partial}{\partial x^i}(\rho \phi u^i) = S, \quad i = 1, \dots, n, \tag{2.61}$$

where ρ and ϵ denote the density and diffusion coefficient of the fluid, respectively. Taking

$$A^i = \rho \phi u^i - \epsilon \frac{\partial \phi}{\partial x^i},$$

we obtain, in accordance with the relation (2.57),

$$\overline{A}^j = \rho \phi \overline{u}^j - \epsilon \frac{\partial \phi}{\partial \xi^k} g^{kj}.$$

Therefore, using (2.56), the convection–diffusion equation (2.61) in the curvilinear coordinates $\xi^1, \dots, \xi^n$ is expressed by the divergent form

$$\frac{\partial}{\partial \xi^j}\left[J\left(\rho \phi \overline{u}^j - \epsilon g^{kj} \frac{\partial \phi}{\partial \xi^k}\right)\right] = JS, \quad j, k = 1, \dots, n. \tag{2.62}$$

2.4.1.3 Laplace Equation

Analogously, the Laplace equation

$$\nabla^2 f = \frac{\partial}{\partial x^j}\frac{\partial f}{\partial x^j} = 0, \quad j = 1, \dots, n, \tag{2.63}$$

has the form (2.55) if we take

$$A^i = \frac{\partial f}{\partial x^i}, \quad i = 1, \dots, n.$$

Using (2.57), we obtain

$$\overline{A}^j = g^{ij}\frac{\partial f}{\partial \xi^i}, \quad i = 1, \dots, n.$$

Therefore the Laplace equation (2.63) results in

$$\nabla^2 f = \frac{1}{J}\frac{\partial}{\partial \xi^i}\left(Jg^{ij}\frac{\partial f}{\partial \xi^j}\right) = 0, \tag{2.64}$$

since (2.56) applies.

2.4.2 Vector Conservation Laws

Many physical problems are also modeled as a system of conservation-law equations in the vector form

$$\frac{\partial A^{ij}}{\partial x^j} = F^i, \quad i, j = 1, \dots, n. \tag{2.65}$$

For the representation of the system (2.65) in new coordinates $\xi^1, \dots, \xi^n$ in a form which includes only coefficients derived from the elements of the metric tensor, it is necessary to make a transition from the original expression for A^{ij} to a new one $\overline{A}^{ij}$. One convenient formula for such a transition from the dependent variables A^{ij} to $\overline{A}^{ij}$, $i, j = 1, \dots, n$,

$$\overline{A}^{ij} = A^{km}\frac{\partial \xi^i}{\partial x^k}\frac{\partial \xi^j}{\partial x^m}, \quad i, j, k, m = 1, \dots, n. \tag{2.66}$$

This relation between A^{ij} and $\overline{A}^{km}$ is in fact composed of transitions of the kind (2.57) for the rows and columns of the tensor A^{ij}. In tensor analysis the quantity $\overline{A}^{ij}$ means the (i, j) component of the second-rank contravariant tensor (A^{ij}) in the coordinates $\xi^1, \dots, \xi^n$.

Multiplying (2.66) by $(\partial x^p/\partial \xi^i)(\partial x^l/\partial \xi^j)$ and summing over i and j we also obtain a formula for the transition from the new dependent variables $\overline{A}^{ij}$ to the original ones A^{ij}:

$$A^{ij} = \overline{A}^{km} \frac{\partial x^i}{\partial \xi^k} \frac{\partial x^j}{\partial \xi^m}, \quad i,j,k,m = 1,\dots,n. \tag{2.67}$$

Therefore we can obtain a system of equations for the new dependent variables $\overline{A}^{ij}$ by replacing the dependent quantities A^{ij} in (2.65) with their expressions (2.67). As a result we obtain

$$\begin{aligned}
\frac{\partial A^{ij}}{\partial x^j} &= \frac{\partial}{\partial x^j}\left(\overline{A}^{km} \frac{\partial x^i}{\partial \xi^k} \frac{\partial x^j}{\partial \xi^m}\right) \\
&= \frac{\partial \overline{A}^{km}}{\partial \xi^m} \frac{\partial x^i}{\partial \xi^k} + \overline{A}^{km} \frac{\partial^2 x^i}{\partial \xi^k \partial \xi^m} + \overline{A}^{km} \frac{\partial x^i}{\partial \xi^k} \frac{\partial}{\partial x^j}\left(\frac{\partial x^j}{\partial \xi^m}\right) \\
&= F^i, \quad i,j,k,m = 1,\dots,n.
\end{aligned}$$

The use of the formula (2.45) for differentiation of the Jacobian in the summation in the equation above yields

$$\frac{\partial A^{ij}}{\partial x^j} = \frac{\partial \overline{A}^{km}}{\partial \xi^m} \frac{\partial x^i}{\partial \xi^k} + \overline{A}^{km} \frac{\partial^2 x^i}{\partial \xi^k \partial \xi^m} + \frac{1}{J} \overline{A}^{km} \frac{\partial x^i}{\partial \xi^k} \frac{\partial J}{\partial \xi^m} = F^i,$$
$$i,j,k,l,m = 1,\dots,n.$$

Multiplying this system by $\partial \xi^p / \partial x^i$ and summing over i we obtain, after simple manipulations,

$$\frac{1}{J} \frac{\partial}{\partial \xi^j} (J \overline{A}^{ij}) + \frac{\partial^2 x^l}{\partial \xi^k \partial \xi^j} \frac{\partial \xi^i}{\partial x^l} \overline{A}^{kj} = \overline{F}^i, \quad i,j,k,l = 1,\dots,n, \tag{2.68}$$

where

$$\overline{F}^i = F^j \frac{\partial \xi^i}{\partial x^j}, \quad i,j = 1,\dots,n,$$

is the ith contravariant component of the vector $\boldsymbol{F} = (F^1,\dots,F^n)$ in the basis $\boldsymbol{x}_{\xi^1},\dots,\boldsymbol{x}_{\xi^n}$. The quantities $(\partial^2 x^l / \partial \xi^k \partial \xi^j)(\partial \xi^i / \partial x^l)$ in (2.68) are the space Christoffel symbols of the second kind Γ^i_{kj}. Thus the system (2.68) has, using the notation Γ^i_{jk}, the form

$$\frac{1}{J} \frac{\partial}{\partial \xi^j} (J \overline{A}^{ij}) + \Gamma^i_{kj} \overline{A}^{kj} = \overline{F}^i, \quad i,j,k = 1,\dots,n. \tag{2.69}$$

We see that all coefficients of (2.69) are derived from the metric tensor (g_{ij}).

Equations of the form (2.69), in contrast to (2.65), do not have a conservative form. The conservative form of (2.65) in new dependent variables is obtained, in analogy with (2.56), from the system

$$\frac{1}{J} \frac{\partial}{\partial \xi^j} (J \overline{A}^j_i) = F^i, \quad i,j = 1,\dots,n, \tag{2.70}$$

where $\overline{A}^j_i$ is the jth component of the vector $\boldsymbol{A}_i = (A^{i1},\dots,A^{in})$ in the basis $\boldsymbol{x}_{\xi^j}$, $j = 1,\dots,n$, i.e.

$$\overline{A}_i^j = A^{ik}\frac{\partial \xi^j}{\partial x^k}, \quad i,j,k = 1,\dots,n. \tag{2.71}$$

In fact, (2.70) is the result of the application of (2.56) to the ith line of (2.65). Therefore in the relations (2.65), (2.70), (2.71) we can assume an arbitrary range for the index i, i.e. the matrix A^{ij} in (2.65) can be a nonsquare matrix with $i = 1,\dots,m,\ j = 1,\dots,n$.

Though the system (2.70) is conservative and more compact than (2.69), it has its drawbacks. In particular, mathematical simulations of fluid flows are generally formulated in the form (2.65) with the tensor A^{ij} represented as

$$A^{ij} = B^{ij} + \rho u^i u^j, \quad i,j = 1,\dots,n,$$

where u^i, $i = 1,\dots,n$, are the Cartesian components of the flow velocity. The transformation of the tensor $\rho u^i u^j$ by the rule (2.71),

$$\rho u^i u^k \frac{\partial \xi^j}{\partial x^k} = \rho u^i \overline{u}^j, \quad i,j,k = 1,\dots,n,$$

results in equations with an increased number of dependent variables, namely u^i and $\overline{u}^j$. The substitution of u^i for $\overline{u}^j$ or vice versa leads to equations whose coefficients are derived from the elements $\partial x^i/\partial \xi^j$ of the Jacobi matrix and not from the elements of the metric tensor (g_{ij}).

2.4.2.1 Example

As an example of (2.65) we consider the stationary equation of a compressible gas flow

$$\frac{\partial}{\partial x^j}(\rho u^i u^j) + \frac{\partial p}{\partial x^i} - \frac{\partial}{\partial x^j}\mu\frac{\partial u^i}{\partial x^j} = \rho F^i, \quad i,j = 1,\dots,n, \tag{2.72}$$

where u^i is the ith Cartesian component of the vector of the fluid velocity $\boldsymbol{u}$, ρ is the density, p is the pressure and, μ is the viscosity. The tensor form of (2.65) is given by

$$A^{ij} = \rho u^i u^j + \delta^i_j p - \mu\frac{\partial u^i}{\partial x^j}, \quad i,j = 1,\dots,n.$$

From (2.66) we obtain in this case

$$\overline{A}^{ij} = \rho \overline{u}^i \overline{u}^j + g^{ij} p - \mu \frac{\partial u^l}{\partial x^k}\frac{\partial \xi^i}{\partial x^l}\frac{\partial \xi^j}{\partial x^k}, \tag{2.73}$$

where $\overline{u}^i$ is the ith component of $\boldsymbol{u}$ in the basis $\boldsymbol{x}_{\xi^i}$, i.e. $\overline{u}^i$ is computed from the formula (2.60). It is obvious that

$$u^l = \overline{u}^j \frac{\partial x^l}{\partial \xi^j}, \quad j,l = 1,\dots,n. \tag{2.74}$$

Therefore

$$\begin{aligned}\frac{\partial u^l}{\partial x^k} &= \frac{\partial}{\partial \xi^m}\left(\overline{u}^p \frac{\partial x^l}{\partial \xi^p}\right)\frac{\partial \xi^m}{\partial x^k} \\ &= \frac{\partial \overline{u}^p}{\partial \xi^m}\frac{\partial x^l}{\partial \xi^p}\frac{\partial \xi^m}{\partial x^k} + \overline{u}^p \frac{\partial^2 x^l}{\partial \xi^p \partial \xi^m}\frac{\partial \xi^m}{\partial x^k}.\end{aligned}$$

Using this equation, we obtain, for the last term of (2.73),

$$\mu \frac{\partial u^l}{\partial x^k}\frac{\partial \xi^i}{\partial x^l}\frac{\partial \xi^j}{\partial x^k} = \mu g^{mj}\left(\frac{\partial \overline{u}^i}{\partial \xi^m} + \Gamma^i_{pm}\overline{u}^p\right), \quad i,j,m,p = 1,\ldots,n,$$

since (2.37) applies. Thus (2.73) has the form

$$\overline{A}^{ij} = \rho \overline{u}^i \overline{u}^j + g^{ij} p - \mu g^{mj}\left(\frac{\partial \overline{u}^i}{\partial \xi^m} + \Gamma^i_{pm}\overline{u}^p\right), \quad i,j,m,p = 1,\ldots,n, \tag{2.75}$$

and, applying (2.69), we obtain the following system of stationary equations (2.72) with respect to the new dependent variables $\rho, \overline{u}^i$, and p and the independent variables ξ^i:

$$\begin{aligned}&\frac{1}{J}\frac{\partial}{\partial \xi^j}\left\{J\left[\rho \overline{u}^i \overline{u}^j + g^{ij} p - \mu g^{mj}\left(\frac{\partial \overline{u}^i}{\partial \xi^m} + \Gamma^i_{pm}\overline{u}^p\right)\right]\right\} \\ &\quad + \Gamma^i_{kj}\left[\rho \overline{u}^k \overline{u}^j + g^{kj} p - \mu g^{mj}\left(\frac{\partial \overline{u}^k}{\partial \xi^m} + \Gamma^k_{pm}\overline{u}^p\right)\right] = \rho \overline{F}^i, \\ &i,j,k,m,p = 1,\ldots,n.\end{aligned} \tag{2.76}$$

The application of (2.70) to (2.72) yields the following system of stationary equations:

$$\begin{aligned}&\frac{1}{J}\frac{\partial}{\partial \xi^j}\left[J\left(\rho u^i \overline{u}^j + \frac{\partial \xi^j}{\partial x^i} p - \mu \frac{\partial u^i}{\partial \xi^k} g^{kj}\right)\right] = \rho F^i, \\ &\overline{u}^j = u^k \frac{\partial \xi^j}{\partial x^k}, \ i,j,k = 1,\ldots,n.\end{aligned} \tag{2.77}$$

Now, as an example of the utilization of the Christoffel symbols of the second kind Γ^i_{kj}, we write out the expression for the transformed elements of the tensor

$$\sigma^{ij} = \mu\left(\frac{\partial u^i}{\partial x^j} + \frac{\partial u^j}{\partial x^i}\right), \quad i,j = 1,\ldots,n, \tag{2.78}$$

in the coordinates $\xi^1,\ldots,\xi^n$, obtained in accordance with the rule (2.66). This tensor is very common and is important in applications simulating deformation in the theory of elasticity and deformation rate in fluid mechanics. Using the notations described above, the tensor $\overline{\sigma}^{ij}$ can be expressed in the coordinates $\xi^1,\ldots,\xi^n$ through the metric elements and the Christoffel symbols of the second kind. For the component $\overline{\sigma}^{ij}$ we have

$$\overline{\sigma}^{ij} = \sigma^{mk}\frac{\partial\xi^i}{\partial x^m}\frac{\partial\xi^j}{\partial x^k}$$
$$= \mu\left(g^{jl}\frac{\partial\overline{u}^i}{\partial\xi^l} + g^{il}\frac{\partial\overline{u}^j}{\partial\xi^l} + (g^{jl}\Gamma^i_{pl} + g^{il}\Gamma^j_{pl})\overline{u}^p\right),$$
$$i, j, l, p = 1, \ldots, n. \tag{2.79}$$

This formula is obtained rather easily. For this purpose one can use the relation (2.74) for the inverse transition from the contravariant components $\overline{u}^i$ to the Cartesian components u^j of the vector $\boldsymbol{u} = (u^1, \ldots, u^n)$ and the formula (2.37). By substituting (2.74) in (2.78), carrying out differentiation by the chain rule, and using the expression (2.37), we obtain (2.79).

2.5 Time-Dependent Transformations

The numerical solution of time-dependent equations requires the application of moving grids and the corresponding coordinate transformations, which are dependent on time. Commonly, such coordinate transformations are determined in the form of a vector-valued time-dependent function

$$\boldsymbol{x}(t, \boldsymbol{\xi}) : \Xi^n \to X^n_t, \quad \boldsymbol{\xi} \in \Xi^n,\ t \in [0, 1], \tag{2.80}$$

where the variable t represents the time and X^n_t is an n-dimensional domain whose boundary points change smoothly with respect to t. It is assumed that $\boldsymbol{x}(t, \boldsymbol{\xi})$ is sufficiently smooth with respect to ξ^i and t and, in addition, that it is invertible for all $t \in [0, 1]$. Therefore there is also the time-dependent inverse transformation

$$\boldsymbol{\xi}(t, \boldsymbol{x}) : X^n_t \to \Xi^n \tag{2.81}$$

for every $t \in [0, 1]$. The introduction of these time-dependent coordinate transformations enables one to compute an unsteady solution on a fixed uniform grid in Ξ^n by the numerical solution of the transformed equations.

2.5.1 Reformulation of Time-Dependent Transformations

Many physical problems are modeled in the form of nonstationary conservation-law equations which include the time derivative. The formulas of Sects. 2.3 and 2.4 can be used directly, by transforming the equations at every value of time t. However, such utilization of the formulas does not influence the temporal derivative, which is transformed simply to the form

$$\frac{\partial}{\partial t} + \frac{\partial\xi^i}{\partial t}\frac{\partial}{\partial\xi^i}, \quad i = 1, \ldots, n,$$

so that does not maintain the property of divergency and its coefficients are not derived from the elements of the metric tensor.

Instead, the formulas of Sects. 2.3 and 2.4 can be more successfully applied to time-dependent conservation-law equations if the set of the functions $\boldsymbol{x}(t, \boldsymbol{\xi})$ is expanded to an $(n+1)$-dimensional coordinate transformation in which the temporal parameter t is considered in the same manner as the spatial variables.

To carry out this process we expand the n-dimensional computational and physical domains in (2.80) to $(n+1)$-dimensional ones, assuming

$$\Xi^{n+1} = I \times \Xi^n, \qquad X^{n+1} = \bigcup_t t \times X_t^n.$$

Let the points of these domains be designated by $\boldsymbol{\xi}_0 = (\xi^0, \xi^1, \dots, \xi^n)$ and $\boldsymbol{x}_0 = (x^0, x^1, \dots, x^n)$, respectively. The expanded coordinate transformation is defined as

$$\boldsymbol{x}_0(\boldsymbol{\xi}_0) : \Xi^{n+1} \to X^{n+1}, \tag{2.82}$$

where $x^0(\boldsymbol{\xi}_0) = \xi^0$ and $x^i(\boldsymbol{\xi}_0)$, $i = 1, \dots, n$, this coincides with (2.80) with $\xi^0 = t$.

The variables x^0 and ξ^0 in (2.82) represent in fact the temporal variable t. For convenience and in order to avoid ambiguity we shall also designate the variable ξ^0 in Ξ^{n+1} by τ and the variable x^0 in X^{n+1} by t. Thus $\boldsymbol{x}_0(\boldsymbol{\xi}_0)$ is the $(n+1)$-dimensional coordinate transformation which is identical to $\boldsymbol{x}(\tau, \boldsymbol{\xi})$ at every section $\xi^0 = \tau$.

The inverted coordinate transformation

$$\boldsymbol{\xi}_0(\boldsymbol{x}_0) : X^{n+1} \to \Xi^{n+1} \tag{2.83}$$

satisfies

$$\xi^0(\boldsymbol{x}_0) = x^0, \qquad \xi^i(\boldsymbol{x}_0) = \xi^i(t, \boldsymbol{x}), \quad i = 1, \dots, n,$$

where $t = x^0$, $\boldsymbol{x} = (x^1, \dots, x^n)$, and $\xi^i(t, \boldsymbol{x})$ is defined by (2.81). Thus (2.83) is identical to (2.81) at each section X_t.

2.5.2 Basic Relations

This subsection discusses some relations and, in particular, identities of the kind (2.45) and (2.47) for the time-dependent coordinate transformations (2.80), using for this purpose the $(n+1)$-dimensional vector functions (2.82) and (2.83) introduced above.

2.5.2.1 Velocity of Grid Movement

The first derivative $\boldsymbol{x}_\tau$, $\boldsymbol{x} = (x^1, x^2, \dots, x^n)$, of the transformation $\boldsymbol{x}(\xi, \tau)$ has a clear physical interpretation as the velocity vector of grid point movement. Let the vector $\boldsymbol{x}_\tau$, in analogy with the flow velocity vector $\boldsymbol{u}$, be designated by $\boldsymbol{w} = (w^1, \dots, w^n)$, i.e. $w^i = x_\tau^i$. The ith component $\overline{w}^i$ of the vector $\boldsymbol{w}^i$ in the tangential bases $\boldsymbol{x}_{\xi^i}$, $i = 1, \dots, n$, is expressed by (2.7) as

$$\overline{w}^i = w^j \frac{\partial \xi^i}{\partial x^j} = \frac{\partial x^j}{\partial \tau}\frac{\partial \xi^i}{\partial x^j}, \quad i,j = 1,\dots,n.$$

Therefore

$$\boldsymbol{w} = \overline{w}^i \boldsymbol{x}_{\xi^i}, \quad i = 1,\dots,n, \tag{2.84}$$

i.e.

$$w^i = \frac{\partial x^i}{\partial \tau} = \overline{w}^j \frac{\partial x^i}{\partial \xi^j}, \quad i,j = 1,\dots,n.$$

Differentiation with respect to ξ^0 of the composition of $\boldsymbol{x}_0(\boldsymbol{\xi}_0)$ and $\boldsymbol{\xi}_0(\boldsymbol{x}_0)$ yields

$$\frac{\partial \xi^i}{\partial x^0}\frac{\partial x^0}{\partial \xi^0} + \frac{\partial \xi^i}{\partial x^j}\frac{\partial x^j}{\partial \xi^0} = 0, \quad i,j = 1,\dots,n.$$

Therefore we obtain the result

$$\frac{\partial \xi^i}{\partial t} = -\frac{\partial x^j}{\partial \tau}\frac{\partial \xi^i}{\partial x^j} = -\overline{w}^i, \quad i,j = 1,\dots,n. \tag{2.85}$$

2.5.2.2 Derivatives of the Jacobian

It is apparent that the Jacobians of the coordinate transformations $\boldsymbol{x}(\tau, \boldsymbol{\xi})$ and $\boldsymbol{x}_0(\boldsymbol{\xi}_0)$ coincide, i.e.

$$\det\left(\frac{\partial x^i}{\partial \xi^j}\right) = \det\left(\frac{\partial x^k}{\partial \xi^l}\right) = J, \quad i,j = 0,1,\dots,n, \ k,l = 1,\dots,n.$$

In the notation introduced above, the formula (2.45) for differentiation of the Jacobian of the transformation

$$\boldsymbol{x}_0(\boldsymbol{\xi}_0) : \Xi^{n+1} \to X^{n+1}$$

is expressed by the relation

$$\frac{1}{J}\frac{\partial}{\partial \xi^i} J = \frac{\partial^2 x^k}{\partial \xi^i \partial \xi^m}\frac{\partial \xi^m}{\partial x^k}, \quad i,k,m = 0,1,\dots,n, \tag{2.86}$$

differing from (2.45) only by the range of the indices. As a result, we obtain from (2.86) for $i = 0$,

$$\frac{1}{J}\frac{\partial}{\partial \tau} J = \frac{\partial}{\partial \xi^m}\left(\frac{\partial x^k}{\partial \tau}\right)\frac{\partial \xi^m}{\partial x^k} = \operatorname{div}_x \frac{\partial \boldsymbol{x}}{\partial \tau}, \quad k,m = 0,1,\dots,n,$$

and, taking into account (2.84),

$$\begin{aligned}\frac{1}{J}\frac{\partial}{\partial \tau} J &= \frac{\partial}{\partial \xi^m}\left(\overline{w}^j \frac{\partial x^k}{\partial \xi^j}\right)\frac{\partial \xi^m}{\partial x^k} \\ &= \frac{\partial \overline{w}^m}{\partial \xi^m} + \overline{w}^i \frac{\partial^2 x^k}{\partial \xi^j \partial \xi^m}\frac{\partial \xi^m}{\partial x^k}, \quad j,k,m = 1,\dots,n.\end{aligned}$$

Now, taking advantage of the formula for differentiation of the Jacobian (2.45), in the last sum of this equation we have

$$\frac{1}{J}\frac{\partial}{\partial\tau}J = \frac{\partial \overline{w}^m}{\partial \xi^m} + \frac{1}{J}\overline{w}^j\frac{\partial J}{\partial \xi^j}, \quad j,m = 1,\dots,n,$$

and consequently

$$\frac{1}{J}\frac{\partial}{\partial\tau}J = \frac{1}{J}\frac{\partial}{\partial \xi^j}(J\overline{w}^j), \quad j = 1,\dots,n. \tag{2.87}$$

2.5.2.3 Basic Identity

Analogously, the system of identities (2.47) has the following form:

$$\frac{\partial}{\partial \xi^j}\left(J\frac{\partial \xi^j}{\partial x^i}\right) = 0, \quad i,j = 0,1,\dots,n. \tag{2.88}$$

Therefore for $i = 0$ we obtain

$$\frac{\partial}{\partial\tau}(J) + \frac{\partial}{\partial \xi^j}\left(J\frac{\partial \xi^j}{\partial t}\right) = 0, \quad j = 1,\dots,n, \tag{2.89}$$

and, taking into account (2.85),

$$\frac{\partial}{\partial\tau}J - \frac{\partial}{\partial \xi^j}(J\overline{w}^j) = 0, \quad j = 1,\dots,n, \tag{2.90}$$

which corresponds to (2.87). For $i > 0$ the identity (2.88) coincides with (2.47), i.e.

$$\frac{\partial}{\partial \xi^j}\left(J\frac{\partial \xi^j}{\partial x^i}\right) = 0, \quad i,j = 1,\dots,n.$$

As a result of (2.89), we obtain, in analogy with (2.50),

$$\frac{\partial f}{\partial t} = \frac{1}{J}\frac{\partial}{\partial \xi^j}\left(J\frac{\partial \xi^j}{\partial t}f\right) = \frac{1}{J}\left(\frac{\partial}{\partial\tau}(Jf) - \frac{\partial}{\partial \xi^k}(J\overline{w}^k f)\right),$$
$$j = 0,1,\dots,n, \ k = 1,\dots,n. \tag{2.91}$$

2.5.3 Equations in the Form of Scalar Conservation Laws

Many time-dependent equations can be expressed in the form of a scalar conservation law in the Cartesian coordinates $t, x^1,\dots,x^n$:

$$\frac{\partial A^0}{\partial t} + \frac{\partial A^i}{\partial x^i} = F, \quad i = 1,\dots,n. \tag{2.92}$$

Using (2.88), in analogy with (2.56), this equation is transformed in the coordinates $\xi^0,\xi^1,\dots,\xi^n$, $\xi^0 = \tau$ to

$$\frac{1}{J}\left(\frac{\partial}{\partial \xi^j}(J\overline{A}_0^j)\right) = F, \quad j = 0, 1, \ldots, n, \tag{2.93}$$

where by $\overline{A}_0^j$ we denote the jth contravariant component of the $(n+1)$-dimensional vector $\boldsymbol{A}_0 = (A^0, A^1, \ldots, A^n)$ in the basis $\partial \boldsymbol{x}_0/\partial \xi^i$, $i = 0, 1, \ldots, n$, i.e.

$$\overline{A}_0^j = A^i \frac{\partial \xi^j}{\partial x^i}, \quad i, j = 0, 1, \ldots, n. \tag{2.94}$$

We can express each component $\overline{A}_0^j$, $j = 1, \ldots, n$, of the vector $\boldsymbol{A}_0$ through the components $\overline{A}^i$ and $\overline{w}^k$, $i, k = 1, \ldots, n$, of the n-dimensional spatial vectors $\boldsymbol{A} = (A^1, \ldots, A^n)$ and $\boldsymbol{w} = (w^1, \ldots, w^n)$ in the coordinates ξ^l, $l = 1, \ldots, n$, where $\boldsymbol{A}$ is a vector obtained by projecting the vector $\boldsymbol{A}_0$ into the space R^n, i.e. $P(A^0, A^1, \ldots, A^n) = (A^1, \ldots, A^n)$. Namely,

$$\overline{A}_0^j = A^0 \frac{\partial \xi^j}{\partial t} + A^i \frac{\partial \xi^j}{\partial x^i} = \overline{A}^j - A^0 \overline{w}^j, \quad i, j = 1, \ldots, n,$$

using (2.85). Further, we have

$$\overline{A}_0^0 = A^k \frac{\partial \xi^0}{\partial x^k} = A^0, \quad k = 0, 1, \ldots, n.$$

Therefore (2.93) implies a conservation law in the variables $\tau, \xi^1, \ldots, \xi^n$ in the conservative form

$$\frac{1}{J}\left(\frac{\partial}{\partial \tau}(JA^0) + \frac{\partial}{\partial \xi^j}[J(\overline{A}^j - A^0 \overline{w}^j)]\right) = F, \quad j = 1, \ldots, n. \tag{2.95}$$

2.5.3.1 Examples of Scalar Conservation-Law Equations

As an illustration of the formula (2.95), we write out some time-dependent scalar conservation law equations presented first in the form (2.92).

Parabolic Equation

For the parabolic equation

$$\frac{\partial f}{\partial t} = \frac{\partial}{\partial x^j}\frac{\partial f}{\partial x^j}, \quad j = 1, \ldots, n, \tag{2.96}$$

we obtain from (2.95), with $A^0 = f$ and $A^i = \partial f/\partial x^i$, $i = 1, \ldots, n$,

$$\frac{\partial J f}{\partial \tau} = \frac{\partial}{\partial \xi^j}\left[J\left(g^{jk}\frac{\partial f}{\partial \xi^k} + f\overline{w}^j\right)\right], \quad j, k = 1, \ldots, n. \tag{2.97}$$

Mass Conservation Law

The scalar mass conservation law for unsteady compressible gas flow

$$\frac{\partial \rho}{\partial t} + \frac{\partial \rho u^i}{\partial x^i} = F, \quad i = 1, 2, 3, \tag{2.98}$$

is expressed in the new coordinates as

$$\frac{\partial J\rho}{\partial t} + \frac{\partial J\rho(\overline{u}^j - \overline{w}^j)}{\partial \xi^j} = JF, \quad j = 1, 2, 3. \tag{2.99}$$

Convection–Diffusion Equation

The unsteady convection–diffusion conservation equation

$$\frac{\partial}{\partial t}(\rho\phi) + \frac{\partial}{\partial x^i}(\rho\phi u^i) - \frac{\partial}{\partial x^i}\left(\epsilon\frac{\partial \phi}{\partial x^i}\right) = S, \quad i = 1, \dots, n, \tag{2.100}$$

has the form in the coordinates $\tau, \xi^1, \dots, \xi^n$

$$\frac{\partial}{\partial \tau}(J\rho\phi) + \frac{\partial}{\partial \xi^j}\left(J\rho\phi(\overline{u}^j - \overline{w}^j) - Jg^{kj}\epsilon\frac{\partial \phi}{\partial \xi^k}\right) = JS, \quad j, k = 1, \dots, n. \tag{2.101}$$

Energy Conservation Law

Analogously, the energy conservation law

$$\frac{\partial}{\partial t}\rho(e + u^2/2) + \frac{\partial}{\partial x^j}\rho u^j(e + u^2/2 + \overline{p}/\rho) = \rho F_j u_j, \quad j = 1, 2, 3, \tag{2.102}$$

where

$$e = e(\rho, p), \qquad \overline{p} = p - \gamma\frac{\partial u_i}{\partial x^i}, \quad i = 1, 2, 3,$$

$$u^2 = \sum_{i=1}^{3}(u_i)^2,$$

is transformed in accordance with (2.95) to

$$\frac{\partial}{\partial \tau}\left[J\rho\left(e + \frac{1}{2}g_{mk}\overline{u}^m\overline{u}^k\right)\right] + \frac{\partial}{\partial \xi^j}\left[J\rho\left(e + \frac{1}{2}g_{mk}\overline{u}^m\overline{u}^k\right)(\overline{u}^j - \overline{w}^j) + J\overline{pu}^j\right]$$
$$= J\rho g_{mk}\overline{f}^m\overline{u}^k, \quad j, m, k = 1, 2, 3, \tag{2.103}$$

where, taking into account (2.56),

$$\overline{p} = p - \frac{\gamma}{J}\frac{\partial}{\partial \xi^i}(J\overline{u}^i), \quad i = 1, 2, 3.$$

Linear Wave Equation

The linear wave equation

$$u_{tt} = c^2 \nabla^2 u \tag{2.104}$$

arises in many areas such as fluid dynamics, elasticity, acoustics, and magnetohydrodynamics. If the coefficient c is constant than (2.104) has a divergent form (2.92) with

$$A^0 = u_t, \qquad A^i = -c^2 \frac{\partial u}{\partial x^i}, \quad i = 1, \ldots, n,$$

or, in the coordinates $\tau, \xi^1, \ldots, \xi^n$,

$$A^0 = u_\tau - \overline{w}^i \frac{\partial u}{\partial \xi^i}, \qquad A^i = -c^2 \frac{\partial u}{\partial \xi^k} \frac{\partial \xi^k}{\partial x^i}, \quad i, k = 1, \ldots, n.$$

Therefore the divergent form (2.93) of (2.104) in the coordinates $\tau, \xi^1, \ldots, \xi^n$ has the form

$$\frac{\partial}{\partial \tau} \left[J \left(u_\tau - \overline{w}^i \frac{\partial u}{\partial \xi^i} \right) \right] + \frac{\partial}{\partial \xi^j} \left[J \left(u_\tau \overline{w}^j + (c^2 g^{mj} - \overline{w}^i \overline{w}^j) \frac{\partial u}{\partial \xi^i} \right) \right] = 0. \tag{2.105}$$

Another representation of the linear wave equation (2.104) in the coordinates $\tau, \xi^1, \ldots, \xi^n$ comes from the formula (2.64) for the Laplace operator and the description of the temporal derivative (2.91). Taking advantage of (2.91), we obtain

$$\begin{aligned} u_{tt} &= \frac{1}{J} \left[\frac{\partial}{\partial \tau} \left(J \frac{\partial u}{\partial t} \right) - \frac{\partial}{\partial \xi^k} \left(J \overline{w}^k \frac{\partial u}{\partial t} \right) \right] \\ &= \frac{1}{J} \frac{\partial}{\partial \tau} \left[J \left(u_\tau - \overline{w}^i \frac{\partial u}{\partial \xi^i} \right) \right] \\ &\quad - \frac{1}{J} \frac{\partial}{\partial \xi^k} \left[J \overline{w}^k \left(u_\tau - \overline{w}^i \frac{\partial u}{\partial \xi^i} \right) \right]. \end{aligned}$$

This equation and (2.64) allow one to derive the following form of (2.104) in the coordinates $\tau, \xi^1, \ldots, \xi^n$:

$$\begin{aligned} &\frac{\partial}{\partial \tau} \left[J \left(u_\tau - \overline{w}^i \frac{\partial u}{\partial \xi^i} \right) \right] \\ &= \frac{\partial}{\partial \xi^k} \left[J \overline{w}^k \left(u_\tau - \overline{w}^j \frac{\partial u}{\partial \xi^i} \right) \right] + c^2 \frac{\partial}{\partial \xi^k} \left(J g^{kj} \frac{\partial u}{\partial \xi^j} \right), \end{aligned} \tag{2.106}$$

which coincides with (2.105) if c^2 is a constant.

2.5.3.2 Lagrangian Coordinates

One of the most popular systems of coordinates in fluid dynamics is the Lagrangian system. A coordinate ξ^i is Lagrangian if the both the ith component of the flow velocity vector $\boldsymbol{u}$ and the grid velocity $\boldsymbol{w}$ in the tangent basis $\boldsymbol{x}_{\xi^j}$, $j = 1, \ldots, m$, coincide, i.e.

$$\overline{u}^i - \overline{w}^i = 0. \tag{2.107}$$

The examples of gas-dynamics equations described above, which include the terms $\overline{w}^i$, allow one to obtain the equations in Lagrange coordinates by substituting $\overline{u}^i$ for $\overline{w}^i$ in the written-out equations in accordance with the relation (2.107). In such a manner, we obtain the equation of mass conservation, for example, in the Lagrangian coordinates $\xi^1, \ldots, \xi^n$ as

$$\frac{\partial J\rho}{\partial \tau} = JF \tag{2.108}$$

from (2.99). Analogously, the convection–diffusion equation (2.101) and the energy conservation law (2.103) have the forms in Lagrangian coordinates ξ^i

$$\frac{\partial}{\partial \tau}(J\rho\phi) + \frac{\partial}{\partial \xi^j}\left(Jg^{kj}\epsilon\frac{\partial \phi}{\partial \xi^k}\right) = JS, \quad j,k = 1, \ldots, n,$$

and

$$\frac{\partial}{\partial \tau}\left[J\rho\left(e + \frac{1}{2}g_{mk}\overline{u}^m\overline{u}^k\right)\right] + \frac{\partial}{\partial \xi^j}(J\overline{pu}^j) = J\rho g_{mk}\overline{f}^m\overline{u}^k, \quad j,m,k = 1,2,3,$$

respectively.

In the same manner, the equations can be written in the Euler–Lagrange form, where some coordinates are Lagrangian while the rest are Cartesian coordinates.

2.5.4 Equations in the Form of Vector Conservation Laws

Now we consider a formula for a vector conservation law with time-dependent physical magnitudes A^{ij}

$$\frac{\partial}{\partial x^j}A^{ij} = F^i, \quad i,j = 0,1,\ldots,n, \tag{2.109}$$

where the independent variable x^0 represents the time variable t, i.e. $x^0 = t$. Let the new independent variables $\xi^0, \xi^1, \ldots, \xi^n$ be obtained by means of (2.82). As well as (2.69), which expresses the vector conservation law (2.65) in the coordinates $\xi^1, \ldots, \xi^n$, we find that the transformation (2.109) has the form of the following system of equations for the new dependent quantities

$\overline{A}_0^{ij}$, $i,j=0,1,\ldots,n$, with respect to the independent variables $\xi^0,\xi^1,\ldots,\xi^n$, $\xi^0=\tau$:

$$\frac{1}{J}\frac{\partial}{\partial\xi^j}(J\overline{A}_0^{ij})+\overline{\Gamma}^i_{kj}\overline{A}_0^{kj}=\overline{F}_0^i,\quad i,j=0,1,\ldots,n, \tag{2.110}$$

where

$$\begin{aligned}
\overline{A}_0^{ij}&=A^{mn}\frac{\partial\xi^i}{\partial x^m}\frac{\partial\xi^j}{\partial x^n}, && i,j,m,n=0,1,\ldots,n,\\
\overline{\Gamma}^i_{kj}&=\frac{\partial^2x^l}{\partial\xi^k\partial\xi^j}\frac{\partial\xi^i}{\partial x^l}, && i,j,k,l=0,1,\ldots,n,\\
\overline{F}_0^i&=F^j\frac{\partial\xi^i}{\partial x^j}, && i,j=0,1,\ldots,n.
\end{aligned}$$

As in the case of the scalar conservation law, we represent all of the terms of (2.110) through A^{00} and the spatial components:

$$\begin{aligned}
\overline{A}^{ij}&=A^{km}\frac{\partial\xi^i}{\partial x^k}\frac{\partial\xi^j}{\partial x^m}, && i,j,m,n=1,\ldots,n,\\
\Gamma^i_{kj}&=\frac{\partial^2x^l}{\partial\xi^k\partial\xi^j}\frac{\partial\xi^i}{\partial x^l}, && i,j,k,l=1,\ldots,n,\\
\overline{F}^i&=F^j\frac{\partial\xi^i}{\partial x^j}, && i,j=1,\ldots,n,\\
\overline{w}^i&=-\frac{\partial\xi^i}{\partial t}=\frac{\partial x^j}{\partial\tau}\frac{\partial\xi^i}{\partial x^j}, && i,j=1,\ldots,n.
\end{aligned}$$

For $\overline{A}_0^{ij}$ we obtain

$$\begin{aligned}
\overline{A}_0^{00}&=A^{00},\\
\overline{A}_0^{0i}&=A^{00}\frac{\partial\xi^i}{\partial t}+A^{0m}\frac{\partial\xi^i}{\partial x^m}=\overline{A}^{0i}-A^{00}\overline{w}^i, && i=1,\ldots,n,\\
\overline{A}_0^{i0}&=\overline{A}^{i0}-A^{00}\overline{w}^i, && i=1,\ldots,n,\\
\overline{A}_0^{ij}&=A^{00}\overline{w}^i\overline{w}^j-\overline{A}^{0j}\overline{w}^i-\overline{A}^{i0}\overline{w}^j+\overline{A}^{ij}, && i,j=1,\ldots,n.
\end{aligned}$$

Analogously, for $\overline{\Gamma}^i_{kj}$ we obtain

$$\begin{aligned}
\overline{\Gamma}^0_{kj}&=0, && k,j=0,1,\ldots,n,\\
\overline{\Gamma}^i_{00}&=\frac{\partial\overline{w}^i}{\partial t}+\overline{w}^l\overline{w}^m\Gamma^i_{lm}, && i,l,m=1,\ldots,n,\\
\overline{\Gamma}^i_{j0}&=\Gamma^i_{0j}=\frac{\partial\overline{w}^i}{\partial\xi^j}+\overline{w}^l\Gamma^i_{jl}, && i,j,l=1,\ldots,n,\\
\overline{\Gamma}^i_{jk}&=\Gamma^i_{jk}, && i,j,k=1,\ldots,n,
\end{aligned}$$

and for $\overline{F}_0^i$,

$$\overline{F}_0^0 = F^0,$$
$$\overline{F}_0^i = \overline{F}^i - A^0\overline{w}^i, \quad i = 1, \dots, n.$$

Using these expression in (2.110), we obtain a system of equations for the vector conservation law in the coordinates $\tau, \xi^1, \dots, \xi^n$ with an explicit expression for the components of the speed of the grid movement:

$$\begin{aligned}
&\frac{\partial}{\partial \tau}(JA^{00}) + \frac{\partial}{\partial \xi^j}[J(\overline{A}^{0i} - A^{00}\overline{w}^j)] = JF^0,\\
&\frac{\partial}{\partial \tau}J(\overline{A}^{i0} - A^{00}\overline{w}^j) + \frac{\partial}{\partial \xi^j}J(\overline{A}^{ij} + A^{00}\overline{w}^i\overline{w}^j - \overline{A}^{0j}\overline{w}^j - \overline{A}^{i0}\overline{w}^i)\\
&\quad + JA^{00}\left(\frac{\partial \overline{w}^i}{\partial \tau} + \overline{w}^l\frac{\partial \overline{w}^i}{\partial \xi^l} + \overline{w}^l\overline{w}^j\Gamma^i_{lj}\right)\\
&\quad + J(\overline{A}^{j0} + \overline{A}^{0j} - 2A^{00}\overline{w}^j)\left(\frac{\partial \overline{w}^i}{\partial \xi^j} + \overline{w}^l\Gamma^i_{jl}\right)\\
&\quad + J(\overline{A}^{lj} + A^{00}\overline{w}^l\overline{w}^j - \overline{A}^{0j}\overline{w}^l - \overline{A}^{0l}\overline{w}^j)\Gamma^i_{lj}\\
&= J(\overline{F}^i - F^0\overline{w}^i), \quad i, j, l = 1, \dots, n.
\end{aligned} \tag{2.111}$$

Another representation of (2.109) in new coordinates can be derived in the form of (2.95) by applying (2.95) to each line of the system (2.109). As a result we obtain

$$\frac{1}{J}\left\{\frac{\partial}{\partial \tau}(JA^{i0}) + \frac{\partial}{\partial \xi^j}\left[J\left(A^{ik}\frac{\partial \xi^j}{\partial x^k} - A^{i0}\overline{w}^j\right)\right]\right\} = F, \quad j, k = 1, \dots, n. \tag{2.112}$$

Recall that this approach is not restricted to a square form of the system (2.109), i.e. the ranges for the indices i and j can be different.

As an illustration of these equations for a vector conservation law in the curvilinear coordinates $\tau, \xi^1, \dots, \xi^n$, we write out a joint system for the conservation of mass and momentum, which in the coordinates t, x^1, x^2, x^3 has the following form:

$$\begin{aligned}
&\frac{\partial \rho}{\partial t} + \frac{\partial}{\partial x^i}\rho u^i = 0, \quad i = 1, 2, 3,\\
&\frac{\partial \rho u^i}{\partial t} + \frac{\partial}{\partial x^j}(\rho u^i u^j + \overline{p}\delta^i_j) = \rho f^i, \quad i, j = 1, 2, 3,
\end{aligned} \tag{2.113}$$

where

$$\overline{p} = p - \gamma\frac{\partial u^i}{\partial x^i}, \qquad \delta^i_j = 0 \ \text{ if } i \neq j \ \text{ and } \ \delta^i_j = 1 \ \text{ if } i = j.$$

This system is represented in the form (2.109) with

$$\begin{aligned}
A^{00} &= \rho,\\
A^{0i} &= A^{i0} = \rho u^i, \quad i = 1, 2, 3,\\
A^{ij} &= \rho u^i u^j + \delta^{ij}p, \quad i, j = 1, 2, 3,
\end{aligned}$$

i.e.

$$(A^{ij}) = \begin{pmatrix} \rho & \rho u^1 & \rho u^2 & \rho u^3 \\ \rho u^1 & \rho u^1 u^1 + \overline{p} & \rho u^1 u^2 & \rho u^1 u^3 \\ \rho u^2 & \rho u^2 u^1 & \rho u^2 u^2 + \overline{p} & \rho u^2 u^3 \\ \rho u^3 & \rho u^3 u^1 & \rho u^3 u^2 & \rho u^3 u^3 + \overline{p} \end{pmatrix}, \qquad i,j = 0,1,2,3 \tag{2.114}$$

and

$$\begin{aligned} F^0 &= 0, \\ F^i &= \rho f^i, \quad i = 1,2,3. \end{aligned}$$

For the coordinate system $\tau, \xi^1, \ldots, \xi^n$ we obtain

$$\begin{aligned} \overline{A}^{00} &= \rho, \\ \overline{A}^{0i} &= \overline{A}^{i0} = \rho(\overline{u}^i - \overline{w}^i), \quad i = 1,2,3, \\ \overline{A}^{ij} &= \rho(\overline{u}^i - \overline{w}^i)(\overline{u}^j - \overline{w}^j) + \overline{p} g^{ij}, \quad i,j = 1,2,3. \end{aligned}$$

Substituting these expressions in (2.111) we obtain a system of equations for the mass and momentum conservation laws in the coordinates $\tau, \xi^1, \ldots, \xi^n$:

$$\begin{aligned} &\frac{\partial}{\partial \tau}(J\rho) + \frac{\partial}{\partial \xi^j}[J\rho(\overline{u}^j - \overline{w}^j)] = 0, \quad i = 1,2,3, \\ &\frac{\partial}{\partial \tau}[J\rho(\overline{u}^i - \overline{w}^i)] + \frac{\partial}{\partial \xi^j}[J\rho(\overline{u}^i - \overline{w}^i)(\overline{u}^j - \overline{w}^j) + J\overline{p} g^{ij}] \\ &\qquad + J\rho \frac{\partial \overline{w}^i}{\partial \tau} + J\rho(2\overline{u}^j - \overline{w}^j)\frac{\partial \overline{w}^i}{\partial \xi^j} + J(\rho \overline{u}^l \overline{u}^j + \overline{p} g^{lj})\Gamma^i_{lj} \\ &\quad = J\rho \overline{f}^i, \quad i,j,l = 1,2,3. \end{aligned} \tag{2.115}$$

If the coordinates ξ^i are the Lagrangian ones, i.e. $\overline{u}^i = \overline{w}^i$, then we obtain from (2.115)

$$\begin{aligned} &\frac{\partial}{\partial \tau}(J\rho) = 0, \\ &J\rho \frac{\partial \overline{u}^i}{\partial \tau} + \frac{\partial}{\partial \xi^j}(J\overline{p} g^{ij}) + J\rho \overline{u}^j \frac{\partial \overline{u}^i}{\partial \xi^j} + J(\rho \overline{u}^l \overline{u}^j + \overline{p} g^{lj})\Gamma^i_{lj} = J\rho \overline{f}^i. \end{aligned} \tag{2.116}$$

Note that the first equation of the system (2.115) coincides with (2.99) if $F = 0$; this was obtained as the scalar mass conservation law.

In the same manner, we can obtain an expression for the general Navier–Stokes equations of mass and momentum conservation by inserting the tensor (σ^{ij}) described by (2.78) in the system (2.113) and the tensor $(\overline{\sigma}^{ij})$ represented by (2.79) in the system (2.115).

A divergent form of (2.113) in arbitrary coordinates $\tau, \xi^1, \ldots, \xi^n$ is obtained by applying (2.112). With this, we obtain the system

$$\begin{aligned}
&\frac{\partial}{\partial \tau}(J\rho) + \frac{\partial}{\partial \xi^j}[J\rho(\overline{u}^j - \overline{w}^j)] = 0, \\
&\frac{\partial}{\partial \tau}(J\rho\overline{u}^i) + \frac{\partial}{\partial \xi^j}\left[J\left(\rho u^i(\overline{u}^j - \overline{w}^j) + \overline{p}\frac{\partial \xi^j}{\partial x^i}\right)\right] = JF^i, \\
&\overline{u}^j = u^k\frac{\partial \xi^j}{\partial x^k}, \quad i, j, k = 1, 2, 3,
\end{aligned} \tag{2.117}$$

and, in the Lagrangian coordinates,

$$\begin{aligned}
&\frac{\partial}{\partial \tau}(J\rho) = 0, \\
&\frac{\partial}{\partial \tau}(J\rho\overline{u}^i) + \frac{\partial}{\partial \xi^j}\left(J\overline{p}\frac{\partial \xi^j}{\partial x^i}\right) = JF^i, \quad i, j = 1, 2, 3.
\end{aligned} \tag{2.118}$$

2.6 Comments

Many of the basic formulations of vector calculus and tensor analysis may be found in the books by Kochin (1951), Sokolnikoff (1964) and Gurtin (1981).

The formulation of general metric and tensor concepts specifically aimed at grid generation was originally performed by Eiseman (1980) and Warsi (1981).

Very important applications of the most general tensor relations to the formulation of unsteady equations in curvilinear coordinates in a strong conservative form were presented by Vinokur (1974). A strong conservation-law form of unsteady Euler equations also was also described by Viviand (1974).

A derivation of various forms of the Navier–Stokes equations in general moving coordinates was described by Ogawa and Ishiguto (1987).

3

Grid Quality Measures

3.1 Introduction

It is very important to develop grid generation techniques which sense grid quality features and possess means to eliminate the deficiencies of the grids. These requirements give rise to the problem of selecting and adequately formulating the necessary grid quality measures and finding out how they affect the solution error and the solution efficiency, in order to control the performance of the numerical analysis of physical problems with grids. Commonly, these quality measures encompass grid skewness, stretching, torsion, cell aspect ratio, cell volume, departure from conformality, and cell deformation.

In this chapter we utilize the notions and relations discussed in Sects. 2.2 and 2.3 to describe some qualitative and quantitative characteristics of structured grids. The structured grid concept allows one to define the grid characteristics through coordinate transformations as features of the coordinate curves, coordinate surfaces, coordinate volumes, etc. In general these features are determined through the elements of the metric tensors and their derivatives. In particular, some grid properties can be described in terms of the invariants of the covariant metric tensor.

The chapter starts with an introduction to the elementary theory of curves and surfaces, necessary for the description of the quality measures of the coordinate curves and coordinate surfaces. It also includes a discussion of the metric invariants. Various grid characteristics are then formulated through quantities which measure the features of the coordinate curves, surfaces, and transformations.

3.2 Curve Geometry

Commonly, the curves lying in the n-dimensional space R^n are represented by smooth nondegenerate parametrizations

V.D. Liseikin, *Grid Generation Methods*, Scientific Computation,
DOI 10.1007/978-90-481-2912-6_3, © Springer Science+Business Media B.V. 2010

$$\boldsymbol{x}(\varphi) : [a, b] \to R^n, \quad \boldsymbol{x}(\varphi) = [x^1(\varphi), \dots, x^n(\varphi)].$$

In our considerations we will use the designation S^{x1} for the curve with the parametrization $\boldsymbol{x}(\varphi)$. In this chapter we discuss the important measures of the local curve quality known as curvature and torsion. These measures are derived by some manipulations of basic curve vectors using the operations of dot and cross products.

3.2.1 Basic Curve Vectors

3.2.1.1 Tangent Vector

The first derivative of the parametrization $\boldsymbol{x}(\varphi)$ is a tangential vector

$$\boldsymbol{x}_\varphi = (x^1_\varphi, \dots, x^n_\varphi)$$

to the curve S^{x1}. The quantity

$$g^{x\varphi} = \boldsymbol{x}_\varphi \cdot \boldsymbol{x}_\varphi = x^i_\varphi x^i_\varphi, \quad i = 1, \dots, n,$$

is the metric tensor of the curve and its square root is the length of the tangent vector $\boldsymbol{x}_\varphi$. Thus the length l of the curve S^{x1} is computed from the integral

$$l = \int_a^b \sqrt{g^{x\varphi}} \mathrm{d}\varphi.$$

The most important notions related to curves are connected with the arc length parameter s defined by the equation

$$s(\varphi) = \int_0^\varphi \sqrt{g^{x\varphi}} \mathrm{d}\varphi. \tag{3.1}$$

The vector $\mathrm{d}\boldsymbol{x}[\varphi(s)]/\mathrm{d}s$, where $\varphi(s)$ is the inverse of $s(\varphi)$, is a tangent vector designated by $\boldsymbol{t}$. From (3.1) we obtain

$$\boldsymbol{t} = \frac{\mathrm{d}}{\mathrm{d}s}\boldsymbol{x}[\varphi(s)] = \frac{\mathrm{d}\varphi}{\mathrm{d}s}\boldsymbol{x}_\varphi = \frac{1}{\sqrt{g^{x\varphi}}}\boldsymbol{x}_\varphi.$$

Therefore $\boldsymbol{t}$ is the unit tangent vector and, after differentiating the relation $\boldsymbol{t} \cdot \boldsymbol{t} = 1$, we find that the derivative $\boldsymbol{t}_s$ is orthogonal to $\boldsymbol{t}$. The vector $\boldsymbol{t}_s$ is called the curvature vector and denoted by $\boldsymbol{k}$. Let $\boldsymbol{n}$ be a unit vector that is parallel to $\boldsymbol{t}_s$; there then exists a scalar k, such that

$$\boldsymbol{t}_s = \boldsymbol{k} = k\boldsymbol{n}, \quad k = (\boldsymbol{t}_s \cdot \boldsymbol{t}_s)^{1/2} = \alpha|\boldsymbol{k}|, \tag{3.2}$$

where $\alpha = 1$ or $\alpha = -1$.

The magnitude k is called the curvature, while the quantity $\rho = 1/k$ is called the radius of curvature of the curve.

Using the identity $\boldsymbol{x}_\varphi = \sqrt{g^{x\varphi}}\boldsymbol{t}$, we obtain from (3.2)

$$\boldsymbol{x}_{\varphi\varphi} = \frac{1}{\sqrt{g^{x\varphi}}}(\boldsymbol{x}_{\varphi\varphi} \cdot \boldsymbol{x}_\varphi)\boldsymbol{t} + g^{x\varphi} k\boldsymbol{n}. \tag{3.3}$$

The identity (3.3) is an analog of the Gauss relations (2.35). This identity shows that the vector $\boldsymbol{x}_{\varphi\varphi}$ lies in the $\boldsymbol{t}$–$\boldsymbol{n}$ plane.

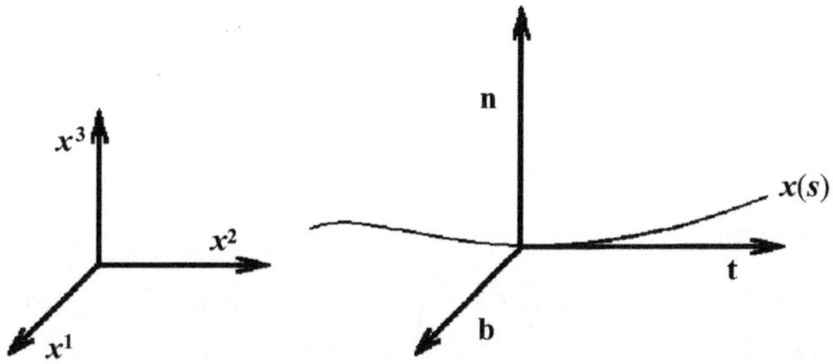

Fig. 3.1. Base curve vectors

3.2.1.2 Curves in Three-Dimensional Space

In three dimensions we can apply the operation of the cross product to the basic tangential and normal vectors. The vector $\boldsymbol{b} = \boldsymbol{t} \times \boldsymbol{n}$ is a unit vector which is orthogonal to both $\boldsymbol{t}$ and $\boldsymbol{n}$. It is called the binormal vector. From (3.3) we find that $\boldsymbol{b}$ is orthogonal to $\boldsymbol{x}_{\varphi\varphi}$.

The three vectors ($\boldsymbol{t}$, $\boldsymbol{n}$, $\boldsymbol{b}$) form a right-handed triad (Fig. 3.1). Note that if the curve lies in a plane, then the vectors $\boldsymbol{t}$ and $\boldsymbol{n}$ lie in the plane as well and $\boldsymbol{b}$ is constant unit vector normal to the plane.

The vectors $\boldsymbol{t}$, $\boldsymbol{n}$, and $\boldsymbol{b}$ are connected by the Serret–Frenet equations

$$\begin{aligned}\frac{\mathrm{d}\boldsymbol{t}}{\mathrm{d}s} &= k\boldsymbol{n},\\ \frac{\mathrm{d}\boldsymbol{n}}{\mathrm{d}s} &= -k\boldsymbol{t} + \tau\boldsymbol{b},\\ \frac{\mathrm{d}\boldsymbol{b}}{\mathrm{d}s} &= -\tau\boldsymbol{n},\end{aligned} \tag{3.4}$$

where the coefficient τ is called the torsion of the curve. The first equation of the system (3.4) is taken from (3.2). The second and third equations are readily obtained from the formula (2.5) by replacing the $\boldsymbol{b}$ in (2.5) by the vectors on the left-hand side of (3.4), while the vectors $\boldsymbol{t}$, $\boldsymbol{n}$, and $\boldsymbol{b}$ substitute for $\boldsymbol{e}_1$, $\boldsymbol{e}_2$, and $\boldsymbol{e}_3$, respectively. The vectors $\boldsymbol{t}$, $\boldsymbol{n}$, and $\boldsymbol{b}$ constitute an orthonormal basis, i.e.

$$a_{ij} = a^{ij} = \delta^i_j, \quad i, j = 1, 2, 3,$$

where, in accordance with Sect. 2.2.4 $a_{ij} = \boldsymbol{e}_i \cdot \boldsymbol{e}_j$, and (a^{ij}) is the inverse of the tensor (a_{ij}). Now, using (2.5) we obtain

$$\frac{\mathrm{d}\boldsymbol{n}}{\mathrm{d}s} = \left(\frac{\mathrm{d}\boldsymbol{n}}{\mathrm{d}s} \cdot \boldsymbol{t}\right)\boldsymbol{t} + \left(\frac{\mathrm{d}\boldsymbol{n}}{\mathrm{d}s} \cdot \boldsymbol{n}\right)\boldsymbol{n} + \left(\frac{\mathrm{d}\boldsymbol{n}}{\mathrm{d}s} \cdot \boldsymbol{b}\right)\boldsymbol{b} = -k\boldsymbol{t} + \left(\frac{\mathrm{d}\boldsymbol{n}}{\mathrm{d}s} \cdot \boldsymbol{b}\right)\boldsymbol{b},$$

since $\boldsymbol{n}_s \cdot \boldsymbol{t} = -\boldsymbol{n} \cdot \boldsymbol{t}_s$, $\boldsymbol{n}_s \cdot \boldsymbol{n} = 0$. Thus we obtain the second equation of (3.4) with $\tau = \boldsymbol{n}_s \cdot \boldsymbol{b}$. Analogously we obtain the last equation of (3.4) by expanding the vector $\boldsymbol{b}_s$ through $\boldsymbol{t}$, $\boldsymbol{n}$, and $\boldsymbol{b}$ using the relation (2.5):

$$\frac{\mathrm{d}\boldsymbol{b}}{\mathrm{d}s} = \left(\frac{\mathrm{d}\boldsymbol{b}}{\mathrm{d}s} \cdot \boldsymbol{t}\right)\boldsymbol{t} + \left(\frac{\mathrm{d}\boldsymbol{b}}{\mathrm{d}s} \cdot \boldsymbol{n}\right)\boldsymbol{n} + \left(\frac{\mathrm{d}\boldsymbol{b}}{\mathrm{d}s} \cdot \boldsymbol{b}\right)\boldsymbol{b} = -\left(\frac{\mathrm{d}\boldsymbol{n}}{\mathrm{d}s} \cdot \boldsymbol{b}\right)\boldsymbol{n} = -\tau\boldsymbol{n},$$

as $\boldsymbol{b}_s \cdot \boldsymbol{t} = -\boldsymbol{b} \cdot \boldsymbol{t}_s = 0,\ \boldsymbol{b}_s \cdot \boldsymbol{b} = 0$.

3.2.2 Curvature

A very important characteristic of a curve which is related to grid generation is the curvature k. This quantity is used as a measure of coordinate line bending.

One way to compute the curvature is to multiply (3.2) by $\boldsymbol{n}$ using the dot product operation. As

$$\frac{\mathrm{d}\boldsymbol{t}}{\mathrm{d}s} = \frac{1}{\sqrt{g^{x\varphi}}} \frac{\mathrm{d}}{\mathrm{d}\varphi}\left(\frac{1}{\sqrt{g^{x\varphi}}}\boldsymbol{x}_\varphi\right) = \frac{1}{g^{x\varphi}}\boldsymbol{x}_{\varphi\varphi} - \frac{1}{(g^{x\varphi})^2}(\boldsymbol{x}_\varphi \cdot \boldsymbol{x}_{\varphi\varphi})\boldsymbol{x}_\varphi,$$

from (3.1), (3.2), the result is

$$k = \frac{1}{g^{x\varphi}}\boldsymbol{x}_{\varphi\varphi} \cdot \boldsymbol{n}. \tag{3.5}$$

The vector $\boldsymbol{n}$ is independent of the curve parametrization, and therefore we find from (3.3), (3.5) that k is an invariant of parametrizations of the curve.

In two dimensions,

$$\boldsymbol{n} = \frac{1}{\sqrt{g^{x\varphi}}}(-x^2_\varphi, x^1_\varphi),$$

therefore in this case we obtain from (3.5),

$$k^2 = \frac{(x_\varphi y_{\varphi\varphi} - y_\varphi x_{\varphi\varphi})^2}{[(x_\varphi)^2 + (y_\varphi)^2]^3} \tag{3.6}$$

with the convention $x = x^1,\ y = x^2$. In particular, when the curve in R^2 is defined by a function $u = u(x)$, we obtain from (3.6), assuming $\boldsymbol{x}(\varphi) = [\varphi, u(\varphi)],\ \varphi = x$,

$$k^2 = (u_{xx})^2/[1 + (u_x)^2]^3.$$

In the case of three-dimensional space the curvature k can also be computed from the relation obtained by multiplying (3.3) by $\boldsymbol{x}_\varphi$ with the cross product operation:

$$\boldsymbol{x}_\varphi \times \boldsymbol{x}_{\varphi\varphi} = g^{x\varphi}k(\boldsymbol{x}_\varphi \times \boldsymbol{n}) = (g^{x\varphi})^{3/2}k\boldsymbol{b}.$$

Thus we obtain

$$k^2 = \frac{|\boldsymbol{x}_\varphi \times \boldsymbol{x}_{\varphi\varphi}|^2}{(g^{x\varphi})^3} \tag{3.7}$$

and, consequently, from (2.25)

$$k^2 = \frac{(x^1_\varphi x^2_{\varphi\varphi} - x^2_\varphi x^1_{\varphi\varphi})^2 + (x^2_\varphi x^3_{\varphi\varphi} - x^3_\varphi x^2_{\varphi\varphi})^2 + (x^3_\varphi x^1_{\varphi\varphi} - x^1_\varphi x^3_{\varphi\varphi})^2}{[(x^1_\varphi)^2 + (x^2_\varphi)^2 + (x^3_\varphi)^2]^3}.$$

3.2.3 Torsion

Another important quality measure of curves in three-dimensional space is the torsion τ. This quantity is suitable for measuring the rate of twisting of the lines of coordinate grids.

In order to figure out the value of τ for a curve in R^3 we use the last relation in (3.4), which yields

$$\tau = -\frac{\mathrm{d}\boldsymbol{b}}{\mathrm{d}s} \cdot \boldsymbol{n}.$$

As $\boldsymbol{b} = \boldsymbol{t} \times \boldsymbol{n}$, we obtain

$$\frac{\mathrm{d}\boldsymbol{b}}{\mathrm{d}s} = \frac{\mathrm{d}\boldsymbol{t}}{\mathrm{d}s} \times \boldsymbol{n} + \boldsymbol{t} \times \frac{\mathrm{d}\boldsymbol{n}}{\mathrm{d}s} = \boldsymbol{t} \times \frac{\mathrm{d}\boldsymbol{n}}{\mathrm{d}s},$$

since $\mathrm{d}\boldsymbol{t}/\mathrm{d}s = k\boldsymbol{n}$. Thus

$$\tau = \left(-\boldsymbol{t} \times \frac{\mathrm{d}\boldsymbol{n}}{\mathrm{d}s}\right) \cdot \boldsymbol{n}. \tag{3.8}$$

From (3.1), (3.2) we have the following obvious relations for the basic vectors $\boldsymbol{t}$ and $\boldsymbol{n}$ in terms of the parametrization $\boldsymbol{x}(\varphi)$ and its derivatives:

$$\begin{aligned}
\boldsymbol{t} &= \frac{1}{\sqrt{g^{x\varphi}}}\boldsymbol{x}_\varphi, \\
\boldsymbol{n} &= \frac{1}{k}\frac{\mathrm{d}\boldsymbol{t}}{\mathrm{d}s} = \frac{1}{k}\left(\frac{1}{g^{x\varphi}}\boldsymbol{x}_{\varphi\varphi} - \frac{\boldsymbol{x}_\varphi \cdot \boldsymbol{x}_{\varphi\varphi}}{(g^{x\varphi})^2}\boldsymbol{x}_\varphi\right), \\
\frac{\mathrm{d}\boldsymbol{n}}{\mathrm{d}s} &= \frac{1}{k}\left(\frac{1}{(g^{x\varphi})^{3/2}}\boldsymbol{x}_{\varphi\varphi\varphi} - 2\frac{\boldsymbol{x}_\varphi \cdot \boldsymbol{x}_{\varphi\varphi}}{(g^{x\varphi})^2}\boldsymbol{x}_{\varphi\varphi}\right. \\
&\quad \left. - \frac{\mathrm{d}}{\mathrm{d}\varphi}\left(\frac{\boldsymbol{x}_\varphi \cdot \boldsymbol{x}_{\varphi\varphi}}{(g^{x\varphi})^2}\right)\boldsymbol{x}_\varphi - \frac{1}{k}\frac{\mathrm{d}k}{\mathrm{d}s}\boldsymbol{n}\right).
\end{aligned} \tag{3.9}$$

Thus

$$\boldsymbol{t} \times \frac{\mathrm{d}\boldsymbol{n}}{\mathrm{d}s} = \frac{1}{k(g^{x\varphi})^2}\boldsymbol{x}_\varphi \times \boldsymbol{x}_{\varphi\varphi\varphi} - 2\frac{\boldsymbol{x}_\varphi \cdot \boldsymbol{x}_{\varphi\varphi}}{k(g^{x\varphi})^{5/2}}\boldsymbol{x}_\varphi \times \boldsymbol{x}_{\varphi\varphi} - \frac{1}{k^2\sqrt{g^{x\varphi}}}\frac{\mathrm{d}k}{\mathrm{d}s}\boldsymbol{x}_\varphi \times \boldsymbol{n}.$$

As $(\boldsymbol{a} \times \boldsymbol{b}) \cdot \boldsymbol{a} = (\boldsymbol{a} \times \boldsymbol{b}) \cdot \boldsymbol{b} = 0$ for arbitrary vectors $\boldsymbol{a}$ and $\boldsymbol{b}$, we obtain from (3.8), (3.9)

$$\tau = -\frac{1}{k^2(g^{x\varphi})^3}(\boldsymbol{x}_\varphi \times \boldsymbol{x}_{\varphi\varphi\varphi}) \cdot \boldsymbol{x}_{\varphi\varphi} = \frac{1}{k^2(g^{x\varphi})^3}(\boldsymbol{x}_\varphi \times \boldsymbol{x}_{\varphi\varphi}) \cdot \boldsymbol{x}_{\varphi\varphi\varphi}. \tag{3.10}$$

And using (2.30) we also obtain

$$\tau = \frac{1}{k^2(g^{x\varphi})^3}\det\begin{pmatrix} x^1_\varphi & x^2_\varphi & x^3_\varphi \\ x^1_{\varphi\varphi} & x^2_{\varphi\varphi} & x^3_{\varphi\varphi} \\ x^1_{\varphi\varphi\varphi} & x^2_{\varphi\varphi\varphi} & x^3_{\varphi\varphi\varphi} \end{pmatrix}.$$

3.3 Surface Geometry

In general, a surface in the three-dimensional space R^3 is assumed to be locally represented by some parametric two-dimensional domain S^2 and a parametrization

$$\boldsymbol{x}(\boldsymbol{s}) : S^2 \to R^3, \quad \boldsymbol{x}(\boldsymbol{s}) = [x^1(\boldsymbol{s}), x^2(\boldsymbol{s}), x^3(\boldsymbol{s})], \ \boldsymbol{s} = (s^1, s^2),$$

where $\boldsymbol{x}(\boldsymbol{s})$ is a smooth nondegenerate vector function. We use the designation S^{x2} for the surface with the parametrization $\boldsymbol{x}(\boldsymbol{s})$. In analogy with domains, the transformation $\boldsymbol{x}(\boldsymbol{s})$ defines the coordinate system s^1, s^2 on the surface as well as the respective base vectors and metric tensors.

For the purpose of adaptive grid generation, the so-called monitor surfaces are very important. These surfaces are defined by the values of some vector-valued function $\boldsymbol{u}(\boldsymbol{s})$, referred to as the height function, over the domain S^2. The natural form of the parametrization of the monitor surface formed with a scalar height function $u(\boldsymbol{x})$ is represented by the formula

$$\boldsymbol{x}(\boldsymbol{s}) = [s^1, s^2, u(s^1, s^2)].$$

3.3.1 Surface Base Vectors

A surface in R^3 has three base vectors: two tangents (one to each coordinate curve) and a normal. The two tangential vectors to the coordinates s^1 and s^2 represented by $\boldsymbol{x}(\boldsymbol{s})$ are, respectively,

$$\boldsymbol{x}_{s^i} = \frac{\partial \boldsymbol{x}}{\partial s^i} = \left(\frac{\partial x^1}{\partial s^i}, \frac{\partial x^2}{\partial s^i}, \frac{\partial x^3}{\partial s^i}\right), \quad i = 1, 2.$$

The unit normal vector to the surface S^{x2} is defined through the cross product of the tangent vectors $\boldsymbol{x}_{s^1}$ and $\boldsymbol{x}_{s^2}$:

$$\boldsymbol{n} = \frac{1}{|\boldsymbol{x}_{s^1} \times \boldsymbol{x}_{s^2}|}(\boldsymbol{x}_{s^1} \times \boldsymbol{x}_{s^2}).$$

Since $(\boldsymbol{x}_{s^1} \times \boldsymbol{x}_{s^2}) \cdot \boldsymbol{n} > 0$, the base surface vectors $\boldsymbol{x}_{s^1}$, $\boldsymbol{x}_{s^2}$, and $\boldsymbol{n}$ comprise a right-handed triad (Fig. 3.2). In accordance with (2.25), the unit normal $\boldsymbol{n}$ can also be expressed as

$$\boldsymbol{n} = \frac{1}{\sqrt{g^{rs}}}\left(\frac{\partial x^{l+1}}{\partial s^1}\frac{\partial x^{l+2}}{\partial s^2} - \frac{\partial x^{l+2}}{\partial s^1}\frac{\partial x^{l+1}}{\partial s^2}\right)\boldsymbol{e}_l, \quad l = 1, 2, 3, \qquad (3.11)$$

where $(\boldsymbol{e}_1, \boldsymbol{e}_2, \boldsymbol{e}_3)$ is the Cartesian basis of R^3. Recall that this formula implies the identification convention for indices in three dimensions, where k is equivalent to $k \pm 3$. If the surface S^{x2} is a monitor surface represented by a height function $u(\boldsymbol{s})$ then we obtain from (3.11)

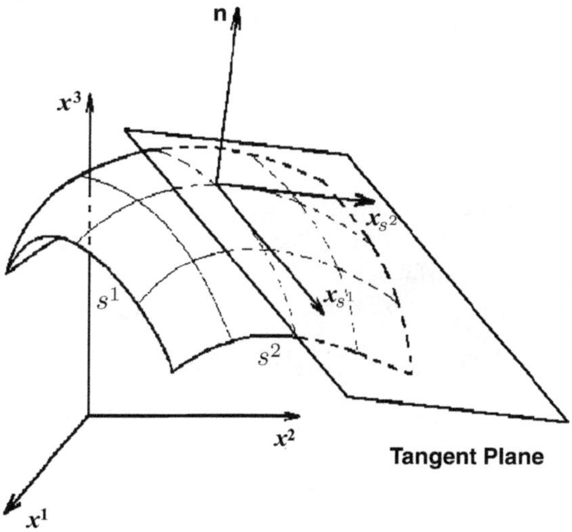

Fig. 3.2. Surface base vectors

$$\boldsymbol{n} = \frac{1}{\sqrt{1 + (u_{s^1})^2 + (u_{s^2})^2}} \left(-\frac{\partial u}{\partial s^1}, -\frac{\partial u}{\partial s^2}, 1 \right).$$

In another particular case, when the surface points are found from the equation $f(\boldsymbol{x}) = c$, we obtain $\nabla f \cdot \boldsymbol{x}_{s^i} = 0, \ i = 1, 2$, and therefore

$$\boldsymbol{n} = l \nabla f, \quad |l| = 1/|\nabla f|.$$

3.3.2 Metric Tensors

The surface metric tensors, like the domain metric tensors, are defined through the operation of the dot product on the vectors tangential to the coordinate lines.

3.3.2.1 Covariant Metric Tensor

We designate the covariant metric tensor of the surface S^{x2} in the coordinates $s^1, \ s^2$ by

$$G^{xs} = (g_{ij}^{xs}), \quad i, j = 1, 2,$$

where

$$g_{ij}^{xs} = \boldsymbol{x}_{s^i} \cdot \boldsymbol{x}_{s^j}, \quad i, j = 1, 2. \tag{3.12}$$

In particular, when a surface is defined by the values of some scalar function $u(\boldsymbol{s})$ over the domain S^2 then

$$g_{ij}^{xs} = \delta_i^j + \frac{\partial u}{\partial s^i} \frac{\partial u}{\partial s^j}, \quad i, j = 1, 2.$$

The quantity $\sqrt{g^{xs}_{ii}}$ in (3.12) for a fixed i has the geometrical meaning of the length of the tangent vector $\boldsymbol{x}_{s^i}$ to the coordinate curve s^i.

The differential quadratic form

$$g^{xs}_{ij}\mathrm{d}s^i\mathrm{d}s^j, \quad i,j=1,2,$$

relating to the line elements in space, is called the first fundamental form of the surface. It represents the value of the square of the length of an elementary displacement $\mathrm{d}\boldsymbol{x}$ on the surface.

Let the Jacobian of G^{xs} be designated by g^{xs}. Since

$$g^{xs}=|\boldsymbol{x}_{s^1}|^2|\boldsymbol{x}_{s^2}|^2(1-\cos^2\theta)=(|\boldsymbol{x}_{s^1}|\cdot|\boldsymbol{x}_{s^2}|\sin\theta)^2=(\boldsymbol{x}_{s^1}\times\boldsymbol{x}_{s^2})^2,$$

where θ is the angle between $\boldsymbol{x}_{s^1}$ and $\boldsymbol{x}_{s^2}$, we find that the quantity g^{xs} is the area squared of the parallelogram formed by the vectors $\boldsymbol{x}_{s^1}$ and $\boldsymbol{x}_{s^2}$. Therefore the area of the surface S^{x2} is computed from the formula

$$S=\int_{S^2}\sqrt{g^{xs}}\mathrm{d}\boldsymbol{s}.$$

3.3.2.2 Contravariant Metric Tensor

Consequently the contravariant metric tensor of the surface S^{x2} in the coordinates s^1, s^2 is the matrix

$$G_{sx}=(g^{ij}_{sx}), \quad i,j=1,2$$

inverse to G^{xs}, i.e.

$$g^{xs}_{ij}g^{jk}_{sx}=\delta^i_k, \quad i,j,k=1,2.$$

Thus, in analogy with (2.20), we obtain

$$\begin{aligned} g^{ij}_{sx}&=(-1)^{i+j}g^{xs}_{3-i\ 3-j}/g^{xs}, \\ g^{xs}_{ij}&=(-1)^{i+j}g^{xs}g^{3-i\ 3-j}_{sx}, \quad i,j=1,2, \end{aligned} \tag{3.13}$$

with fixed indices i and j. The diagonal elements g^{11}_{sx} and g^{22}_{sx} of the contravariant metric tensor G_{sx} are connected with the natural geometric quantities of the parallelogram defined by the tangent vectors $\boldsymbol{x}_{s^1}$ and $\boldsymbol{x}_{s^2}$ (see Fig. 3.3).

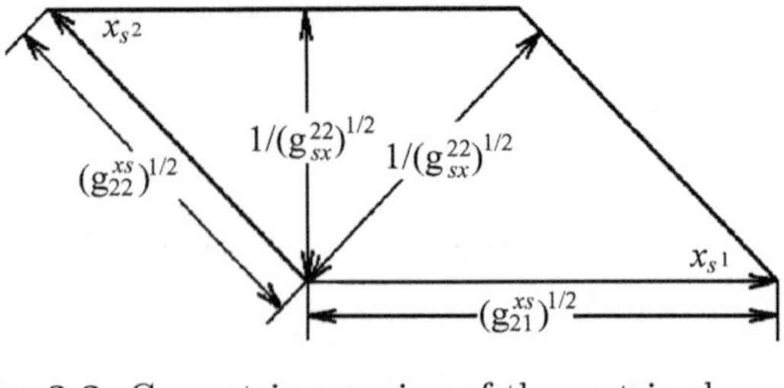

Fig. 3.3. Geometric meaning of the metric elements

Namely, taking into account the relation $g^{xs} = g^{xs}_{11}/g^{22}_{sx}$, we find that $\sqrt{g^{22}_{sx}}$ is the inverse of the value of the distance between the parallel edges of the parallelogram formed by the vector $\boldsymbol{x}_{s^1}$. Analogously, $\sqrt{g^{11}_{sx}}$ is the inverse of the distance between the other pair of parallelogram edges, i.e. those formed by $\boldsymbol{x}_{s^2}$.

3.3.3 Second Fundamental Form

The coefficients of the second fundamental form

$$b_{ij}\mathrm{d}s^i\mathrm{d}s^j, \quad i,j = 1,2,$$

of the surface S^{x2} are defined by the dot products of the second derivatives of the vector function $\boldsymbol{x}(\boldsymbol{s})$ and the unit normal vector $\boldsymbol{n}$ to the surface at the point $\boldsymbol{s}$ under consideration:

$$b_{ij} = \boldsymbol{x}_{s^is^j} \cdot \boldsymbol{n}, \quad i,j = 1,2. \tag{3.14}$$

Thus, from (3.11), (3.14) we obtain for b_{ij}, $i,j = 1,2$,

$$b_{ij} = \frac{1}{\sqrt{g^{xs}}}\left[\frac{\partial^2 x^l}{\partial s^i \partial s^j}\left(\frac{\partial x^{l+1}}{\partial s^1}\frac{\partial x^{l+2}}{\partial s^2} - \frac{\partial x^{l+2}}{\partial s^1}\frac{\partial x^{l+1}}{\partial s^2}\right)\right], \quad l = 1,2,3, \tag{3.15}$$

with the identification convention for the superscripts that k is equivalent to $k \pm 3$. Correspondingly, for the monitor surface with the height function $u(\boldsymbol{s})$, we obtain

$$b_{ij} = \frac{1}{\sqrt{1 + (u_{s^1})^2 + (u_{s^2})^2}}\, u_{s^is^j}, \quad i,j = 1,2.$$

The tensor (b_{ij}) reflects the local warping of the surface, namely its deviation from the tangent plane at the point under consideration. In particular, if $(b_{ij}) \equiv 0$ at all points of S^2 then the surface is a plane.

3.3.4 Surface Curvatures

3.3.4.1 Principal Curvatures

Let a curve on the surface be defined by the intersection of a plane containing the normal $\boldsymbol{n}$ with the surface. It is obvious that either $\boldsymbol{n}$ or $-\boldsymbol{n}$ is also the curve normal vector. Taking into account (3.5), we obtain for the curvature of this curve

$$k = \frac{b_{ij}\mathrm{d}s^i\mathrm{d}s^j}{g^{xs}_{ij}\mathrm{d}s^i\mathrm{d}s^j}, \quad i,j = 1,2. \tag{3.16}$$

Here $(\mathrm{d}s^1, \mathrm{d}s^2)$ is the direction of the curve, i.e. $\mathrm{d}s^i = c(\mathrm{d}s^i/\mathrm{d}\varphi)$, where $\boldsymbol{s}(\varphi)$ is a curve parametrization. The two extreme quantities K_{I} and K_{II} of the values of k are called the principal curvatures of the surface at the point under consideration. In order to compute the principal curvatures, we consider the following relation for the value of the curvature:

$$(b_{ij} - k g^{xs}_{ij}) \mathrm{d}s^i \mathrm{d}s^j = 0, \quad i, j = 1, 2, \tag{3.17}$$

which follows from (3.16). In order to find the maximum and minimum values of k, the usual method of equating to zero the derivative with respect to $\mathrm{d}s^i$ is applied. Thus the components of the $(\mathrm{d}s^1, \mathrm{d}s^2)$ direction giving an extreme value of k are subject to the restriction

$$(b_{ij} - k g^{xs}_{ij}) \mathrm{d}s^j = 0, \quad i, j = 1, 2,$$

which, in fact, is the eigenvalue problem for curvature. One finds the eigenvalues k by setting the determinant of this equation equal to zero, obtaining thereby the secular equation for k:

$$\det(b_{ij} - k g^{xs}_{ij}) = 0, \quad i, j = 1, 2.$$

This equation, written out in full, is a quadratic equation

$$k^2 - g^{ij}_{sx} b_{ij} k + [b_{11} b_{22} - (b_{12})^2]/g^{xs} = 0,$$

with two roots, which are the maximum and minimum values K_{I} and K_{II} of the curvature k:

$$K_{\mathrm{I,II}} = \frac{1}{2} g^{ij}_{sx} b_{ij} \pm \sqrt{\frac{1}{4}(g^{ij}_{sx} b_{ij})^2 - \frac{1}{g^{xs}}[b_{11} b_{22} - (b_{12})^2]}. \tag{3.18}$$

3.3.4.2 Mean Curvature

One half of the sum of the principal curvatures is referred to as the mean surface curvature. Taking advantage of (3.18), the mean curvature, designated by K_{m}, is defined through the coefficients of the second fundamental form and elements of the contravariant metric tensor by

$$K_{\mathrm{m}} = \frac{1}{2}(K_{\mathrm{I}} + K_{\mathrm{II}}) = \frac{1}{2} g^{ij}_{sx} b_{ij}, \quad i, j = 1, 2. \tag{3.19}$$

In the case of the monitor surface represented by the function $u(s^1, s^2)$, we obtain

$$K_{\mathrm{m}} = \frac{u_{s^1 s^1}[1 + (u_{s^2})^2] + u_{s^2 s^2}[1 + (u_{s^1})^2] - 2 u_{s^1} u_{s^2} u_{s^1 s^2}}{2[1 + (u_{s^1})^2 + (u_{s^2})^2]^{3/2}}.$$

Now we consider the tensor

$$(K^i_j) = (g^{ik}_{sx} b_{kj}), \quad i, j, k = 1, 2.$$

It is easy to see that (K^i_j) is a mixed tensor contravariant with respect to the upper index and covariant with respect to the lower one. From (3.19) we find that the mean curvature is defined as the trace of the tensor, namely,

$$2K_{\mathrm{m}} = \mathrm{tr}(K^i_j), \quad i,j = 1,2. \tag{3.20}$$

A surface whose mean curvature is zero, i.e. $K_{\mathrm{I}} = -K_{\mathrm{II}}$, possesses the following unique property. Namely, if a surface bounded by a specified contour has a minimum area then its mean curvature is zero. Conversely, of all the surfaces bounded by a curve whose length is sufficiently small, the minimum area is possessed by the surface whose mean curvature is zero.

3.3.4.3 Gaussian Curvature

The determinant of the tensor (K^i_j) represents the Gaussian curvature of the surface

$$K_{\mathrm{G}} = \det(K^i_j) = \frac{1}{g^{xs}}[b_{11}b_{22} - (b_{12})^2]. \tag{3.21}$$

Taking into account (3.18), we readily see that the Gaussian curvature is the product of the two principal curvatures K_{I} and K_{II}, i.e.

$$K_{\mathrm{G}} = K_{\mathrm{I}}K_{\mathrm{II}}.$$

In terms of the height function $u(\boldsymbol{s})$ representing the monitor surface S^{x2}, we have

$$K_{\mathrm{G}} = \frac{u_{s^1s^1}u_{s^2s^2} - (u_{s^1s^2})^2}{[1 + (u_{s^1})^2 + (u_{s^2})^2]^2}.$$

A surface point is called elliptic if $K_{\mathrm{G}} > 0$, i.e. both K_{I} and K_{II} are both negative or both positive at the point of consideration. A saddle or hyperbolic point has principal curvatures of opposite sign, and therefore has negative Gaussian curvature. A parabolic point has one principal curvature vanishing and, consequently, a vanishing Gaussian curvature. This classification of points is prompted by the form of the curve which is obtained by the intersection of the surface with a slightly offset tangent plane. For an elliptic point the curve is an ellipse; for a saddle point it is a hyperbola. It is a pair of lines (degenerate conic) at a parabolic point, and it vanishes at a planar point, where both principal curvatures are zero.

It is easily shown that both the mean and the Gaussian curvatures are invariant of surface parametrizations.

3.4 Metric-Tensor Invariants

The coordinate transformation $\boldsymbol{x}(\boldsymbol{\xi}) : \Xi^n \to X^n$ of a physical n-dimensional domain X^n applied to generate grids can be locally interpreted as some deformation of a uniform cell in the computational domain Ξ^n into the corresponding cell in the domain X^n. The local deformation of any cell is approximated by a linear transformation represented by the Jacobi matrix $(\partial x^i/\partial \xi^j)$. This deformation is not changed if any orthogonal transformation is applied to the cell in X^n. The deformation is also preserved if the orientation of the computational domain Ξ^n is changed. Therefore it is logical to formulate the features of the grid cells in terms of the invariants of the orthogonal transformations of the covariant metric tensor (g_{ij}).

3.4.1 Algebraic Expressions for the Invariants

According to the theory of matrices a symmetric nondegenerate $(n \times n)$ matrix (a_{ij}) has n independent invariants I_i, $i = 1, \dots, n$, of its orthogonal transformations. The ith invariant I_i is defined by summing all of the principal minors of order i of the matrix. Recall that the principal minors of a square matrix are the determinants of the square submatrices of the matrix. Thus, for example,

$$\begin{aligned} I_1 &= \sum_{i=1}^{n} a_{ii} = \mathrm{tr}(a_{ij}), \\ I_{n-1} &= \sum_{i=1}^{n} \text{cofactor } a_{ii} = \det(a_{ij}) \sum_{i=1}^{n} a^{ii} = \det(a_{ij})\, \mathrm{tr}(a^{ij}), \\ I_n &= \det(a_{ij}), \end{aligned} \tag{3.22}$$

where (a^{ij}) is the inverse of (a_{ij}).

When we use for (a_{ij}) the covariant metric tensor g_{ij}, $g_{ij} = \boldsymbol{x}_{\xi^i} \cdot \boldsymbol{x}_{\xi^j}$, of a domain X^n, then, taking advantage of (3.22), the invariants I_1 and I_2 in two dimensions are expressed as

$$\begin{aligned} I_1 &= g_{11} + g_{22}, \\ I_2 &= g_{11}\, g_{22} - (g_{12})^2 = g = J^2, \end{aligned} \tag{3.23}$$

where $J = \det(\partial x^i/\partial \xi^j)$. The invariants of the three-dimensional metric tensor (g_{ij}) are expressed as follows:

$$\begin{aligned} I_1 &= g_{11} + g_{22} + g_{33}, \\ I_2 &= g(g^{11} + g^{22} + g^{33}), \\ I_3 &= \det(g_{ij}) = g, \quad i, j = 1, 2, 3, \end{aligned} \tag{3.24}$$

where $g^{ij} = \nabla \xi^i \cdot \nabla \xi^j$. Analogously, the invariants of the surface metric tensor G^{xs}, represented in the coordinates s^1, s^2 by (3.12), are written out as

$$\begin{aligned} I_1 &= g_{11}^{xs} + g_{22}^{xs}, \\ I_2 &= g^{xs}. \end{aligned} \tag{3.25}$$

The notion of an invariant can be helpful to identity conformal coordinate transformations. For example, in two dimensions we know that a conformal mapping $\boldsymbol{x}(\boldsymbol{\xi})$ satisfies the Cauchy–Riemann equations

$$\frac{\partial x^1}{\partial \xi^1} = \frac{\partial x^2}{\partial \xi^2}, \qquad \frac{\partial x^1}{\partial \xi^2} = -\frac{\partial x^2}{\partial \xi^1}.$$

Therefore a zero value of the quantity

$$Q = \left(\frac{\partial x^1}{\partial \xi^1} - \frac{\partial x^2}{\partial \xi^2}\right)^2 + \left(\frac{\partial x^1}{\partial \xi^2} + \frac{\partial x^2}{\partial \xi^1}\right)^2$$

is an indication of the conformality of $\boldsymbol{x}(\boldsymbol{\xi})$. We obtain

$$Q = g_{11} + g_{22} - 2J = I_1 - 2\sqrt{I_2},$$

using (3.23). Thus the two-dimensional coordinate transformation $\boldsymbol{x}(\boldsymbol{\xi})$ is conformal if only if the invariants I_1 and I_2 satisfy the restriction $I_1/\sqrt{I_2} = 2$. In Sect. 3.7.6 it will be shown that an analogous relation is valid for an arbitrary dimension $n \geq 2$.

We also can see that the mean and Gaussian curvatures described by (3.20) and (3.21), respectively, are defined through the invariants of the tensor (K^i_j), namely,

$$K_{\mathrm{m}} = \frac{1}{2} I_1, \qquad K_{\mathrm{G}} = I_2.$$

3.4.2 Geometric Interpretation

The invariants of the covariant metric tensor (g_{ij}) can also be described in terms of some geometric characteristics of the n-dimensional parallelepiped (parallelogram in two dimensions) determined by the tangent vectors $\boldsymbol{x}_{\xi^i}$, thus giving a relationship between the grid cell characteristics and the invariants. For example, we see from (3.23), (3.25) in two dimensions that the invariant I_1 equals the sum squares of the parallelogram edge lengths, while I_2 is equal to the parallelogram area squared. In three-dimensional space we find from (3.24) that I_1 equals the sum of the squares of the lengths of the base vectors $\boldsymbol{x}_{\xi^i}$, $i = 1, 2, 3$, which are the edges of the parallelepiped. The invariant I_2 is the sum of the squares of the areas of the faces of the parallelepiped, while the invariant I_3 is its volume squared.

These geometric interpretations can be extended to arbitrary dimensions by the following consideration. Every principal minor of order m is the determinant of an m-dimensional square matrix A^m obtained from the covariant tensor (g_{ij}) by crossing out $n - m$ rows and columns that intersect pairwise on the diagonal. Therefore the elements of A^m are the dot products of m particular vectors of the base tangential vectors $\boldsymbol{x}_{\xi^i}$, $i = 1, \ldots, n$. Thus, geometrically, the determinant of A^m equals the square of the m-dimensional volume of the m-dimensional parallelepiped constructed by the vectors of the basic set $\boldsymbol{x}_{\xi^i}$, $i = 1, \ldots, n$, whose dot products form the matrix A^m. Therefore I_i, $i = 1, \ldots, n$, is geometrically the sum of the squares of the i-dimensional volumes of the i-dimensional sides of the n-dimensional parallelepiped spanned by the base vectors $\boldsymbol{x}_{\xi^i}$, $i = 1, \ldots, n$.

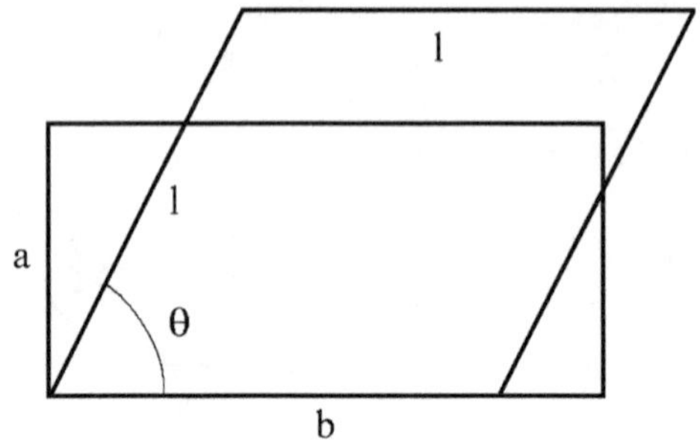

Fig. 3.4. Quadrilaterals with the same invariants

We note that the invariants do not describe all of the geometric features of the grid cells. In the two-dimensional case, the invariants I_1 and I_2 given by (3.23) can be the same for parallelepipeds that are not similar. For example, if we take a transformation $\boldsymbol{x}(\boldsymbol{\xi})$ whose tangential vectors $\boldsymbol{x}_{\xi^1}$ and $\boldsymbol{x}_{\xi^2}$ define a rectangle with sides of different lengths a and b, then we obtain

$$I_1 = a^2 + b^2, \qquad I_2 = (ab)^2.$$

However, as demonstrated in Fig. 3.4, the same invariants are produced by a transformation $\boldsymbol{x}(\boldsymbol{\xi})$, whose tangent vectors yield a rhombus with a side length l equal to $\sqrt{(a^2+b^2)/2}$ and an angle θ defined by

$$\theta = \arcsin \frac{2ab}{a^2+b^2},$$

since

$$\begin{aligned} I_1 &= 2l^2 = a^2 + b^2, \\ I_2 &= l^2 \sin^2 \theta = (ab)^2. \end{aligned}$$

Thus, a knowledge of only the values of the invariants I_1 and I_2 is not sufficient to distinguish the rectangle from the rhombus. However, the value of the quantity $I_1/\sqrt{I_2}$ imposes restriction on the maximal angle between the parallelogram edges and on the maximum cell aspect ratio. These bounds will be evaluated in Sect. 3.7.6. In particular, if $I_1 = 2\sqrt{I_2}$, then we can definitely state that the parallelogram is a square.

3.5 Characteristics of Grid Lines

This section describes some characteristics of curvilinear coordinate lines in domains specified by the parametrization $\boldsymbol{x}(\boldsymbol{\xi}) : \Xi^n \to X^n$. These characteristics can be used for the evaluation of the grid properties and for the formulation of grid generation techniques through the calculus of variations.

All considerations in this section are concerned with a selected coordinate line ξ^i for a specified i, and therefore summation is not carried out over the repeated index i here.

3.5.1 Sum of Squares of Cell Edge Lengths

The length l_i of any cell edge along the coordinate curve ξ^i is expressed through the element g_{ii} of the covariant metric tensor (g_{ij}):

$$l_i \approx \sqrt{g_{ii}}h.$$

The sum of the squares of the cell edge lengths equals $Q_l h^2$, where

$$Q_l = \sum_{j=1}^{n} g_{jj} = \mathrm{tr}(g_{ij}). \tag{3.26}$$

The quantity Q_l is one of the important characteristics of the grid cell. This characteristic is the first invariant I_1 of the tensor matrix (g_{ij}).

3.5.2 Eccentricity

The ratio between two adjacent grid steps along any coordinate curve ξ^i is a quantity which characterizes the change of the length of the cell edge in the ξ^i direction. This quantity is designated as ϵ^i, and at the point $\boldsymbol{\xi}$ it is expressed as follows:

$$\epsilon^i \approx \frac{|\boldsymbol{x}_{\xi^i}(\boldsymbol{\xi} + h\boldsymbol{e}^i)|}{|\boldsymbol{x}_{\xi^i}(\boldsymbol{\xi})|}.$$

We also find that

$$\epsilon^i \approx \frac{\sqrt{g_{ii}(\boldsymbol{\xi} + h\boldsymbol{e}^i)} - \sqrt{g_{ii}(\boldsymbol{\xi})}}{\sqrt{g_{ii}(\boldsymbol{\xi})}} + 1 \approx h\frac{1}{\sqrt{g_{ii}}}\frac{\partial}{\partial \xi^i}\sqrt{g_{ii}} + 1,$$

for a fixed i, since $|\boldsymbol{x}_{\xi^i}| = \sqrt{g_{ii}}$. The quantity

$$Q^i_\epsilon = \left(\frac{1}{\sqrt{g_{ii}}}\frac{\partial}{\partial \xi^i}\sqrt{g_{ii}}\right)^2 = \left(\frac{\partial}{\partial \xi^i}\ln\sqrt{g_{ii}}\right)^2 \tag{3.27}$$

obtain from the expression for ϵ^i is a measure of the relative eccentricity. When $Q_\epsilon = 0$ then the length of the cell edge does not change in the ξ^i direction. With the Christoffel symbol notation (2.39), we also obtain

$$Q^i_\epsilon = \left(\frac{1}{g_{ii}}\frac{\partial \boldsymbol{x}}{\partial \xi^i}\cdot\frac{\partial^2 \boldsymbol{x}}{\partial \xi^i \partial \xi^i}\right)^2 = \left(\frac{1}{g_{ii}}[ii,i]\right)^2. \tag{3.28}$$

3.5.3 Curvature

The relative eccentricity Q^i_ϵ describes the change of the length of the cell edge along the coordinate curve ξ^i, however, it fails to describe the change of its direction. The quantity which characterizes this grid quality is derived through a curvature vector.

In accordance with (3.2), the curvature vector $\boldsymbol{k}_i$ of the coordinate line ξ^i for a fixed i is defined by the relation $\boldsymbol{k}_i = \boldsymbol{x}_{ss}$, where s is the arc length parametrization of the coordinate line ξ^i, i.e. the variable s is defined by the transformation $s(\xi^i)$ satisfying the equation

$$\frac{\mathrm{d}s}{\mathrm{d}\xi^i} = \sqrt{g_{ii}}.$$

Therefore

$$\frac{\partial}{\partial s} = \frac{1}{\sqrt{g_{ii}}}\frac{\partial}{\partial \xi^i}$$

and consequently

$$\boldsymbol{k}_i = \frac{1}{g_{ii}}\boldsymbol{x}_{\xi^i\xi^i} - \frac{\boldsymbol{x}_{\xi^i}}{(g_{ii})^2}\boldsymbol{x}_{\xi^i}\cdot\boldsymbol{x}_{\xi^i\xi^i}. \tag{3.29}$$

3.5.3.1 Local Straightness of the Coordinate Line

Equation (3.29) shows that if the curvature vector $\boldsymbol{k}_i$ equals zero ($\boldsymbol{k}_i = \boldsymbol{0}$) then the vector $\boldsymbol{x}_{\xi^i\xi^i}$ is parallel to the vector $\boldsymbol{x}_{\xi^i}$, i.e. the tangential vector does not change its direction. Therefore the coordinate line ξ^i is locally straight at a point of zero curvature. From (3.29), we obtain in this case

$$\boldsymbol{x}_{\xi^i\xi^i} = \frac{(\boldsymbol{x}_{\xi^i\xi^i}\cdot\boldsymbol{x}_{\xi^i})}{g_{ii}}\boldsymbol{x}_{\xi^i}.$$

Using the Gauss relations (2.35), we also obtain

$$\boldsymbol{x}_{\xi^i\xi^i} = \Gamma^l_{ii}\boldsymbol{x}_{\xi^l}, \quad l = 1, \dots, n.$$

Comparing these two expansions of $\boldsymbol{x}_{\xi^i\xi^i}$ we see that the vector $\boldsymbol{x}_{\xi^i\xi^i}$ is parallel to $\boldsymbol{x}_{\xi^i}$ if

$$\Gamma^l_{ii} = 0 \quad \text{for all } l \neq i. \tag{3.30}$$

The relation (3.30) is a criterion of local straightness of the coordinate curve ξ^i. A measure of the deviation of the curve ξ^i from a straight line may, therefore, be determined as

$$Q^i_{\mathrm{st}} = d_{lm}\Gamma^l_{ii}\Gamma^m_{ii}, \quad l, m \neq i, \tag{3.31}$$

where d_{lm} is a positive $(n-1)\times(n-1)$ tensor.

3.5.3.2 Expansion of the Curvature Vector in the Normal Vectors

We know that the curvature vector $\boldsymbol{k}_i$ is orthogonal to the unit tangential vector $\boldsymbol{x}_s$. On the other hand, the normal base vectors $\nabla\xi^j$, $j \neq i$, are also orthogonal to the tangent vector $\boldsymbol{x}_{\xi^i}$ and therefore to $\boldsymbol{x}_s$. Thus the curvature vector $\boldsymbol{k}_i$ of the coordinate curve ξ^i can be expanded in the $n-1$ normal vectors $\nabla\xi^j$, $j \neq i$. In order to find such an expansion we first recall that in accordance with (2.40),

$$\boldsymbol{x}_{\xi^i\xi^i} = [ii, m]\nabla\xi^m, \quad m = 1, \dots, n,$$

with summation over m, where

$$[ii, m] = \boldsymbol{x}_{\xi^i\xi^i} \cdot \boldsymbol{x}_{\xi^m} = \frac{\partial g_{im}}{\partial \xi^i} - \frac{1}{2}\frac{\partial g_{ii}}{\partial \xi^m},$$

from (2.44). Further, from (2.22),

$$\boldsymbol{x}_{\xi^i} = g_{im}\nabla\xi^m, \quad m = 1, \dots, n.$$

Therefore the relation (3.29) is equivalent to

$$\begin{aligned} \boldsymbol{k}_i &= \frac{1}{g_{ii}}\left([ii, m]\nabla\xi^m - \frac{1}{g_{ii}}[ii, i]\right) g_{im}\nabla\xi^m \\ &= \frac{1}{(g_{ii})^2}(g_{ii}[ii, l] - g_{il}[ii, i])\nabla\xi^l, \\ m &= 1, \dots, n, \ l = 1, \dots, n, \ l \neq i, \ i \text{ fixed.} \end{aligned} \tag{3.32}$$

This equation represents the curvature vector $\boldsymbol{k}_i$ through the $n-1$ normal base vectors $\nabla\xi^l$, $l \neq i$.

In particular, in two dimensions the relation (3.32) for $i = 1$ becomes

$$\boldsymbol{k}_1 = \frac{1}{(g_{11})^2}(g_{11}[11, 2] - g_{12}[11, 1])\nabla\xi^2. \tag{3.33}$$

And, from, (2.20)

$$\boldsymbol{k}_1 = \frac{g}{(g_{11})^2}(g^{22}[11, 2] + g^{21}[11, 1])\nabla\xi^2.$$

Therefore, using (2.42), we obtain

$$\boldsymbol{k}_1 = \frac{g}{(g_{11})^2}\Gamma^2_{11}\nabla\xi^2. \tag{3.34}$$

Analogously, the curvature vector $\boldsymbol{k}_2$ along the coordinate ξ^2 is expressed as follows:

$$\boldsymbol{k}_2 = \frac{g}{(g_{22})^2}\Gamma^1_{22}\nabla\xi^1. \tag{3.35}$$

In the same way the curvature vector of the coordinate curves in the case of three-dimensional space R^3 is computed. For example, in accordance with (3.32), the vector $\boldsymbol{k}_1$ can be expanded in the normal vectors $\boldsymbol{\nabla}\xi^2$ and $\boldsymbol{\nabla}\xi^3$ as

$$\begin{aligned}\boldsymbol{k}_1 = &\frac{1}{(g_{11})^2}\{(g_{11}[11,2] - g_{12}[11,1])\boldsymbol{\nabla}\xi^2 \\ &+ (g_{11}[11,3] - g_{13}[11,1])\boldsymbol{\nabla}\xi^3\}.\end{aligned} \tag{3.36}$$

3.5.3.3 Measure of Coordinate Line Curvature

The length of the vector $\boldsymbol{k}_i$ is the modulus of the curvature and denoted by $|k_i|$. Thus, for the curvature k_1 of the coordinate line ξ^1 in the two-dimensional domain X^2, we obtain from (3.34)

$$|k_1| = \frac{g\sqrt{g^{22}}}{(g_{11})^2}|\Gamma^2_{11}| = \frac{g\sqrt{g^{22}}}{(g_{11})^2}\left|\frac{\partial^2 x^1}{\partial\xi^1\partial\xi^1}\frac{\partial\xi^2}{\partial x^1} + \frac{\partial^2 x^2}{\partial\xi^1\partial\xi^1}\frac{\partial\xi^2}{\partial x^2}\right|. \tag{3.37}$$

Taking into account the two-dimensional relation (2.3)

$$\frac{\partial\xi^i}{\partial x^j} = (-1)^{i+j}\frac{1}{J}\frac{\partial x^{3-j}}{\partial\xi^{3-i}}, \quad i,j = 1,2, \ J = \sqrt{g},$$

with i, j fixed, we find that

$$\Gamma^2_{11} = \frac{1}{J}\left(\frac{\partial x^1}{\partial\xi^1}\frac{\partial^2 x^2}{\partial\xi^1\partial\xi^1} - \frac{\partial x^2}{\partial\xi^1}\frac{\partial^2 x^1}{\partial\xi^1\partial\xi^1}\right).$$

Therefore, for the curvature of the coordinate ξ^1, we also obtain from (2.20) and (3.37)

$$|k_1| = \frac{1}{(g_{11})^{3/2}}\left|\frac{\partial x^1}{\partial\xi^1}\frac{\partial^2 x^2}{\partial\xi^1\partial\xi^1} - \frac{\partial x^2}{\partial\xi^1}\frac{\partial^2 x^1}{\partial\xi^1\partial\xi^1}\right|. \tag{3.38}$$

Analogously, using the relation (3.35) we get for the curvature of the coordinate curve ξ^2

$$|k_2| = \frac{1}{(g_{22})^{3/2}}\left|\frac{\partial x^2}{\partial\xi^2}\frac{\partial^2 x^1}{\partial\xi^2\partial\xi^2} - \frac{\partial x^1}{\partial\xi^2}\frac{\partial^2 x^2}{\partial\xi^2\partial\xi^2}\right|. \tag{3.39}$$

In the case of three-dimensional space, the curvature measure of the coordinate line ξ^i is computed from the relation (3.7):

$$|k_i| = \frac{1}{\sqrt{(g_{ii})^3}}|\boldsymbol{x}_{\xi^i} \times \boldsymbol{x}_{\xi^i\xi^i}|, \quad i = 1,2,3. \tag{3.40}$$

The curvature representation can provide various measures of the curvature of the coordinate line ξ^i. The simplest measure may be described in the common manner as the square of the curvature

$$Q^i_k = (k_i)^2. \tag{3.41}$$

In analogy with (3.31), the quantity Q^i_k is also a measure of the departure of the coordinate line ξ^i from a straight line.

3.5.4 Measure of Coordinate Line Torsion

The square of the torsion is another measure of a coordinate line ξ^i lying in three-dimensional space. This measure is computed in accordance with (3.10) from the relation

$$\begin{aligned} Q^i_\tau &= \frac{1}{(k_i)^4(g_{ii})^6}[(\boldsymbol{x}_{\xi^i} \times \boldsymbol{x}_{\xi^i\xi^i}) \cdot \boldsymbol{x}_{\xi^i\xi^i\xi^i}]^2 \\ &= \frac{1}{(k_i)^4(g_{ii})^6}\det{}^2 \begin{pmatrix} \boldsymbol{x}_{\xi^i} \\ \boldsymbol{x}_{\xi^i\xi^i} \\ \boldsymbol{x}_{\xi^i\xi^i\xi^i} \end{pmatrix}. \end{aligned} \tag{3.42}$$

The condition $Q^i_\tau \equiv 0$ means that the coordinate line ξ^i lies in a plane. Thus the quantity Q^i_τ is a measure of the departure of the coordinate line ξ^i from a plane line.

3.6 Characteristics of Faces of Three-Dimensional Grids

A structured coordinate grid in a three-dimensional domain X^3 is composed of three-dimensional curvilinear hexahedral cells which are images of elementary cubes obtained through a coordinate transformation

$$\boldsymbol{x}(\boldsymbol{\xi}) : \Xi^3 \to X^3.$$

The boundary of each cell is segmented into six curvilinear quadrilaterals, through which some characteristics of the cell can be defined. This section describes some important quality measures of the faces of three-dimensional coordinate cells.

3.6.1 Cell Face Skewness

The skewness of a cell face is described through the angle between the two tangent vectors defining the cell face. Let the cell face lie in the surface $\xi^l =$ const; the tangent vectors of the surface are then the vectors $\boldsymbol{x}_{\xi^i}$ and $\boldsymbol{x}_{\xi^j}$, $i = l+1,\ j = l+2$, with the identification convention for the index m that m is equivalent to $m \pm 3$. One of the cell face skewness characteristics can be determined as the square of the cosine of the angle between the vectors. Thus, for a fixed l,

$$Q^l_{\mathrm{sk},1} = \cos^2\theta = \frac{(g_{ij})^2}{g_{ii}g_{jj}}, \quad i = l+1,\ j = l+2. \tag{3.43}$$

Another expression for the cell face skewness is specified by the cotangent squared of the angle θ:

$$Q^l_{\mathrm{sk},2} = \cot^2\theta = \frac{(g_{ij})^2}{g_{ii}g_{jj} - (g_{ij})^2}, \quad i = l+1,\ j = l+2. \tag{3.44}$$

Taking into account the relations (2.28) and (2.32), this can also be written in the form

$$Q^l_{\mathrm{sk},2} = \frac{(g_{ij})^2}{(\boldsymbol{x}_{\xi^i} \times \boldsymbol{x}_{\xi^j})^2} = \frac{(g_{ij})^2}{gg^{ll}}, \quad i = l+1,\ j = l+2.$$

Since $(g_{ij})^2 = g_{ii}g_{jj}(1 - \sin^2\theta)$, we also obtain

$$Q^l_{\mathrm{sk},2} = \frac{g_{ii}g_{jj}}{(\boldsymbol{x}_{\xi^i} \times \boldsymbol{x}_{\xi^j})^2} - 1 = \frac{g_{ii}g_{jj}}{gg^{ll}}, \quad i = l+1,\ j = l+2.$$

The quantities for the grid face skewness introduced above equal zero when the edges of the cell face are orthogonal. Therefore these quantities characterize the departure of the cell face from a rectangle. One more characteristic of the cell face nonorthogonality is defined as square of the dot product of the vectors $\boldsymbol{x}_{\xi^i}$ and $\boldsymbol{x}_{\xi^j}$:

$$Q^l_{\mathrm{o},1} = (g_{ij})^2, \quad i = l+1,\ j = l+2. \tag{3.45}$$

3.6.2 Face Aspect-Ratio

A measure of the aspect-ratio of the cell face formed by the tangent vectors $\boldsymbol{x}_{\xi^i}$ and $\boldsymbol{x}_{\xi^j}$ is defined through the diagonal elements g_{ii} and g_{jj} of the covariant metric tensor (g_{km}), $k, m = 1, 2, 3$. One form of this measure is given by the expression

$$Q^l_{\mathrm{as}} = \frac{g_{ii}}{g_{jj}} + \frac{g_{jj}}{g_{ii}} = \frac{(g_{ii} + g_{jj})^2}{g_{ii}g_{jj}} - 2, \tag{3.46}$$

where $i = l+1$, $j = l+2$, and $m + 3$ is equivalent to $\pm m$. We have the inequality $Q^l_{\mathrm{as}} \geq 2$, which is an equality if and only if $g_{ii} = g_{jj}$, i.e. the parallelogram formed by the vectors $\boldsymbol{x}_{\xi^i}$ and $\boldsymbol{x}_{\xi^j}$ is a rhombus. Thus (3.46) is a measure of the departure of the cell from a rhombus.

3.6.3 Cell Face Area Squared

The square of the area of the face of the basic parallelepiped formed by the two tangential vectors $\boldsymbol{x}_{\xi^i}$ and $\boldsymbol{x}_{\xi^j}$ is expressed as follows:

$$Q^l_{\mathrm{ar}} = |\boldsymbol{x}_{\xi^i}|^2 |\boldsymbol{x}_{\xi^j}|^2 \sin^2\theta = g_{ii}g_{jj} - (g_{ij})^2, \quad i = l+1,\ j = l+2, \tag{3.47}$$

where θ is the angle of intersection of the vectors and i and j are chosen to satisfy the condition $l \neq i$ and $l \neq j$. Taking advantage of (2.28) and (2.32), we see that

$$Q^l_{\mathrm{ar}} = |\boldsymbol{x}_{\xi^i} \times \boldsymbol{x}_{\xi^j}|^2 = g|\boldsymbol{\nabla}\boldsymbol{\xi}^l|^2 = gg^{ll}, \quad l \text{ fixed}. \tag{3.48}$$

As the square of the area of the coordinate cell face which corresponds to the parallelogram defined by the vectors $\boldsymbol{x}_{\xi^i}$ and $\boldsymbol{x}_{\xi^j}$ equals $h^2 Q_{\mathrm{ar}} + O(h^3)$, the quantity Q^l_{ar} can be applied to characterize the area of the cell face.

3.6.4 Cell Face Warping

Measures of the cell face warping are obtained through the curvatures of the coordinate surface on which the face lies. Let this be the coordinate surface $\xi^3 = \xi_0^3$. Then a natural parametrization $\boldsymbol{x}(\boldsymbol{\xi}) : \Xi^2 \to R^3$, $\boldsymbol{\xi} = (\xi^1, \xi^2)$ of the surface is represented by $\boldsymbol{x}(\xi^1, \xi^2, \xi_0^3)$.

3.6.4.1 Mean Curvature of the Coordinate Surface

Twice the mean curvature of the coordinate surface is defined through the formula (3.19) or (3.20) as

$$2K_{3,m} = g_{\xi x}^{ij} b_{ij}, \quad i,j = 1,2, \tag{3.49}$$

where $b_{ij} = \boldsymbol{x}_{\xi^i \xi^j} \cdot \boldsymbol{n}$. It is obvious that the contravariant metric tensor $(g_{\xi x}^{ij})$ of the surface $\xi^3 = \xi_0^3$ in the coordinates ξ^1, ξ^2 is the inverse of the 2×2 matrix $(g_{ij}^{x\xi})$ whose elements are the elements of the volume metric tensor (g_{ij}) with the indices $i,j = 1,2$, i.e.

$$g_{ij}^{x\xi} = g_{ij} = \boldsymbol{x}_{\xi^i} \cdot \boldsymbol{x}_{\xi^j}, \quad i,j = 1,2.$$

Therefore, using (3.13) and (2.32), we have

$$g_{\xi x}^{ij} = (-1)^{i+j} g_{3-i\ 3-j} / (\boldsymbol{x}_{\xi^1} \times \boldsymbol{x}_{\xi^2})^2 = \frac{(-1)^{i+j} g^{33}}{g} g_{3-i\ 3-j}, \quad i,j = 1,2,$$

without summation over i or j. Also, it is clear that

$$\boldsymbol{n} = \frac{1}{\sqrt{g^{33}}} \nabla \xi^3,$$

and consequently the coefficients of the second fundamental form of the coordinate surface $\xi^3 = \xi_0^3$ are expressed as follows:

$$b_{ij} = \frac{1}{\sqrt{g^{33}}} \boldsymbol{x}_{\xi^i \xi^j} \cdot \nabla \xi^3 = \frac{1}{\sqrt{g^{33}}} \Gamma_{ij}^3.$$

Thus, (3.49) results in

$$2K_{3,m} = \frac{(-1)^{i+j} \sqrt{g^{33}}}{g} g_{3-i\ 3-j} \Gamma_{ij}^3, \quad i,j = 1,2.$$

Analogously, we obtain a general formula for the coefficients of the second fundamental form of the coordinate surface $\xi^l = \xi_0^l$, $l = 1,2,3$:

$$b_{ij} = \frac{1}{\sqrt{g^{ll}}} \Gamma_{l+i\ l+j}^l, \quad i,j = 1,2, \tag{3.50}$$

with l fixed and where m is equivalent to $m \pm 3$. Thus twice the mean curvature of the coordinate surface $\xi^l = \xi^l_0$, $l = 1, 2$, is expressed by

$$2K_{l,m} = \frac{(-1)^{i+j}\sqrt{g^{ll}}}{g} g_{l-i\ l-j}\Gamma^l_{l+i\ l+j}, \quad i, j = 1, 2, \tag{3.51}$$

with l fixed.

3.6.4.2 Gaussian Curvature of the Coordinate Surface

Taking advantage of (3.21) and (3.50), the Gaussian curvature of the coordinate surface $\xi^l = \xi^l_0$ can be expressed as follows:

$$K_{l,\mathrm{G}} = \frac{\sqrt{g^{ll}}}{g}[\Gamma^l_{l+1\ l+1}\Gamma^l_{l+2\ l+2} - (\Gamma^l_{l+1\ l+2})^2], \tag{3.52}$$

with the index l fixed.

3.6.4.3 Measures of Face Warping

The quantities which measure the warping of the face of a three-dimensional cell are obtained through the coefficients of the second fundamental form or through the mean and Gaussian curvatures of a coordinate surface containing the face. Let this be the surface $\xi^l = \xi^l_0$. Then, taking advantage of (3.51) and (3.52), the measures may be expressed as follows:

$$\begin{aligned} Q^l_{\mathrm{w},1} &= (K_{l,m})^2 = \frac{g^{ll}}{g^2}[(-1)^{i+j}g_{l-i\ l-j}\Gamma^l_{l+i\ l+j}]^2, \\ Q^l_{\mathrm{w},2} &= (K_{l,g})^2 = \frac{g^{ll}}{g^2}[\Gamma^l_{l+1\ l+1}\Gamma^l_{l+2\ l+2} - (\Gamma^l_{l+1\ l+2})^2], \end{aligned} \tag{3.53}$$

with l fixed.

Equation (3.50) for the second fundamental form of the surface $\xi^l = \xi^l_0$ also gives an expression for the third measure of the cell face warpness:

$$Q^l_{\mathrm{w},3} = \sum_{i,j=1}^{2}(b_{ij})^2 = \frac{1}{g^{ll}}\sum_{i,j=1}^{2}(\Gamma^l_{l+i\ l+j})^2, \quad l \text{ fixed}. \tag{3.54}$$

3.7 Characteristics of Grid Cells

Cell features are described by the cell volume (area in two dimensions) and by the characteristics of the cell edges and faces.

3.7.1 Cell Aspect-Ratio

A measure of the aspect-ratio of a three-dimensional cell is formulated through the measures of the aspect-ratio of its faces described by (3.46). The simplest formulation is provided by summing these measures, which results in

$$Q_{\mathrm{sk}} = \sum_{l=1}^{3} Q_{\mathrm{sk}}^{l}. \tag{3.55}$$

3.7.2 Square of Cell Volume

The characteristic related to the square of the cell volume is

$$Q_V = g = \det(g_{ij}) = I_n. \tag{3.56}$$

In three dimensions, we also obtain from (2.31)

$$Q_V = [\boldsymbol{x}_{\xi^1} \cdot (\boldsymbol{x}_{\xi^2} \times \boldsymbol{x}_{\xi^3})]^2.$$

3.7.3 Cell Area Squared

We denote by Q_{ar} the sum of the quantities Q_{ar}^{ij}, $i \neq j$, from (3.48). These quantities are the area characteristics of the faces of a three-dimensional cell; thus, in accordance with (3.24), the magnitude Q_{ar} coincides with the invariant I_2:

$$Q_{\mathrm{ar}} = \sum_{i=1}^{3} g g^{ii} = I_2. \tag{3.57}$$

3.7.4 Cell Skewness

One way to describe the cell skewness characteristics of three-dimensional grids utilizes the angles between the tangential vectors in the forms of the corresponding expressions (3.43) and (3.44) introduced for the formulation of the face skewness. For example, summation of these quantities gives the following expressions for the cell skewness measures:

$$\begin{aligned} Q_{\mathrm{sk},1} &= \frac{(g_{12})^2}{g_{11}g_{22}} + \frac{(g_{23})^2}{g_{22}g_{33}} + \frac{(g_{13})^2}{g_{11}g_{33}}, \\ Q_{\mathrm{sk},2} &= \frac{(g_{12})^2}{g_{11}g_{22} - (g_{12})^2} + \frac{(g_{13})^2}{g_{11}g_{33} - (g_{13})^2} + \frac{(g_{23})^2}{g_{22}g_{33} - (g_{23})^2} \\ &= \frac{1}{g}\left(\frac{(g_{12})^2}{g^{33}} + \frac{(g_{13})^2}{g^{22}} + \frac{(g_{23})^2}{g^{11}}\right). \end{aligned} \tag{3.58}$$

Here $Q_{\mathrm{sk},1}$ is the sum of the squares of the cosines of the angles between the edges of the cell, while $Q_{\mathrm{sk},2}$ is the sum of the squares of the cotangents of the angles.

Other quantities to express the three-dimensional cell skewness can be defined through the angles between the normals to the coordinate surfaces. Any normal to the coordinate surface $\xi^i = \xi^i_0$ is parallel to the normal vector $\boldsymbol{\nabla}\xi^i$. Therefore the cell skewness can be derived through the angles between the base normal vectors $\boldsymbol{\nabla}\xi^i$. The quantity

$$\frac{(\boldsymbol{\nabla}\xi^i \cdot \boldsymbol{\nabla}\xi^j)^2}{g^{ii}g^{jj}} = \frac{(g^{ij})^2}{g^{ii}g^{jj}}, \quad i, j \text{ fixed}$$

is the cosine squared of the angle between the respective faces of the coordinate cell. This characteristic is a dimensionless magnitude. The sum of such quantities is the third characteristic of the three-dimensional cell skewness:

$$Q_{\text{sk},3} = \frac{(g^{12})^2}{g^{11}g^{22}} + \frac{(g^{13})^2}{g^{11}g^{33}} + \frac{(g^{23})^2}{g^{22}g^{33}}. \tag{3.59}$$

Another dimensionless quantity which characterizes the mutual skewness of two faces of the cell is the cotangent squared of the angle between the normal vectors $\boldsymbol{\nabla}\xi^i$ and $\boldsymbol{\nabla}\xi^j$:

$$\frac{(\boldsymbol{\nabla}\xi^i \cdot \boldsymbol{\nabla}\xi^j)^2}{|\boldsymbol{\nabla}\xi^i \times \boldsymbol{\nabla}\xi^j|^2} = \frac{g(g^{ij})^2}{g_{kk}} = \frac{(g^{ij})^2}{g^{ii}g^{jj} - (g^{ij})^2},$$

where (i, j, k) are cyclic and fixed. The summation of this over k defines the fourth grid skewness characteristic

$$\begin{aligned} Q_{\text{sk},4} &= \frac{(g^{12})^2}{g^{11}g^{22} - (g^{12})^2} + \frac{(g^{13})^2}{g^{11}g^{33} - (g^{13})^2} + \frac{(g^{23})^2}{g^{22}g^{33} - (g^{23})^2} \\ &= g\left(\frac{(g^{12})^2}{g_{33}} + \frac{(g^{13})^2}{g_{22}} + \frac{(g^{23})^2}{g_{11}}\right). \end{aligned} \tag{3.60}$$

Note that the three-dimensional cell skewness quantities $Q_{\text{sk},1}$ and $Q_{\text{sk},3}$ can be readily extended to arbitrary dimensions $n \geq 2$.

3.7.5 Characteristics of Nonorthogonality

The quantities $Q_{\text{sk},i}$, $i = 1, 2, 3, 4$, from (3.58)–(3.60) reach their minimum values equal to zero, only when the three-dimensional transformation $\boldsymbol{x}(\boldsymbol{\xi})$ is orthogonal at the respective point, and vice-versa. Therefore these quantities, which provide the possibility to detect orthogonal grids, may be considered as some measures of grid nonorthogonality.

Other quantities characterizing the departure of a three-dimensional grid from an orthogonal one are as follows:

$$\begin{aligned} Q_{\text{o},1} &= \frac{g_{11}g_{22}g_{33}}{g}, \\ Q_{\text{o},2} &= g\,(g^{11}g^{22}g^{33}). \end{aligned} \tag{3.61}$$

Obviously, these quantities $Q_{o,1}$ and $Q_{o,2}$ are dimensionless and reach their minimum equal to 1, if and only if the coordinate transformation $\boldsymbol{x}(\boldsymbol{\xi})$ is orthogonal.

The sum of the squares of the nondiagonal elements of the covariant metric tensor (g_{ij}) yields another characteristic of cell nonorthogonality,

$$Q_{o,3} = (g_{12})^2 + (g_{13})^2 + (g_{23})^2. \tag{3.62}$$

An analogous formulation is given through the elements of the contravariant metric tensor,

$$Q_{o,4} = (g^{12})^2 + (g^{13})^2 + (g^{23})^2. \tag{3.63}$$

Note that, in contrast to $Q_{o,1}$ and $Q_{o,2}$, the quantities $Q_{o,3}$ and $Q_{o,4}$ are dimensionally heterogeneous.

3.7.6 Grid Density

The invariants of the tensor (g_{ij}) can be useful for specifying some characteristics of grid quality. For example, one important characteristic describing the concentration of grid nodes can be derived from the ratio I_{n-1}/I_n.

In order to show this we first note that in accordance with the geometrical interpretation of the invariants given in Sect. 3.4.2 we can write

$$\frac{I_{n-1}}{I_n} = \sum_{m=1}^{n} \left(V_m^{n-1}\right)^2 / \left(V^n\right)^2, \tag{3.64}$$

where V_m^{n-1} is the space of the boundary segment ξ^m = const of the basic parallelepiped defined by the tangential vectors $\boldsymbol{x}_{\xi^i}$, $i = 1, \ldots, n$.

It is evident that

$$V^n = d_m V_m^{n-1}, \quad m = 1, \ldots, n,$$

where d_m is the distance between the vertex of the tangential vector $\boldsymbol{x}_{\xi^m}$ and the $(n-1)$-dimensional plane P^{n-1} spanned by the vectors $\boldsymbol{x}_{\xi^i}$, $i \neq m$. Hence, from (3.64),

$$\frac{I_{n-1}}{I_n} = \sum_{m=1}^{n} (1/d^m)^2. \tag{3.65}$$

Now let us consider two grid surfaces $\xi^m = c$ and $\xi^m = c + h$ obtained by mapping a uniform rectangular grid with a step size h in the computational domain Ξ^n onto X^n. Let us denote by l_m the distance between a node on the coordinate surface $\xi^m = c$ and the nearest node on the surface $\xi^m = c + h$ (Fig. 3.5). We have

$$l_m = d_m h + O(h)^2$$

and, therefore, from (3.65),

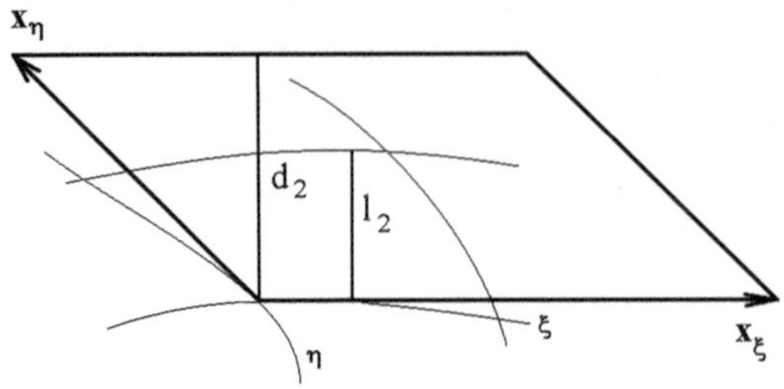

Fig. 3.5. Illustration of invariants

$$\frac{I_{n-1}}{I_n} = \sum_{m=1}^{n} (h/l_m)^2 + O(h).$$

The quantity $(h/l_m)^2$ increases if the grid nodes cluster in the direction normal to the surface $\xi^m = c$. Therefore this quantity can be considered as some measure of the grid concentration in the normal direction and, consequently, the magnitude $1/d_m$ means the density of the grid concentration in the $\nabla\xi^m$ direction. In particular, we readily see that $1/d_m = \sqrt{g^{mm}}$, with m fixed. Thus the expression (3.65) defines a measure of the grid density in all directions. We denote this quantity by Q_{cn}, where the subscript "cn" represents "concentration". Note that, in accordance with (3.22), this measure can be expressed as follows:

$$Q_{\text{cn}} = \frac{I_{n-1}}{I_n} = g^{11} + \cdots + g^{nn}. \tag{3.66}$$

3.7.7 Characteristics of Deviation from Conformality

Conformal coordinate transformations are distinguished by the fact that the Jacobi matrix J is orthonormal and consequently the metric tensor (g_{ij}) is a multiple of the unit matrix:

$$(g_{ij}) = g(\boldsymbol{\xi})I = g(\boldsymbol{\xi})(\delta^i_j), \quad i, j = 1, \ldots, n.$$

The cells of the coordinate grid derived from the conformal mapping $\boldsymbol{x}(\boldsymbol{\xi})$ are close to n-dimensional cubes (squares in two dimensions). Grids with such cells are attractive from the computational point of view. Therefore it is desirable to define simple grid quantities which can allow one to detect grids whose cells are close to n-dimensional cubes. It is clear that the condition of conformality can be described by the system

$$g_{ij} = 0, \quad i \neq j,$$
$$g_{11} = g_{22} = \cdots = g_{nn}.$$

These relations give a rise to a natural quantity

$$Q = \sum_{i \neq j} (g_{ij})^2 + \sum_{i=2}^{n} (g_{ii} - g_{11})^2,$$

which is zero if and only if the coordinate transformation $\boldsymbol{x}(\boldsymbol{\xi})$ is conformal. So this quantity can help one to detect when the grid is conformal. However, the above formula is too cumbersome and it is dimensionally heterogeneous. More compact expressions for the analysis of the conformality or nonconformality of grid cells and for the formulation of algorithms to construct nearly conformal grids are obtained by the use of the metric-tensor invariants.

3.7.7.1 Two-Dimensional Space

The departure from conformality of the two-dimensional transformation $\boldsymbol{x}(\xi) : \Xi^2 \to X^2$ is expressed by the quantity

$$Q_{\mathrm{cf},1} = \frac{I_1}{\sqrt{I_2}} = \frac{|\boldsymbol{x}_{\xi^1}|^2 + |\boldsymbol{x}_{\xi^2}|^2}{|\boldsymbol{x}_{\xi^1}||\boldsymbol{r}_{\xi^2}||\sin\theta|} = \frac{g_{11} + g_{22}}{\sqrt{g_{11}}\sqrt{g_{22}}|\sin\theta|}, \qquad (3.67)$$

where θ is the angle between the tangent vectors $\boldsymbol{x}_{\xi^1}$ and $\boldsymbol{x}_{\xi^2}$. Since

$$Q_{\mathrm{cf},1} \geq \frac{g_{11} + g_{22}}{\sqrt{g_{11}g_{22}}},$$

it is clear that the value of $I_1/\sqrt{I_2}$ exceeds 2. The minimum value 2 is achieved only if $g_{11} = g_{22}$ and $\theta = \pi/2$, i.e. when the parallelogram with sides defined by the vectors $\boldsymbol{x}_{\xi^1}$ and $\boldsymbol{x}_{\xi^2}$ is a square. Thus the characteristic $Q_{\mathrm{cf},1}$ allows one to state with certainty when the coordinate transformation $\boldsymbol{x}(\boldsymbol{\xi})$ is conformal at a point $\boldsymbol{\xi}$, namely when $Q_{\mathrm{cf},1}(\boldsymbol{\xi}) = 2$. Therefore in the two-dimensional case the quantity

$$Q_{\mathrm{cf},1} - 2 = I_1/\sqrt{I_2} - 2$$

reflects some measure of the deviation of the cell from a square. We see that the quantity $Q_{\mathrm{cf},1}$ given by (3.67) is dimensionally homogeneous.

Through the quantity $Q_{\mathrm{cf},1}$ we can also estimate the bounds of the aspect ratio of the two-dimensional cell and the angle between the edges of this cell.

Evaluation of the Cell Angles

First, we obtain an estimate of the angle between the cell edges. From (3.67) we have

$$\sin^2\theta = \frac{(F^2+1)^2}{F^2}/Q^2_{\mathrm{cf},1}, \qquad (3.68)$$

where $F^2 = g_{11}/g_{22}$. As $(F^2+1)^2/F^2 \geq 4$, we have from (3.68) that

$$\sin^2\theta \geq 4/Q^2_{\mathrm{cf},1} \qquad (3.69)$$

and, accordingly, we obtain the following estimate for the angle θ:

$$\pi - \arcsin(2/Q_{\mathrm{cf},1}) \geq \theta \geq \arcsin(2/Q_{\mathrm{cf},1}). \tag{3.70}$$

From (3.68), we find that the minimum value $4/Q^2_{\mathrm{cf},1}$ of $\sin^2\theta$ for a fixed value of $Q_{\mathrm{cf},1}$ is achieved when $F = 1$, i.e. when the parallelogram is the rhombus.

Although it is desirable to generate orthogonal grids, a departure from orthogonality is practically inevitable when grid adaptation is performed. Commonly, this departure is required to be restricted to 45°. Beyound this range the contribution of the grid skewness to the truncation error may become unacceptable. The inequality (3.70) shows that this barrier of 45° is not broken if $Q_{\mathrm{cf},1} \leq 2\sqrt{2}$.

Evaluation of the Cell Aspect Ratio

Now we estimate the quantity $F = \sqrt{g_{11}/g_{22}}$. The quantity F, called the cell aspect ratio, is the ratio of the lengths of the edges of the cell. By computing F from (3.68) we obtain

$$F = \frac{\alpha}{2} - 1 \pm \sqrt{\frac{\alpha^2}{4} - \alpha}, \quad \alpha = Q^2_{\mathrm{cf},1}\sin^2\theta. \tag{3.71}$$

Equation (3.71) gives two values of the cell aspect ratio,

$$F_1 = \frac{\alpha}{2} - 1 + \sqrt{\frac{\alpha^2}{4} - \alpha} \quad \text{and} \quad F_2 = \frac{\alpha}{2} - 1 - \sqrt{\frac{\alpha^2}{4} - \alpha},$$

satisfying the relation $F_1F_2 = 1$. We find that

$$F_1 = \max(\sqrt{g_{11}/g_{22}},\ \sqrt{g_{22}/g_{11}})$$

and

$$F_2 = \min(\sqrt{g_{11}/g_{22}},\ \sqrt{g_{22}/g_{11}}).$$

Thus

$$\frac{\alpha}{2} - 1 - \sqrt{\frac{\alpha^2}{4} - \alpha} \leq F_i \leq \frac{\alpha}{2} - 1 + \sqrt{\frac{\alpha^2}{4} - \alpha}, \quad i = 1, 2, \tag{3.72}$$

and consequently

$$2 \leq F_i + 1/F_i \leq \alpha - 2, \quad i = 1, 2. \tag{3.73}$$

As $Q^2_{\mathrm{cf},1} \geq \alpha \geq 4$, from (3.69), we also obtain from (3.72) and (3.73) the following upper and lower estimates of the aspect ratios F_i, $i = 1, 2$, which depend only on the quantity $Q_{\mathrm{cf},1}$:

$$\frac{Q^2_{\mathrm{cf},1}}{2} - 1 - Q_{\mathrm{cf},1}\sqrt{\frac{Q^2_{\mathrm{cf},1}}{4} - 1} \leq F_i \leq \frac{Q^2_{\mathrm{cf},1}}{2} - 1 + Q_{\mathrm{cf},1}\sqrt{\frac{Q^2_{\mathrm{cf},1}}{4} - 1}, \tag{3.74}$$

and

$$2 \leq F_i + 1/F_i \leq Q^2_{\mathrm{cf},1} - 2, \quad i = 1, 2. \tag{3.75}$$

The maximum value of F_i for a given value of $Q_{\mathrm{cf},1}$ is realized when $\sin^2\theta = 1$, i.e. the parallelogram is a rectangle.

3.7.7.2 Three-Dimensional Space

In three-dimensional space the deviation from conformality can be described by the dimensionless magnitude

$$Q_{\mathrm{cf},1} = (g)^{1/3}(g^{11} + g^{22} + g^{33}), \tag{3.76}$$

which, in accordance with (3.24), is expressed by means of the invariants I_2 and I_3 as follows:

$$Q_{\mathrm{cf},1} = I_2/(I_3)^{2/3}. \tag{3.77}$$

The value of (3.77) reaches its minimum only if

$$g^{11} = g^{22} = g^{33} \quad \text{and} \quad g^{-1} = g^{11}g^{22}g^{33}, \tag{3.78}$$

i.e. when the parallelogram defined by the basic normal vectors $\boldsymbol{\nabla}\xi^i$ is a cube. To prove this fact we note that

$$\frac{1}{g} \le g^{11}g^{22}g^{33}.$$

Therefore, from (3.76),

$$Q_{\mathrm{cf},1} \ge \frac{g^{11} + g^{22} + g^{33}}{\sqrt[3]{g^{11}g^{22}g^{33}}}$$

and, taking into account the general inequality for arbitrary positive numbers $a_1, \ldots, a_n$

$$\frac{1}{n}\sum_{i=1}^{n} a_i \ge \sqrt[n]{\prod_{i=1}^{n} a_i},$$

we find that $Q_{\mathrm{cf},1} \ge 3$. Obviously, $Q_{\mathrm{cf},1} = 3$ when the relations (3.78) are satisfied. From (2.34),

$$\frac{1}{g} = |\boldsymbol{\nabla}\xi^1 \cdot \boldsymbol{\nabla}\xi^2 \times \boldsymbol{\nabla}\xi^3|^2$$

and therefore (3.78) is satisfied only when the normal vectors $\boldsymbol{\nabla}\xi^i$, $i = 1, 2, 3$, are orthogonal to each other and have the same length. But then this is valid for the base tangential vectors $\boldsymbol{x}_{\xi^i}$, $i = 1, 2, 3$, as well. Thus (3.78) is satisfied only when the transformation $\boldsymbol{x}(\boldsymbol{\xi})$ is conformal.

In the same manner as in the two-dimensional case, one can derive bounds on the angles of the parallelepiped and on the ratio of the lengths of its edges that depend on the quantity $Q_{\mathrm{cf},1}$.

3.7.7.3 Generalization to Arbitrary Dimensions

Analogously, in the n-dimensional case a local measure of the deviation of the transformation $\boldsymbol{x}(\boldsymbol{\xi})$ from a conformal one is expressed by the quantity $Q_{\mathrm{cf},1} - n$, where

$$Q_{\mathrm{cf},1} = I_{n-1}/(I_n)^{1-1/n} = g^{1/n}(g^{11} + \cdots + g^{nn}). \tag{3.79}$$

The quantity $Q_{\mathrm{cf},1}$ equals n if and only if the mapping $\boldsymbol{x}(\boldsymbol{\xi})$ is conformal.

Another local characteristic of the deviation from conformality is described by the quantity $Q_{\mathrm{cf},2} - n$, where

$$Q_{\mathrm{cf},2} = I_1/(I_n)^{1/n}. \tag{3.80}$$

As for $Q_{\mathrm{cf},1}$, we can show that $Q_{\mathrm{cf},2} \geq n$ and that $Q_{\mathrm{cf},2} = n$ if the transformation $\boldsymbol{x}(\boldsymbol{\xi})$ is conformal at the point under consideration. Note also that $Q_{\mathrm{cf},1} = Q_{\mathrm{cf},2}$ in two dimensions.

3.7.8 Grid Eccentricity

One grid eccentricity characteristic is defined by summing the squares of the coordinate-line eccentricities (3.27). Thus the quantity

$$Q_{\epsilon,1} = \sum_{i=1}^{n} \left(\frac{\partial}{\partial \xi^i} \ln \sqrt{g_{ii}} \right)^2 \tag{3.81}$$

is a measure of the change of the lengths of all of the grid cell edges.

A similar characteristic of eccentricity can be formulated through the terms g^{ii}, namely

$$Q_{\epsilon,2} = \sum_{i=1}^{n} \left(\frac{\partial}{\partial x^i} \ln \sqrt{g^{ii}} \right)^2. \tag{3.82}$$

3.7.9 Measures of Grid Warping and Grid Torsion

In the same way as for grid eccentricity, we may formulate measures of grid warping by summing the surface-coordinate characteristics (3.53) and (3.54). As a result we obtain

$$\begin{aligned}
Q_{\mathrm{w},1} &= \frac{1}{g^2} \sum_{l=1}^{3} g^{ll} \left((-1)^{i+j} g_{l-i\ l-j} \Gamma^l_{l+i\ l+j} \right)^2, \\
Q_{\mathrm{w},2} &= \frac{1}{g^2} \sum_{l=1}^{3} g^{ll} \left[\Gamma^l_{l+1\ l+1} \Gamma^l_{l+2\ l+2} - \left(\Gamma^l_{l+1\ l+1} \right)^2 \right], \\
Q_{\mathrm{w},3} &= \sum_{l=1}^{3} \sum_{i,j=1}^{2} \frac{1}{g^{ll}} \left(\Gamma^l_{l+i\ l+j} \right)^2.
\end{aligned} \tag{3.83}$$

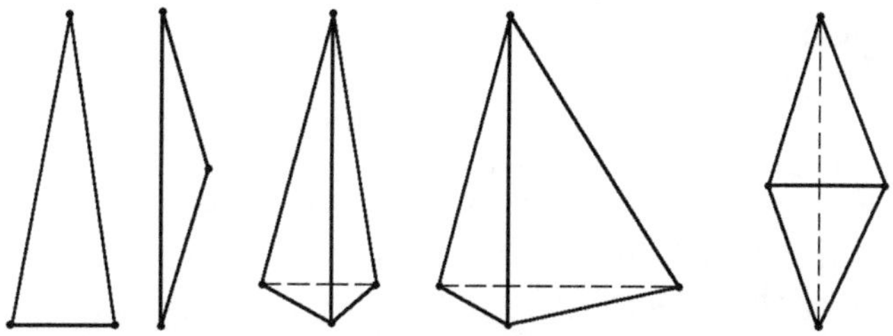

Fig. 3.6. Examples of poorly shaped triangles (**a**, **b**) and tetrahedrons (**c**, **d**, **e**)

The measure of grid torsion is formulated by summing the torsion measures (3.42) of the coordinate lines ξ^i, $i = 1, 2, 3$:

$$Q_\tau = \sum_{i=1}^{3} Q_\tau^i. \tag{3.84}$$

3.7.10 Quality Measures of Simplexes

The quantities which are applied to measure the quality of triangles and tetrahedrons are the following:

(1) the maximum edge length H,
(2) the minimum edge length h,
(3) the circum-radius R,
(4) the inradius r.

There are four deformation measures that allow one to characterize the quality of triangular and tetrahedral cells:

$$Q_{\mathrm{d},1} = \frac{H}{r}, \qquad Q_{\mathrm{d},2} = \frac{R}{H}, \qquad Q_{\mathrm{d},3} = \frac{H}{h}, \qquad Q_{\mathrm{d},4} = \frac{R}{r}.$$

The uniformity condition for a cell is satisfied when $Q_{\mathrm{d},1} = O(1)$ or $Q_{\mathrm{d},4} = O(1)$.

Examples of poorly shaped cells are shown in Fig. 3.6. Cases (a) and (c) correspond to needle-shaped cells. Figure 3.6(d) shows a wedge-shaped cell, while Figs. 3.6(b), (e) show sliver-shaped cells.

The cell is excessively deformed if $Q_{\mathrm{d},1} \gg 1$. In this case the cell has either a very acute or a very obtuse angle. The former case corresponds to $Q_{\mathrm{d},2} = O(1)$, $Q_{\mathrm{d},3} \gg 1$ (Figs. 3.6(a), (c), (d)), while the latter corresponds to $Q_{\mathrm{d},4} \gg 1$, $Q_{\mathrm{d},3} = O(1)$ (Figs. 3.6(b), (e)). The condition $Q_{\mathrm{d},2} = O(1)$ precludes obtuse angles.

3.8 Comments

Various aspects of mesh quality were surveyed by Knupp (2001, 2007). The introduction of metric-tensor invariants to describe some of the qualitative properties of grids was originally proposed by Jacquotte (1987). The grid measures in terms of the invariants and their relations described in this chapter were obtained by the author.

Prokopov (1989) introduced the dimensionless characteristics of two-dimensional cells.

Triangular elements were extensively analyzed by Field (2000).

Some questions concerned with the assessment of the contribution of the grid quality properties to the accuracy of solutions obtained using the grid were discussed by Kerlic and Klopfer (1982), Mastin (1982), Lee and Tsuei (1992), and Huang and Prosperetti (1994).

Discrete length, area, and orthogonality grid measures using averages and deviations were formulated by Steinberg and Roache (1992).

Babuŝka and Aziz (1976) have shown that the minimum-angle condition in a planar triangulation is too restrictive and can be replaced by a condition that limits the maximum allowable angle. Also the influence of grid quality measures on solution accuracy was discussed by Knupp (2007) and Shewchuck (2002).

Measures to quantify the shape of triangles and tetrahedrons were introduced by Field (1986), Baker (1989), Cougny, Shephard, and Georges (1990), and Dannelongue and Tanguy (1991).

A brief overview of tetrahedron quality measures with a comparison of the fidelity of these measures to a set of distortion sensitivity tests, as well as a comparison of the computational expense of such measures was given by Parthasarathy, Graichen, and Hathaway (1993).

An overview of several element quality metrics was given by Field (2000). Chen, Tristano, and Kwok (2003) extended the angle-based quality metric originally defined by Lee and Lo (1994) for use in the optimization of meshes consisting of triangles and quadrilaterals.

A special tetrahedron shape measure was given by Liu and Joe (1994a). It is based on eigenvalues of the metric tensor for the transformation between a tetrahedron and a regular reference tetrahedron. The geometric explanation of this measure is that it characterizes the shape of the inscribed ellipsoid. Another shape regularity quality of a triangle was given Bank and Xu (1996) and Bank and Smith (1997). They showed that the quality has circular level sets, when considered a function of the location of one vertex of triangle with the other two vertices fixed. Three tetrahedron measures—the minimum solid angle, the radius ratio, and the mean ratio and their relations were discussed by Liu and Joe (1994b).

An algorithm for construction of solution-adapted triangular meshes within an optimization framework was considered by Buscaglia and Dari (1997).

Here, the optimized quality measure is a product of "shape" quality and a function of mesh size.

A local cell quality measure as a function of Jacobian matrix and combined element-shape and size-control metrics for different cell types was analyzed by Garanzha (2000) and Branets and Carey (2005).

Dompierre et al. (2005) analyzed and generalized several simplex shape measures documented in the literature and used them for mesh adaptation and mesh optimization. Conclusions were drawn on the choice of simplex shape measures to control mesh optimization.

4

Stretching Method

4.1 Introduction

The stretching approach for generating structured grids is applied widely in the numerical solution of partial differential equations. Its major advantage is the rapidity of grid generation and direct control of grid spacing, while the main disadvantage is the necessity to explicitly select the zones where the stretching is needed. Of central importance in the method are intermediate transformations constructed on the basis of some standard stretching functions which provide the required spacing between the coordinate lines in selected zones. This chapter is concerned with the specification of the stretching mappings and of the intermediate transformations for the generation of grids with node clustering in the areas of solution singularities.

For this purpose some basic univariate, nonuniform coordinate transformations are described. These transformations can smooth the singularities arising in boundary value problems whose solutions undergo large variations in narrow zones. The grids generated through the use of such functions, each of which transforms an individual coordinate, appear to be well adapted to the expected physical features.

The basic functions incorporated into the method allow the grid to adjust automatically to solution singularities arising from the physical parameters, e.g. viscosity, high Reynolds number, or shell thickness, while a practical problem is solved. Such automation is one of the requirements imposed on comprehensive grid codes. The grids obtained by such methods enable users to obtain numerical solutions of singularly perturbed equations which converge uniformly to the exact solution with respect to the parameter. They also provide uniform interpolation of the numerical solution over the entire region, including boundary and interior layers.

A stretching method utilizing the standard stretching functions supplies one with a very simple means to cluster the nodes of the computational grid within the regions of steep gradients without an increase in the total number of grid nodes. This grid concentration improves the spatial resolution in

V.D. Liseikin, *Grid Generation Methods*, Scientific Computation,
DOI 10.1007/978-90-481-2912-6_4,

the regions of large variation, thus enhancing the accuracy of the algorithms applied to the numerical solution of partial differential equations.

The stretching mappings can also be used successively to derive the blending functions in algebraic methods of transfinite interpolation. The algebraic techniques are usually contained in large, multipurpose grid generation codes in combination with more sophisticated elliptic and parabolic methods, where their major task is to provide an initial grid which serves to start the iterative process of the grid generators. The blending functions implemented through the stretching mappings ease the process of the generation of the elliptic and parabolic grids by taking part of the solution adjustment on themselves.

4.2 Formulation of the Method

The stretching method is one of the simplest and fastest approaches applied to generate nonuniform grids. As a preliminary step it requires the introduction of some specified curvilinear coordinates in the physical region X^n. The coordinates are chosen by a parametrization

$$\boldsymbol{x}(\boldsymbol{q}) : Q^n \to X^n, \quad \boldsymbol{q} = (q^1, \dots, q^n), \ \boldsymbol{x} = (x^1, \dots, x^n),$$

from a domain $Q^n \subset R^n$ with a system of Cartesian coordinates q^i, $i = 1, \dots, n$. This system is selected in such a way that it includes the coordinates along which the grid nodes are to be redistributed by the stretching technique. Then, in the zones where the nodes are to be concentrated, every required variable q^i is replaced by some stretching variable ξ^i using a specified separate univariate transformation $\xi^i(q^i)$. To provide stretching of the coordinate q^i, the function $\xi^i(q^i)$ must have a large first derivative with respect to q^i. The inverse transformation $q^i(\xi^i)$, having, in contrast, a small first derivative with respect to ξ^i, is a contraction transformation in these zones. A smooth or continuous expansion of these separate local contraction functions $q^i(\xi^i)$ to produce a new coordinate system $\xi^1, \dots, \xi^n$ in the whole region Q^n provides an intermediate transformation

$$\boldsymbol{q}(\boldsymbol{\xi}) : \Xi^n \to Q^n$$

from some parametric domain $\Xi^n \subset R^n$. The composition $\boldsymbol{x}[\boldsymbol{q}(\boldsymbol{\xi})]$ defines a coordinate transformation which yields a numerical grid with nodal clustering in the required parts of the domain X^n.

Analogously, the transformation $\boldsymbol{q}(\boldsymbol{\xi})$ can be obtained as the inverse to a mapping

$$\boldsymbol{\xi}(\boldsymbol{q}) : Q^n \to \Xi^n,$$

which is an expansion of the local stretching functions $\xi^i(q^i)$ over the whole domain Q^n.

Without losing generality, we assume that the domain Q^n, called the intermediate domain, as well as the domain Ξ^n, called the logical or computational

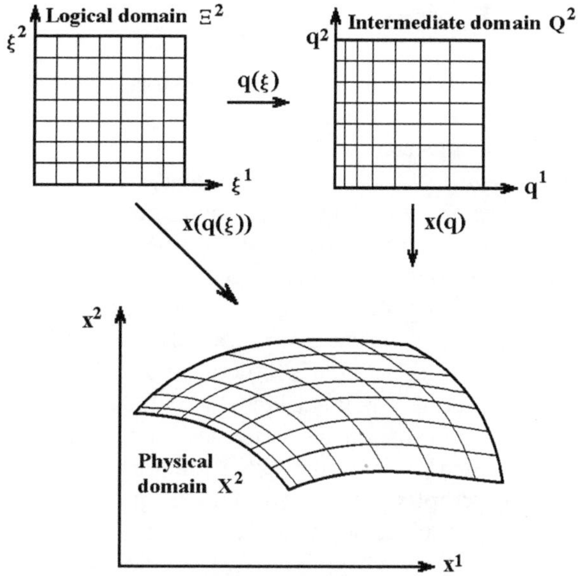

Fig. 4.1. Illustration of the stretching method

domain is the unit n-dimensional cube. So the coordinate transformation from the unit logical cube Ξ^n onto the physical region X^n is defined as the composition of two transformations: $\boldsymbol{q}(\boldsymbol{\xi})$ from Ξ^n onto Q^n and $\boldsymbol{x}(\boldsymbol{q})$ from Q^n onto X^n (Fig. 4.1), i.e.

$$\boldsymbol{x}(\boldsymbol{q})(\boldsymbol{\xi}) : \Xi^n \to Q^n \to X^n.$$

The splitting of the sought coordinate transformation $\boldsymbol{x}[\boldsymbol{q}(\boldsymbol{\xi})]$ into the two transformations $\boldsymbol{q}(\boldsymbol{\xi})$ and $\boldsymbol{x}(\boldsymbol{q})$ enables one to divide the task of grid generation into two steps: one performed by the intermediate transformation $\boldsymbol{q}(\boldsymbol{\xi}) : \Xi^n \to Q^n$, obtained with the help of some specified contraction functions $q^i(\xi^i)$ and responsible for the control of the grid, and another one performed by the parametric mapping $\boldsymbol{x}(\boldsymbol{q}) : Q^n \to X^n$, which is concerned with the specification of the coordinates requiring stretching in some zones of the physical domain X^n. These two steps can be considered as separate and distinct operations and as such can be developed in an independent and modular way.

The intermediate coordinate transformations $\boldsymbol{q}(\boldsymbol{\xi})$ are designed to be between the standard unit cubes Ξ^n and Q^n, which are fixed regardless of the physical domain X^n and the physical solution. Therefore there is an opportunity to create a kind of database of reference functions which can be used as elements to construct comprehensive intermediate transformations $\boldsymbol{q}(\boldsymbol{\xi})$.

4.3 Theoretical Foundation

The construction of the basic intermediate transformations $\boldsymbol{q}(\boldsymbol{\xi}) : \Xi^n \to Q^n$ starts with the definition of the basic univariate functions $q^i(\xi^i)$ which are, as was mentioned, the inverses of the basic univariate stretching functions $\xi^i(q^i)$. The functions $q^i(\xi^i)$ should be suitable for providing adequate grid clustering in the necessary zones through their implementation in formulas for the intermediate transformations.

The form of the univariate transformation $\xi^i(q^i)$ which stretches the coordinate q^i in the zones of large derivatives depends on the qualitative behavior of the solution. Thus, for effective use of the stretching method in the numerical solution of multidimensional problems, one needs both to select the directions q^i in the region X^n along which the solution has large derivatives and to have some information on its structure along these particular directions. Information about the qualitative solution structure is obtained from a theoretical analysis of simpler model equations, in particular, ordinary differential equations which simulate the qualitative features of the solutions, or it can be obtained from a preliminary numerical calculation for similar problems on coarse grids.

One set of stretching functions can be formed by local nonuniform mappings applied to the numerical solution of equations with a small parameter ϵ affecting the higher derivatives. Equations with a small parameter ϵ before the higher derivatives are widespread in practical applications. For example, such equations can model flows with small viscosity or high Reynolds number, describe problems of elasticity where the small parameter represents the shell thickness, or simulate flows of liquid in regions having orifices with a small diameter. These problems have narrow boundary and interior layers where the derivatives of the solutions with respect to the coordinates orthogonal to the layers reach very large magnitudes when the parameter ϵ is small. In the center of such a layer these derivatives have values of order ϵ^{-k}, $k > 0$.

Problems with a small parameter affecting the higher derivatives have been studied thoroughly by analytical and numerical methods. At present, there is a lot of analytical information related to the qualitative features of the solutions of these problems in the layers, which can be efficiently applied to the development of well-behaved numerical methods, in particular, to the generation of grids with nodal clustering in narrow zones of large variation of the physical quantities.

The analysis of these problems has revealed new forms of local stretching functions in addition to the well-known ones aimed chiefly at the treatment of exponential-type layers. These new stretching functions are very suitable for coping with power-type and mixed layers as well, which are common in practical applications. A stretching technique based on the new functions provides efficient concentration of coordinate lines and control of the generation of the coordinate system.

In this section some theoretical facts concerning qualitative features of solutions in boundary and interior layers are outlined. These facts serve to justify the forms of the basic stretching functions applied to generate grids by the stretching method.

4.3.1 Model Problems

The stretching method has efficient application to the numerical solution of ordinary differential equations and multidimensional problems with boundary and interior layers, where the solutions may have large variations along the coordinate lines intersecting the layers. For the numerical calculation of a two-dimensional viscous gas flow, for instance, the stretching method can be used successfully to generate grids with nonuniform clustering in the region of the boundary layer, where the longitudinal component u of the velocity $\boldsymbol{u} = (u, v)$ has its highest gradient near a solid boundary, in the direction x orthogonal to the boundary, in the case of a laminar flow. Some information about the qualitative behavior of the tangential velocity u in the direction x can be gained from the study of a model two-point boundary value semilinear problem with a small parameter ϵ:

$$\begin{aligned} &\epsilon u'' + a(x)u' = f(x, u), \quad 0 < x < 1, \\ &u(0) = u_0, \qquad u(1) = u_1. \end{aligned} \tag{4.1}$$

The model equation (4.1) is derived from the steady equation for the tangential velocity component u of the Navier–Stokes system. With respect to the independent transverse and longitudinal variables x and y, it can be written in the form

$$\frac{\partial}{\partial x}\left(\mu\frac{\partial u}{\partial x}\right) - \rho v\frac{\partial u}{\partial x} = g\left[u, \frac{\partial u}{\partial y}, \frac{\partial^2 u}{\partial y^2}, \frac{\partial v}{\partial y}, \frac{\partial}{\partial x}\left(\mu\frac{\partial v}{\partial y}\right), \frac{\partial}{\partial y}\left(\mu\frac{\partial v}{\partial x}\right)\right]. \tag{4.2}$$

The model two-point boundary value problem (4.1) is obtained from this equation with the assumptions that the dynamic viscosity μ is a constant, the longitudinal coordinate y is a parameter, and the right-hand side is uniformly bounded with respect to μ. Therefore $a(x)$ in (4.1) corresponds to $-\rho v$, ϵ to μ, and $f(x, u)$ to the right-hand side of (4.2).

The boundary-layer behavior of the solution $u(x, \epsilon)$ of the two-point boundary value problem (4.1) for $f_u(x, u) > 0$ is of three types, depending on the values of $a = a(0)$ and $a'(0)$, and is characterized by estimates for the derivatives of $u(x, \epsilon)$ with respect to x.

Another model equation to investigate the behavior of the solutions in boundary and interior layers is obtained from a problem simulating the shock wave structure of steady heat-conducting gas flow:

$$
\begin{aligned}
&\frac{\mathrm{d}\rho u}{\mathrm{d}x} = 0,\\
&\rho u \frac{\mathrm{d}u}{\mathrm{d}x} + \frac{\mathrm{d}p}{\mathrm{d}x} - \epsilon \frac{\mathrm{d}^2 u}{\mathrm{d}x^2} = 0,\\
&\rho u \frac{\mathrm{d}e}{\mathrm{d}x} + p \frac{\mathrm{d}u}{\mathrm{d}x} - \epsilon \left(\frac{\mathrm{d}u}{\mathrm{d}x}\right)^2 - \frac{\mathrm{d}}{\mathrm{d}x}\left(\chi \frac{\mathrm{d}T}{\mathrm{d}x}\right) = 0, \quad 0 < x < 1,\\
&(\rho, u, e)(0) = (\rho_0, u_0, e_0), \qquad (\rho, u, e)(1) = (\rho_1, u_1, e_1),
\end{aligned}
\tag{4.3}
$$

where ρ is the density, u the velocity, p the pressure, T the temperature, e the energy, ϵ the coefficient of viscosity of the gas, and χ the coefficient of thermal conductivity.

In the case

$$
\begin{aligned}
e &= c_v T,\\
p &= (\nu - 1)\rho e,
\end{aligned}
$$

we obtain from the system (4.3)

$$
\begin{aligned}
&-\epsilon u'' + c[u + (\nu - 1)e/u]' = 0, \quad 0 < x < 1,\\
&u(0) = u_0, \qquad u(1) = u_1,
\end{aligned}
\tag{4.4}
$$

$$
\begin{aligned}
&-(\epsilon_1 e')' + c\left(e - \frac{u^2}{2} + \frac{c_2}{c} u\right)' = 0, \quad 0 < x < 1,\\
&e(0) = e_0, \qquad e(1) = e_1,
\end{aligned}
\tag{4.5}
$$

where

$$
\begin{aligned}
c &= \rho_0 u_0, \qquad \epsilon_1 = \chi / c_v,\\
c_2 &= \{-\epsilon u' + c[u + (\nu - 1)e/u]\}|_{x=0}.
\end{aligned}
$$

The functions $u(x)$ and $e(x)$ are monotonic in the layer of their rapid variation. Hence the dependent variables u and e are connected by some relations

$$
e = E(u), \qquad u = U(e).
$$

Therefore the problem (4.4) can be presented as a two-point boundary value problem of a very simple, standard, autonomous quasilinear form

$$
\begin{aligned}
&-\epsilon u'' + a(u)u' = 0, \quad 0 < x < 1,\\
&u(0) = u_0, \qquad u(1) = u_1,
\end{aligned}
\tag{4.6}
$$

which represents a model problem to study the qualitative features of solutions with singularities in interior layers. An analogous expression can be obtained for the problem (4.5) if ϵ_1 is a constant.

One more model suitable for investigating the qualitative features of solutions in layers is the boundary value problem of a gas flow near a round hole with a small radius $r = \epsilon$ or a corresponding problem of electron motion. The behavior of the solution to these problems in the vicinity of the boundary layer is simulated qualitatively by a semilinear two-point boundary value problem

$$\begin{aligned} &(\epsilon + x)^p u'' + a(x)u' = f(x,u), \quad p > 0,\ 0 < x < 1, \\ &u(0) = u_0, \qquad u(1) = u_1. \end{aligned} \tag{4.7}$$

The problems (4.1), (4.6), (4.7) are amenable to analytical study. Though they represent highly idealized cases, they nevertheless give a rather profound understanding of the variety and complexity of the singularities arising in practical applications. The study of these two-point boundary value problems has provided solid knowledge about the possible qualitative features of solutions in boundary and interior layers.

The next considerations of this section are concerned with some results related to the qualitative behavior of the solutions to the problems (4.1), (4.6), (4.7). The results mainly apply to estimates of the derivatives of the solution appropriate to specifying the stretching functions. Any analytical proof of the facts outlined below is beyond the scope of this book. But we note that the principal technique used to analyze the asymptotic behavior of the solutions and to provide estimates of the solutions and of their derivatives employs the theory of differential inequalities developed by Nagumo (1937). For the Dirichlet problem

$$\begin{aligned} &u'' = f(x,u,u'), \quad 0 < x < 1, \\ &u(0) = u_0, \qquad u(1) = u_1, \end{aligned} \tag{4.8}$$

where f is a continuous function of the arguments x, u, u', the Nagumo inequality theory states that if there exist continuous, twice differentiable functions $\alpha(x)$ and $\beta(x)$ with the properties

$$\begin{aligned} &\alpha(x) \le \beta(x), && 0 \le x \le 1, \\ &\alpha(0) \le u(0) \le \beta(0), && \alpha(1) \le u(1) \le \beta(1), \\ &\alpha'' \ge f(x,\alpha,\alpha'), && 0 < x < 1, \\ &\beta'' \le f(x,\beta,\beta''), && 0 < x < 1, \end{aligned}$$

then the problem (4.8) with the condition $f(x,u,z) = O(z^2)$ has a solution $u(x)$ and

$$\alpha(x) \le u(x) \le \beta(x).$$

The functions $\alpha(x)$ and $\beta(x)$ are called the bounding functions. Estimates of the solutions to the problems (4.1), (4.6), (4.7) and of their derivatives are obtained rather readily by selecting the appropriate bounding functions $\alpha(x)$ and $\beta(x)$.

4.3.2 Basic Majorants

This subsection presents some estimates of the first and higher derivatives of the solutions to the problems (4.1), (4.6), (4.7). The solutions of these problems can have highly localized regions of rapid variation. The first derivative of the solutions may reach a magnitude of ϵ^{-k}, $1 \geq k > 0$, and therefore it tends to infinity when ϵ approaches zero. Outside the layers the derivative is estimated by a constant M independent of the parameter ϵ. These estimates are used to define an optimum coordinate ξ with a transformation $x(\xi)$.

4.3.2.1 Relation Between Optimal Univariate Transformations and Majorants of the First Derivative

The optimum univariate transformation $x(\xi)$ for a monotonic univariate function $u(x)$ would be one for which $u[x(\xi)]$ varied linearly with respect to ξ, since this would result in zero truncation errors for any approximation. However, if the function $u(x)$ is not monotonic and is found from a solution of a particular problem, this formulation of the optimal transformation is too good to be realized in practice. With regard to the problems (4.1), (4.6), (4.7) with the small parameter ϵ, the optimum transformation $x(\xi)$ would be one that eliminated the layers of singularity of the solution $u(x, \epsilon)$ of these problems, i.e., in particular, one in which the first derivative of the transformation $u[x(\xi), \epsilon]$ was limited by a constant M independent of ϵ. Such a transformation $x(\xi)$ eliminates the singularities of $u(x, \epsilon)$, as a result the function $u[x(\xi), \epsilon]$ does not have large variations, and therefore the transformed problem with respect to the independent variable ξ can be efficiently solved on the uniform grid

$$\xi_i = ih, \quad i = 1, \dots, N, \ h = 1/N.$$

The univariate transformations eliminating the singularities inherent in the solutions of the problems (4.1), (4.6), (4.7) depend inevitably on the small parameter ϵ. Nevertheless, for simplicity, we use the notation $x(\xi)$ for such functions.

Let the ranges of the variables x and ξ be normalized, say

$$0 \leq x \leq 1, \qquad 0 \leq \xi \leq 1.$$

Then the optimum transformation $x(\xi)$ exists if the derivative of the function $u(x, \epsilon)$ is bounded by a strictly positive function $\psi(x, \epsilon)$ whose total integral is limited by a constant M independent of ϵ, i.e.

$$\left| \frac{\mathrm{d}u}{\mathrm{d}x} \right| \leq \psi(x, \epsilon),$$

with

$$\int_0^1 \psi(x, \epsilon) \mathrm{d}x \leq M. \tag{4.9}$$

Equation (4.9) means that $u(x,\epsilon)$ is a function with a uniformly limited total variation on the interval $[0,1]$, i.e.

$$\int_0^1 \left|\frac{\mathrm{d}u}{\mathrm{d}x}\right| \mathrm{d}x \le M.$$

The required function $x(\xi)$ eliminating the singularity of first order of $u(x,\epsilon)$ is obtained, for example, as the inverse of the solution $\xi(x)$ of the initial-value problem

$$\begin{aligned} &\frac{\mathrm{d}\xi}{\mathrm{d}x} = c\psi(x,\epsilon), \quad x > 0, \\ &\xi(0) = 0, \end{aligned} \tag{4.10}$$

where c is a scaling constant providing the condition $\xi(1) = 1$. After integrating (4.10) we obtain

$$c = 1 \Big/ \int_0^1 \psi(x,\epsilon)\mathrm{d}x$$

and hence we have

$$\left|\frac{\mathrm{d}u}{\mathrm{d}\xi}\right| = \left|\frac{\mathrm{d}u}{\mathrm{d}x}\right|\frac{\mathrm{d}x}{\mathrm{d}\xi} \le \int_0^1 \psi(x,\epsilon)\mathrm{d}x \le M. \tag{4.11}$$

So the function $u[x(\xi),\epsilon]$ does not have layers of rapid variation in the interval $[0,1]$ of the independent variable ξ.

The initial-value problem (4.10) can also be replaced by an equivalent linear two-point boundary value problem for an ordinary equation of the second order,

$$\begin{aligned} &\frac{\mathrm{d}}{\mathrm{d}x}\left(\frac{\mathrm{d}\xi}{\mathrm{d}x}/\psi(x,\epsilon)\right) = 0, \quad 0 < x < 1, \\ &\xi(0) = 0, \qquad \xi(1) = 1, \end{aligned} \tag{4.12}$$

or by a nonlinear problem for an equation with ξ as the independent variable and x as the dependent variable,

$$\begin{aligned} &\frac{\mathrm{d}}{\mathrm{d}\xi}\left(\frac{\mathrm{d}x}{\mathrm{d}\xi}\psi(x,\epsilon)\right) = 0, \quad 0 < \xi < 1, \\ &x(0) = 0, \qquad x(1) = 1. \end{aligned} \tag{4.13}$$

So, the singular functions $u(x,\epsilon)$ whose total variation is limited on the interval $[0,1]$ by a constant M independent of the parameter ϵ can be transformed to the function $u[x(\xi),\epsilon]$ with a uniformly limited first derivative with respect to ξ on the interval $[0,1]$.

The one-dimensional grid derived through a transformation $x(\xi)$ which satisfies the relations (4.10), (4.12), (4.13) is optimal in the above sense. Taking into account (4.11), which shows that the first derivative of the function

$u[x(\xi),\epsilon]$ with respect to ξ is uniformly bounded, we find that the variation of the function $u[x(\xi),\epsilon]$ on the neighboring points x_{i+t} and x_i of the grid derived by the transformation $x(\xi)$, where

$$x_i = x(ih), \quad h = 1/N, \; i = 0, \dots, N,$$

is uniformly limited as well, i.e.

$$|u_{i+1} - u_i| \leq Mh, \quad u_i = u(x_i, \epsilon), \; i = 0, \dots, N-1,$$

where M is independent of ξ. Therefore the values of u_i can be uniformly interpolated over the whole interval $[0,1]$ by a piecewise function $P(x)$ which uniformly approximates $u(x,\epsilon)$ over the entire interval $[0,1]$:

$$|u(x,\xi) - P(x)| \leq Mh, \quad 0 \leq x \leq 1.$$

Analytical results guarantee the existence of a majorant $\psi(x,\epsilon)$ satisfying the condition (4.9) for the solutions to the problems (4.1), (4.7) if the function $f(x,u)$ satisfies the condition of strong ellipticity, i.e.

$$f_u(x,u) \geq m > 0,$$

which also ensures uniqueness of the solution. Note, however, that the problem

$$\begin{aligned} \epsilon u'' &= -u = 0, \quad 0 < x < 1, \\ u(0) &= 0, \qquad u(1) = 1, \end{aligned}$$

for example, does not satisfy the above condition of strong ellipticity and as a result the total variation of its solution is not uniformly limited with respect to ϵ.

The solution $u(x,\epsilon)$ to the problem (4.6) is always a monotonic function, and therefore its total variation equals $|u_1 - u_0|$.

Now we present four positive basic singular functions $\psi_i(x,\epsilon)$, with a uniformly limited total integral, whose combinations bound the first derivative of the solutions to the problems (4.1), (4.6), (4.7) in the boundary layers.

4.3.2.2 Exponential Functions

The most popular function used to demonstrate a boundary singularity is the exponential function

$$u(x,\epsilon) = \exp(-bx/\epsilon^k), \quad 0 < x < 1,$$

$k > 0$, $b > 0$, whose first derivative yields an expression for the basic majorant $M\psi_1(x,b,\epsilon)$, where

$$\psi_1(x,b,\epsilon) = \epsilon^{-k} \exp(-bx/\epsilon^k), \tag{4.14}$$

satisfying the condition (4.9).

An exponential singularity of the solution $u(x,\epsilon)$ to the problem (4.1) can occur only near the boundary point as follows:

(1) at $x = 0$, when $a(0) \geq m > 0$, or $a(0) = 0,\ a'(0) = 0$;
(2) at $x = 1$, when $-a(0) \geq m > 0$, or $a(1) = 0,\ a'(1) = 0$.

The condition $a(0) > 0$ in the gas flow simulation (4.2) corresponds to $v(0) < 0$, which means the physical situation of the gas being sucked through a side wall.

The first derivative of the solution of (4.7) is also estimated by the majorant $M\psi_1(x, b, \xi)$, with $k = p$, when $p > 1$.

4.3.2.3 Power Singularities

The most common condition for viscous gas flows is that of adhesion of the gas to a solid wall, which, for two-dimensional flows (4.2), is expressed mathematically by the equation

$$u(0) = 0, \qquad v(0) = 0,$$

corresponding to $a(0) = 0$ in (4.1). In this case the nature of the boundary layer singularity of the solution to the problem (4.1) depends on the sign of the first derivative of $a(x)$ at the point $x = 0$. The relations

$$a(0) = 0, \qquad a'(0) < 0, \quad \text{or} \quad a'(0) > 0,$$

express physically attraction or repulsion of the gas to or from the wall, respectively. Therefore the singularities of the gas flow are directly connected with the direction of the transverse velocity near the solid wall.

For $a(0) = 0,\ a'(0) < 0$, the first derivative of the solution $u(x, \epsilon)$ to the problem (4.1) in the vicinity of the boundary $x = 0$ when $f_u(x, u) \geq m > 0$ is estimated by the majorant $M\psi_2(x, b, \epsilon)$, where

$$\psi_2(x, b, \epsilon) = \epsilon^{kb}/(\epsilon^k + x)^{b+1}, \quad b > 0,\ k = 1/2. \tag{4.15}$$

The function $M\psi_2(x, b, \epsilon)$ also estimates the first derivative of the solution to the problem (4.7) when $p = 1$ and $a(0) < 1$.

Another power function near the boundary $x = 0$ is expressed by the majorant $M\psi_3(x, b, \epsilon)$:

$$\psi_3(x, b, \epsilon) = (\epsilon^k + x)^{b-1}, \quad 1 > b > 0. \tag{4.16}$$

The combination of this function for $k = 1/2$ with the majorant $M\psi_2(x, b, \epsilon)$ estimates the first derivative of the solution to the problem (4.1) in the vicinity of $x = 0$ when $a(0) = 0,\ a'(0) > 0$. So a viscous flow in the direction of the repulsion of the gas from the wall may have a combined boundary layer. The function $M\psi_3(x, b, \epsilon)$ also bounds the first derivative of the solution to the problem (4.7) when $p = 1,\ a(0) > 1$.

4.3.2.4 Logarithmic Function

One more important majorant function satisfying (4.9) appears in an estimate of the first derivative of the solution to the problem (4.7) with $p = 1,\ a(0) = 1$. Qualitatively, the solution $u(x, \epsilon)$ in the boundary layer is described in this case by a logarithmic function

$$c(x)\frac{\ln(\epsilon^k + x)}{\ln \epsilon^k}$$

with $k = 1$. The first derivative of $u(x, \epsilon)$ is estimated by the basic majorant $M\psi_4(x, \epsilon)$, where

$$\psi_4(x, \epsilon) = \frac{1}{(\epsilon^k + x)|\ln \epsilon|}. \tag{4.17}$$

4.3.2.5 Relations Among Basic Majorants

For the majorants $\psi_i(x, b, \epsilon),\ i = 1, 2, 3$, and $\psi_4(x, \epsilon)$ the following relations, expressed by inequalities, apply:

$$\begin{aligned} &\psi_i(x, b_1, \epsilon) \leq M\psi_i(x, b_2, \epsilon), \quad b_1 \geq b_2 > 0,\ i = 1, 2, 3, \\ &\psi_1(x, b, \epsilon) \leq M\psi_2(x, d, \epsilon) \quad \text{for arbitrary } d > 0, \\ &\psi_i(x, b, \epsilon) \leq M|\ln \epsilon^k|\psi_4(x, \epsilon), \quad b > 0,\ i = 1, 2. \end{aligned} \tag{4.18}$$

These relations are readily proved. For example, the confirmation of the second inequality follows from

$$\begin{aligned} \frac{\psi_1(x, b, \epsilon)}{\psi_2(x, d, \epsilon)} &= \frac{(\epsilon^k + x)^{d+1}}{\epsilon^{k(1+d)}} \exp(-bx/\epsilon^k) \\ &\leq 2^{d+1} \exp(-bx/\epsilon^k) + \left(\frac{x}{\epsilon^k}\right)^{d+1} \exp(-bx/\epsilon^k). \end{aligned}$$

As

$$x^n \exp(-cx) \leq M, \quad n > 0,\ c > 0,\ 0 \leq x < \infty,$$

where the constant M is dependent only on n and c, we find that

$$\frac{\psi_1(x, b, \epsilon)}{\psi_2(x, d, \epsilon)} \leq M,$$

i.e. the second inequality of (4.18) is proved.

4.3.2.6 Interior Layers

The solutions to the problems (4.1) and (4.6) also have interior layers with large variations. Moreover, the problem (4.6) models qualitatively the wave tracks and shocks of many gas dynamic flows with layers of sharp variation away from the boundaries. The problems (4.1), (4.6) are remarkable in that the first derivatives of their solutions in the interior layers can be estimated by combinations of the same majorants $\psi_i(x, b, \epsilon)$, $i = 1, 2, 3$, in which the independent variable x is replaced by $|x - x_0|$, namely

$$\psi_i(|x - x_0|, b, \xi), \quad i = 1, 2, 3,$$

where x_0 is the center of the layer, i.e. x_0 is the point of the fastest local variation. Thus the estimates near the boundary point $x = 0$ serve also to estimate the first derivative in the interior layer near the point x_0. For example, for the derivative of the solution $u(x, \xi)$ to the problem (4.1), we have

$$\left|\frac{\mathrm{d}u}{\mathrm{d}x}(x, \epsilon)\right| \leq M\psi_3(|x - x_0|, b, \xi), \quad |x - x_0| \leq m,$$

where x_0 is defined by the condition

$$a(x_0) = 0, \qquad a'(x_0) \geq f_u(u, x).$$

The location of the center of the interior layer of the solution to the problem (4.6) is dependent on the properties of the function

$$b(u) = \int_{u_0}^{u} a(\eta)\mathrm{d}\eta.$$

An interior layer of the solution exists if

$$b(u_0) = b(u_1), \qquad b(u) > b(u_0), \tag{4.19}$$

and its center point x_0 is defined by the first nonzero coefficients of the Taylor expansions of the function $b(u)$ in the vicinity of the points u_0 and u_1.

For example, if in addition to the condition (4.19) the condition

$$b'(u_i) = a(u_i) \neq 0, \quad i = 0, 1,$$

is satisfied, then

$$x_0 = a_1/(a_1 - a_0), \qquad a_i = a(u_i), \quad i = 0, 1,$$

and

$$|u'(x)| \leq M\psi_1(|x - x_0|, b, \epsilon), \quad 0 \leq x \leq 1,$$

where the constant b is defined by a_0 and a_1.

An instructive example is that of an interior layer of the solution to the problem (4.6) that moves unlimitedly to the boundary as the parameter ϵ approaches zero. Such a layer is realized in a solution to the problem (4.6) if the function $b(u)$ satisfies the condition (4.19) and

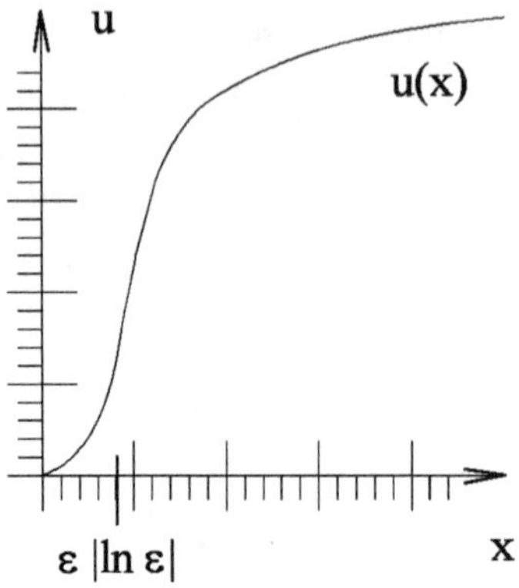

Fig. 4.2. Function with a mixed interior layer approaching the boundary unlimitedly

$$a(u_0) \neq 0, \qquad \frac{\mathrm{d}^k}{\mathrm{d}u^k} b(u_1) = 0, \quad k \leq p, \; p \geq 1,$$
$$\frac{\mathrm{d}^{p+1}}{\mathrm{d}u^{p+1}} b(u_1) \neq 0.$$

For the derivative of the solution $u(x, \epsilon)$ in this case, we have

$$|u'(x,\epsilon)| \leq \begin{cases} M[1+\epsilon^{-1}\{\exp[a_0(x-x_0)/\epsilon]\}], & 0 \leq x \leq x_0, \\ M[1+\epsilon^b(\epsilon + x - x_0)^{(-1-b)}], & x_0 \leq x \leq 1, \end{cases} \tag{4.20}$$

where

$$b = 1/p, \quad x_0 = (1+b)\frac{\epsilon}{a_0} \ln \epsilon^{-1}.$$

So the derivative in the left-hand part of the layer is estimated by the majorant $M\psi_1(x - x_0, a_0, \epsilon)$ and in the right-hand part it is bounded by $M\psi_2(x - x_0, b, \epsilon)$. Therefore the stretching of the variable x should be different in the left- and right-hand parts of the boundary layer. As the center point x_0 approaches the boundary unlimitedly as $\epsilon \to 0$, the solution with an interior layer tends to the solution of the reduced problem ($\epsilon = 0$) with a boundary layer (Fig. 4.2). Thus this example shows a drawback of the analysis of the locations of layers by means of reduced problems.

4.3.2.7 Estimates of the Higher Derivatives

The basic majorants of the higher derivatives of the solutions to the problems (4.1), (4.6), (4.7) in the layers have the form of the derivatives of the majorants (4.14)–(4.17). Namely, they are expressed by the following functions $\psi_i^n(x, b, \epsilon)$, $i = 1, 2, 3$, and $\psi_4^n(x, \epsilon)$:

$$
\begin{aligned}
\psi_1^n(x,b,\epsilon) &= \epsilon^{-kn}\exp(-b|x-x_0|/\epsilon^k),\\
\psi_2^n(x,b,\epsilon) &= \epsilon^{kb}/(\epsilon^k+|x-x_0|)^{b+n},\\
\psi_3^n(x,b,\epsilon) &= (\epsilon^k+|x-x_0|)^{b-n},\\
\psi_4^n(x,\epsilon) &= \frac{1}{(\epsilon^k+|x-x_0|)^n|\ln\epsilon|},
\end{aligned}
\tag{4.21}
$$

where x_0 is the point at the center of the layer.

The general estimate of the nth derivative of $u(x,\epsilon)$ has the form

$$
\left|\frac{\mathrm{d}^n u}{\mathrm{d}x^n}\right| \le M[1+\psi^n(x,\epsilon)], \quad 0\le x\le 1, \tag{4.22}
$$

where ψ^n is a combination of the functions described by the formulas (4.21). The distinctive feature of these estimates is that they guarantee that the local transformation $x(\xi)$ obtained from equations of the forms (4.10), (4.12), (4.13) is suitable for smoothing the higher-order singularities, i.e. the following estimate is valid:

$$
\left|\frac{\mathrm{d}^n}{\mathrm{d}\xi^n}u[x(\xi),\epsilon]\right| \le M, \quad 0\le x\le 1. \tag{4.23}
$$

4.3.2.8 Invariants of Equations

It is apparent that the boundary layers of the singularly perturbed solutions do not vanish when the coordinate x is replaced by a coordinate q with a one-to-one smooth transformation $x(q)$ of the interval $[0,1]$ if $x(q)$ is independent of ϵ. And it is apparent that (4.1) has some invariants under such transformations which determine the qualitative behavior of the solutions to the equations. For example, (4.1) in the new independent variable q and dependent variable $u_1(q,\epsilon)=u[x(q),\epsilon]$ has the form

$$
\begin{aligned}
&\epsilon u_1'' + a_1(q)u_1' = f_1(q,u_1), \quad 0<q<1,\\
&a_1(q) = x'a[x(q)] + \epsilon x''/(x')^2,\\
&f_1(q_1,u_1) = (x')^2 f[x(a),u_1].
\end{aligned}
$$

So the invariants are

(1) the sign of the coefficient $a_1(q)$ of the first derivative for $q=0,\ \epsilon=0$,
(2) the expression

$$
a_1'(0)/f_{u_1}[0,u_1(0)]
$$

for $\epsilon=0$ when $a_1(0)=0$.

As was mentioned above, the estimate of the first derivative of the solution to this problem is defined through these two invariants.

The invariant defining the structure of the solution to the problem (4.7) in the boundary layer is also the value of the coefficient $a(x)$ for $x=0$.

The qualitative behavior of the solution to the problem (4.6) is determined by the values of the derivatives of the function $b(u) = \int_{u_0}^{u} a(u)du$ at the points u_0 and u_1. These values are the invariants of transformations $v = f(u)$ of the dependent variable u.

The preceding remarks show the importance of the study of the invariants of equations and their connection with the qualitative features of the solutions in the layers.

4.4 Basic Intermediate Transformations

This section gives a detailed description of the basic univariate stretching functions and, consequently, the contraction functions, which are applied to construct the intermediate transformations that provide grid clustering in boundary and interior layers.

4.4.1 Basic Local Stretching Functions

The derivatives of the solution of a singularly perturbed equation are large in the center of a layer and decrease towards its boundary. Outside the layers the derivatives are estimated by a constant M independent of the small parameter ϵ, while within a layer the derivatives of any singular solution with respect to the coordinate x transverse to the layer can be bounded by the derivatives of one or a combination of the basic functions $\psi_i^n(x, \epsilon)$, $i = 1, 2, 3$, defined by (4.31). These basic functions generate four basic univariate transformations $\varphi_i(x, \epsilon)$ which stretch the layers. The introduction of these functions to stretch the coordinate transverse to a layer nonuniformly allows one to build a new local coordinate system with respect to which the solution has no layers with large derivatives.

Local coordinate transformations $\varphi_i(x, \epsilon)$ which nonuniformly stretch the coordinate lines within the boundary layers have already been utilized to generate grids for the numerical solution of some singularly perturbed problems. Analytical and numerical analyses have demonstrated that the grids generated in the layers by these coordinate transformations allow one to obtain a numerical solution to a singularly perturbed problems which converges uniformly with respect to the small parameter to the exact solution. Also, the solution can be interpolated uniformly over the entire region, including the layers. Therefore the incorporation of stretching functions into formulas for intermediate transformations is a promising way to develop grid techniques.

The four standard, local, stretching, coordinate transformations denoted by $\varphi_i(x, \epsilon)$, $i = 1, 2, 3, 4$, where x is a scalar-valued independent variable interpreted here as a coordinate orthogonal to a layer and ϵ is a small parameter, are have been designed only to stretch the boundary layer at the point $x = 0$. These functions are defined by integrating the basic majorants (4.14)–(4.17). In reality, these local stretching transformations are boundary layer functions which describe the qualitative behavior of the physical solutions across the boundary layers. The functions which stretch the interior layers are derived from these basic transformations by the procedures described in Sect. 4.4.4.

The boundary layer functions corresponding to the majorants described above (4.14)–(4.17) are computed by solving an initial-value problem of the type (4.10):

$$\frac{d\varphi}{dx} = d\psi(x, \epsilon), \quad x > 0,$$
$$\varphi(0) = 0,$$

where $\psi(x, \epsilon)$ is the majorant of the first derivative. For convenience the local stretching functions are written in a form that satisfies the following conditions:

$$\varphi(0, \epsilon) = 0, \qquad \frac{d}{dx}\varphi(x, \epsilon) > 0.$$

The first function is the well-known exponential mapping

$$\varphi_1(x, \epsilon) = \frac{1 - \exp(-bx/\epsilon^k)}{c}, \quad k > 0, \ b > 0, \ c > 0. \tag{4.24}$$

The next two local stretching mappings are power functions,

$$\varphi_2(x, \epsilon) = \frac{1 - [\epsilon^k/(\epsilon^k + x)]^b}{c}, \quad k > 0, \ b > 0, \ c > 0, \tag{4.25}$$

and

$$\varphi_3(x, \epsilon) = \frac{(\epsilon^k + x)^b - \epsilon^{kb}}{c}, \quad k > 0, \ 1 > b > 0, \ c > 0. \tag{4.26}$$

The fourth local stretching function is a logarithmic map

$$\varphi_4(x, \epsilon) = \frac{\ln(1 + x\epsilon^{-k})}{c\ln(1 + \epsilon^{-k})}, \quad k > 0, \ c > 0. \tag{4.27}$$

The numbers k, b, and c in these expressions for the stretching functions $\varphi_i(x, \epsilon)$ are positive constants. The number k shows the scale of a layer. It is easily computed analytically. For example, for problems of viscous flows, $k = 1/2$ in a boundary layer and $k = 1$ in a shock wave. The constant c serves to control the length of the interval of the new stretching coordinate φ that is transformed into the layer. The constant b controls the type of stretching nonuniformity and the width of the layer. The parameter ϵ provides the major contribution to determining the slopes of the stretching functions in the vicinity of the point $x = 0$.

The stretching functions $\varphi_i(x, \epsilon)$, $i = 1, 2, 3, 4$, for $\epsilon^k = 1/30$ are shown in Fig. 4.3. The symbols $+, \times, \Diamond$, and $\Box$ identify the functions $\varphi_1(x, \epsilon)$, $\varphi_2(x, \epsilon)$, $\varphi_3(x, \epsilon)$, and $\varphi_4(x, \epsilon)$, respectively. The constant c is selected to satisfy the restriction $\varphi_i(1, \epsilon) = 1$, $i = 1, 2, 3, 4$.

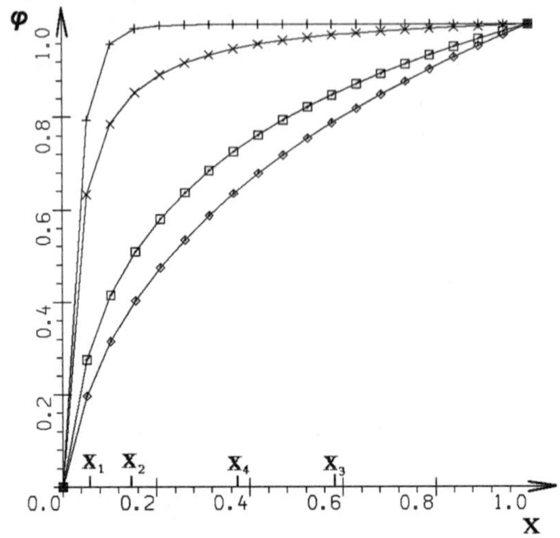

Fig. 4.3. Basic local stretching functions

4.4.1.1 Width of Boundary Layers

The interval where any function $\varphi_i(x, \epsilon)$ provides a stretching of the coordinate x coincides with the interval where the first derivative with respect to x of this function $\varphi_i(x, \epsilon)$ is large. The first derivatives of the basic stretching transformations $\varphi_i(x, \epsilon)$, $i = 1, 2, 3, 4$, are

$$\begin{aligned}
\frac{\mathrm{d}\varphi_1}{\mathrm{d}x}(x, \epsilon) &= \frac{b\epsilon^{-k}}{c}\exp(-bx/\epsilon^k), \quad k > 0,\ b > 0, \\
\frac{\mathrm{d}\varphi_2}{\mathrm{d}x}(x, \epsilon) &= \frac{b\epsilon^{kb}}{c(\epsilon^k + x)^{b+1}}, \quad k > 0,\ b > 0, \\
\frac{\mathrm{d}\varphi_3}{\mathrm{d}x}(x, \epsilon) &= \frac{b}{c}(\epsilon^k + x)^{b-1}, \quad k > 0,\ 1 > b > 0, \\
\frac{\mathrm{d}\varphi_4}{\mathrm{d}x}(x, \epsilon) &= \frac{1}{c\ln(1 + \epsilon^{-k})(\epsilon^k + x)}, \quad k > 0.
\end{aligned} \tag{4.28}$$

For the first derivative $\mathrm{d}\varphi_i(x, \epsilon)/\mathrm{d}x$ of the stretching functions φ_i, $i = 1, 2, 4$, one can readily obtain the following relations:

$$\begin{aligned}
\frac{\mathrm{d}\varphi_1}{\mathrm{d}x}(x, \epsilon) &\leq M\frac{\mathrm{d}\varphi_2}{\mathrm{d}x}(x, \epsilon), \quad 0 \leq x \leq 1, \\
\frac{\mathrm{d}}{\mathrm{d}x}\varphi_2(x, \epsilon) &\leq M\frac{\mathrm{d}}{\mathrm{d}x}\varphi_4(x, \epsilon^p), \quad p > 1,\ 0 \leq x \leq 1,
\end{aligned}$$

where the constant M does not depend on ϵ. Therefore the stretching transformation $\varphi_2(x, \epsilon)$ can be used to eliminate both exponential and power layers, while the mapping $\varphi_4(x, \epsilon)$ is suitable for smoothing exponential and power layers and also the singularities of the type described by $\varphi_4(x, \epsilon)$.

The derivative $\mathrm{d}\varphi_i(x,\epsilon)/\mathrm{d}x$ of each local stretching mapping $\varphi_i(x,\epsilon)$ is large in the vicinity of the point $x=0$ when the parameter ϵ is small, and decreases as x increases. The boundary of the layer for the function $\varphi_i(x,\epsilon)$ is defined to be at the point x_i where the modulus of the first derivative $(\mathrm{d}/\mathrm{d}x)\varphi_i(x,\epsilon)$ is limited by a constant $M>0$ independent of the parameter ϵ, i.e.

$$\left|\frac{\mathrm{d}\varphi_i}{\mathrm{d}x}(x_i,\epsilon)\right| \le M.$$

The value of x_i coincides with the width of the layer, denoted by $\Delta_i(x,\epsilon)$, of the function $\varphi_i(x,\epsilon)$. So from (4.28),

$$\begin{aligned}
x_1 &= \Delta_1(x,\epsilon) = \frac{\epsilon^k}{b}\ln\epsilon^{-k}, \quad \frac{\mathrm{d}\varphi_1}{\mathrm{d}x}(x_1,\epsilon) = b/c,\\
x_2 &= \Delta_2(x,\epsilon) = \epsilon^{kb/(b+1)} - \epsilon^k, \quad \frac{\mathrm{d}\varphi_2}{\mathrm{d}x}(x_2,\epsilon) = b/c,\\
x_3 &= \Delta_3(x,\epsilon) = d\epsilon^0 - \epsilon^k, \quad \frac{\mathrm{d}\varphi_3}{\mathrm{d}x}(x_3,\epsilon) = d^{(b-1)}b/c,\\
x_4 &= \Delta_4(x,\epsilon) = \frac{1}{\ln(1+\epsilon^{-k})} - \epsilon^k, \quad \frac{\mathrm{d}\varphi_4}{\mathrm{d}x}(x_4,\epsilon) = 1/c.
\end{aligned} \tag{4.29}$$

These expressions evidently provide a rule for controlling, with the constant b, the width of the layers where the grid nodes are to be clustered. In order to make the layer wider this constant needs to be reduced.

Also, from (4.29) one can obtain the maximum value $m_i>0$ of the parameter ϵ for each stretching function $\varphi_i(x,\epsilon)$. The value of m is obtained from the obvious condition $x_i<1,\ i=1,2,3,4$. This value defines the range for ϵ for the application of the stretching, $0<\epsilon\le m_i$, and consequently the contraction functions for the construction of the intermediate transformations. In the following discussion we consider only those values of the parameter ϵ which are subject to the restriction $x_i<1$.

The formulas for $\Delta_2(x,\epsilon)$, $\Delta_3(x,\epsilon)$, and $\Delta_4(x,\epsilon)$ contain the quantity $-\epsilon^k$, which, asymptotically, does not influence the width of the layers, but is included purely to simplify the expression for the first derivative of $\varphi_i(x,\epsilon)$ at the point x_i. Equations (4.29) clearly show that there exists a number $\epsilon_0>0$ such that

$$\Delta_3(x,\epsilon) > \Delta_4(x,\epsilon) > \Delta_2(x,\epsilon) > \Delta_1(x,\epsilon)$$

for all positive $\epsilon<\epsilon_0$.

The equations in (4.28) indicate that the length of the central part of the layer, where the first derivative reaches the maximum values $M\epsilon^{-k}$ for $\varphi_i(x,\epsilon)$, $i=1,2,3$, and $M\epsilon^{-k}/\ln\epsilon^{-k}$ for $\varphi_4(x,\epsilon)$, is similar for all functions $\varphi_i(x,\epsilon)$ and equals $m\epsilon^k$. However, the relations (4.29) state that the transitional part of the layer, between the center and the boundary, is very much larger than $m\epsilon^k$, especially for the functions $\varphi_2(x,\epsilon)$, $\varphi_3(x,\epsilon)$, and $\varphi_4(x,\epsilon)$. The first derivative in the transitional part of the layer is also large when the parameter ϵ is small, and therefore stretching of the central part is required as well, though in lesser degree.

In contrast, the first derivative of each function $\varphi_i(x,\epsilon)$, $i=1,2,4$, is very small outside a layer when the parameter ϵ is small. Namely, from (4.29), for a point $x=\text{const}$ lying outside a layer,

$$\begin{aligned}
\frac{\mathrm{d}\varphi_1}{\mathrm{d}x}(x,\epsilon) &\sim \epsilon^{-k}\exp(-b_1/\epsilon^k),\\
\frac{\mathrm{d}\varphi_2}{\mathrm{d}x}(x,\epsilon) &\sim \epsilon^{km},\\
\frac{\mathrm{d}\varphi_4}{\mathrm{d}x}(x,\epsilon) &\sim \frac{1}{\ln \epsilon^{-k}}.
\end{aligned}$$

Of these three expressions for the first derivative, the last one has the least tendency to become zero outside the layer and, when the parameter ϵ is not too small, the function $\varphi_4(x,\epsilon)$ can be used over the whole interval $[0,1]$ to introduce a new coordinate variable to stretch the coordinate x in the layer.

The boundary point x_i of the layer for the transformation $\varphi_i(x,\epsilon)$ corresponds to the value $\varphi_i(x_i,\epsilon)=\varphi_i$ of the dependent variable $\varphi_i(x,\epsilon)$. For generality, we denote by φ the dependent variable. Then φ_i defines the interval $[0,\varphi_i]$ which is transformed into the layer by the function inverse to $\varphi_i(x,\epsilon)$. The values of these points $\varphi_i(x_i,\epsilon)=\varphi_i$ corresponding to the values of x_i specified by (4.29) are given by

$$\begin{aligned}
\varphi_1 &= \frac{1-\epsilon^k}{c},\\
\varphi_2 &= \frac{1-\epsilon^{kb/(b+1)}}{c},\\
\varphi_3 &= \frac{d^b-\epsilon^{kb}}{c},\\
\varphi_4 &= \frac{\ln(\epsilon^{-k})-\ln[\ln(1+\epsilon^{-k})]}{c\ln(1+\epsilon^{-k})}.
\end{aligned} \tag{4.30}$$

These expressions imply that

$$\begin{aligned}
\varphi_i &\to 1/c, \quad i=1,2,4,\\
\varphi_3 &\to d^b/c,
\end{aligned}$$

when ϵ tends to 0.

4.4.2 Basic Boundary Contraction Functions

The functions $\varphi_i(x,\epsilon)$ stretch the coordinate x within the narrow layers $[0,x_i]$, therefore the mappings that are inverse to $\varphi_i(x,\epsilon)$ provide a contraction of the coordinate φ in the interval $[0,\varphi_i]$. Thus these inverse functions can be used as the univariate local transformations $q^j(\xi^j)$, where $q^j = x$, $\xi^j = \varphi$, to build the intermediate n-dimensional transformations

$$\boldsymbol{q}(\boldsymbol{\xi}) : \Xi^n \to Q^n$$

which generate nodal clustering in the layers along the selected coordinates q^j.

Taking into account (4.24)–(4.27), the local inverse transformations $x_i(\varphi, \epsilon)$ of the corresponding stretching functions $\varphi_i(x, \epsilon)$ have the following form:

$$\begin{aligned} x_1(\varphi, \epsilon) &= -\frac{\epsilon^k}{b} \ln(1 - c\varphi), \quad k > 0,\ b > 0, \\ x_2(\varphi, \epsilon) &= \epsilon^k \left((1 - c\varphi)^{-1/b} - 1\right), \quad k > 0,\ b > 0, \\ x_3(\varphi, \epsilon) &= (\epsilon^{kb} + c\varphi)^{1/b} - \epsilon^k, \quad k > 0,\ 1 > b > 0, \\ x_4(\varphi, \epsilon) &= \epsilon^k \left((1 + \epsilon^{-k})^{c\varphi} - 1\right), \quad k > 0. \end{aligned} \tag{4.31}$$

The first derivative of any of the functions $x_i(\varphi, \epsilon)$ is small at the points where $0 \le \varphi \le \varphi_i(x_i, \epsilon) = \varphi_i$, and therefore the magnitude of the grid spacing in the x direction of the grid generated by the mapping $x_i(\varphi, \epsilon)$ is also small in the layer; it is approximately of the order of $(\mathrm{d}/\mathrm{d}\varphi)[x_i(\varphi, \epsilon)]h$. The degree of grid clustering at the center of the layer reaches a value of ϵ^k and increases at the points near the boundary x_i of the layer.

The stretching functions $\varphi_i(x, \epsilon)$ themselves describe the qualitative behavior of solutions within their zones of large gradients along the coordinate lines normal to the layers. Therefore, as any mapping $x_i(\varphi, \epsilon)$, $i = 1, 2, 3, 4$, is the inverse of the corresponding function $\varphi_i(x, \epsilon)$, the grids derived from the transformations $x_i(\varphi, \epsilon)$ provide the optimum nonuniform resolution of the physical solution in the layers with an economy of nodal points. However, as mentioned above, these functions, excluding the mapping $x_3(\varphi, \epsilon)$, produce excessively sparse grids outside the layers, since their first derivative, satisfying the equation

$$\frac{\mathrm{d}x_i}{\mathrm{d}\varphi}(\varphi, \epsilon) = 1 \Big/ \frac{\mathrm{d}\varphi_i}{\mathrm{d}x}[x(\varphi), \epsilon]$$

is, according to (4.28), very large when $\varphi > \varphi_i$ and tends to infinity as the parameter ϵ nears zero. Therefore the contraction functions $x_i(\varphi, \epsilon)$ can only be used to provide grid clustering in the layers; outside the layers the grids must be generated through other mappings producing less coarse grids.

4.4.2.1 Basic Univariate Transformations

An intermediate tansformation

$$\boldsymbol{q}(\boldsymbol{\xi}) : \Xi^n \to Q^n$$

can be constructed through the use of separate univariate mappings $q^i(\xi^i)$: $[0,1] \to [0,1]$ such that $q^i(\boldsymbol{\xi}) = q^i(\xi^i)$. Therefore in order to define a nonuniform intermediate transformation $\boldsymbol{q}(\boldsymbol{\xi})$ through the local univariate mappings $x_i(\varphi, \epsilon)$, $i = 1, 2, 3, 4$, specified on the corresponding intervals $[0, \varphi_i]$, to provide adequate clustering of grid points where necessary, these mappings need to be extended continuously or smoothly over the whole interval $[0,1]$ to map this interval monotonically onto the unit interval $[0,1]$. This can be done by "gluing" these local nonuniform transformations $x_i(\varphi, \epsilon)$ to other mappings in the interval $[\varphi^i, 1]$ that are more uniform than the basic functions $x_i(\varphi, \epsilon)$, for example, linear or polynomial functions. The glued transformation extending $x_i(\varphi, \epsilon)$ must be smooth, or at least continuous.

Continuous Mappings

Nonsmooth continuous univariate mappings, denoted here as $x_{i,\mathrm{c}}(\varphi, \epsilon)$, can be defined as

$$x_{i,\mathrm{c}}(\varphi, \epsilon) = \begin{cases} x_i(\varphi, \epsilon), & 0 \le \varphi \le \varphi_i, \\ x_i + \frac{(1-x_i)(\varphi - \varphi_i)}{1-\varphi_i}, & \varphi_i \le \varphi \le 1. \end{cases} \tag{4.32}$$

These functions are monotonically increasing, given a suitable choice of the interval $[0, m_i]$ for the parameter ϵ, and vary from 0 to 1. Therefore they generate the individual univariate transformations $q^i(\xi^i)$, assuming

$$q^i(\xi^i) = x_{j,\mathrm{c}}(\xi^i, \epsilon), \quad j = 1, 2, 3, 4.$$

The first derivative of the function $x_3(\varphi, \epsilon)$ is limited uniformly, with respect to the parameter ϵ, for all $0 < \epsilon \le m_3$, so the matching of this function to any other one to transform the interval $[0,1]$ onto the interval $[0,1]$ is, in general, not necessary. The proper functions, monotonically increasing and varying from 0 to 1, denoted by $x_{3,\mathrm{s}}(\varphi, \epsilon)$, is obtained by adjusting the constant c in (4.31):

$$x_{3,\mathrm{s}}(\varphi, \epsilon) = (\epsilon^{kb} + c\varphi)^{1/b} - \epsilon^k, \quad c = (\epsilon^k + 1)^b - \epsilon^{kb}.$$

The length of any interval $[0, \varphi_i]$, $i = 1, 2, 4$, transformed into the corresponding layer by the corresponding function $x_{i,\mathrm{c}}(\varphi, \epsilon)$ defined by (4.32) approaches the constant $1/c$ as the parameter ϵ tends to zero. This quantity $1/c$ specifies that part of the uniform grid in the interval $[0,1]$ of the independent variable φ which is transformed into the layer. Obviously, the value of the constant c must be more than 1.

Smooth Mappings

Smooth basic univariate global transformations of the interval $[0,1]$ can be defined by matching the basic local transformations $x_i(\varphi, \epsilon)$ at the corresponding points φ_i to polynomials of the second order or to tangent straight lines emerging from the point $(1,1)$ in the plane x,φ. The latter matching is made at the cost of the constant c in the formulas for $x_i(\varphi,\epsilon)$. These very smooth transformations, obtained by matching smoothly the local contraction functions $x_i(\varphi,\epsilon)$, $i=1,2,4$, with the tangent lines and denoted below by $x_{i,\mathrm{s}}(\varphi,\epsilon)$, are given here.

The local basic stretching function $x_1(\varphi,\epsilon)$ is extended by the procedure of smooth matching to

$$x_{1,\mathrm{s}}(\varphi,\epsilon)=\begin{cases}-\frac{\epsilon^k}{b}\ln(1-c\varphi), & 0\le\varphi\le\varphi_1,\\ x_1+\frac{c}{b}(\varphi-\varphi_1), & \varphi_1\le\varphi\le 1,\end{cases} \tag{4.33}$$

with

$$\varphi_1=(1-\epsilon^k)/c, \qquad x_1=\frac{1}{b}\epsilon^k\ln\epsilon^{-k}, \qquad c=1-\epsilon^k+b(1-x_1).$$

The use of the local stretching function $x_2(\varphi,\epsilon)$ yields

$$x_{2,\mathrm{s}}(\varphi,\epsilon)=\begin{cases}\epsilon^k((1-c\varphi)^{(-1/b)}-1), & 0\le\varphi\le\varphi_2,\\ x_2+\frac{c}{b}(\varphi-\varphi_2), & \varphi_2\le\varphi\le 1,\end{cases} \tag{4.34}$$

where

$$\varphi_2=\frac{1}{c}(1-\epsilon^{kb/(b+1)}), \qquad x_2=\epsilon^{kb/(b+1)}-\epsilon^k,$$
$$c=1-\epsilon^{kb/(b+1)}+b(1-x_2).$$

When $b=1$ this transformation has a simpler form:

$$x_{2,\mathrm{s}}(\varphi,\epsilon)=\begin{cases}\epsilon^k\frac{c\varphi}{1-c\varphi}, & 0\le\varphi\le\varphi_2,\\ x_2+c(\varphi-\varphi_2), & \varphi_2\le\varphi\le 1,\end{cases} \tag{4.35}$$

where

$$\varphi_2=\frac{1}{c}(1-\epsilon^{k/2}), \qquad x_2=\epsilon^{k/2}-\epsilon^k, \qquad c=2-2\epsilon^{k/2}+\epsilon^k.$$

Finally, for the fourth basic mapping, we obtain

$$x_{4,\mathrm{s}}(\varphi,\epsilon)=\begin{cases}\epsilon^k((1+\epsilon^{-k})^{c\varphi}-1), & 0\le\varphi\le\varphi_4,\\ x_4+c(\varphi-\varphi_4), & \varphi_4\le\varphi\le 1,\end{cases} \tag{4.36}$$

where

$$\varphi_4=\frac{\ln(\epsilon^{-k})-\ln[\ln(1+\epsilon^{-k})]}{c\ln(1+\epsilon^{-k})}, \qquad x_4=\frac{1}{\ln(1+\epsilon^{-k})}-\epsilon^k,$$
$$c=1-x_4+\frac{\ln(\epsilon^{-k})-\ln[\ln(1+\epsilon^{-k})]}{\ln(1+\epsilon^{-k})}.$$

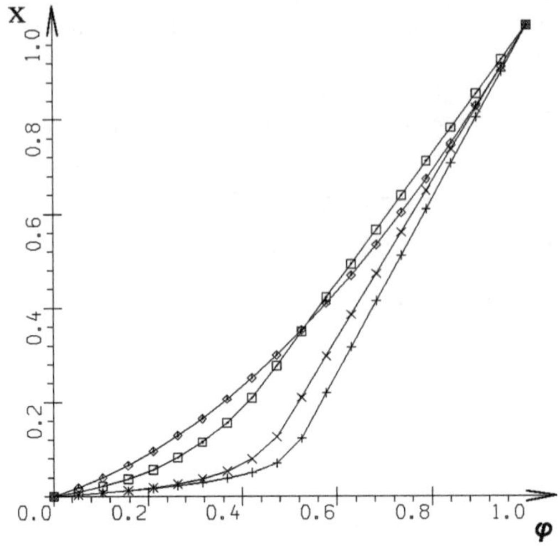

Fig. 4.4. Basic local contraction functions

Figure 4.4 illustrates these functions for $\epsilon = 1/30$. The symbols $+$, $\times$, $\Diamond$, and $\Box$ correspond to the functions $x_{1,\mathrm{s}}(\varphi, \epsilon)$, $x_{2,\mathrm{s}}(\varphi, \epsilon)$, $x_{3,\mathrm{s}}(\varphi, \epsilon)$, and $x_{4,\mathrm{s}}(\varphi, \epsilon)$, respectively.

The first derivative of the local function

$$x_4(\varphi, \epsilon) = \epsilon^k \big((1 + \epsilon^{-k})^{c\varphi} - 1\big)$$

tends to $\ln \epsilon^{-k}$ as $\varphi > \varphi_4$. This quantity is large when the parameter ϵ is very small; however, if ϵ is not too small, the magnitude of $\ln \epsilon^{-k}$, which characterizes the grid spacing, is not very large and may be tolerable for grid generation. In this case it is reasonable to use the local basic transformation $x_4(\varphi, \epsilon)$ as a global one from $[0, 1]$ to $[0, 1]$ without matching it to any other one to generate a grid. By adjusting the constant c, we obtain the form

$$x_5(\varphi, \epsilon) = \epsilon^k \big((1 + \epsilon^{-k})^{\varphi} - 1\big).$$

The length of the interval $[0, \varphi_1]$ transformed into the corresponding layer $[0, x_1]$ by the smooth function $x_{1,\mathrm{s}}(\varphi, \epsilon)$ tends to $1/(1 + b)$ as the parameter ϵ tends to zero. For the function $x_{2,\mathrm{s}}(\varphi, \epsilon)$, the length of the interval $[0, \varphi_2]$ also tends to $1/(1 + b)$ and, for the function $x_{4,\mathrm{s}}(\varphi, \epsilon)$, the length of the interval $[0, \varphi_4]$ tends to $1/2$. Consequently, this part of the uniform grid on the interval $[0, 1]$ is transformed into the corresponding layer. If there is a need for a larger proportion of the grid points to be distributed by smooth mappings into layers, the basic local contraction functions $x_i(\varphi, \epsilon)$, $i = 1, 2, 4$, should be matched smoothly to polynomials. In this case the point of matching can be chosen with less restriction and in this case it will not be completely prescribed, unlike the functions $x_{i,\mathrm{s}}(\varphi, \epsilon)$.

4.4.3 Other Univariate Transformations

Besides the univariate transformations described above, which depend on the parameter ϵ and are directly connected with the solutions to singularly perturbed equations, there are other monotonic functions, e.g. polynomials, hyperbolic functions, sines, and tangents which are used as local contraction mappings to yield grid clustering in boundary and interior layers.

4.4.3.1 Eriksson Function

One such reference function was introduced by Eriksson (1982):

$$x_6(\varphi) = \frac{e^{d\varphi} - 1}{e^d - 1}, \qquad d > 0, \ 0 \le \varphi \le 1. \tag{4.37}$$

This function provides a concentration of the grid towards the boundary $\varphi = 0$.

There is a direct correspondence between $x_6(\varphi)$ and the basic transformation $x_5(\varphi, \epsilon)$. Namely, if d in (4.37) is equal to $\ln(1 + \epsilon^{-k})$ then the Eriksson function (4.37) coincides with the contraction transformation $x_5(\varphi, \epsilon)$, i.e.

$$x_6(\varphi) = \epsilon^k \big((1 + \epsilon^{-k})^{\varphi} - 1\big).$$

This relation shows clearly how to adjust the grid spacing automatically to the physical small parameter ϵ by means of the Eriksson basic function.

Other functions, based on the inverse hyperbolic sines and tangents, were introduced by Vinokur (1983) to treat exponential singularities. Note that hyperbolic sines and tangents are defined through exponential functions and therefore, in the case of narrow layers, are locally similar to the exponential function $\varphi_1(x, \epsilon)$.

4.4.3.2 Tangent Function

The basic function

$$y(\varphi) = \tan \varphi \tag{4.38}$$

is very popular for generating grid clustering. Using two parameters α and β, a monotonic function transforming the interval $[0, 1]$ onto $(0, 1]$ with an opportunity to control the contraction near the boundary $\varphi = 0$ can be defined by

$$x(\varphi) = \alpha \tan(\beta\varphi). \tag{4.39}$$

The condition $x(1) = 1$ implies

$$\alpha = 1/\tan\beta.$$

For the derivative of the function $x(\varphi)$ with respect to φ, we have

$$x'(\varphi) = \frac{\beta}{\tan\beta\cos^2\beta\varphi}.$$

In order to cope with any boundary layer quantity whose derivative with respect to x reaches values of ϵ^{-k} at the point $x = 0$, the function $x'(\varphi)$ must have a value of the order of ϵ^k at the point $\varphi = 0$. This condition implies

$$\beta = \arctan\epsilon^{-k}$$

and consequently

$$\alpha = \epsilon^k,$$
$$x'(0) = \epsilon^k\arctan\epsilon^{-k} \sim \frac{\pi}{2}\epsilon^k, \quad \text{for } 0 < \epsilon \ll 1.$$

Thus the required expression for the local contraction function (4.39) eliminating the boundary layer is

$$x(\varphi) = \epsilon^k\tan[\arctan(\epsilon^{-k})\varphi]. \tag{4.40}$$

The corresponding inverse local stretching function $\varphi(x)$ has the form

$$\varphi(x) = \frac{\arctan(\epsilon^{-k}x)}{\arctan(\epsilon^{-k})}.$$

We have

$$\varphi'(x) = \frac{\epsilon^k}{\arctan\epsilon^{-1}(\epsilon^{2k} + x^2)}$$

and thus

$$M_1\frac{\epsilon^k}{(\epsilon^k + x)^2} \le |\varphi'(x)| \le M_2\frac{\epsilon^k}{(\epsilon^k + x)^2}, \tag{4.41}$$

where $0 < \epsilon \le 1/2$, and the constants M_1 and M_2 are independent of the parameter ϵ. A comparison of the inequality (4.41) and the relations (4.28) shows that the local stretching function $\varphi(x)$ is qualitatively equivalent to the function $\varphi_2(x, \epsilon)$ described by (4.25) with $b = 1$. The function (4.40) is therefore suitable to cope with solutions that are close to step functions in layers of exponential and power types.

4.4.3.3 Procedure for the Construction of Local Contraction Functions

The features of the tangent function (4.38) and the procedure described give a clue as to how to build new functions which can generate local grid clustering near a boundary point. These functions are derived from some basic univariate mappings $y(\varphi)$, satisfying, in analogy with $\tan\varphi$, the following conditions:

(1) $y'(0) = 1$,

(2) $y(\varphi)$ is a monotonical increasing function for $0 \le \varphi < a$ for some $a > 0$,
(3) $y(\varphi) \to \infty$ when $\varphi \to a$.

If $y(\varphi)$ is a function satisfying these properties then, assuming in analogy with (4.39) and (4.40)

$$x(\varphi) = \frac{y(\beta\varphi)}{y(\beta)} = \epsilon^k y[y^{-1}(\epsilon^{-k})\varphi],$$

where y^{-1} is the inverse of $y(\varphi)$, we obtain a monotonic transformation of the interval $[0,1]$ onto $[0,1]$ with a contraction of the order of ϵ^k at the point $\varphi = 0$.

Note that the transformations

$$x_1(\varphi,\epsilon) = -\frac{\epsilon^k}{b}\ln(1-c\varphi),$$
$$x_2(\varphi,\epsilon) = \epsilon^k\left[(1-c\varphi)^{(-1/b)} - 1\right],$$

from (4.31) can be obtained in accordance with this scheme from the functions

$$y_1(\varphi) = -\ln(1-\varphi)$$

and

$$y_2(\varphi) = b\left[(1-\varphi)^{-1/b} - 1\right],$$

respectively.

The original function $y(\varphi)$ can be formed as a ratio

$$y(\varphi) = b_1(\varphi)/b_2(\varphi)$$

of two functions $b_1(\varphi)$ and $b_2(\varphi)$ which are strongly positive on the interval $[0,a]$. In addition, the function $b_1(\varphi)$ must be monotonically increasing, while $b_2(\varphi)$ is monotonically decreasing and satisfies the condition $b_2(a) = 0$. For example, the function

$$y(\varphi) = \frac{\varphi}{1-\varphi}$$

generates the local contraction transformation

$$x(\varphi) = \frac{\epsilon^k\varphi}{1-(1-\epsilon^k)\varphi},$$

which coincides with the transformation $x_2(\varphi,\epsilon)$ from (4.31) for $b=1$, $c = (1-\epsilon^k)$.

4.4.4 Construction of Basic Intermediate Transformations

The basic functions $x_{i,\mathrm{c}}(\varphi,\epsilon)$ and $x_{i,\mathrm{s}}(\varphi,\epsilon)$ described above can be considered as construction elements for building intermediate transformations $\boldsymbol{q}(\boldsymbol{\xi}) : \Xi^n \to Q^n$ that serve to provide adequate grid clustering where necessary.

Firstly, these basic functions can be used as separate transformations $q^i(\xi^i)$ of the coordinates ξ^i. The first derivatives of the basic mappings $x_{i,\mathrm{c}}(\varphi,\epsilon)$ and $x_{i,\mathrm{s}}(\varphi,\epsilon)$ are small near the point $\varphi=0$, and therefore the derived intermediate transformations produce grid clustering in the vicinity of the selected boundary surfaces $\xi^i=0$.

Functions which provide grid clustering near arbitrary coordinate surfaces can be derived from these basic univariate mappings. For this purpose it is sufficient to define monotonic scalar functions having a small first derivative near arbitrary boundary or interior points in the interval $[0,1]$. Such mappings can be defined by simple procedures of scaling, shifting, and matching with the basic functions $x_{i,\mathrm{c}}(\varphi,\epsilon)$ and $x_{i,\mathrm{s}}(\varphi,\epsilon)$, as described below.

4.4.4.1 Functions with Boundary Contraction

For example, let $x(0,\varphi)$ be one of these basic monotonically increasing functions varying from 0 to 1 and having a small value of the first derivative near the point $\varphi=0$, thus exhibiting a grid contraction near the point $x=0$. Then the mapping

$$x(1,\varphi)=1-x(0,1-\varphi)$$

is also a monotonically increasing function transforming the interval $[0,1]$ onto itself and having the same small first derivative, but near the boundary point $x=1$. Therefore it performs a nodal concentration near the point $x=1$.

Grid clustering near two boundary points 0 and 1 can be produced by the mapping $x(0,1,\varphi)$ that is a composite of the two functions $x(0,\varphi)$ and $x(1,\varphi)$, say

$$x(0,1,\varphi)=x[0,x(1,\varphi)],$$

or can be obtained by the following formula of scaling and matching of the functions $x(0,\varphi)$ and $x(1,\varphi)$:

$$x(0,1,\varphi)=\begin{cases} x_0x(0,\varphi/x_0), & 0\le\varphi\le x_0,\\ 1-(1-x_0)x(0,\frac{1-\varphi}{1-x_0}), & x_0\le\varphi\le 1,\end{cases}$$

where x_0 is an interior matching point of the interval $[0,1]$.

4.4.4.2 Functions with Interior Contraction

Further, if x_0 is an inner point of the interval $[0,1]$ then the mapping

$$x(x_0,\varphi)=\begin{cases} x_0[1-x(0,1-\varphi/x_0)], & 0\le\varphi\le x_0,\\ x_0+(1-x_0)x(0,\frac{\varphi-x_0}{1-x_0}), & x_0\le\varphi\le 1,\end{cases}$$

is a monotonically increasing function from the interval $[0,1]$ onto the interval $[0,1]$ which provides grid clustering in the vicinity of the point $x=x_0$.

The function $x(0, x_0, 1, \varphi)$, defined as the composition of the two mappings $x(0, 1, \varphi)$ and $x(x_0, \varphi)$ introduced above, namely

$$x(0, x_0, 1, \varphi) = x[x_0, x(0, 1, \varphi)],$$

provides a concentration of grid nodes in the vicinity of the boundary points 0, 1 and of the interior point x_0.

A monotonically increasing function $x(x_0, x_1, \varphi)$ performing grid clustering near two interior points x_0 and x_1, $x_0 < x_1$, can be defined as a composition of two functions of the type $x(x_0, \varphi)$ or can be given by the formula

$$x(x_0, x_1, \varphi) = \begin{cases} x_0[1 - x(0, 1 - \varphi/x_0)], & 0 \le \varphi \le x_0, \\ x_0 + (d - x_0)x(0, \frac{\varphi - x_0}{d - x_0}), & x_0 \le \varphi \le b, \\ x_1 + (d - x_1)x(0, \frac{x_1 - \varphi}{x_1 - d}), & b \le \varphi \le x_1, \\ x_1 + (1 - x_1)\, x(0, \frac{\varphi - x_1}{1 - x_1}), & x_1 \le \varphi \le 1, \end{cases}$$

where d is a specified number satisfying $x_0 < b < x_1$. The same procedures allow one to construct monotonically increasing functions providing grid clustering near an arbitrary number of points. If the original mapping $x(0, \varphi)$ is smooth then the functions derived by these procedures are smooth as well.

Analogously, there can be defined monotonically decreasing functions providing grid concentration near an arbitrary number of points using the basic decreasing transformations $1 - x(0, \varphi)$. So there is a broad range of possibilities in using the basic transformations to generate effective grid clustering.

4.4.4.3 Clustering near Arbitrary Surfaces

The intermediate transformations $\boldsymbol{q}(\boldsymbol{\xi})$ constructed by the above approach through these modifications of the basic scalar functions provide grid clustering near the coordinate surfaces $\xi^i = \xi^i_l$. One drawback of such intermediate transformations is that they generate grid clustering only near these coordinate surfaces, with the same spacing in the vicinity of each of them. Therefore some procedures are needed to construct intermediate functions with a broader range of possibilities.

In three-dimensional domains, for instance, there is often a need to define an intermediate transformation $\boldsymbol{q}(\boldsymbol{\xi})$ providing grid clustering near an arbitrary surface intersecting a coordinate direction, say ξ^3. Let the surface be prescribed by the function

$$\xi^3 = g(\xi^1, \xi^2).$$

The required mapping $\boldsymbol{q}(\xi)$, providing grid concentration near this surface, is given by the formula

$$q^1(\boldsymbol{\xi}) = \xi^1,$$
$$q^2(\boldsymbol{\xi}) = \xi^2,$$
$$q^3(\boldsymbol{\xi}) = \begin{cases} g(\xi^1, \xi^2)\{1 - f[1 - \xi^3/g(\xi^1, \xi^2)]\}, & 0 \le \xi^3 \le g(\xi^1, \xi^2), \\ g(\xi^1, \xi^2) + [1 - g(\xi^1, \xi^2)]f(\frac{\xi^3 - g(\xi^1, \xi^2)}{1 - g(\xi^1, \xi^2)}), & g(\xi^1, \xi^2) \le \xi^3 \le 1, \end{cases}$$

where $f(\varphi)$ is a basic monotonically increasing function with a small first derivative near the point $\varphi = 0$. Compositions of such transformations produce maps that provide grid clustering near a number of surfaces intersecting different directions.

4.4.4.4 Nonuniform Clustering

The procedures described above provide adequate grid clustering in the vicinity of arbitrary surfaces, but with the same grid spacing around any one specified surface. However, in some cases, for example when gridding a flow region around a body, there is a need for nonuniform grid clustering in the transverse direction with respect to different parts of the surface of the body. Such grid clustering can be realized by intermediate transformations constructed along the coordinate surface (line in two dimensions) through a combination of basic functions with different values of the parameter ϵ. For example, a two-dimensional intermediate transformation $\boldsymbol{q}(\boldsymbol{\xi})$ can be defined as

$$q^1(\xi^1, \xi^2) = \xi^1,$$
$$q^2(\xi^1, \xi^2) = \xi^1 x_1(\xi^2, \epsilon^k) + (1 - \xi^1) x_2(\xi^2, \epsilon^d),$$

where $x_i(\xi^2, \epsilon^m)$ is one of the basic functions. The mapping $\boldsymbol{q}(\boldsymbol{\xi})$ provides a nonuniform grid spacing along the coordinate $\xi^2 = \xi^2_0$ in the ξ^1 direction.

The procedures presented here can be applied to other mappings as well to construct intermediate transformations generating nonuniform grid clustering in the desired zones of the physical domain.

4.5 Comments

There are three basic approaches to treating problems with boundary and interior layers. The classical approach relies on expansion of the solution in a series of singular and slowly changing functions. The second technique applies special approximations of equations. The third one is based on the implementation of local stretching functions to stretch the coordinates and, correspondingly, provide clustered grids.

The approach using stretching functions appears to be more effective in comparison with the other techniques because it requires only rough information about the qualitative properties of the solution and enables one to interpolate the solution uniformly over the entire physical region. The application of interactive procedures using the basic intermediate transformations allows one to generate efficient grids in arbitrary zones even without preliminary information about the qualitative features of the solution.

Estimates of the derivatives of the solution to the problem of the type (4.8) with exponential layers were obtained by Brish (1954). Investigation of the qualitative properties of the solution to the linear problem (4.1) in interior layers was carried out by Berger, Han, and Kellog (1984).

The asymptotic location of the interior layers of the solution to the problem (4.6) was found by Lorenz (1982, 1984). The asymptotic expansion of the linear version of the problem (4.7) for $a(0) > 1$ was considered by Lomov (1964). A qualitative investigation of the solutions to the problems (4.1), (4.6), (4.7) in arbitrary boundary and interior layers was carried out by Liseikin (1984, 1986, 1993b). In these papers, estimates of the derivatives of the solution were obtained. A detailed description of the estimates of derivatives of the solutions of singularly perturbed equations is presented in the monograph by Liseikin and Petrenko (1989) and Liseikin (2001a).

The logarithmic transformation $x_1(\varphi, \epsilon)$ in (4.31) was introduced by Bahvalov (1969) for the generation of clustered grids in the vicinity of exponential boundary layers of singularly perturbed equations. The mappings $x_i(\varphi, \epsilon)$, $i = 2, 3, 4$, were proposed by Liseikin (1984, 1986) for the construction of nonuniform clustered grids within power and combined boundary and interior layers. A particular shape of the basic contraction mapping $x_2(\varphi, \epsilon)$ for $b = 1$, having the form

$$x_2(\varphi, \epsilon) = \epsilon^k \frac{c\varphi}{1 - c\varphi},$$

was also proposed independently by Vulanovic (1984) to generate grids within some exponential layers for boundary layer functions of the type described by (4.24).

Stretching functions based on inverse hyperbolic sines were employed by Thomas et al. (1972) in a numerical solution of inviscid supersonic flow.

A two-sided stretching function of the logarithmic type (4.27) was introduced by Roberts (1971) to study boundary layer flows.

A family of tangent mappings of the form

$$x = x_0 + \alpha \tan[(s - s_0)\beta\varphi],$$

suitable primarily for internal layers, was introduced by Vinokur (1983) to generate grids. These mappings were also employed by Bayliss and Garbey (1995) as part of the adaptive pseudospectral method.

Physical quantities were used as new coordinates for stretching boundary and interior layers by Tolstykh (1973).

5

Algebraic Grid Generation

5.1 Introduction

The algebraic grid generation approach relies chiefly on an explicit construction of coordinate transformations through the formulas of transfinite interpolation. Of central importance in the method are blending functions (univariate quantities each depending on one chosen coordinate only). These provide matching of the grid distribution on, and grid directions from, the boundaries and specified interior surfaces of an arbitrary domain. Direct control of the essential properties of the coordinate transformations in the vicinity of the boundaries and interior surfaces is carried out by the specification of the out-of-surface-direction derivatives and blending functions.

The purpose of this chapter is to describe common techniques of algebraic grid generation.

Nearly all of the formulas of transfinite interpolation include both repeated indices over which a summation is carried out and one repeated index, usually i, that is fixed. Therefore in this chapter we do not use the convention of summation of repeated indices but instead use the common notation $\sum$ to indicate summation.

5.2 Transfinite Interpolation

This section describes some general three-dimensional formulas of transfinite interpolation which are used to define algebraic coordinate transformations from a standard three-dimensional cube Ξ^3 with Cartesian coordinates ξ^i, $i = 1, 2, 3$, onto a physical domain X^3 with Cartesian coordinates x^i, $i = 1, 2, 3$. The formulation of the three-dimensional interpolation is based on a particular operation of Boolean summation over unidirectional interpolations. So, first, the general formulas of unidirectional interpolation are reviewed.

V.D. Liseikin, *Grid Generation Methods*, Scientific Computation,
DOI 10.1007/978-90-481-2912-6_5, © Springer Science+Business Media B.V. 2010

5.2.1 Unidirectional Interpolation

5.2.1.1 General Formulas

For the unit cube Ξ^3, let there be chosen one coordinate direction ξ^i and some sections of the cube orthogonal to this direction, defined by the planes $\xi^i = \xi^i_l$, $l = 1, \dots, L_i$. Furthermore, on each section $\xi^i = \xi^i_l$, let there be given the values of some vector-valued function $\boldsymbol{r}(\boldsymbol{\xi})$, $\boldsymbol{\xi} = (\xi^1, \xi^2, \xi^3)$, and of its derivatives with respect to ξ^i up to order P^i_l. Then the unidirectional interpolation of the function $\boldsymbol{r}(\boldsymbol{\xi})$ is a vector-valued function $\boldsymbol{P}^i[\boldsymbol{r}](\boldsymbol{\xi})$ from Ξ^3 into R^3 defined by the formula

$$\boldsymbol{P}_i[\boldsymbol{r}](\boldsymbol{\xi}) = \sum_{l=1}^{L^i} \sum_{n=0}^{P^i_l} \alpha^i_{l,n}(\xi^i) \frac{\partial^n}{(\partial \xi^i)^n} \boldsymbol{r}(\boldsymbol{\xi}|_{\xi^i = \xi^i_l}). \tag{5.1}$$

Here the smooth scalar functions $\alpha^i_{l,n}(\xi^i)$, depending on one independent variable ξ^i are subject to the following restrictions:

$$\frac{\mathrm{d}^m}{(\mathrm{d}\xi^i)^m} \alpha^i_{l,n}(\xi^i_k) = \delta^l_k \delta^n_m, \quad l, k = 1, \dots, L^i, \ m, n = 0, 1, \dots, P^i_l, \tag{5.2}$$

where δ^j_i is the Kronecker delta function, i.e.

$$\delta^j_i = \begin{cases} 1, & i = j, \\ 0, & i \neq j. \end{cases}$$

The expression $(\boldsymbol{\xi}|_{\xi^i = \xi^i_l})$ in (5.1) designates a point that is a projection of $\boldsymbol{\xi} = (\xi^1, \xi^2, \xi^3)$ on the section $\xi^i = \xi^i_l$, i.e. the ith coordinate ξ^i of $\boldsymbol{\xi}$ is fixed and equal to ξ^i_l; for example,

$$(\boldsymbol{\xi}|_{\xi^1 = \xi^1_l}) = (\xi^1_l, \xi^2, \xi^3).$$

It is also assumed in (5.1) and below that the operator for the zero-order derivative is the identity operator, i.e.

$$\frac{\partial^0}{(\partial \xi^i)^0} f(\boldsymbol{\xi}) = f(\boldsymbol{\xi}), \qquad \frac{\mathrm{d}^0}{(\mathrm{d}\xi^i)^0} g(\xi^i) = g(\xi^i).$$

The coefficients $\alpha^i_{l,n}(\xi^i)$ in (5.1) are referred to as the blending functions. They serve to propagate the values of the vector-valued function $\boldsymbol{r}(\boldsymbol{\xi})$ from the specified sections of the cube $\boldsymbol{\Xi}^3$ into its interior. It is easily shown that the conditions (5.2) imposed on the blending functions $\alpha^i_{l,n}(\xi^i)$ provide matching at the sections $\xi^i = \xi^i_l$ of the values of the function $\boldsymbol{P}_i[\boldsymbol{r}](\boldsymbol{\xi})$ and $\boldsymbol{r}(\boldsymbol{\xi})$, as well as the values of their derivatives with respect to ξ^i, namely,

$$\frac{\partial^n \boldsymbol{P}_i[\boldsymbol{r}]}{(\partial \xi^i)^n}(\boldsymbol{\xi}|_{\xi^i = \xi^i_l}) = \frac{\partial^n \boldsymbol{r}}{(\partial \xi^i)^n}(\boldsymbol{\xi}|_{\xi^i = \xi^i_l}), \quad n = 0, \dots, P^i_l.$$

5.2.1.2 Two-Boundary Interpolation

A very important interpolation for grid generation applications is the one which matches the values of the vector-valued function $\boldsymbol{r}(\boldsymbol{\xi})$ and of its derivatives exclusively at the boundary planes of the cube Ξ^3. In this case $L^i = 2$, $\xi^i_1 = 0$, and $\xi^i_2 = 1$, and the relations (5.1) and (5.2) have the form

$$\boldsymbol{P}_i[\boldsymbol{r}](\boldsymbol{\xi}) = \sum_{n=0}^{P^i_1} \alpha^1_{l,n}(\xi^i)\frac{\partial^n}{(\partial\xi^i)^n}\boldsymbol{r}(\boldsymbol{\xi}|_{\xi^i=0}) + \sum_{n=0}^{P^i_2} \alpha^i_{2,n}(\xi^i)\frac{\partial^n}{(\partial\xi^i)^n}\boldsymbol{r}(\boldsymbol{\xi}|_{\xi^i=1}), \quad (5.3)$$

$$\frac{\mathrm{d}^m}{(\mathrm{d}\xi^i)^m}\alpha^i_{l,n}(\xi^i_k) = \delta^l_k\delta^n_m, \quad l,k = 1,2, \ m,n = 0,1,\ldots,P^i_l. \quad (5.4)$$

The interpolation described by (5.3) is referred to as the two-boundary interpolation.

5.2.2 Tensor Product

The composition of two unidirectional interpolations $\boldsymbol{P}_i[\boldsymbol{r}](\boldsymbol{\xi})$ and $\boldsymbol{P}_j[\boldsymbol{r}](\boldsymbol{\xi})$ of $\boldsymbol{r}(\boldsymbol{\xi})$ in the directions ξ^i and ξ^j, respectively, is called their tensor product. This operation is denoted by $\boldsymbol{P}_i[r] \otimes \boldsymbol{P}_j[r](\xi)$ and, in accordance with (5.1), we obtain

$$\begin{aligned}\boldsymbol{P}_i[\boldsymbol{r}] \otimes \boldsymbol{P}_j[\boldsymbol{r}](\boldsymbol{\xi}) &= \boldsymbol{P}_i[\boldsymbol{P}_j[\boldsymbol{r}]](\boldsymbol{\xi}) \\ &= \sum_{l=1}^{L^i}\sum_{n=0}^{P^i_l} \alpha^i_{l,n}(\xi^i)\frac{\partial^n \boldsymbol{P}_j[\boldsymbol{r}]}{(\partial\xi^i)^n}(\boldsymbol{\xi}|_{\xi^i=\xi^i_l}) \\ &= \sum_{k=1}^{L^j}\sum_{m=0}^{P^j_k}\sum_{l=1}^{L^i}\sum_{n=0}^{P^i_l} \alpha^i_{l,n}(\xi^i)\alpha^j_{k,m}(\xi^j) \\ &\quad \times \frac{\partial^{n+m}\boldsymbol{r}}{(\partial\xi^i)^n(\partial\xi^j)^m}(\boldsymbol{\xi}|_{\xi^i=\xi^i_l,\xi^j=\xi^j_k}). \end{aligned} \quad (5.5)$$

Here by the notation $(\boldsymbol{\xi}|_{\xi^i=\xi^i_l,\xi^j=\xi^j_k})$ we mean the point which is the projection of $\boldsymbol{\xi}$ on the intersection of the planes $\xi^i = \xi^i_l$ and $\xi^j = \xi^j_k$, e.g.

$$(\boldsymbol{\xi}|_{\xi^1=\xi^1_l,\xi^3=\xi^3_k}) = (\xi^1_l, \xi^2, \xi^3_k).$$

Equation (5.5) shows clearly that the tensor product is a commutative operation, i.e.

$$\boldsymbol{P}_i[\boldsymbol{r}] \otimes \boldsymbol{P}_j[\boldsymbol{r}] = \boldsymbol{P}_j[\boldsymbol{r}] \otimes \boldsymbol{P}_i[\boldsymbol{r}].$$

Using the relations (5.1), (5.2), and (5.5) we obtain

$$\frac{\partial}{\partial \xi^i} \boldsymbol{P}_i[\boldsymbol{r}] \otimes \boldsymbol{P}_j[\boldsymbol{r}](\boldsymbol{\xi}|_{\xi^i=\xi^i_s, \xi^j=\xi^j_t})$$

$$= \sum_{m=1}^{L^j} \sum_{k=0}^{P^j_k} \sum_{l=1}^{L^i} \sum_{p=0}^{P^i_l} \frac{\mathrm{d}}{\mathrm{d}\xi^i} \alpha^i_{m,k}(\xi^i_s) \alpha^j_{l,p}(\xi^j_t) \frac{\partial^{k+p} \boldsymbol{r}}{(\partial \xi^i)^k (\partial \xi^j)^p} (\boldsymbol{\xi}|_{\xi^i=\xi^i_m, \xi^j=\xi^j_l})$$

$$= \frac{\partial \boldsymbol{r}}{\partial \xi^i} (\boldsymbol{\xi}|_{\xi^i=\xi^i_s, \xi^j=\xi^j_t}).$$

Analogously,

$$\frac{\partial^{k+p}}{(\partial \xi^i)^k (\partial \xi^j)^p} (\boldsymbol{P}_i[\boldsymbol{r}] \otimes \boldsymbol{P}_j[\boldsymbol{r}])(\boldsymbol{\xi}|_{\xi^i=\xi^i_s, \xi^j=\xi^j_t}) = \frac{\partial^{k+p}}{(\partial \xi^i)^k (\partial \xi^j)^p} \boldsymbol{r}(\boldsymbol{\xi}|_{\xi^i=\xi^i_s, \xi^j=\xi^j_t}).$$

Thus the derivatives of the tensor product $\boldsymbol{P}_i[\boldsymbol{r}] \otimes \boldsymbol{P}_j[\boldsymbol{r}]$ with respect to ξ^i and ξ^j match the derivatives of the function $\boldsymbol{r}(\boldsymbol{\xi})$ at the intersections of the planes $\xi^i = \xi^i_s$ and $\xi^j = \xi^j_t$.

5.2.3 Boolean Summation

5.2.3.1 Bidirectional Interpolation

The bidirectional interpolation matching the values of the function $\boldsymbol{r}(\boldsymbol{\xi})$ and of its derivatives at the sections in the directions ξ^i and ξ^j is defined through the Boolean summation $\oplus$:

$$\boldsymbol{P}_i[\boldsymbol{r}] \oplus \boldsymbol{P}_j[\boldsymbol{r}](\boldsymbol{\xi}) = \boldsymbol{P}_i[\boldsymbol{r}](\boldsymbol{\xi}) + \boldsymbol{P}_j[\boldsymbol{r}](\boldsymbol{\xi}) - \boldsymbol{P}_i[\boldsymbol{r}] \otimes \boldsymbol{P}_j[\boldsymbol{r}](\boldsymbol{\xi}). \tag{5.6}$$

Using (5.1) and (5.5), we obtain

$$\boldsymbol{P}_i[\boldsymbol{r}] \oplus \boldsymbol{P}_j[\boldsymbol{r}](\boldsymbol{\xi})$$

$$= \sum_{l=1}^{L^i} \sum_{n=0}^{P^i_l} \alpha^i_{l,n}(\xi^i) \frac{\partial^n \boldsymbol{r}}{(\partial \xi^i)^n} (\boldsymbol{\xi}|_{\xi^i=\xi^i_l}) + \sum_{k=1}^{L^j} \sum_{m=0}^{P^j_k} \alpha^j_{k,m}(\xi^i) \frac{\partial^m \boldsymbol{r}}{(\partial \xi^j)^m} (\boldsymbol{\xi}|_{\xi^j=\xi^j_k})$$

$$- \sum_{k=1}^{L^j} \sum_{m=0}^{P^j_k} \sum_{l=1}^{L^i} \sum_{n=0}^{P^i_l} \alpha^i_{l,n}(\xi^i) \alpha^j_{k,m}(\xi^j) \frac{\partial^{n+m} \boldsymbol{r}}{(\partial \xi^i)^n (\partial \xi^j)^m} (\boldsymbol{\xi}|_{\xi^i=\xi^i_l, \xi^j=\xi^j_k}). \tag{5.7}$$

Taking into account the relation

$$\boldsymbol{P}_j[\boldsymbol{r}] - \boldsymbol{P}_i[\boldsymbol{r}] \otimes \boldsymbol{P}_j[\boldsymbol{r}] = \boldsymbol{P}_j[\boldsymbol{r} - \boldsymbol{P}_i[\boldsymbol{r}]] \tag{5.8}$$

we obtain the result that the formulas (5.6), and (5.7) for the Boolean summation can be written as the ordinary sum of two unidirectional interpolants $\boldsymbol{P}_i[\boldsymbol{r}]$ and $\boldsymbol{P}_j[\boldsymbol{r} - \boldsymbol{P}_i[\boldsymbol{r}]]$. Thus, using (5.1), we obtain

$$\boldsymbol{P}_i[\boldsymbol{r}] \oplus \boldsymbol{P}_j[\boldsymbol{r}](\boldsymbol{\xi}) = \boldsymbol{P}_j[\boldsymbol{r}](\boldsymbol{\xi}) + \sum_{l=1}^{L^i} \sum_{n=0}^{P^i_l} \alpha^i_{l,n}(\xi^i) \frac{\partial^n (\boldsymbol{r} - \boldsymbol{P}_j[\boldsymbol{r}])}{(\partial \xi^i)^n} (\boldsymbol{\xi}|_{\xi^i=\xi^i_l}). \tag{5.9}$$

From (5.7) it is evident that

$$\boldsymbol{P}_i[\boldsymbol{r}] \oplus \boldsymbol{P}_j[\boldsymbol{r}] = \boldsymbol{P}_j[\boldsymbol{r}] \oplus \boldsymbol{P}_i[\boldsymbol{r}],$$

so the indices i and j in (5.7), (5.9) can be interchanged.

The Boolean summation (5.6) matches $\boldsymbol{r}(\boldsymbol{\xi})$ and its derivatives at all sections $\xi^i = \xi^i_k$ and $\xi^j = \xi^j_l$, i.e.

$$\frac{\partial^{k+p}}{(\partial\xi^i)^k(\partial\xi^j)^p}(\boldsymbol{P}_i[\boldsymbol{r}] \otimes \boldsymbol{P}_j[\boldsymbol{r}])(\boldsymbol{\xi}|_{\xi^t=\xi^t_l}) = \frac{\partial^{k+p}}{(\partial\xi^i)^k(\partial\xi^j)^p}\boldsymbol{r}(\boldsymbol{\xi}|_{\xi^t=\xi^t_l}),$$

where either $t = i$ or $t = j$.

5.2.3.2 Three-Dimensional Interpolation

A multidirectional interpolation $\boldsymbol{P}[\boldsymbol{r}](\boldsymbol{\xi})$ of $\boldsymbol{r}(\boldsymbol{\xi})$, which matches the values of the function $\boldsymbol{r}(\boldsymbol{\xi})$ and of its derivatives at the sections $\xi^i = \xi^i_l$, $l = 1, \ldots, L_i$, in all directions $\xi^i, i = 1, 2, 3$, is defined through the Boolean summation of all unidirectional interpolations $\boldsymbol{P}_i[\boldsymbol{r}], i = 1.2, 3$:

$$\boldsymbol{P}[\boldsymbol{r}] = \boldsymbol{P}_1[\boldsymbol{r}] \oplus \boldsymbol{P}_2[\boldsymbol{r}] \oplus \boldsymbol{P}_3[\boldsymbol{r}]. \tag{5.10}$$

Taking into account (5.6), we obtain

$$\begin{aligned}\boldsymbol{P}[\boldsymbol{r}] = {} & \boldsymbol{P}_1[\boldsymbol{r}] + \boldsymbol{P}_2[\boldsymbol{r}] + \boldsymbol{P}_3[\boldsymbol{r}] \\ & - \boldsymbol{P}_1[\boldsymbol{r}] \otimes \boldsymbol{P}_2[\boldsymbol{r}] - \boldsymbol{P}_1[\boldsymbol{r}] \otimes \boldsymbol{P}_3[\boldsymbol{r}] - \boldsymbol{P}_2[\boldsymbol{r}] \otimes \boldsymbol{P}_3[\boldsymbol{r}] \\ & + \boldsymbol{P}_1[\boldsymbol{r}] \otimes \boldsymbol{P}_2[\boldsymbol{r}] \otimes \boldsymbol{P}_3[\boldsymbol{r}]. \end{aligned} \tag{5.11}$$

5.2.3.3 Recursive Form of Transfinite Interpolation

Using the relation (5.8) we can easily show that (5.11) is also equal to the following equation:

$$\boldsymbol{P}[\boldsymbol{r}] = \boldsymbol{P}_1[\boldsymbol{r}] + \boldsymbol{P}_2[\boldsymbol{r} - \boldsymbol{P}_1[\boldsymbol{r}]] + \boldsymbol{P}_3\big[\boldsymbol{r} - \boldsymbol{P}_1[\boldsymbol{r}] - \boldsymbol{P}_2[\boldsymbol{r} - \boldsymbol{P}_1[\boldsymbol{r}]]\big]. \tag{5.12}$$

This represents the formula (5.10) for multidirectional interpolation as the sum of the three unidirectional interpolations $\boldsymbol{P}_1[\boldsymbol{r}]$, $\boldsymbol{P}_2[\boldsymbol{r} - \boldsymbol{P}_1[\boldsymbol{r}]]$, and $\boldsymbol{P}_3[\boldsymbol{r} - \boldsymbol{P}_1[\boldsymbol{r}] - \boldsymbol{P}_2[\boldsymbol{r} - \boldsymbol{P}_1[\boldsymbol{r}]]]$. Therefore the expression (5.12) for $\boldsymbol{P}[\boldsymbol{r}]$ gives a recursive form of the interpolation (5.10) through a sequence of the unidirectional interpolations (5.1):

$$\begin{aligned}\boldsymbol{F}_1[\boldsymbol{r}] &= \boldsymbol{P}_1[\boldsymbol{r}], \\ \boldsymbol{F}_2[\boldsymbol{r}] &= \boldsymbol{F}_1[\boldsymbol{r}] + \boldsymbol{P}_2[\boldsymbol{r} - \boldsymbol{F}_1[\boldsymbol{r}]], \\ \boldsymbol{P}[\boldsymbol{r}] &= \boldsymbol{F}_2[\boldsymbol{r}] + \boldsymbol{P}_3[\boldsymbol{r} - \boldsymbol{F}_2[\boldsymbol{r}]]\end{aligned}$$

which is usually applied in constructing algebraic coordinate transformations. Using (5.1) we obtain

$$
\begin{aligned}
\boldsymbol{F}_1[\boldsymbol{r}](\boldsymbol{\xi}) &= \sum_{l=1}^{L^1}\sum_{n=0}^{P_l^1} \alpha_{l,n}^1(\xi^1)\frac{\partial^n \boldsymbol{r}}{(\partial \xi^1)^n}(\xi_l^1,\xi^2,\xi^3),\\
\boldsymbol{F}_2[\boldsymbol{r}](\boldsymbol{\xi}) &= \boldsymbol{F}_1[\boldsymbol{r}](\boldsymbol{\xi}) + \sum_{l=1}^{L^2}\sum_{n=0}^{P_l^2} \alpha_{l,n}^2(\xi^2)\frac{\partial^n (\boldsymbol{r}-\boldsymbol{F}_1[\boldsymbol{r}])}{(\partial \xi^2)^n}(\xi^1,\xi_l^2,\xi^3),\\
\boldsymbol{P}[\boldsymbol{r}](\boldsymbol{\xi}) &= \boldsymbol{F}_2[\boldsymbol{r}](\boldsymbol{\xi}) + \sum_{l=1}^{L^3}\sum_{n=0}^{P_l^3} \alpha_{l,n}^3(\xi^3)\frac{\partial^n (\boldsymbol{r}-\boldsymbol{F}_2[\boldsymbol{r}])}{(\partial \xi^3)^n}(\xi^1,\xi^2,\xi_l^3).
\end{aligned}
\tag{5.13}
$$

It is easy to see, taking advantage of (5.2), that the multiple summation matches the function $\boldsymbol{r}(\boldsymbol{\xi})$ and its derivatives with respect to ξ^1, ξ^2, and ξ^3 on all sections $\xi^i = \xi_l^i$, $i = 1, 2, 3$, of the cube Ξ^3.

5.2.3.4 Outer Boundary Interpolation

Equation (5.13) shows that the outer boundary interpolation based on the two-boundary unidirectional interpolations described by (5.4) has the following form:

$$
\begin{aligned}
\boldsymbol{F}_1[\boldsymbol{r}](\boldsymbol{\xi}) =& \sum_{n=0}^{P_1^1} \alpha_{1,n}^1(\xi^1)\frac{\partial^n \boldsymbol{r}}{(\partial \xi^1)^n}(0,\xi^2,\xi^3)\\
&+\sum_{n=0}^{P_2^1} \alpha_{2,n}^1(\xi^1)\frac{\partial^n \boldsymbol{r}}{(\partial \xi^1)^n}(1,\xi^2,\xi^3),\\
\boldsymbol{F}_2[\boldsymbol{r}](\boldsymbol{\xi}) =& \boldsymbol{F}_1[\boldsymbol{r}](\boldsymbol{\xi}) + \sum_{n=0}^{P_1^2} \alpha_{1,n}^2(\xi^2)\frac{\partial^n (\boldsymbol{r}-\boldsymbol{F}_1[\boldsymbol{r}])}{(\partial \xi^2)^n}(\xi^1,0,\xi^3)\\
&+\sum_{n=0}^{P_2^2} \alpha_{2,n}^2(\xi^2)\frac{\partial^n (\boldsymbol{r}-\boldsymbol{F}_1[\boldsymbol{r}])}{(\partial \xi^2)^n}(\xi^1,1,\xi^3),\\
\boldsymbol{P}[\boldsymbol{r}](\boldsymbol{\xi}) =& \boldsymbol{F}_2[\boldsymbol{r}](\boldsymbol{\xi}) + \sum_{n=0}^{P_1^3} \alpha_{1,n}^3(\xi^3)\frac{\partial^n (\boldsymbol{r}-\boldsymbol{F}_2[\boldsymbol{r}])}{(\partial \xi^3)^n}(\xi^1,\xi^2,0)\\
&+\sum_{n=0}^{P_2^3} \alpha_{2,n}^3(\xi^3)\frac{\partial^n (\boldsymbol{r}-\boldsymbol{F}_2[\boldsymbol{r}])}{(\partial \xi^3)^n}(\xi^1,\xi^2,1).
\end{aligned}
\tag{5.14}
$$

5.2.3.5 Two-Dimensional Interpolation

The formulas for two-dimensional transfinite interpolation of a two-dimensional vector-valued function $\boldsymbol{x}(\boldsymbol{\xi}) : \Xi^2 \to X^2$ are obtained from (5.13) and (5.14) by assuming $\boldsymbol{F}_2(\boldsymbol{r}) = \boldsymbol{P}(\boldsymbol{r})$, $\alpha_{l,k}^3 = 0$, and omitting ξ^3. For example, we obtain from (5.13) the following formula for two-dimensional transfinite interpolation:

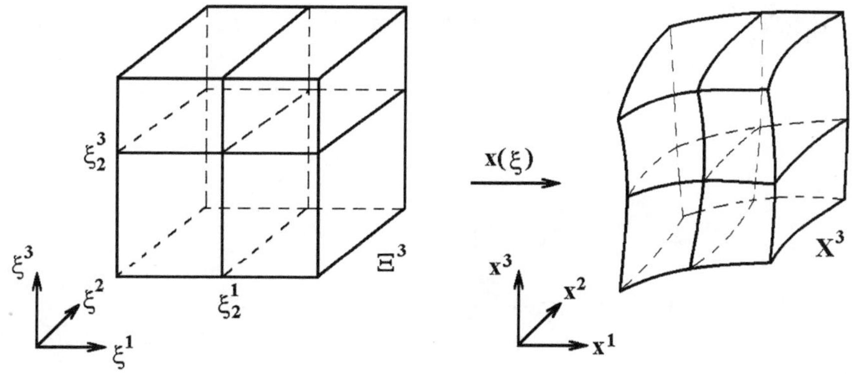

Fig. 5.1. Coordinate transformation

$$
\begin{aligned}
\boldsymbol{F}_1[\boldsymbol{r}](\xi^1,\xi^2) &= \sum_{l=1}^{L^1}\sum_{k=0}^{P_l^1} \alpha_k^1(\xi^1)\frac{\partial^k \boldsymbol{r}}{(\partial\xi^1)^k}(\xi_l^1,\xi^2),\\
\boldsymbol{P}[\boldsymbol{r}](\xi^1,\xi^2) &= \boldsymbol{F}_1[\boldsymbol{r}](\xi^1,\xi^2)\\
&\quad+\sum_{l=1}^{L^2}\sum_{m=0}^{P_l^2} \alpha_{l,m}^2(\xi^2)\frac{\partial^m(\boldsymbol{r}-\boldsymbol{F}_1[\boldsymbol{r}])}{(\partial\xi^2)^m}(\xi^1,\xi_l^2).
\end{aligned}
\tag{5.15}
$$

5.3 Algebraic Coordinate Transformations

This section sets out the definitions of the algebraic coordinate transformations appropriate for the generation of structured grids through the formulas of transfinite interpolation.

5.3.1 Formulation of Algebraic Coordinate Transformation

The formulas of transfinite interpolation described above give clear guidance on how to define an algebraic coordinate transformation

$$\boldsymbol{x}(\boldsymbol{\xi}) : \Xi^3 \to X^3, \quad \boldsymbol{x}(\boldsymbol{\xi}) = (x^1(\boldsymbol{\xi}), x^2(\boldsymbol{\xi}), x^3(\boldsymbol{\xi})),\ \boldsymbol{\xi} = (\xi^1,\xi^2,\xi^3)$$

from the cube Ξ^3 onto a domain $X^3 \subset R^3$ which matches, at the boundary and some chosen intermediate coordinate planes of the cube, the prescribed values and the specified derivatives of $\boldsymbol{x}(\boldsymbol{\xi})$ along the coordinate directions emerging from the coordinate surfaces (Fig. 5.1).

Let there be chosen, in each direction ξ^i, some coordinate planes $\xi^i = \xi_l^i$, $l = 1,\dots,L^i$, of the cube Ξ^3, including two opposite boundary planes $\xi^i = \xi_0^i = 0$, $\xi^i = \xi_{L^i}^i = 1$. Furthermore, let there be specified, at each section $\xi^i = \xi_l^i$, a smooth three-dimensional vector-valued function denoted by $\boldsymbol{A}^i_{l,0}(\boldsymbol{\xi}|_{\xi^i=\xi^i_l})$, which is assumed to represent the values of the function $\boldsymbol{x}(\boldsymbol{\xi})$ being constructed at the points of this section. Also, let there be specified, at this section, three-dimensional vector-valued functions denoted by $\boldsymbol{A}^i_{l,n}(\boldsymbol{\xi}|_{\xi^i=\xi^i_l})$ which represent derivatives with respect to ξ^i of the function $\boldsymbol{x}(\boldsymbol{\xi})$ on the respective sections $\xi^i = \xi^i_l$. Thus it is assumed that

$$\boldsymbol{A}^i_{l,0}(\boldsymbol{\xi}|_{\xi^i=\xi^i_l}) = \frac{\partial^0}{(\partial \xi^i)^0}\boldsymbol{x}(\boldsymbol{\xi}|_{\xi^i=\xi^i_l}) = \boldsymbol{x}(\boldsymbol{\xi}|_{\xi^l=\xi^i_l}), \quad l = 1, \dots, L_i,$$

$$\boldsymbol{A}^i_{l,n}(\boldsymbol{\xi}|_{\xi^i=\xi^i_l}) = \frac{\partial^n}{(\partial \xi^i)^n}\boldsymbol{x}(\boldsymbol{\xi}|_{\xi^i=\xi^i_l}), \quad n = 1, \dots, P^i_l.$$

Since

$$\frac{\partial^m}{(\partial \xi^j)^m}\left(\frac{\partial^n \boldsymbol{x}}{(\partial \xi^i)^n}\right) = \frac{\partial^n}{(\partial \xi^j)^n}\left(\frac{\partial^m \boldsymbol{x}}{(\partial \xi^j)^m}\right),$$

we find that the vector functions $\boldsymbol{A}^i_{l,n}(\boldsymbol{\xi}|_{\xi^i=\xi^i_l})$ and $\boldsymbol{A}^j_{k,m}(\boldsymbol{\xi}|_{\xi^j=\xi^j_k})$ specifying the corresponding derivatives on the planes $\xi^i = \xi^i_l$ and $\xi^j = \xi^j_k$, respectively, must be compatible at the intersection of these planes, i.e.

$$\frac{\partial^m}{(\partial \xi^j)^m}\boldsymbol{A}^i_{l,n}(\boldsymbol{\xi}|_{\xi^i=\xi^i_l,\xi^j=\xi^j_k}) = \frac{\partial^n}{(\partial \xi^j)^n}\boldsymbol{A}^j_{k,m}(\boldsymbol{\xi}|_{\xi^i=\xi^i_l,\xi^j=\xi^j_k}),$$
$$n = 0, \dots, P^i_l, \; m = 0, \dots, P^j_k. \tag{5.16}$$

When the vector-valued functions $\boldsymbol{A}^i_{l,k}$ satisfying (5.16) are specified, the transformation $\boldsymbol{x}(\boldsymbol{\xi})$ is obtained by substituting the functions $\boldsymbol{A}^i_{l,0}$ and $\boldsymbol{A}^i_{l,n}$ for the values of $\boldsymbol{r}(\boldsymbol{\xi})$ and of its derivatives $\partial^n \boldsymbol{r}/\partial(\xi^i)^n(\boldsymbol{\xi}|_{\xi^i=\xi^i_l})$, respectively in the above formulas for transfinite interpolation. Hence the transformation based on the unidirectional interpolation given by (5.1) has the form

$$\boldsymbol{P}_i(\boldsymbol{\xi}) = \sum_{l=1}^{L^i}\sum_{n=0}^{P^i_l} \alpha^i_{l,n}(\xi^i)\boldsymbol{A}^i_{l,n}(\boldsymbol{\xi}|_{\xi^i=\xi^i_l}). \tag{5.17}$$

This mapping matches the values of $\boldsymbol{A}^i_{l,n}$ only at the coordinate planes $\xi^i = \xi^i_l$ crossing the chosen coordinate ξ^i.

The formula (5.5) for the tensor product $\otimes$ of the two mappings $\boldsymbol{P}_i(\boldsymbol{\xi})$ and $\boldsymbol{P}_j(\boldsymbol{\xi})$ obtained from (5.17) gives then the transformation

$$\boldsymbol{P}_i \otimes \boldsymbol{P}_j(\boldsymbol{\xi})$$
$$= \sum_{k=1}^{L^j}\sum_{m=0}^{P^j_k}\sum_{l=1}^{L^i}\sum_{n=0}^{P^i_l} \alpha^i_{l,n}(\xi^i)\alpha^j_{k,m}(\xi^j)\frac{\partial^n}{(\partial \xi^i)^n}\boldsymbol{A}^j_{k,m}(\boldsymbol{\xi}|_{\xi^i=\xi^i_l,\xi^j=\xi^j_k}), \tag{5.18}$$

which matches the values of $\boldsymbol{A}^i_{l,n}$ and $\boldsymbol{A}^j_{k,m}$ at the intersection of the planes $\xi^i = \xi^i_l$ and $\xi^j = \xi^j_k$. According to the consistency conditions (5.16), the operation of the tensor product is commutative, i.e.

$$\boldsymbol{P}_i \otimes \boldsymbol{P}_j(\boldsymbol{\xi}) = \boldsymbol{P}_j \otimes \boldsymbol{P}_i(\boldsymbol{\xi}),$$

which is indispensable for an appropriate definition of the coordinate transformation $\boldsymbol{x}(\boldsymbol{\xi})$.

5.3.2 General Algebraic Transformations

The general formula for the three-dimensional coordinate transformation $\boldsymbol{x}(\boldsymbol{\xi})$ that provides a matching with $\boldsymbol{A}^i_{l,n}$ in all directions and at all chosen coordinate planes $\xi^i = \xi^i_l$ is given by the replacement of the values of the function $\boldsymbol{r}(\boldsymbol{\xi})$ and of its derivatives in the recursive formula (5.13) by the functions $\boldsymbol{A}^i_{l,n}$. Thus we obtain

$$\begin{aligned} \boldsymbol{F}_1(\boldsymbol{\xi}) &= \sum_{l=1}^{L^1}\sum_{n=0}^{P^1_l} \alpha^1_{l,n}(\xi^1)\boldsymbol{A}^1_{l,n}(\xi^1_l,\xi^2,\xi^3), \\ \boldsymbol{F}_2(\boldsymbol{\xi}) &= \boldsymbol{F}_1(\boldsymbol{\xi}) + \sum_{l=1}^{L^2}\sum_{n=0}^{P^2_l} \alpha^2_{l,n}(\xi^2)\left(\boldsymbol{A}^2_{l,n} - \frac{\partial^n \boldsymbol{F}_1}{(\partial\xi^2)^n}\right)(\xi^1,\xi^2_l,\xi^3), \\ \boldsymbol{x}(\boldsymbol{\xi}) &= \boldsymbol{F}_2(\boldsymbol{\xi}) + \sum_{l=1}^{L^3}\sum_{n=0}^{P^3_l} \alpha^3_{l,n}(\xi^3)\left(\boldsymbol{A}^3_{l,n} - \frac{\partial^n \boldsymbol{F}_2}{(\partial\xi^3)^n}\right)(\xi^1,\xi^2,\xi^3_l). \end{aligned} \tag{5.19}$$

As the specified functions $\boldsymbol{A}^i_{l,n}$ are consistent on the intersections of the planes $\xi^i = \xi^i_l$ and $\xi^j = \xi^j_k$ and, therefore, the tensor product of the transformations $\boldsymbol{P}_i(\boldsymbol{\xi})$ and $\boldsymbol{P}_j(\boldsymbol{\xi})$ is commutative, the result (5.19) is independent of the specific ordering of the successive interpolation directions ξ^i.

The formula for the two-dimensional algebraic coordinate transformation is obtained in a corresponding way from (5.15):

$$\begin{aligned} \boldsymbol{F}_1(\boldsymbol{\xi}) &= \sum_{l=1}^{L^1}\sum_{n=0}^{P^1_l} \alpha^1_{l,n}(\xi^1)\boldsymbol{A}^1_{l,n}(\xi^1_l,\xi^2), \\ \boldsymbol{x}(\boldsymbol{\xi}) &= \boldsymbol{F}_1(\boldsymbol{\xi}) + \sum_{l=1}^{L^2}\sum_{n=0}^{P^2_l} \alpha^2_{l,n}\left(\boldsymbol{A}^2_{l,n} - \frac{\partial^n \boldsymbol{F}_1}{(\partial\xi^2)^n}\right)(\xi^1,\xi^2_l), \end{aligned} \tag{5.20}$$

where $\boldsymbol{A}^i_{l,n}$ are two-dimensional vector-valued functions representing $\boldsymbol{x}(\boldsymbol{\xi})$ for $n = 0$ and its derivatives for $P^i_l \geq n > 0$ at the sections

$$\xi^i = \xi^i_l, \quad i = 1,2, \ l = 1,\dots,L^i.$$

These functions must satisfy the relations (5.16) at the points $(\xi^1_l,\ \xi^2_m)$, $l = 1,\dots,L^1$, $m = 1,\dots,L^2$.

The vector-valued function $\boldsymbol{x}(\boldsymbol{\xi})$ defined by (5.19) maps the unit cube Ξ^3 onto the physical region X^3 bounded by the six coordinate surfaces specified by the parametrizations $\boldsymbol{A}^i_{1,0}(\boldsymbol{\xi}|_{\xi^i=0})$ and $\boldsymbol{A}^i_{L^i,0}(\boldsymbol{\xi}|_{\xi^i=1})$, $i = 1,2,3$, from the respective boundary intervals of Ξ^3. The introduction of the intermediate planes $\xi^i = \xi^i_l$, $0 < \xi^i_l < 1$, into the formulas of transfinite interpolation allows one to control the grid distribution and grid spacing in the vicinity of some selected interior surfaces of the domain X^3. A similar result is achieved by joining, at the selected boundary surfaces, a series of transformations $\boldsymbol{x}(\boldsymbol{\xi})$ constructed using the above described outer boundary interpolation equation (5.14):

$$
\begin{aligned}
\boldsymbol{F}_1(\boldsymbol{\xi}) &= \sum_{n=0}^{P_1^1} \alpha^1_{1,n}(\xi^1)\boldsymbol{A}^1_{1,n}(0,\xi^2,\xi^3) \\
&\quad + \sum_{n=0}^{P_2^1} \alpha^1_{2,n}(\xi^1)\boldsymbol{A}^1_{2,n}(1,\xi^2,\xi^3), \\
\boldsymbol{F}_2(\xi) &= \boldsymbol{F}_1(\xi) + \sum_{n=0}^{P_1^2} \alpha^2_{1,n}(\xi^2)\left(\boldsymbol{A}^2_{1,n} - \frac{\partial^n \boldsymbol{F}_1}{(\partial \xi^2)^n}(\xi^1,0,\xi^3)\right) \\
&\quad + \sum_{n=0}^{P_2^2} \alpha^2_{2,n}(\xi^2)\left(\boldsymbol{A}^2_{2,n} - \frac{\partial^n \boldsymbol{F}_1}{(\partial \xi^2)^n}\right)(\xi^1,1,\xi^3), \\
\boldsymbol{x}(\boldsymbol{\xi}) &= \boldsymbol{F}_2(\xi) + \sum_{n=0}^{P_1^3} \alpha^3_{1,n}(\xi^3)\left(\boldsymbol{A}^3_{1,n} - \frac{\partial^n \boldsymbol{F}_2}{(\partial \xi^3)^n}\right)(\xi^1,\xi^2,0) \\
&\quad + \sum_{n=0}^{P_2^3} \alpha^3_{2,n}(\xi^3)\left(\boldsymbol{A}^3_{2,n} - \frac{\partial^n \boldsymbol{F}_2}{(\partial \xi^3)^n}\right)(\xi^1,\xi^2,1).
\end{aligned}
\tag{5.21}
$$

This boundary interpolation transformation $\boldsymbol{x}(\boldsymbol{\xi})$ is widely applied to generate grids in regions around bodies. These domains cannot be successfully gridded by one global mapping $\boldsymbol{x}(\boldsymbol{\xi})$ from the unit cube Ξ^3 because of the inevitable singularities pertinent to such global maps. An approach based on the matching of a series of boundary-interpolated transformations is thus preferable. It only requires the consistent specification of the parametrizations and coordinate directions at the corresponding boundary surfaces.

Equations (5.18)–(5.21) use the same set of blending functions $\alpha^i_{l,n}(\xi^i)$ to define each component $x^i(\xi)$ of the transformation $\boldsymbol{x}(\boldsymbol{\xi})$. These formulas can be generalized by introducing an individual set of blending functions $\alpha^i_{l,n}(\xi^i)$ for the definition of each component $x^i(\boldsymbol{\xi})$ of the map $\boldsymbol{x}(\boldsymbol{\xi})$ being built. Such a generalization gives broader opportunities to define appropriate algebraic coordinate transformations $\boldsymbol{x}(\boldsymbol{\xi})$ and, therefore, to generate grids more successfully.

5.4 Lagrange and Hermite Interpolations

The recursive formula (5.19) represents a general form of transfinite interpolation which includes the prescribed values of the constructed coordinate transformation $\boldsymbol{x}(\boldsymbol{\xi})$ and of its derivatives up to order P^i_l at the sections $\xi^i = \xi^i_l$ of the cube Ξ^3. However, most grid generation codes require, as a rule, the specification of only the values of the function $\boldsymbol{x}(\boldsymbol{\xi})$ being sought and sometimes, in addition, the values of its first derivatives at the selected sections. Such sorts of algebraic coordinate transformation are described in this section.

5.4.1 Coordinate Transformations Based on Lagrange Interpolation

A Lagrange interpolation matches only the values of the function $\boldsymbol{r}(\boldsymbol{\xi})$ at some prescribed sections $\xi^i = \xi^i_l$, $l = 1, \ldots, L^i$, of the cube Ξ^3. So, in accordance with (5.1), the unidirectional Lagrange interpolation has the following form:

$$\boldsymbol{P}_i[\boldsymbol{r}](\boldsymbol{\xi}) = \sum_{l=1}^{L^i} \alpha^i_l(\xi^i)\boldsymbol{r}(\boldsymbol{\xi}|_{\xi^i=\xi^i_l}).$$

The blending function $\alpha^i_l(\xi^i)$ in this equation corresponds to $\alpha^i_{l,0}(\xi^i)$ in the formula (5.1). Taking into account (5.2), the blending functions $\alpha^i_l(\xi^i)$, $l = 1, \ldots, L^i$, depending on one independent variable ξ^i, must be subject to the following restrictions:

$$\alpha^i_l(\xi^i_k) = \delta^l_k, \quad l, k = 1, \ldots, L^i. \tag{5.22}$$

These restrictions imply that the blending function α^i_l for a fixed l equals 1 at the point $\xi^i = \xi^i_l$ and equals zero at all other points ξ^i_m, $m \neq l$. The formula for the construction of a three-dimensional coordinate mapping $\boldsymbol{x}(\boldsymbol{\xi})$ based on the Lagrangian interpolation is obtained from (5.19) as

$$\begin{aligned}
\boldsymbol{F}_1(\boldsymbol{\xi}) &= \sum_{l=1}^{L^1} \alpha^1_l(\xi^1)\boldsymbol{A}^1_l(\boldsymbol{\xi}|_{\xi^1=\xi^1_l}),\\
\boldsymbol{F}_2(\boldsymbol{\xi}) &= \boldsymbol{F}_1(\boldsymbol{\xi}) + \sum_{l=1}^{L^2} \alpha^2_l(\xi^2)(\boldsymbol{A}^2_l - \boldsymbol{F}_1)(\boldsymbol{\xi}|_{\xi^2=\xi^2_l}),\\
\boldsymbol{x}(\boldsymbol{\xi}) &= \boldsymbol{F}_2(\boldsymbol{\xi}) + \sum_{l=1}^{L^3} \alpha^3_l(\xi^3)(\boldsymbol{A}^3_l - \boldsymbol{F}_2)(\boldsymbol{\xi}|_{\xi^3=\xi^3_l}),
\end{aligned} \tag{5.23}$$

where the blending functions $\alpha^i_l(\xi^i)$ satisfy (5.22), and the functions $\boldsymbol{A}^i_l(\xi|_{\xi^i=\xi^i_l})$ corresponding to $\boldsymbol{A}^i_{l,0}$ in (5.22) specify the values of the mapping $\boldsymbol{x}(\boldsymbol{\xi})$ being sought. In accordance with (5.16), the specified functions $\boldsymbol{A}^i_l$ must coincide at the intersection of their respective coordinate planes $\xi^i = \xi^i_l$, i.e.

$$\boldsymbol{A}^i_l(\boldsymbol{\xi}|_{\xi^i=\xi^i_l,\xi^j=\xi^j_k}) = \boldsymbol{A}^j_k(\boldsymbol{\xi}|_{\xi^i=\xi^i_l,\xi^j=\xi^j_k}).$$

Now we consider some examples of the blending functions used in Lagrange interpolations.

5.4.1.1 Lagrange Polynomials

The best-known blending functions $\alpha^i_l(\xi^i)$ satisfying (5.22) are defined as Lagrange polynomials applied to the points $\xi^i_1, \dots, \xi^i_{L^i}$:

$$\alpha^i_l(\xi^i) = \prod_{j=1}^{L^i} \frac{\xi^i - \xi^i_j}{\xi^i_l - \xi^i_j}, \quad j \neq l. \tag{5.24}$$

For example, when $L^i = 2$, then from (5.24)

$$\alpha^i_1(\xi^i) = \frac{\xi^i - \xi^i_2}{\xi^i_1 - \xi^i_2}, \qquad \alpha^i_2(\xi^i) = \frac{\xi^i - \xi^i_1}{\xi^i_2 - \xi^i_1} = 1 - \alpha^i_1(\xi^i). \tag{5.25}$$

Therefore, for the boundary interpolation, we obtain

$$\alpha^i_1(\xi^i) = 1 - \xi^i, \qquad \alpha^i_2(\xi^i) = \xi^i. \tag{5.26}$$

5.4.1.2 Spline Functions

The Lagrange polynomials become polynomials of a high-order when a large number of intermediate sections $\xi^i = \xi^i_l$ is applied to control the grid distribution in the interior of the domain X^3. These polynomials of high order may cause oscillations. One way to overcome this drawback is to use splines as blending functions $\alpha^i_l(\xi^i)$. The splines are defined as polynomials of low-order between each of the specified points $\xi^i = \xi^i_{L^i}$, with continuity of some derivatives at the interior points.

Piecewise-continuous splines satisfying (5.22) can be derived by means of linear polynomials. The simplest pattern of such blending functions in the form of splines consists of piecewise linear functions:

$$\alpha_l(\xi^i) = \begin{cases} 0, & \xi^i \le \xi^i_{l-1}, \\ \frac{\xi^i - \xi^i_{l-1}}{\xi^i_l - \xi^i_{l-1}}, & \xi^i_{l-1} \le \xi^i \le \xi^i_l, \\ \frac{\xi^i_{l+1} - \xi^i}{\xi^i_{l+1} - \xi^i_l}, & \xi^i_l \le \xi^i \le \xi^i_{l+1}, \\ 0, & \xi^i \ge \xi^i_{l+1}. \end{cases}$$

However, the use of these blending functions results in a nonsmooth point distribution since they themselves are not smooth.

Continuity of the first derivative of a spline blending function can be achieved with polynomials of the third-order, regardless of the number of interior sections.

5.4.1.3 Construction Based on General Functions

The application of polynomials in the Lagrange interpolation gives only a poor opportunity to control the grid spacing near the selected boundary and interior surfaces. In this subsection we describe a general approach to constructing the blending functions $\alpha_l^i(\xi^i)$ by the use of a wide range of basic functions, which provides a real opportunity to control the grid point distribution.

The formulation of the blending functions on the interval $0 \le \xi^i \le 1$, with L^i specified points,

$$0 = \xi_1^i < \cdots < \xi_{L^i}^i = 1,$$

requires only the specification of some univariate smooth positive function

$$\phi(x) : [0, \infty) \to [0, \infty),$$

satisfying the restrictions $\phi(0) = 0, \phi(1) = 1$. This function can be used as a basic element to derive the blending functions satisfying (5.22) through the following standard procedure.

First we define two series of functions

$$\phi_l^f(\xi^i) \quad \text{and} \quad \phi_l^b(\xi^i), \quad l = 1, \dots, L^i.$$

The functions $\phi_l^f(\xi^i)$ are defined for $l = 1$ by

$$\phi_1^f(\xi^i) = \phi(1 - \xi^i), \quad 0 \le \xi \le 1,$$

and for $1 < l \le L^i$ by

$$\phi_l^f(\xi^i) = \begin{cases} 0, & 0 \le \xi^i \le \xi_{l-1}^i, \\ \phi(\frac{\xi^i - \xi_{l-1}^i}{\xi_l^i - \xi_{l-1}^i}), & \xi_{l-1}^i \le \xi^i \le 1. \end{cases}$$

The functions $\phi_l^b(\xi^i)$ are determined similarly:

$$\phi_{L^i}^b(\xi^i) = \phi(\xi^i)$$

and for $1 \le l < L^i$,

$$\phi_l^b(\xi^i) = \begin{cases} 0, & 1 \ge \xi^i \ge \xi_{l+1}^i, \\ \phi(\frac{\xi_{l+1}^i - \xi^i}{\xi_{l+1}^i - \xi_l^i}), & 0 \le \xi^i \le \xi_{l+1}^i. \end{cases}$$

Using the functions $\phi_l^f(\xi^i)$ and $\phi_l^b(\xi^i)$, the blending coefficients $\alpha_l^i(\xi^i)$ satisfying (5.22) are defined by

$$\alpha_l^i(\xi^i) = \phi_l^f(\xi^i)\phi_l^b(\xi^i), \quad l = 1, \dots, L^i. \tag{5.27}$$

Each of these blending functions vanishes outside some interval and thus it affects the interpolation function only locally.

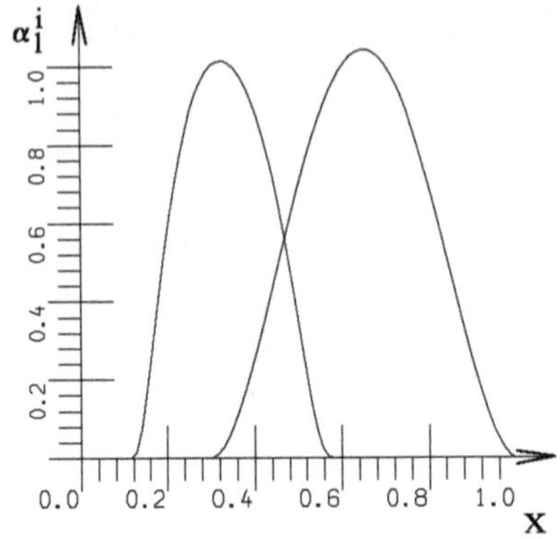

Fig. 5.2. Smooth blending functions

Note that this procedure for constructing blending functions for the Lagrange interpolations will yield splines if the original function ϕ is a polynomial. This construction may be extended by using various original functions for the terms ϕ_l^f and ϕ_l^b in (5.27).

The simplest example of the basic function is $\phi(x) = x$. However, this function generates nonsmooth blending coefficients $\alpha_l^i(\xi^i)$ at the points ξ_{l-1}^i and ξ_{l+1}^i, since $\alpha_l^i(\xi^i) \equiv 0$ outside the interval $(\xi_{l-1}^i, \xi_{l+1}^i)$. If the derivative of $\phi(x)$ at the point $x = 0$ is zero then the blending functions derived by the procedure described are smooth. One example of such a function is $\phi(x) = x^2$. It can readily be shown that in this case the blending functions α_l^i are of the class $C^1[0, 1]$.

Continuity of the higher-order derivatives of the blending functions (5.27) is obtained when the basic function $\phi(x)$ satisfies the condition $\phi^{(k)}(0) = 0,\ k > 1$, in particular, if $\phi(x) = x^{k+1}$. The function $\phi(x) = \varphi(x)$, where

$$\varphi(x) = \begin{cases} 0, & x = 0, \\ a^{1-1/x}, & x > 0, \end{cases}$$

with $a > 1$, generates an infinitely differentiable blending function $\alpha_l^i(\xi^i)$ on the interval $[0, 1]$. Figure 5.2 demonstrates the blending functions constructed for $\phi(x) = \varphi(x)$ (left) and $\phi(x) = x^2$ (right).

5.4.1.4 Relations Between Blending Functions

Now, we point out some relations between blending functions which can be useful for their construction. If the functions $\alpha_l^i(\xi^i)$ are blending functions for Lagrangian interpolation, namely, they are subject to the restrictions (5.22), then the functions $\beta_l^i(\xi^i)$ defined below satisfy the condition (5.22) as well:

$$
\begin{aligned}
&(1) \quad \beta_l^i(\xi^i) = \alpha_l^i(\xi^i) f(\xi) \quad \text{if } f(\xi_l^i) = 1, \\
&(2) \quad \beta_l^i(\xi^i) = \alpha_l^i[f(\xi^i)] \quad \text{if } f(\xi_l^i) = \xi_l^i, \\
&(3) \quad \beta_l^i(\xi^i) = f[\alpha_l^i(\xi^i)] \quad \text{if } f(0) = 0,\ f(1) = 1, \\
&(4) \quad \beta_l^i(\xi^i) = \alpha_l^i(\xi^i) + f(\xi^i) \quad \text{if } f(\xi_l^i) = 0, \\
&(5) \quad \beta_l^i(\xi^i) = 0.5[\alpha_l^i(\xi^i) + \gamma_l^i(\xi^i)] \quad \text{if } \gamma_l^i(\xi) \text{ satisfies (5.22).}
\end{aligned}
\tag{5.28}
$$

5.4.2 Transformations Based on Hermite Interpolation

Hermite interpolation matches the values of both the function $\boldsymbol{r}(\boldsymbol{\xi})$, and its first derivatives $\partial \boldsymbol{r}/\partial \xi^i(\boldsymbol{\xi}|_{\xi^i=\xi_l^i})$ at each section $\xi^i = \xi_l^i$, $l = 1, \ldots, L^i$, therefore the unidirectional interpolation (5.1) takes the following form:

$$
\boldsymbol{P}_i[\boldsymbol{r}](\boldsymbol{\xi}) = \sum_{l=1}^{L^i} \left(\alpha_{l,0}^i(\xi^i) \boldsymbol{r}(\boldsymbol{\xi}|_{\xi^i=\xi_l^i}) + \alpha_{l,1}^i(\xi^i) \frac{\partial \boldsymbol{r}}{\partial \xi^i} (\boldsymbol{\xi}|_{\xi^i=\xi_l^i}) \right). \tag{5.29}
$$

The formula (5.19) in the case of a Hermite coordinate mapping $\boldsymbol{x}(\boldsymbol{\xi})$ which matches the specified values of $\boldsymbol{x}(\boldsymbol{\xi})$, denoted by $\boldsymbol{A}_{l,0}^i$, and of its first derivatives, denoted by $\boldsymbol{A}_{l,1}^i$, at all sections $\xi^i = \xi_l^i$, $l = 1, \ldots, L^i$, and in all directions ξ^i, $i = 1, 2, 3$, is thus reduced to

$$
\begin{aligned}
\boldsymbol{F}_1(\boldsymbol{\xi}) &= \sum_{l=1}^{L^1} \left(\alpha_{l,0}^1(\xi^1) \boldsymbol{A}_{l,0}^1(\xi_l^1, \xi^2, \xi^3) + \alpha_{l,1}^1(\xi^1) \boldsymbol{A}_{l,1}^1(\xi_l^1, \xi^2, \xi^3) \right), \\
\boldsymbol{F}_2(\boldsymbol{\xi}) &= \boldsymbol{F}_1(\boldsymbol{\xi}) + \sum_{l=1}^{L^2} \Bigg(\alpha_{l,0}^2(\xi^2) (\boldsymbol{A}_{l,0}^2 - \boldsymbol{F}_1)(\xi^1, \xi_l^2, \xi^3) \\
&\quad + \alpha_{l,1}^2(\xi^2) \left(\boldsymbol{A}_{l,1}^2 - \frac{\partial \boldsymbol{F}_1}{\partial \xi^2} \right) (\xi^1, \xi_l^2, \xi^3) \Bigg), \\
\boldsymbol{x}(\boldsymbol{\xi}) &= \boldsymbol{F}_2(\boldsymbol{\xi}) + \sum_{l=1}^{L^3} \Bigg(\alpha_{l,0}^3(\xi^3) (\boldsymbol{A}_{l,0}^3 - \boldsymbol{F}_2)(\xi^1, \xi^2, \xi_l^3) \\
&\quad + \alpha_{l,1}^3(\xi^3) \left(\boldsymbol{A}_{l,1}^3 - \frac{\partial \boldsymbol{F}_2}{\partial \xi^3} \right) (\xi^1, \xi^2, \xi_l^3) \Bigg),
\end{aligned}
\tag{5.30}
$$

where, in accordance with (5.2), the blending functions $\alpha_{l,0}^i$, $\alpha_{l,1}^i$ satisfy the conditions

$$
\begin{aligned}
&\alpha_{l,0}^i(\xi_k^i) = \delta_k^l, \qquad \alpha_{l,1}^i(\xi_k^i) = 0, \\
&\frac{\mathrm{d}}{\mathrm{d}\xi^i} \alpha_{l,1}^i(\xi_k^i) = \delta_k^l, \qquad \frac{\mathrm{d}}{\mathrm{d}\xi^i} \alpha_{l,0}^i(\xi_k^i) = 0, \\
&l, k = 1, \ldots, L^i, \ i = 1, 2, 3,
\end{aligned}
\tag{5.31}
$$

and the vector-valued functions $\boldsymbol{A}_{l,n}^i(\xi|_{\xi^i=\xi_l^i})$ satisfy the consistency conditions (5.16):

$$\begin{aligned} \boldsymbol{A}^i_{l,0}(\boldsymbol{\xi}|_{\xi^i=\xi^i_l,\xi^j=\xi^j_k}) &= \boldsymbol{A}^j_{k,0}(\boldsymbol{\xi}|_{\xi^i=\xi^i_l,\xi^j=\xi^j_k}), \\ \frac{\partial}{\partial \xi^j}\boldsymbol{A}^i_{l,0}(\boldsymbol{\xi}|_{\xi^i=\xi^i_l,\xi^j=\xi^j_k}) &= \boldsymbol{A}^j_{k,1}(\boldsymbol{\xi}|_{\xi^i=\xi^i_l,\xi^j=\xi^j_k}). \end{aligned} \tag{5.32}$$

5.4.2.1 Construction of Blending Functions

The blending functions $\alpha^i_{l,m}(\xi^i)$, $m = 0, 1$, for Hermite interpolations can be obtained from the smooth blending functions defined for Lagrange interpolations. Namely, let $\alpha^i_l(\xi^i)$, $l = 1, \dots, L^i$, be some smooth scalar functions meeting the conditions (5.22). The functions $\alpha^i_{l,m}$, $m = 0, 1$, determined by the relations

$$\begin{aligned} \alpha^i_{l,0} &= \left(1 - 2(\xi^i - \xi^i_l)\frac{d\alpha^i_l}{d\xi^i}(\xi_i)\right)[\alpha^i_l(\xi^i)]^2, \\ \alpha^i_{l,1} &= (\xi^i - \xi^i_l)[\alpha^i_l(\xi^i)]^2, \end{aligned} \tag{5.33}$$

then satisfy (5.31) and, therefore, are the blending functions for the Hermite interpolations. For example, if $L^i = 2$ and the Lagrangian blending functions are defined through (5.25), then from (5.33),

$$\begin{aligned} \alpha^i_{1,0}(\xi^i) &= \left(1 - 2\frac{\xi^i - \xi^i_1}{\xi^i_1 - \xi^i_2}\right)\left(\frac{\xi^i - \xi^i_2}{\xi^i_1 - \xi^i_2}\right)^2, \\ \alpha^i_{2,0}(\xi^i) &= \left(1 - 2\frac{\xi^i - \xi^i_2}{\xi^i_2 - \xi^i_1}\right)\left(\frac{\xi^i - \xi^i_1}{\xi^i_2 - \xi^i_1}\right)^2, \\ \alpha^i_{1,1}(\xi^i) &= (\xi^i - \xi^i_1)\left(\frac{\xi^i - \xi^i_2}{\xi^i_1 - \xi^i_2}\right)^2, \\ \alpha^i_{2,1}(\xi^i) &= (\xi^i - \xi^i_2)\left(\frac{\xi^i - \xi^i_1}{\xi^i_2 - \xi^i_1}\right)^2. \end{aligned} \tag{5.34}$$

So if $\xi^i_1 = 0$, $\xi^i_2 = 1$, then from these relations,

$$\begin{aligned} \alpha^i_{1,0}(\xi^i) &= (1 + 2\xi^i)(\xi^i - 1)^2, \\ \alpha^i_{2,0}(\xi^i) &= (3 - 2\xi^i)(\xi^i)^2 = 1 - \alpha^i_{1,0}(\xi^i), \\ \alpha^i_{1,1}(\xi^i) &= \xi^i(1 - \xi^i)^2, \\ \alpha^i_{2,1}(\xi^i) &= (\xi^i - 1)(\xi^i)^2. \end{aligned} \tag{5.35}$$

If the blending functions for Lagrange interpolation satisfy the condition

$$\frac{d\alpha^i_l}{d\xi^i}(\xi^i) \equiv 0, \quad \text{if } \xi^i \geq \xi^i_{l+1} \text{ and } \xi^i \leq \xi^i_{l-1}, \tag{5.36}$$

then the blending functions $\alpha^i_{l,n}(\xi^i)$ for the Hermite interpolation can be derived from $\alpha^i_l(\xi^i)$ by the relations

$$\alpha^i_{l,0}(\xi^i) = \left(1 + (\xi^i - \xi^i_l)\frac{\mathrm{d}\alpha^i_l}{\mathrm{d}\xi^i}(\xi^i_l)\right)\alpha^i_l(\xi^i),$$
$$\alpha^i_{l,1}(\xi^i) = (\xi^i - \xi^i_l)\alpha^i_l(\xi^i). \tag{5.37}$$

It is readily shown that the blending functions $\alpha^i_{l,n}(\xi^i)$, $n = 0, 1$, satisfy the restriction (5.31). Note that the approach described above for the general construction of the blending functions for Lagrange interpolation yields the smooth blending functions $\alpha^i_l(\xi^i)$, $l = 1, \ldots, L^i$, in the form (5.27), which, in addition to (5.22), are also subject to (5.36).

5.4.2.2 Deficient Form of Hermite Interpolation

Often it is not reasonable to specify the values of the first derivative with respect to ξ^i of the sought coordinate transformation $\boldsymbol{x}(\boldsymbol{\xi})$ at all sections $\xi^i = \xi^i_l$, $l = 1, \ldots, L^i$, but only at some selected ones. By omitting the corresponding terms

$$\alpha^1_{l,1}(\xi^1)\boldsymbol{A}^1_{l,1}(\boldsymbol{\xi}|_{\xi^1=\xi^1_l})$$

and/or

$$\alpha^i_{l,1}(\xi^i)\left(\boldsymbol{A}^i_{l,1} - \frac{\partial \boldsymbol{F}_{i-1}}{\partial \xi^i}\right)(\boldsymbol{\xi}|_{\xi^i=\xi^i_l}), \quad i = 2, 3,$$

in (5.30), a deficient form of Hermite interpolation is obtained which matches the values of the first derivatives at the selected sections only. For example, the outer boundary interpolation which contains the outer boundary specifications on all boundaries but the outward derivative with respect to ξ^1 on the boundary $\xi^1 = 0$ only has, in accordance with (5.30), the form

$$\begin{aligned}
\boldsymbol{F}_1(\boldsymbol{\xi}) &= \alpha^1_{1,0}(\xi^1)\boldsymbol{A}^1_{1,0}(0, \xi^2, \xi^3) + \alpha^1_{2,0}(\xi^1)\boldsymbol{A}^1_{2,0}(1, \xi^2, \xi^3) \\
&\quad + \alpha^1_{1,1}(\xi^1)\boldsymbol{A}^1_{1,1}(0, \xi^2, \xi^3), \\
\boldsymbol{F}_2(\boldsymbol{\xi}) &= \boldsymbol{F}_1(\boldsymbol{\xi}) + \alpha^2_{1,0}(\xi^2)(\boldsymbol{A}^2_{1,0} - \boldsymbol{F}_1)(\xi^1, 0, \xi^3) \\
&\quad + \alpha^2_{2,0}(\xi^2)(\boldsymbol{A}^2_{2,0} - \boldsymbol{F}_1)(\xi^1, 1, \xi^3), \\
\boldsymbol{x}(\boldsymbol{\xi}) &= \boldsymbol{F}_2(\boldsymbol{\xi}) + \alpha^3_{1,0}(\xi^3)(\boldsymbol{A}^3_{1,0} - \boldsymbol{F}_2)(\xi^1, \xi^2, 0) \\
&\quad + \alpha^3_{2,0}(\xi^3)(\boldsymbol{A}^3_{2,0} - \boldsymbol{F}_2)(\xi^1, \xi^2, 1).
\end{aligned} \tag{5.38}$$

5.4.2.3 Specification of Normal Directions

In the outer boundary interpolation technique the outward derivatives $\boldsymbol{A}^i_{1,1}(\xi|_{\xi^i=0})$, $\boldsymbol{A}^i_{2,1}(\xi|_{\xi^i=1})$ along the lines emerging from the boundary surfaces are usually required to be performed as normals to the corresponding boundary surfaces in order to generate orthogonal grids near the boundaries. The boundary surfaces are parametrized by the specified boundary transformations $\boldsymbol{A}^i_{1,0}(\xi|_{\xi^i=0})$ and $\boldsymbol{A}^i_{2,0}(\xi|_{\xi^i=1})$, respectively. Therefore, these normals can be computed from the cross product of the vectors tangential to the boundary surfaces. For example, the ξ^1 coordinate direction $\boldsymbol{A}^1_{l,1}(\xi^1_l, \xi^2, \xi^3)$ can be specified as

$$\boldsymbol{A}^1_{1,1}(0, \xi^2, \xi^3) = g(\xi^2, \xi^3)\left(\frac{\partial}{\partial \xi^2}\boldsymbol{A}^1_{1,0}(0, \xi^2, \xi^3) \times \frac{\partial}{\partial \xi^3}\boldsymbol{A}^1_{1,0}(0, \xi^2, \xi^3)\right),$$

where $g(\xi^2, \xi^3)$ is a scalar function that can be used to control the spacing of the grid lines emerging from the boundary surface represented by the parametrization $\boldsymbol{A}^1_{1,0}(0, \xi^2, \xi^3)$. Such a specification of the first derivatives can be chosen on the interior sections as well.

5.5 Control Techniques

Commonly, all algebraic schemes are computationally efficient but require a significant amount of user interaction and control techniques to define workable meshes. This section delineates some control approaches applied to algebraic grid generation.

The spacing between the grid points and the skewness of the grid cells in the physical domain is controlled in the algebraic method, primarily by the blending functions $\alpha^i_{l,n}(\xi^i)$, by the representations of the boundary and intermediate surfaces $\boldsymbol{A}^i_{l,0}(\xi|_{\xi^i=\xi^i_l})$, and by the values of the first derivatives $\boldsymbol{A}^i_{l,1}(\xi|_{\xi^i=\xi^i_l})$ in the interpolation equations.

As was stated in Chap. 4, an effective approach which significantly simplifies the control of grid generation relies on the introduction of an intermediate control domain between the computational and the physical regions. The control domain is a unit cube Q^3 with the Cartesian coordinates q^i, $i = 1, 2, 3$. In this approach the coordinate transformation $\boldsymbol{x}(\boldsymbol{\xi})$ from the unit cube Ξ^3 onto the physical region X^3 is defined as a composition of two transformations: $\boldsymbol{q}(\boldsymbol{\xi})$ from Ξ^3 onto Q^3 and $\boldsymbol{g}(\boldsymbol{q})$, $\boldsymbol{q} = (q^1, q^2, q^3)$, from Q^3 onto X^3, that is,

$$\boldsymbol{x}(\boldsymbol{\xi}) = \boldsymbol{g}[\boldsymbol{q}](\boldsymbol{\xi}) : \Xi^3 \to X^3.$$

The functions $\boldsymbol{g}(\boldsymbol{q})$ and $\boldsymbol{q}(\boldsymbol{\xi})$ can be constructed through the formulas of transfinite interpolation or by other techniques. As both the computational domain Ξ^3 and the intermediate domain Q^3 are the standard unit cubes, the formulas of transfinite interpolation for the generation of the intermediate transformations $\boldsymbol{q}(\boldsymbol{\xi})$ are somewhat simpler then the original expressions. In these formulas it can be assumed, without any loss of generality, that the boundary planes $\xi^i = 0$ and $\xi^i = 1$ for each $i = 1, 2, 3$ are transformed by the function $\boldsymbol{q}(\boldsymbol{\xi})$ onto the boundary planes $q^i = 0$ and $q^i = 1$, respectively, so that

$$\boldsymbol{q}(\boldsymbol{\xi}|_{\xi^1=0}) = [0, q^2(0, \xi^2, \xi^3), q^3(0, \xi^2, \xi^3)],$$
$$\boldsymbol{q}(\boldsymbol{\xi}|_{\xi^1=1}) = [1, q^2(1, \xi^2, \xi^3), q^3(1, \xi^2, \xi^3)].$$

Therefore the first component $q^1(\xi)$ of the Lagrangian boundary interpolation for the intermediate mapping $\boldsymbol{q}(\boldsymbol{\xi})$ has the form

$$\begin{aligned}
F_1(\boldsymbol{\xi}) &= \alpha_2^1(\xi^1),\\
F_2(\boldsymbol{\xi}) &= F_1(\boldsymbol{\xi}) + \alpha_1^2(\xi^2)\big(u^1(\xi^1,0,\xi^3) - \alpha_2^1(\xi^1)\big)\\
&\quad + \alpha_2^2(\xi^2)\big(u^1(\xi^1,1,\xi^3) - \alpha_2^1(\xi^1)\big),\\
q^1(\boldsymbol{\xi}) &= F_2(\boldsymbol{\xi}) + \alpha_1^3(\xi^3)\big(u^1(\xi^1,\xi^2,0) - F_2(\xi^1,\xi^2,0)\big)\\
&\quad + \alpha_2^3(\xi^3)\big(u^1(\xi^1,\xi^2,1) - F_2(\xi^1,\xi^2,1)\big).
\end{aligned} \tag{5.39}$$

Analogous equations can be defined for the other components of the intermediate transformation $\boldsymbol{q}(\boldsymbol{\xi})$.

The functions based on the reference univariate transformations $x_{i,\mathrm{c}}(\varphi,\epsilon)$ and $x_{i,\mathrm{s}}(\varphi,\epsilon)$ introduced in Chap. 4 can be used very successfully as blending functions to construct intermediate transformations by Lagrange and Hermite interpolations in the two-boundary technique. In the case of Lagrange interpolation the blending function $\alpha_{1,0}^i(\xi^i)$ satisfies the conditions $\alpha_{1,0}^i(0) = 1$, $\alpha_{1,0}^i(1) = 0$. Therefore any monotonically decreasing function derived by applying the procedures described in Sect. 4.4 to the reference univariate functions can be used as the blending function $\alpha_{1,0}^i(\xi^i)$. Analogously, the blending function $\alpha_{2,0}^i(\xi^i)$ can be represented by any monotonically increasing mapping based on one of the standard local contraction functions $x_i(\varphi,\epsilon)$. The blending functions $\alpha_{i,1}^i(\xi^i)$ for Hermite interpolations can also use these standard transformations through applying the operation described by (5.33) to the blending functions $\alpha_{1,0}^i(\xi^i)$. By choosing the proper functions, one has an opportunity to construct intermediate transformations that provide adequate grid clustering in the zones where it is necessary.

5.6 Transfinite Interpolation from Triangles and Tetrahedrons

The formulas of transfinite interpolation define a coordinate transformation from the unit cube Ξ^3 (the square Ξ^2 in two dimensions and line Ξ^1 in one dimension) onto a physical domain X^3 (or X^2 or X^1). The application of this interpolation may lead to singularities of the type pertaining to polar transformations when any boundary segment of the physical domain, corresponding to a boundary segment of the computational domain, is contracted into a point. An example is when the boundary of a physical two-dimensional domain X^2 is composed of three smooth segments as shown in Fig. 5.3. One way to treat such regions is to use coordinate transformations from triangular computational domains in two dimensions and tetrahedral domains in three dimensions. It can be seen that the transfinite interpolation approach can be modified to generate triangular or tetrahedral grids by mapping a standard triangular or tetrahedral domain, respectively. The formulation of a transfinite

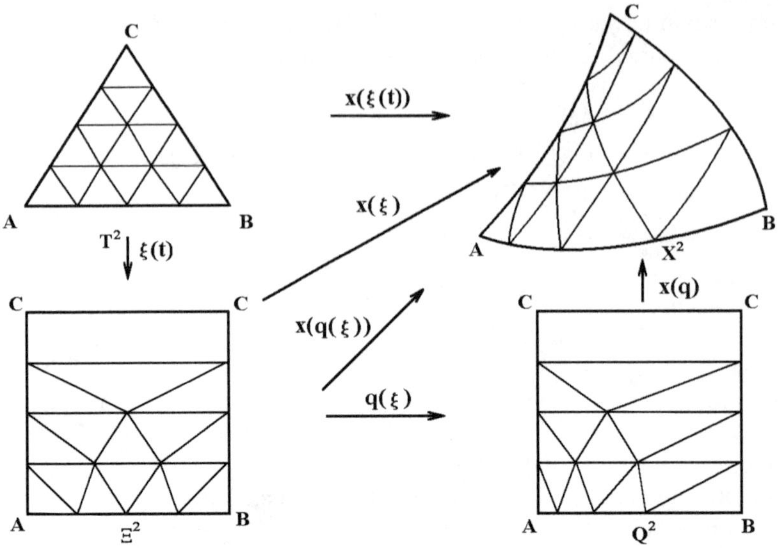

Fig. 5.3. Scheme for gridding triangular curvilinear domains with triangles

interpolation to obtain these transformations from the standard unit tetrahedron (triangle in two dimensions) is based on the composition of an operation of scaling (stretching) the coordinates to deform the tetrahedron into the unit cube Ξ^3 and an algebraic transformation constructed by the equations given above.

This procedure is readily clarified in two dimensions by the scheme depicted in Fig. 5.3. Suppose that the boundary segments AB, BC, and CD of the unit triangle T^2 are mapped onto the corresponding boundary segments AB, BC, and CD of the domain X^2. Then, in this procedure, the standard triangle T^2 with a uniform triangular grid is expanded to a square by a deformation $\boldsymbol{\xi}(\boldsymbol{t})$ uniformly stretching each horizontal line of the triangle to make it a rectangle, and afterwards the rectangle is uniformly stretched in the vertical direction to make it the unit square Ξ^2 as shown in Fig. 5.3. This operation is the inverse of the contraction $\boldsymbol{t}(\boldsymbol{\xi})$ of the square along the horizontal and vertical lines to transform it to the triangle. As a result we obtain a square Ξ^2 with triangular cells on all horizontal levels except the top one. The number of these cells in each horizontal band reduces from the lower levels to the upper ones. The top level consists of one rectangular cell. With this deformation of T^2 the transformation between the boundaries of T^2 and X^2 generates the transformation

$$\boldsymbol{x}(\boldsymbol{\xi}) : \partial\Xi^2 \to \partial X^2,$$

which is the composition of $\boldsymbol{t}(\boldsymbol{\xi})$ and the assumed mapping of the boundary of T^2 onto the boundary of X^2. This boundary transformation maps the top segment of Ξ^2 onto the point C in X^2. Now, applying the formulas of transfinite interpolation to a square Ξ^2 with such grid cells, and the specified boundary transformation one generates the algebraic transformation

$$\boldsymbol{x}(\boldsymbol{\xi}) : \Xi^2 \to X^2$$

and consequently

$$\boldsymbol{x}[\boldsymbol{\xi}(\boldsymbol{t})] : T^2 \to X^2$$

from the triangle to the physical region X^2 with the prescribed values of the transformation at the boundary segments of the triangle. Note that the composition $\boldsymbol{x}[\boldsymbol{\xi}(\boldsymbol{t})]$ is continuous as the upper segment of Ξ^2 is transformed by $\boldsymbol{x}(\boldsymbol{\xi})$ onto one point C in X^2.

In fact, such a triangular grid in the physical domain can be generated directly by mapping the nonuniform grid constructed in the unit square Ξ^2 as described above onto X^2 with a standard algebraic coordinate transformation defined by transfinite interpolation.

The generation of grids by this approach is very well justified for regions which shaped like curvilinear triangles, i.e. their boundaries are composed of three smooth curves intersecting at angles θ less than π. By dividing an arbitrary domain into triangular curvilinear domains one can generate a composite triangular grid in the entire domain by the procedure described above.

An analogous procedure using transfinite interpolation, is readily formulated for generating tetrahedral grids in regions with shapes similar to that of a tetrahedron.

The approach for generating triangular or tetrahedral meshes described above can be extended to include grid adaptation by adding to the scheme an intermediate domain and intermediate transformation $\boldsymbol{q}(\boldsymbol{\xi})$ as illustrated in Fig. 5.3 and special blending functions, as in the case of generating hexahedral (or quadrilateral) grids. Here, an adaptive triangular grid is generated through the composition of the transformations $\boldsymbol{q}(\boldsymbol{\xi})$ and $\boldsymbol{x}(\boldsymbol{q})$, where $\boldsymbol{q}(\boldsymbol{\xi})$ is an intermediate mapping providing grid adaptation and $\boldsymbol{x}(\boldsymbol{q})$ is an algebraic transformation. One example of a two-dimensional adaptive triangular grid, with the intermediate grid generated in such a manner through the basic stretching functions of Chap. 4, is presented in Fig. 5.4.

Note that the procedure described above for generating triangular grids (tetrahedral and prismatic ones in three dimensions) can be realized analogously in other techniques based on coordinate transformations from the unit cube.

5.7 Comments

The standard formulas of multivariate transfinite interpolation using Boolean operations were described by Gordon (1969, 1971), although a two-dimensional interpolation formula with the simplest blending functions for the construction of the boundaries of hexahedral patches from CAD data was proposed by

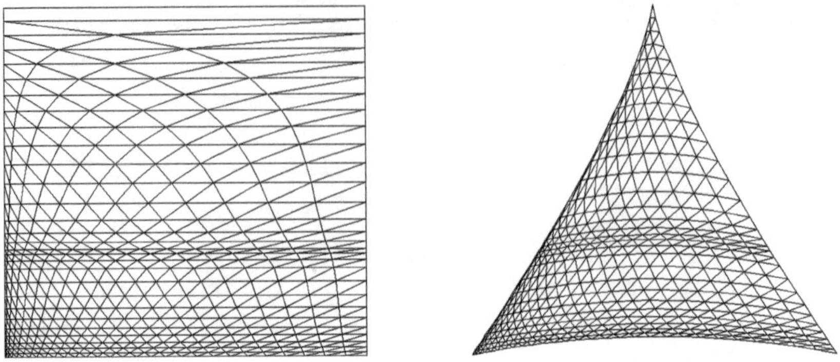

Fig. 5.4. Example of an adaptive algebraic triangular grid (*right*) and the corresponding grid on the intermediate domain (*left*) generated by the algebraic method

Coons (1967) and Ahuja and Coons (1968). The construction of coordinate transformations through the formulas of transfinite interpolation was formulated by Gordon and Hall (1973) and Gordon and Thiel (1982). The Hermite interpolation was presented by Smith (1982).

The multisurface method was described by Eiseman (1980) and was, in its original form, a univariate formula for grid generation based on the specification of two boundary surfaces and an arbitrary number of interior control surfaces. The blending functions were implicitly derived from global and/or local interpolants which result from an expression for the tangential derivative spanning between the exterior boundary surfaces. The multisurface transformation can be described in the context of transfinite interpolation.

A two-boundary technique was introduced by Smith (1981). It is based on the description of two opposite boundary surfaces, tangential derivatives on the boundary surfaces which are used to compute normal derivatives, and Hermite cubic blending functions.

The construction of some special blending functions aimed at grid clustering at boundaries was performed by Eriksson (1982) and Smith and Eriksson (1987). A detailed description of various forms of blending functions with the help of splines was presented in a monograph by Thompson, Warsi, and Mastin (1985).

The procedures described above for generating smooth blending functions and algebraic triangulations were developed by the author of the present book.

6

Grid Generation Through Differential Systems

6.1 Introduction

Grid techniques based on using systems of partial differential equations to derive coordinate transformations are very popular in structured grid generation. The choice of the systems of equations relies on numerical principles and careful analysis of the required properties of the equations. They must have intrinsic abilities to cope with complex geometries and to produce grids which are locally compressed by large factors compared with uniform grids. The equations should be computationally efficient, i.e. easy to model numerically and solve. Therefore the task of formulating satisfactory grid equations is not simple.

The present chapter describes the most typical systems of equations for grid generation: elliptic, hyperbolic, and parabolic.

At present, elliptic methods of grid generation have widespread applications. The formulation of an elliptic method for grid generation relies on the utilization of an elliptic system whose solution defines a coordinate transformation

$$\boldsymbol{x}(\boldsymbol{\xi}) : \Xi^n \to X^n, \quad \boldsymbol{x}(\boldsymbol{\xi}) = \left[x^1(\boldsymbol{\xi}), \ldots, x^n(\boldsymbol{\xi})\right]$$

from the computational domain $\Xi^n \subset R^n$ onto the physical one $X^n \subset R^n$. The values of the vector-valued function $\boldsymbol{x}(\boldsymbol{\xi})$ at the points of a reference grid in Ξ^n define the nodes of the elliptic grid in X^n. However, in practice the grid nodes are obtained by the numerical solution of a boundary value problem for the elliptic system on the reference grid in the domain Ξ^n.

Elliptic equations are attractive for generating curvilinear coordinates because of some of their properties. First, elliptic equations which obey the extremum principle, i.e. the extrema of solutions cannot be within the domain, are readily formulated and numerically implemented. With this property, there is less tendency for folding of the resulting grid cells. Another important property of any elliptic system is the inherent smoothness of its solution and consequently of the resulting coordinate curves in the interior of the

V.D. Liseikin, *Grid Generation Methods*, Scientific Computation,
DOI 10.1007/978-90-481-2912-6_6, © Springer Science+Business Media B.V. 2010

domain and even on smooth segments of the boundary. Moreover, the smoothness can be propagated over the whole boundary with slope discontinuities if the boundary conditions are consistent with the equations of the elliptic system. The third advantageous feature of elliptic systems is that they allow one to specify the coordinate points (and/or coordinate-line slopes) on the whole boundary of the domain. Finally, well-established methods are available to solve elliptic equations.

A disadvantage of elliptic systems is the cost of the numerical solution, especially because the commonly applied equations, considered in the transformed space, are nonlinear and require iteration.

Commonly, the elliptic equations for generating curvilinear coordinates are formulated in two ways:

(1) in the computational domain with the physical Cartesian coordinates x^i as the dependent variables;
(2) in the physical domain with the curvilinear coordinates ξ^i as the dependent variables.

In the second way, the coordinate mapping $\boldsymbol{x}(\boldsymbol{\xi})$ is constructed by solving the elliptic system obtained by a transformation of the original system so as to interchange the dependent and independent variables.

The initial elliptic systems for the generation of grids are generally chosen in the form

$$L_1(x^i) \equiv a_1^{kj} \frac{\partial^2 x^i}{\partial \xi^k \partial \xi^j} + b_1^j \frac{\partial x^i}{\partial \xi^j} + c_1 x^i = f_1^i, \quad i, j, k = 1, \dots, n, \tag{6.1}$$

and

$$L_2(\xi^i) \equiv a_2^{kj} \frac{\partial^2 \xi^i}{\partial x^k \partial x^j} + b_2^j \frac{\partial \xi^i}{\partial x^j} + c_2 \xi^i = f_2^i, \quad i, j, k = 1, \dots, n. \tag{6.2}$$

Recall that repeated indices in formulas mean a summation over them unless otherwise noted. The condition of ellipticity puts a restriction on the coefficients a_l^{ij}:

$$a_l^{ij} b^i b^j \geq c_l b^k b^k, \quad c_l > 0, \ i, j, k = 1, \dots, n, \ l = 1, 2,$$

for an arbitrary vector $\boldsymbol{b} = (b^1, \dots, b^n)$.

Hyperbolic and parabolic methods of grid generation imply the numerical solution of hyperbolic and parabolic differential equations, respectively. Both types of system of equations are solved by marching in the direction of one selected curvilinear coordinate. These procedures are much faster than an elliptic scheme, producing a grid in an order of magnitude less computational time.

6.2 Laplace Systems

The most simple elliptic systems for generating grids are represented by the uncoupled Laplace equations, either in the computational domain,

$$\nabla^2 x^i \equiv \frac{\partial}{\partial \xi^j}\frac{\partial x^i}{\partial \xi^j} = 0, \quad i,j = 1,\ldots,n, \tag{6.3}$$

with the dependent variables x^i, or in the physical domain,

$$\nabla^2 \xi^i \equiv \frac{\partial}{\partial x^j}\frac{\partial \xi^i}{\partial x^j} = 0, \quad i,j = 1,\ldots,n, \tag{6.4}$$

with the dependent variables ξ^i.

Multiplying the system (6.4) by $\partial x^p/\partial \xi^i$ and summing over i, we readily obtain the inverse elliptic system with the dependent and independent quantities interchanged, in the form

$$g^{ij}\frac{\partial^2 x^p}{\partial \xi^i \partial \xi^j} = 0, \quad i,j,p = 1,\ldots,n, \tag{6.5}$$

where, in accordance with (2.17)

$$g^{ij} = \nabla \xi^i \cdot \nabla \xi^j = \frac{\partial \xi^i}{\partial x^k}\frac{\partial \xi^j}{\partial x^k}, \quad i,j,k = 1,\ldots,n.$$

These equations shape the coupled quasilinear elliptic system in the computational domain Ξ^n. A grid in the domain X^n is generated by solving (6.3) or (6.5), with the Cartesian values of $\boldsymbol{x}$ on the physical boundaries used as the boundary conditions along the corresponding boundary segments of Ξ^n.

The maximum principle is valid for the Laplace equations (6.3) and (6.4). In the case of the system (6.3) it guarantees that the image of Ξ^n produced by the coordinate transformation $\boldsymbol{x}(\boldsymbol{\xi})$ will be contained in X^n if the domain X^n is convex. Analogously, the image of X^n produced by the transformation $\boldsymbol{\xi}(\boldsymbol{x})$ satisfying the Laplace system (6.4) will be contained in Ξ^n if Ξ^n is a convex domain. In the latter case the restriction of convexity is not imposed on the physical domain X^n. As the shape of the computational domain Ξ^n can be specified by the user, the system (6.4), and correspondingly (6.5), has been more favored in applications than the system (6.3) for generating grids in general regions. Equations (6.4) also are preferred because the physical-space formulation provides direct control of grid spacing and orthogonality. For these reasons, the formulation of many other elliptic grid generators is also commonly performed in terms of the inverse of the coordinate transformation $\boldsymbol{x}(\boldsymbol{\xi})$.

The main problem in grid generation is to make the coordinate transformation $\boldsymbol{x}(\boldsymbol{\xi}) : \Xi^n \to X^n$ a diffeomorphism, i.e. one-to-one mapping with the Jacobian J not vanishing. In the case $n = 2$ the mathematical foundation of the technique, based on the Laplace system (6.4) with a convex computational domain Ξ^2, is solid. It is founded on the following result, derived from a theorem of Rado.

Let X^2 be a simply connected bounded domain in R^2. In this case, the Jacobian of the transformation $\boldsymbol{\xi}(\boldsymbol{x})$ generated by the system (6.4) does not vanish in the interior of X^2, if Ξ^2 is a rectangle and $\boldsymbol{\xi}(\boldsymbol{x}) : \partial X^2 \to \partial \Xi^2$ is a homeomorphism.

The system (6.4) was introduced by Crowley (1962) and Winslow (1967) and, owing to the properties noted above, it has been the most widely used system for generating fixed grids in general regions.

Some features of the coordinate transformations and corresponding grids derived from the system (6.4) and, correspondingly, (6.5) are considered in the next two subsections.

6.2.1 Two-Dimensional Equations

In this subsection we discuss the qualitative behavior of the coordinate lines generated by the two-dimensional Laplace system (6.4) near the boundary curves. We assume that Ξ^2 is a unit square, X^2 is a simply connected bounded domain, and the coordinate transformation $\boldsymbol{x}(\boldsymbol{\xi})$ is defined as a solution to the Dirichlet problem for the system (6.5) with a specified one-to-one boundary transformation

$$\boldsymbol{x}(\boldsymbol{\xi}) : \partial \Xi^2 \to \partial X^2 .$$

It is obvious from the theorem above that the mapping $\boldsymbol{x}(\boldsymbol{\xi})$ is the inverse of the transformation $\boldsymbol{\xi}(\boldsymbol{x})$, that is, a solution to the Laplace system (6.4) with the Dirichlet boundary conditions

$$\boldsymbol{\xi}(\boldsymbol{x}) : \partial X^2 \to \partial \Xi^2 .$$

From (2.20), the two-dimensional contravariant metric elements g^{ij} in (6.5) are connected with the covariant elements $g_{ij} = \boldsymbol{x}_{\xi^i} \cdot \boldsymbol{x}_{\xi^j}$ by the relation

$$g^{ij} = (-1)^{i+j} \frac{g_{3-i\ 3-j}}{g}, \quad i = 1, 2,$$

with fixed i and j. Therefore the system (6.5) for $n = 2$ is equivalent to

$$g_{22} \frac{\partial^2 x^i}{\partial \xi^1 \partial \xi^1} - 2 g_{12} \frac{\partial^2 x^i}{\partial \xi^1 \partial \xi^2} + g_{11} \frac{\partial^2 x^i}{\partial \xi^2 \partial \xi^2} = 0, \quad i = 1, 2. \tag{6.6}$$

We now demonstrate that the spacing between coordinate lines, say $\xi^2 =$ const., in the vicinity of the respective boundary curve $\xi^2 = \xi^2_0$, increases toward it if the boundary line is convex and, conversely, the spacing decreases when the boundary line is concave.

Let us consider, for clarity, a family of the coordinate curves $\xi^2 =$ const. Then the boundary curve of this family is defined by the relation $\xi^2 = \xi^2_0$ with $\xi^2_0 = 0$ or $\xi^2_0 = 1$.

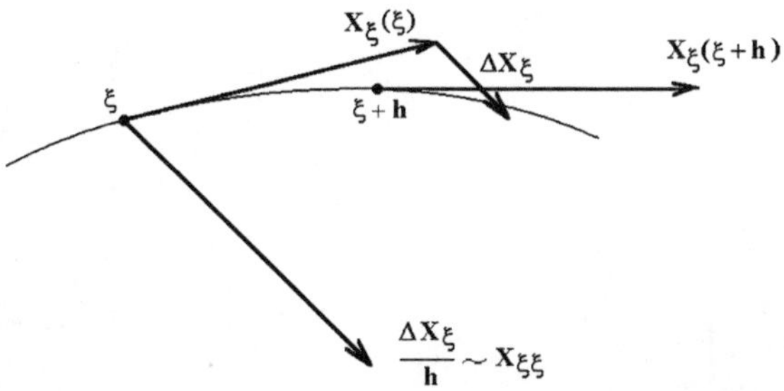

Fig. 6.1. Direction of the derivative of the tangential vector

First we note that the vector $\boldsymbol{x}_{\xi^1\xi^1}$, which is the derivative with respect to ξ^1 of the tangential vector $\boldsymbol{x}_{\xi^1}$, $\boldsymbol{x} = (x^1, x^2)$, is directed, as shown in Fig. 6.1, toward the concavity of the coordinate line $\xi^2 = \xi_0^2$. Another important gradient vector of $\xi^2(\boldsymbol{x})$,

$$\boldsymbol{\nabla}\xi^2 = \left(\frac{\partial \xi^2}{\partial x^1}, \frac{\partial \xi^2}{\partial x^2}\right),$$

is orthogonal to the tangent vector $\boldsymbol{x}_{\xi^1}$. The dot product of the vector $\boldsymbol{\nabla}\xi^2$ and the tangential vector $\boldsymbol{x}_{\xi^2}$ equals 1. Therefore, these vectors are always directed to one side of the line $\xi^2 = \xi_0^2$ (see Figs. 2.2 and 6.2); in particular, they are directed into the domain X^2 if this coordinate line is the boundary curve $\xi^2 = 0$. Thus the sign of the quantity

$$Q = \boldsymbol{\nabla}\xi^2 \cdot \boldsymbol{x}_{\xi^1\xi^1} = \Gamma_{11}^2$$

can serve as a criterion of the local shape of the boundary $\xi^2 = 0$. Namely, if $Q < 0$, which means the vectors $\boldsymbol{x}_{\xi^1\xi^1}$ and $\boldsymbol{\nabla}\xi^2$ are directed toward different sides of the coordinate curve $\xi^2 = 0$, then the domain X^2 is concave (if $Q > 0$ the domain is convex) near that part of the boundary $\xi^2 = 0$ where this inequality is satisfied (see Fig. 6.2).

We have

$$\begin{aligned} Q &= \boldsymbol{x}_{\xi^1\xi^1} \cdot \boldsymbol{\nabla}\xi^2 = -\boldsymbol{x}_{\xi^1} \cdot \frac{\partial}{\partial \xi^1}(\boldsymbol{\nabla}\xi^2) \\ &= -\left(\frac{\partial x^1}{\partial \xi^1}\right)^2 \frac{\partial^2 \xi^2}{\partial x^1 \partial x^1} - 2\frac{\partial x^1}{\partial \xi^1}\frac{\partial x^2}{\partial \xi^1}\frac{\partial^2 \xi^2}{\partial x^1 \partial x^2} - \left(\frac{\partial x^2}{\partial \xi^1}\right)^2 \frac{\partial^2 \xi^2}{\partial x^2 \partial x^2}. \end{aligned} \tag{6.7}$$

The vector $\boldsymbol{n} = \boldsymbol{\nabla}\xi^2/|\boldsymbol{\nabla}\xi^2|$ is the unit normal to the tangential vector $\boldsymbol{x}_{\xi^1}$. It is valid that

$$\boldsymbol{n} \cdot \boldsymbol{x}_{\xi^2} = 1/|\boldsymbol{\nabla}\xi^2| = 1/\sqrt{g^{22}},$$

where

$$g^{22} = \boldsymbol{\nabla}\xi^2 \cdot \boldsymbol{\nabla}\xi^2 = \frac{\partial \xi^2}{\partial x^i}\frac{\partial \xi^2}{\partial x^i}, \quad i = 1, 2.$$

Let us denote by l_h the distance between the two coordinate lines $\xi^2 = 0$ and $\xi^2 = h$. Using the above equation for $\boldsymbol{n} \cdot \boldsymbol{x}_{\xi^2}$, we have

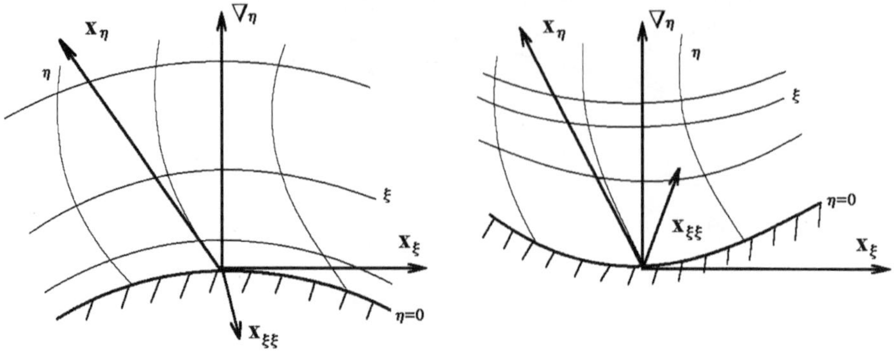

Fig. 6.2. Grid concentration near a concave boundary curve (*left*) and grid rarefaction near a convex part of the boundary (*right*)

$$l_h = (\boldsymbol{n} \cdot \boldsymbol{x}_{\xi^2})h + O(h)^2 = h/\sqrt{g^{22}} + O(h)^2.$$

So the quantity

$$s_2 = 1/\sqrt{g^{22}}$$

reflects the relative spacing between the coordinate grid lines $\xi^2 = \text{const}$.

The vector $\boldsymbol{n}$ is orthogonal to the boundary coordinate line $\xi^2 = 0$, and therefore the rate of change of the relative spacing s_2 of the coordinate curves $\xi^2 = \text{const}$ near this boundary line is computed in the $\boldsymbol{n}$ direction. Since

$$\boldsymbol{n} = \frac{1}{\sqrt{g^{22}}}\left(\frac{\partial \xi^2}{\partial x^1}, \frac{\partial \xi^2}{\partial x^2}\right),$$

we obtain

$$\begin{aligned}\frac{\partial s_2}{\partial \boldsymbol{n}} &= \frac{1}{\sqrt{g^{22}}}\left(\frac{\partial s_2}{\partial x^1}\frac{\partial \xi^2}{\partial x^1} + \frac{\partial s_2}{\partial x^2}\frac{\partial \xi^2}{\partial x^2}\right)\\ &= -\frac{1}{2(g^{22})^2}\left(\left(\frac{\partial \xi^2}{\partial x^1}\right)^2 \frac{\partial^2 \xi^2}{\partial x^1 \partial x^1} + 2\frac{\partial \xi^2}{\partial x^1}\frac{\partial \xi^2}{\partial x^2}\frac{\partial^2 \xi^2}{\partial x^1 \partial x^2}\right.\\ &\quad \left. + \left(\frac{\partial \xi^2}{\partial x^2}\right)^2 \frac{\partial^2 \xi^2}{\partial x^2 \partial x^2}\right).\end{aligned}$$

Using in this equation the relation

$$\frac{\partial \xi^i}{\partial x^j} = (-1)^{i+j}\frac{\partial x^{3-j}}{\partial \xi^{3-i}}\Big/J, \quad i, j, = 1, 2,$$

with fixed i and j, we obtain for the rate of change of the relative spacing s_2

$$\begin{aligned}\frac{\partial s_2}{\partial \boldsymbol{n}} = &-\frac{1}{2J^2(g^{22})^2}\left[\left(\frac{\partial x^2}{\partial \xi^1}\right)^2 \frac{\partial^2 \xi^2}{\partial x^1 \partial x^1} - 2\frac{\partial x^2}{\partial \xi^1}\frac{\partial x^1}{\partial \xi^1}\frac{\partial^2 \xi^2}{\partial x^1 \partial x^2}\right.\\ &\left. + \left(\frac{\partial x^1}{\partial \xi^1}\right)^2 \frac{\partial^2 \xi^2}{\partial x^2 \partial x^2}\right].\end{aligned} \tag{6.8}$$

Equation (6.4), for $n = 2$, implies

$$\frac{\partial^2 \xi^2}{\partial x^2 \partial x^2} = -\frac{\partial^2 \xi^2}{\partial x^1 \partial x^1},$$

and therefore we see from (6.7) and (6.8) that

$$\frac{\partial s_2}{\partial \boldsymbol{n}} = -\frac{1}{2J^2(g^{22})^2} Q.$$

Thus, we find that the quantities $\partial s_2/\partial \boldsymbol{n}$ and Q have different signs. So if the boundary line $\xi^2 = 0$ is convex or concave at some point $\boldsymbol{\xi}_0$, i.e. $Q > 0$ or $Q < 0$, respectively, then

$$\frac{\partial s_2}{\partial \boldsymbol{n}} < 0 \quad \text{or} \quad \frac{\partial s_2}{\partial \boldsymbol{n}} > 0,$$

at this point. These inequalities mean that the spacing of the grid lines $\xi^2 =$ const decreases or increases, respectively, from the boundary curve $\xi^2 = 0$ in the vicinity of the boundary point $\boldsymbol{\xi}_0$.

Analogous computations are readily carried out for the coordinates near the boundary curves $\xi^2 = 1$ and $\xi^1 = 0$ or $\xi^1 = 1$, which result in the conclusion that the grid lines obtained from the Laplace system (6.4) for $n = 2$ are attracted to the concave part of the boundary and repelled near the convex part (Fig. 6.2).

6.2.2 Three-Dimensional Equations

In contrast to two-dimensional domains, the problem of generating one-to-one three-dimensional transformations through the Laplace system (6.4) has not yet been solved theoretically. One of the reasons is the fact that the technique used for the two-dimensional case cannot extended to higher dimensions. This observation was made by Liao (1991). However, we may assume that the transformation $\boldsymbol{x}(\boldsymbol{\xi})$ obtained as a solution to the Dirichlet problem for the system (6.5) with $n = 3$ on the unit cube Ξ^3 is a diffeomorphism, and hence the inverse transformation $\boldsymbol{\xi}(\boldsymbol{x})$ is a solution to the Laplace system (6.4). In this case the analogous property of the concentration of the coordinate surfaces toward the concave part of the boundary and their rarefaction toward the convex part is valid. This subsection gives a detailed proof of this fact.

First, we call that in the three-dimensional case the gradient vector

$$\boldsymbol{\nabla}\xi^3 = \left(\frac{\partial \xi^3}{\partial x^1}, \frac{\partial \xi^3}{\partial x^2}, \frac{\partial \xi^3}{\partial x^3}\right)$$

is orthogonal to the tangent vectors $\boldsymbol{x}_{\xi^1}$ and $\boldsymbol{x}_{\xi^2}$, $\boldsymbol{x} = (x^1, x^2, x^3)$. The vectors $\boldsymbol{\nabla}\xi^3$ and $\boldsymbol{x}_{\xi^3}$ are directed toward one side of the coordinate surface $\xi^3 = \xi^3_0$. And the quantity

$$s_3 = 1/\sqrt{g^{33}},$$

where

$$g^{33} = \nabla\xi^3 \cdot \nabla\xi^3 = \frac{\partial \xi^3}{\partial x^i}\frac{\partial \xi^3}{\partial x^i}, \quad i = 1, 2, 3,$$

means, in the two-dimensional case, the relative grid spacing between the coordinate surfaces $\xi^3 = \text{const}$ in the normal direction $\boldsymbol{n}$, where

$$\boldsymbol{n} = \nabla\xi^3/|\nabla\xi^3| = \nabla\xi^3/\sqrt{g^{33}}.$$

The rate of change of the relative spacing in this direction $\boldsymbol{n}$ equals

$$\begin{aligned} \frac{\partial s_3}{\partial \boldsymbol{n}} &= \frac{1}{\sqrt{g^{33}}}\left(\frac{\partial s_3}{\partial x^1}\frac{\partial \xi^3}{\partial x^1} + \frac{\partial s_3}{\partial x^2}\frac{\partial \xi^3}{\partial x^2} + \frac{\partial s_3}{\partial x^3}\frac{\partial \xi^3}{\partial x^3}\right) \\ &= -\frac{1}{2(g^{33})^2}\left(\frac{\partial \xi^3}{\partial x^i}\frac{\partial \xi^3}{\partial x^j}\frac{\partial^2 \xi^3}{\partial x^i \partial x^j}\right), \quad i, j = 1, 2, 3. \end{aligned} \tag{6.9}$$

Using the general identity (2.48),

$$\frac{\partial^2 \xi^i}{\partial x^k \partial x^m} = -\frac{\partial^2 x^p}{\partial \xi^l \partial \xi^j}\frac{\partial \xi^j}{\partial x^k}\frac{\partial \xi^l}{\partial x^m}\frac{\partial \xi^i}{\partial x^p},$$
$$i, j, k, l, m, p = 1, 2, 3,$$

for $i = 3$, between the second derivatives of the coordinate transformation $\boldsymbol{x}(\boldsymbol{\xi}) : \Xi^3 \to X^3$ and $\boldsymbol{\xi}(\boldsymbol{x}) : X^3 \to \Xi^3$ in (6.9), we obtain

$$\frac{\partial s_3}{\partial \boldsymbol{n}} = \frac{1}{2(g^{33})^2} g^{3j} g^{3l} \frac{\partial^2 x^p}{\partial \xi^l \partial \xi^j}\frac{\partial \xi^3}{\partial x^p}, \quad j, l, p = 1, 2, 3.$$

Now we write out the right-hand side of this equation as the sum of two parts, one of which contains all terms of the kind

$$\frac{\partial^2 x^p}{\partial \xi^l \partial \xi^j}, \quad i, j = 1, 2, \ p = 1, 2, 3,$$

namely,

$$\begin{aligned} \frac{\partial s_3}{\partial \boldsymbol{n}} &= \frac{1}{2(g^{33})^2}(Q_1 + Q_2), \\ Q_1 &= \left((g^{31})^2 \frac{\partial^2 x^p}{\partial \xi^1 \partial \xi^1} + 2g^{31}g^{32}\frac{\partial^2 x^p}{\partial \xi^1 \partial \xi^2} + (g^{32})^2 \frac{\partial^2 x^p}{\partial \xi^2 \partial \xi^2}\right)\frac{\partial \xi^3}{\partial x^p}, \\ Q_2 &= g^{33}\left(2g^{31}\frac{\partial^2 x^p}{\partial \xi^1 \partial \xi^3} + 2g^{32}\frac{\partial^2 x^p}{\partial \xi^2 \partial \xi^3} + g^{33}\frac{\partial^2 x^p}{\partial \xi^3 \partial \xi^3}\right)\frac{\partial \xi^3}{\partial x^p}, \end{aligned} \tag{6.10}$$

where $p = 1, 2, 3$. Multiplying the elliptic system (6.5) by $\partial \xi^m/\partial x^p$ and summing the result over p, we obtain, for $m = 3$,

$$g^{ij}\frac{\partial^2 x^p}{\partial\xi^i\partial\xi^j}\frac{\partial\xi^3}{\partial x^p}=0,\quad i,j,p=1,2,3.$$

Using this equation in the expression (6.10) for the quantity Q_2, we readily obtain

$$Q_2=-g^{33}\left(g^{11}\frac{\partial^2 x^p}{\partial\xi^1\partial\xi^1}+2g^{12}\frac{\partial^2 x^p}{\partial\xi^1\partial\xi^2}+g^{22}\frac{\partial^2 x^p}{\partial\xi^2\partial\xi^2}\right)\frac{\partial\xi^3}{\partial x^p},\quad p=1,2,3.$$

Therefore,

$$\begin{aligned}Q_1+Q_2=\Bigg(&[(g^{31})^2-g^{33}g^{11}]\frac{\partial^2 x^p}{\partial\xi^1\partial\xi^1}\\&+[(g^{32})^2-g^{33}g^{22}]\frac{\partial^2 x^p}{\partial\xi^2\partial\xi^2}\\&+2[g^{12}g^{33}-g^{13}g^{23}]\frac{\partial^2 x^p}{\partial\xi^1\partial\xi^2}\Bigg)\frac{\partial\xi^3}{\partial x^p}.\end{aligned}$$

And, in accordance with the relation

$$g_{ij}=J^2(g^{i+1\ j+1}g^{i+2\ j+2}-g^{i+1\ j+2}g^{i+2\ j+1}),\quad i,j=1,2,3,$$

from (2.21), where any superscript index k can be identified with $k-3$, we have

$$Q_1+Q_2=-\frac{g_{3-i\ 3-j}}{J^2}\frac{\partial^2 x^p}{\partial\xi^i\partial\xi^j}\frac{\partial\xi^3}{\partial x^p},\quad i,j=1,2,\ p=1,2,3. \tag{6.11}$$

Now we consider the value of Q_1+Q_2 at the boundary surface $\xi^3=0$. Let $\boldsymbol{\xi}_0$ be a point at this surface. The derivative of the vector

$$\boldsymbol{b}=a_1\boldsymbol{x}_{\xi^1}+a_2\boldsymbol{x}_{\xi^2},\quad a_i=\text{const},\ i=1,2,$$

along the direction $\boldsymbol{t}=\boldsymbol{b}(\boldsymbol{\xi}_0)$ is the vector

$$\frac{\partial\boldsymbol{b}}{\partial\boldsymbol{t}}=(a_1)^2\boldsymbol{x}_{\xi^1\xi^1}+2a_1a_2\boldsymbol{x}_{\xi^1\xi^2}+(a_2)^2\boldsymbol{x}_{\xi^2\xi^2}.$$

If $\boldsymbol{\xi}_0$ is a point of local convexity of the boundary surface $\xi^3=0$ then, in analogy with the vector $\boldsymbol{x}_{\xi^1\xi^1}$ considered previously in the two-dimensional case, the vector $\partial\boldsymbol{b}/\partial\boldsymbol{t}(\boldsymbol{\xi}_0)$ is directed into the domain X^3. The vector $\nabla\xi^3$ at the point $\boldsymbol{\xi}_0$ and the vector $\boldsymbol{x}_{\xi^3}$ are directed into the domain X^3 also. Therefore the dot product of the vectors $\nabla\xi^3$ and $\partial\boldsymbol{b}/\partial\boldsymbol{t}$ is positive at the point under consideration, i.e.

$$\nabla\xi^3\cdot\frac{\partial\boldsymbol{b}}{\partial\boldsymbol{t}}=\left((a_1)^2\frac{\partial^2 x^p}{\partial\xi^1\partial\xi^1}+2a_1a_2\frac{\partial^2 x^p}{\partial\xi^1\partial\xi^2}+(a_2)^2\frac{\partial^2 x^p}{\partial\xi^2\partial\xi^2}\right)\frac{\partial\xi^3}{\partial x^p}>0, \tag{6.12}$$

where $p=1,2,3$. Considering the three cases

$$(a_1, a_2) = (1, 0), \qquad (a_1, a_2) = (0, 1), \qquad (a_1, a_2) = (1, 1)$$

in (6.12) we find that at the point $\boldsymbol{\xi}_0$

$$\frac{\partial^2 x^p}{\partial \xi^1 \partial \xi^1} \frac{\partial \xi^3}{\partial x^p} > 0, \quad p = 1, 2, 3,$$
$$\frac{\partial^2 x^p}{\partial \xi^2 \partial \xi^2} \frac{\partial \xi^3}{\partial x^p} > 0, \quad p = 1, 2, 3,$$

and

$$\left| \frac{\partial^2 x^p}{\partial \xi^1 \partial \xi^2} \frac{\partial \xi^3}{\partial x^p} \right| < \left(\frac{\partial^2 x^p}{\partial \xi^1 \partial \xi^1} \frac{\partial \xi^3}{\partial x^p} \frac{\partial^2 x^p}{\partial \xi^2 \partial \xi^2} \frac{\partial \xi^3}{\partial x^p} \right)^{1/2}, \quad p = 1, 2, 3.$$

As the quadratic form (g_{ij}), $i, j = 1, 2$, is positive, we obtain from (6.11)

$$Q_1 + Q_2 < 0,$$

and, correspondingly, $\partial s_3 / \partial \boldsymbol{n} < 0$, i.e. the spacing between the coordinate surfaces $\xi^3 =$ const is decreases from the convex part of the boundary $\xi^3 = 0$.

Analogously, we have $Q_1 + Q_2 > 0$ at a point on a concave part of the boundary surface $\xi^3 = 0$, which implies the observation that the grid surface spacing increases locally from a concave part of the boundary surface.

The same facts are obviously true for the corresponding grid spacings near the boundary surfaces $\xi^3 = 1$, $\xi^i = 0$, and $\xi^i = 1$, $i = 1, 2$.

Thus we find, that the coordinate surfaces of the coordinate system derived from the Laplace equations (6.4) are clustered near the concave parts of the boundary and coarser near convex parts of it.

6.3 Poisson Systems

The Laplace system provides little opportunity to control the properties of the grid, in particular, to adapt the mesh to the geometry of the boundary or to the features of the solution of the physical equations in regions of the domain where this is necessary. Only one opportunity is given, by the specification of the boundary conditions. However, the grid point distribution on the boundaries affects noticeably only the disposition of the nearby interior grid nodes. The distribution of the nodes over most of the interior is influenced more by the form of the elliptic equations than by the boundary values.

Therefore, in order to provide global control of the grid node distribution, the Laplace system is replaced by a more general elliptic system with variable coefficients. The simplest way to obtain such a generalization, suggested by Godunov and Prokopov (1972) for the generation of two-dimensional grids, consists of adding right-hand terms to the Laplace system (6.4), thus making it a Poisson system.

The actual generation of the grid is done by the numerical solution of the transformed Poisson system in the computational domain Ξ^n, where the curvilinear coordinates ξ^i are the independent variables and the Cartesian coordinates x^i are the dependent variables.

An elliptic method of grid generation based on the numerical solution of a system of inverted Poisson equations is being used in broad range of practical applications. The method allows the users to generate numerical grids in fairly complicated domains and on surfaces that arise while analyzing multidimensional fluid-flow problems. Practically all big grid generation codes incorporate it as a basic tool to generate structured grids. Other techniques (algebraic, hyperbolic, etc.) play an auxiliary role in the codes, serving as an initial guess for the elliptic solver, or as a technique for generating grids in regions with simple geometry.

6.3.1 Formulation of the System

The system of a Poisson equations for generating grids has the form

$$\nabla^2 \xi^i \equiv \frac{\partial}{\partial x^j}\frac{\partial \xi^i}{\partial x^j} = P^i, \quad i, j = 1, \ldots, n. \tag{6.13}$$

The quantities P^i are called either control functions or source terms. The source terms are essential for providing an effective control of the grid point distribution, although the choice of the proper control functions P^i is difficult, especially for multicomponent geometries.

Since

$$\frac{\partial}{\partial x^j}\left(\frac{\partial \xi^i}{\partial x^j}\right)\frac{\partial x^k}{\partial \xi^i} = -\frac{\partial^2 x^k}{\partial \xi^i \partial \xi^m}\frac{\partial \xi^i}{\partial x^j}\frac{\partial \xi^m}{\partial x^j} = -g^{im}\frac{\partial^2 x^k}{\partial \xi^i \partial \xi^m},$$

by multiplying the Poisson system (6.13) by $\partial x^k / \partial \xi^i$ and summing over i an inverted system of the equations (6.13) is obtained:

$$g^{ij}\frac{\partial^2 x^k}{\partial \xi^i \partial \xi^j} = -P_i \frac{\partial x^k}{\partial \xi^i}, \quad k, i, j = 1, \ldots, n. \tag{6.14}$$

Note that the left-hand part of these equations comprises the left hand part of the system of inverted Laplace equations (6.5). The system (6.14) can also be represented in the following vector notation:

$$g^{ij}\boldsymbol{x}_{\xi^i \xi^j} = -P^i \boldsymbol{x}_{\xi^i}, \quad i, j = 1, \ldots, n. \tag{6.15}$$

For one-dimensional space we obtain from (6.15)

$$\frac{\mathrm{d}^2 x}{\mathrm{d}\xi^2} = -P\left(\frac{\mathrm{d}x}{\mathrm{d}\xi}\right)^3.$$

Assuming

$$P = \left(\frac{\mathrm{d}\xi}{\mathrm{d}x}\right)^2 \frac{\partial w}{\partial \xi} \Big/ w,$$

where w is some positive function, playing the role of a weight in applications, we have

$$w\frac{\mathrm{d}}{\mathrm{d}\xi}\left(w\frac{\mathrm{d}x}{\mathrm{d}\xi}\right) = 0.$$

This equation is related to the equation in (4.13), thus giving a clue as to how to generate univariate grid clustering with the control function P.

Using the relations (2.20) and (2.21) we obtain from (6.15) two- and three-dimensional systems of inverted Poisson grid equations:

$$B_n[\boldsymbol{x}](\boldsymbol{\xi}) = -gP^i \boldsymbol{x}_{\xi^i}, \quad i = 1, \ldots, n, \tag{6.16}$$

where

$$B_n[y] = gg^{ij}\frac{\partial^2 y}{\partial \xi^i \partial \xi^j}, \quad i, j = 1, \ldots, n,$$

i.e. for $n = 2$ and $n = 3$

$$\begin{aligned}
B_2[y] &= g_{22}\frac{\partial^2 y}{\partial \xi^1 \partial \xi^1} - 2g_{12}\frac{\partial^2 y}{\partial \xi^1 \partial \xi^2} + g_{11}\frac{\partial^2 y}{\partial \xi^2 \partial \xi^2},\\
B_3[y] &= [g_{22}g_{33} - (g_{23})^2]\frac{\partial^2 y}{\partial \xi^1 \partial \xi^1} + [g_{11}g_{33} - (g_{13})^2]\frac{\partial^2 y}{\partial \xi^2 \partial \xi^2}\\
&\quad + [g_{11}g_{22} - (g_{12})^2]\frac{\partial^2 y}{\partial \xi^3 \partial \xi^3} + 2[g_{23}g_{13} - g_{12}g_{33}]\frac{\partial^2 y}{\partial \xi^1 \partial \xi^2}\\
&\quad + 2[g_{12}g_{23} - g_{22}g_{13}]\frac{\partial^2 y}{\partial \xi^1 \partial \xi^3} + 2[g_{13}g_{12} - g_{23}g_{11}]\frac{\partial^2 y}{\partial \xi^2 \partial \xi^3}.
\end{aligned}$$

6.3.2 Justification for the Poisson System

The idea of using the Poisson system to provide efficient control of grid generation was justified by the fact that the system of the Poisson type is obtained from the Laplace system, for intermediate coordinates which are transformed to other coordinates. Let every component q^i, $i = 1, \ldots, n$, of the coordinate transformation $\boldsymbol{q}(\boldsymbol{x})$ satisfy the Laplace equation

$$\nabla^2 q^i = \frac{\partial}{\partial x^j}\left(\frac{\partial q^i}{\partial x^j}\right) = 0, \quad i, j = 1, \ldots, n.$$

Futher, let $\boldsymbol{\xi}(\boldsymbol{q})$ be a new intermediate one-to-one smooth coordinate transformation. Then every new coordinate ξ^i will satisfy the inhomogeneous elliptic system

$$\begin{aligned}
\nabla^2 \xi^i &= \frac{\partial}{\partial x^j}\left(\frac{\partial \xi^i}{\partial q^k}\frac{\partial q^k}{\partial x^j}\right) = \frac{\partial^2 \xi^i}{\partial q^k \partial q^m}\overline{g}^{km} + \frac{\partial \xi^i}{\partial q^k}\frac{\partial^2 q^k}{\partial x^j \partial x^j}\\
&= \overline{g}^{km}\frac{\partial^2 \xi^i}{\partial q^k \partial q^m}, \quad i, j, k, m = 1, \ldots, n,
\end{aligned} \tag{6.17}$$

where $\overline{g}^{km}$ is the (k,m) element of the contravariant metric tensor of the domain X^n in the coordinates $q^1,\ldots,q^n$, i.e.

$$\overline{g}^{km} = \frac{\partial q^k}{\partial x^j}\frac{\partial q^m}{\partial x^j}, \quad j,k,m = 1,\ldots,n.$$

The elements g^{ij}, $i,j=1,\ldots,n$, of the contravariant metric tensor of the domain X^n in the coordinates $\xi^1,\ldots,\xi^n$ are connected with $\overline{g}^{ij}$, $i,j=1,\ldots,n$, by

$$\overline{g}^{km} = g^{lj}\frac{\partial q^k}{\partial \xi^l}\frac{\partial q^m}{\partial \xi^j}, \quad k,l,m,j = 1,\ldots,n.$$

Thus, taking into account this relation, the system (6.17) has the form (6.13), i.e.

$$\nabla^2\,\xi^i = P^i, \quad i=1,\ldots,n,$$

where

$$P^i = g^{lj}\frac{\partial q^k}{\partial \xi^l}\frac{\partial q^m}{\partial \xi^j}\frac{\partial^2 \xi^i}{\partial q^k \partial q^m}, \quad i,j,k,l,m = 1,\ldots,n. \tag{6.18}$$

From the identity

$$\frac{\partial q^k}{\partial \xi^l}\frac{\partial q^m}{\partial \xi^j}\frac{\partial^2 \xi^i}{\partial q^k \partial qm} \equiv -\frac{\partial \xi^i}{\partial q^m}\frac{\partial^2 q^m}{\partial \xi^l \partial \xi^j}$$

we also have

$$P^i = -g^{lj}\frac{\partial \xi^i}{\partial q^m}\frac{\partial^2 q^m}{\partial \xi^l \partial \xi^j}. \tag{6.19}$$

Thus by applying the intermediate coordinate transformation $\boldsymbol{q}(\boldsymbol{\xi})$ to a grid generated as a solution of the Laplace system, we obtain a grid which could have been generated directly as the solution of the Poisson system (6.13) with the appropriate control functions defined by (6.18) and (6.19).

The general Poisson system (6.13) does not obey the maximum principle. And, in contrast to the two-dimensional Laplace system (6.4), there is no guarantee that the generated grid is not folded. In fact, any smooth but folded coordinate transformation $\boldsymbol{\xi}(\boldsymbol{x})$ can be obtained from the system (6.13) by computing P^i directly from the Laplacian of $\xi^i(\boldsymbol{x})$. If these P^i are used in the Poisson system (6.13) then the folded transformation $\boldsymbol{\xi}(\boldsymbol{x})$ will be reproduced.

One way to make the Poisson system satisfy the maximum principle is to replace the control functions P^i with other functions which guarantee the maximum principle. One appropriate approach is to define the control functions P^i in the form

$$P^i = g^{jk}P^i_{jk}, \quad i,j,k = 1,\ldots,n.$$

Such an expression for P^i is prompted by (6.18) and (6.19) with

$$P^i_{jk} = \frac{\partial \xi^i}{\partial q^m}\frac{\partial^2 q^m}{\partial \xi^j \partial \xi^k}, \quad i,j,k,m = 1,\ldots,n, \tag{6.20}$$

defined by the transformation from the intermediate coordinates q^i to the final computational coordinates ξ^i. According to the theory of elliptic equations the factors g^{ij} in the expressions for P^i guarantee the maximum principle for the Poisson system (6.13).

Thus an appropriate Poisson system can be defined by the equations

$$\nabla^2 \xi^i = g^{jk} P^i_{jk}, \quad i,j,k = 1,\dots,n, \tag{6.21}$$

where the control functions P^i_{jk} are considered to be specified. The inverse of (6.21) has then the form

$$g^{ij}(\boldsymbol{x}_{\xi^{ij}} + P^k_{ij}\boldsymbol{x}_{\xi^k}) = 0. \tag{6.22}$$

When the intermediate transformation $\boldsymbol{q}(\boldsymbol{\xi})$ is composed of separate one-dimensional mappings $q^i(\xi^i)$ for each coordinate direction ξ^i, then, from (6.20),

$$P^i_{jk} = \delta^i_j \delta^i_k P^i$$

so that the generation system (6.21) becomes

$$\nabla^2 \xi^i = g^{ii} P^i \tag{6.23}$$

for each fixed $i = 1,\dots,n$. The inverted system has then the form

$$g^{ij}\boldsymbol{x}_{\xi^i\xi^j} + g^{ii}P^i\boldsymbol{x}_{\xi^i} = 0. \tag{6.24}$$

The selection of the control functions P^i_{jk} is a difficult task. Equations (6.20) show that these functions are not independent if the coordinate transformation $\boldsymbol{x}(\boldsymbol{\xi})$ is defined as the composition of an intermediate mapping $\boldsymbol{q}(\boldsymbol{\xi})$ and an exterior mapping $\boldsymbol{x}(\boldsymbol{q})$ which satisfies the inverted Laplace equation (6.5). Some forms of the control functions suitable for grid adaptation will be demonstrated in Sect. 7.4.

6.3.3 Equivalent Forms of the Poisson System

Taking into account the general identity (2.56) for arbitrary smooth functions A^i, $i = 1,\dots,n$,

$$\frac{\partial}{\partial x^j}(A^j) \equiv \frac{1}{J}\frac{\partial}{\partial \xi^j}\left(JA^m \frac{\partial \xi^j}{\partial x^m}\right), \quad j,m = 1,\dots,n,$$

we obtain, assuming $A^j = \partial \xi^i / \partial x^j$,

$$\nabla^2 \xi^i = \frac{\partial}{\partial x^j}\left(\frac{\partial \xi^i}{\partial x^j}\right) \equiv \frac{1}{J}\frac{\partial}{\partial \xi^j}\left(J\frac{\partial \xi^i}{\partial x^m}\frac{\partial \xi^j}{\partial x^m}\right) \equiv \frac{1}{J}\frac{\partial}{\partial \xi^j}(Jg^{ij}), \tag{6.25}$$

where $i,j,m = 1,\dots,n$. Therefore the Poisson system (6.13) is equivalent to the following system of equations:

$$\frac{1}{J}\frac{\partial}{\partial \xi^j}(J g^{ij}) = P^i, \quad i,j = 1,\dots,n, \tag{6.26}$$

which is derived from the elements of the metric tensors only. The left-hand part of (6.26) can be expressed through the Christoffel symbols. For this purpose we consider the identity

$$\frac{1}{J}\frac{\partial}{\partial \xi^j}(J g^{ij} \boldsymbol{x}_{\xi^i}) \equiv 0, \tag{6.27}$$

which is a result of (2.47), since $g^{ij}\boldsymbol{x}_{\xi^i} = \boldsymbol{\nabla}\xi^j$. Performing the differentiation in the left-hand part of (6.27), we obtain

$$g^{ij}\boldsymbol{x}_{\xi^i\xi^j} + \frac{1}{J}\frac{\partial}{\partial \xi^j}(J g^{ij})\boldsymbol{x}_{\xi^i} \equiv 0, \quad i,j = 1,\dots,n. \tag{6.28}$$

The dot product of (6.28) and $\boldsymbol{\nabla}\xi^k$ results in

$$\nabla^2 \xi^k \equiv -g^{ij}\Gamma^k_{ij}, \quad i,j,k = 1,\dots,n, \tag{6.29}$$

using (6.25) and (2.38). The identity (6.29) demonstrates that the value of $\nabla^2\xi^k$ is expressed through the metric elements and the space Christoffel symbols of the second kind.

The utilization of (6.29) generates the following equivalent form of the Poisson system (6.13):

$$-g^{ij}\Gamma^k_{ij} = P^k, \quad i,j,k = 1,\dots,n. \tag{6.30}$$

In order to define the value of the forcing terms on the boundaries we use an alternative, equivalent system of equations

$$P^k = -g^{ij}g^{lk}[ij,l], \quad i,j,k,l = 1,\dots,n, \tag{6.31}$$

which is obtained from (6.30) and (2.42), with

$$[ij,l] = \boldsymbol{x}_{\xi^i\xi^j}\cdot\boldsymbol{x}_{\xi^l} = \frac{1}{2}\left(\frac{\partial g_{il}}{\partial \xi^j} + \frac{\partial g_{jl}}{\partial \xi^i} - \frac{\partial g_{ij}}{\partial \xi^l}\right),$$
$$i,j,l = 1,\dots,n.$$

In particular, when the coordinate system ξ^i is orthogonal, then (6.31) results in

$$P^k = -g^{ii}g^{kk}[ii,k] = g^{kk}\left(\frac{1}{2}g^{ii}\frac{\partial g_{ii}}{\partial \xi^k} - g^{kk}\frac{\partial g_{kk}}{\partial \xi^k}\right),$$
$$i = 1,\dots,n, \ k \text{ fixed}.$$

6.3.4 Orthogonality at Boundaries

The grid point distribution in the immediate neighborhood of the boundaries of two-dimensional and three-dimensional regions has a strong influence on the accuracy of the algorithms developed for the numerical solution of partial differential equations. In particular, it is often desirable to have orthogonal or nearly orthogonal grid lines emanating from some boundary segments.

Consider, for example, the evaluation of the outward normal derivative of an arbitrary function f at the boundary of a two-dimensional region X^2:

$$\frac{\partial f}{\partial \boldsymbol{n}} = \boldsymbol{n} \cdot \boldsymbol{\nabla} f.$$

If the boundary is a line of constant ξ^1, then

$$\boldsymbol{n} = \frac{1}{\sqrt{g^{22}}} \boldsymbol{\nabla} \xi^2,$$

and so

$$\frac{\partial f}{\partial \boldsymbol{n}} = \frac{1}{\sqrt{g^{22}}} \frac{\partial f}{\partial x^i} \frac{\partial \xi^2}{\partial x^i} = \frac{g^{2k}}{\sqrt{g^{22}}} \frac{\partial f}{\partial \xi^k}, \quad k = 1, 2. \tag{6.32}$$

If the coordinates ξ^1, ξ^2 are orthogonal, (6.32) reduces to just

$$\frac{\partial f}{\partial \boldsymbol{n}} = \frac{1}{\sqrt{g_{22}}} \frac{\partial f}{\partial \xi^2}.$$

Obviously, this equation is much simpler than (6.32) and is to be preferred for most analytical purposes. Less obviously, but of importance to numerical schemes, (6.32) couples the ξ^1 and ξ^2 variations of the function f, and thus the application of a Neumann boundary condition to f may involve the difference of two large numbers, with a possible loss of numerical accuracy.

The Poisson system provides two opportunities to satisfy the requirement of orthogonality or near orthogonality of the coordinate lines emanating from the boundary segments, either by imposing Neumann boundary conditions or by specifying the source terms P^i through the boundary values of the coordinate transformation.

The commonly used approach to the specification of the source terms P^i to provide boundary orthogonality relies on the computation of the values of $\nabla^2 \xi^i$ on boundary segments, provided the coordinate lines ξ^i are orthogonal to these segments. These computed data generate the boundary conditions for P^i. Expansion of the boundary values of P^i over the whole region by algebraic or differential approaches produces the specification of the control functions. The coincidence of P^i and the computed values of $\nabla^2 \xi^i$ on the boundary provides some grounds for the expectation that the solution of the Poisson system with the specified P^i will yield a coordinate system which is nearly orthogonal in the vicinity of the boundary segments.

In this subsection we find some necessary conditions for the boundary values of the control functions P^i to generate coordinates which emanate orthogonally or nearly orthogonally from the respective boundary segments.

6.3.4.1 Two-Dimensional Equations

Now we consider a two-dimensional case. Let a coordinate curve $\xi^2 = \xi_0^2$ be orthogonal to the opposite family of coordinate lines $\xi^1 = \text{const}$. In this case

$$\begin{aligned} g_{12} &= 0, \qquad J = \sqrt{g_{11}g_{22}}, \\ g^{12} &= 0, \qquad g^{11} = 1/g_{11}, \qquad g^{22} = 1/g_{22}, \end{aligned}$$

along this coordinate line $\xi^2 = \xi_0^2$. With these equations, the relations (6.31) for the definition of P^i, $i = 1, 2$, on the coordinate line $\xi^2 = \xi_0^2$ have the form

$$\begin{aligned} P^1 &= -(g^{11})^2[11,1] - g^{22}g^{11}[22,1] \\ &= -\frac{1}{g_{11}}\left(\frac{1}{2g_{11}}\frac{\partial g_{11}}{\partial \xi^1} + \frac{1}{g_{22}}(\boldsymbol{x}_{\xi^2\xi^2} \cdot \boldsymbol{x}_{\xi^1})\right). \end{aligned} \tag{6.33}$$

Analogously, if the coordinate curve $\xi^1 = \xi_0^1$ is orthogonal to the opposite family of coordinate curves, then for the source term P^2 on the curve $\xi^1 = \xi_0^1$ we obtain

$$\begin{aligned} P^2 &= -\frac{1}{2(g_{22})^2}\frac{\partial g_{22}}{\partial \xi^2} - \frac{1}{J^2}\frac{\partial^2 x^i}{\partial \xi^1\, \partial \xi^1}\frac{\partial x^i}{\partial \xi^2} \\ &= -\frac{1}{g_{22}}\left(\frac{1}{2g_{22}}\frac{\partial g_{22}}{\partial \xi^2} + \frac{1}{g_{11}}[11,2]\right). \end{aligned} \tag{6.34}$$

If the curve $\xi^2 = \xi_0^2$ is the boundary segment then all of the quantities in (6.33) are known except g_{22} and $\boldsymbol{x}_{\xi^2\xi^2}$. The metric term g_{22} is connected with the relative grid spacing $|\boldsymbol{x}_{\xi^2}|$ of the coordinate lines $\xi^2 = \text{const}$ by the relation $g_{22} = |\boldsymbol{x}_{\xi^2}|^2$. If the spacing $|\boldsymbol{x}_{\xi^2}|$ is specified on the boundary curve $\xi^2 = \xi_0^2$, then only $\boldsymbol{x}_{\xi^2\xi^2}$ is an unknown quantity in the specification of the control function P^1 on this boundary. In the same way, on the boundary segment $\xi^1 = \xi_0^1$, only $\boldsymbol{x}_{\xi^1\xi^1}$ is an unknown quantity in (6.34) for P^2. One way to define $\boldsymbol{x}_{\xi^1\xi^1}$ and $\boldsymbol{x}_{\xi^2\xi^2}$ and consequently P^1 and P^2 on the respective boundary segments is to apply an iterative procedure which utilizes the equation (6.16) with the term $-2g_{12}\boldsymbol{x}_{\xi^1\xi^2}$ omitted because of the orthogonality condition:

$$g_{22}\boldsymbol{x}_{\xi^1\xi^1} + g_{11}\boldsymbol{x}_{\xi^2\xi^2} = -g_{11}g_{22}P^i\boldsymbol{x}_{\xi^i}. \tag{6.35}$$

Every step allows one to evaluate the control function P^1 on the boundary curves $\xi^2 = \xi_0^2$ and the control function P^2 on the boundary lines $\xi^1 = \xi_0^1$. By expansion from the boundary values, the control functions P^1 and P^2 are evaluated in the domain X^2. By solving the system (6.35) with the obtained control functions P^i, the grid corresponding to the next step is generated in the domain X^2. If convergence is achieved, the final grid is generated satisfying the condition of orthogonality and the specified spacing at the boundary.

Local Straightness at the Boundary

The equations (6.33) and (6.34), which serve to define the control functions P^i on the boundary, are simplified if an additional requirement of local straightness of coordinate lines is imposed. To demonstrate this we note that the vector

$$\boldsymbol{b} = \left(\frac{\partial x^2}{\partial \xi^1}, -\frac{\partial x^1}{\partial \xi^1}\right)$$

is orthogonal to the tangential vector $\boldsymbol{x}_{\xi_1}$. From the assumed condition of orthogonality of the coordinate system along the curve $\xi^2 = \xi_0^2$, we find that the vector $\boldsymbol{x}_{\xi_2}$ is parallel to the vector $\boldsymbol{b}$:

$$\boldsymbol{x}_{\xi^2} = F\boldsymbol{b},$$

i.e.

$$\begin{aligned} \frac{\partial x^1}{\partial \xi^2} &= -F\frac{\partial x^2}{\partial \xi^1}, \\ \frac{\partial x^2}{\partial \xi^2} &= F\frac{\partial x^1}{\partial \xi^1}. \end{aligned} \tag{6.36}$$

Let $\partial x^2/\partial \xi^2 \neq 0$ at the boundary point $\boldsymbol{\xi}_0$. After squaring every equation of the system (6.36) and summing them, we find that $F = \sqrt{g_{22}/g_{11}}$. Therefore,

$$\begin{aligned} [22,1] &= \frac{\partial x^i}{\partial \xi^1}\frac{\partial^2 x^i}{\partial \xi^2\,\partial \xi^2} \\ &= \frac{1}{F}\left(\frac{\partial x^2}{\partial \xi^2}\frac{\partial^2 x^1}{\partial \xi^2\,\partial \xi^2} - \frac{\partial x^1}{\partial \xi^2}\frac{\partial^2 x^2}{\partial \xi^2\,\partial \xi^2}\right) \\ &= \frac{1}{F}\left(\frac{\partial x^2}{\partial \xi^2}\right)^2\frac{\partial}{\partial \xi^2}\left(\frac{\partial x^1}{\partial \xi^2}\Big/\frac{\partial x^2}{\partial \xi^2}\right) \end{aligned} \tag{6.37}$$

at the point $\boldsymbol{\xi}_0$. The substitution of this relation in (6.33) yields

$$P^1 = -\frac{1}{2(g_{11})^2}\frac{\partial g_{11}}{\partial \xi^1} - \frac{1}{J^2 F}\left(\frac{\partial x^2}{\partial \xi^2}\right)^2\frac{\partial}{\partial \xi^2}\left(\frac{\partial x^1}{\partial \xi^2}\Big/\frac{\partial x^2}{\partial \xi^2}\right). \tag{6.38}$$

The ratio $(\partial x^1/\partial \xi^2)/(\partial x^2/\partial \xi^2)$ is merely the slope $\mathrm{d}x^1/\mathrm{d}x^2$ of the family of the coordinate curves $\xi^1 = \text{const}$, which are transverse to the coordinate $\xi^2 = \xi_0^2$. The imposition of the condition that these transverse coordinate lines $\xi^1 = \text{const}$ are locally straight (i.e. have zero curvature) in the neighborhood of the coordinate $\xi^2 = \xi_0^2$ leads to the equation

$$\frac{\partial}{\partial \xi^2}\left(\frac{\partial x^1}{\partial \xi^2}\Big/\frac{\partial x^2}{\partial \xi^2}\right) = 0 \tag{6.39}$$

on the coordinate line $\xi^2 = \xi_0^2$. So in this case we obtain from (6.33) the following expression for P^1,

$$P^1 = -\frac{1}{2(g_{11})^2}\frac{\partial}{\partial \xi^1} g_{11}, \tag{6.40}$$

which the source term P^1 must satisfy along the coordinate curve $\xi^2 = \xi_0^2$ if it is orthogonal to the family of locally straight coordinate lines $\xi^1 = \text{const}$. This equation can be used to compute the numerical value of P^1 at each grid point on the horizontal boundaries where the transformation $\boldsymbol{x}(\boldsymbol{\xi})$ and consequently the metric element g_{11} is specified.

Analogously, if the coordinate line $\xi^1 = \xi_0^1$ is orthogonal to the family of locally straight coordinate curves $\xi^2 = \text{const}$, we have

$$P^2 = -\frac{1}{2(g_{22})^2}\frac{\partial}{\partial \xi^2} g_{22} \tag{6.41}$$

along this coordinate.

Once the control function P^1 is defined at each mesh point of the horizontal boundaries $\xi^2 = 0$ and $\xi^2 = 1$, its value at the interior mesh points can be computed by unidirectional interpolation along the vertical mesh lines $\xi^1 = \text{const}$ between the horizontal boundaries. Similarly, the control function P^2 can be computed by unidirectional interpolation along the horizontal mesh lines $\xi^2 = \text{const}$.

6.3.4.2 Three-Dimensional Equations

Now we find the values of the system (6.25) on a coordinate surface, say $\xi^3 = \xi_0^3$, when it is orthogonal to the family of coordinates ξ^3. These values define the specification of the control functions P^i, $i = 1, 2$, on the coordinate surface to obtain three-dimensional grids nearly orthogonal about this surface through the system (6.14).

From the condition of orthogonality we have the following relations on the surface $\xi^3 = \xi_0^3$:

$$\begin{aligned} g_{13} = g_{23} = 0, \qquad & g^{33} = 1/g_{33}, \\ J = \sqrt{g_{33}\overline{g}}, \quad & \overline{g} = \det(g_{ij}),\ i,j = 1,2. \end{aligned} \tag{6.42}$$

It is also clear that the orthogonality condition on the coordinate surface $\xi^3 = \xi_0^3$ implies that the matrix (g^{ij}), $i, j = 1, 2$, is inverse to the tensor (g_{ij}), $i, j = 1, 2$. In fact the matrix (g_{ij}), $i, j = 1, 2$, is the covariant metric tensor of the surface $\xi^3 = \xi_0^3$ in the coordinates ξ^1, ξ^2 represented by the parametrization

$$\boldsymbol{r}(\boldsymbol{\xi}) : \Xi^2 \to R^3, \quad \boldsymbol{\xi} = (\xi^1, \xi^2),\ \boldsymbol{r} = (x^1, x^2, x^3),$$

where

$$\boldsymbol{r}(\xi^1, \xi^2) = \boldsymbol{x}(\xi^1, \xi^2, \xi_0^3).$$

Correspondingly, the matrix (g^{ij}), $i, j = 1, 2$, is the contravariant metric tensor of the surface $\xi^3 = \xi_0^3$ in the coordinates ξ^1, ξ^2.

The forcing terms P^i, $i=1,2$, are expressed by the system of equations (6.26). We will write out the equations for $i=1,2$ as a sum of two parts. The first part contains only the terms with the superscripts 1 and 2, which thus are related to the coordinate surface $\xi^3=\xi^3_0$. The second part includes the terms with the superscript 3. Thus we assume

$$\begin{aligned} P^i &= P^i_1 + P^i_2, \\ P^i_1 &= \frac{1}{J}\frac{\partial}{\partial \xi^j}(Jg^{ij}), \quad i,j=1,2, \\ P^i_2 &= \frac{1}{J}\frac{\partial}{\partial \xi^3}(Jg^{i3}), \quad i=1,2. \end{aligned} \tag{6.43}$$

Let us consider the case $i=1$. In accordance with the formula (2.21), we obtain for the element g^{13} in (6.43)

$$g^{13} = \frac{g_{21}g_{32}-g_{31}g_{22}}{g}.$$

So, taking into account the relations (6.42) valid on the surface $\xi^3=\xi^3_0$, we obtain

$$g^{13} = -\frac{1}{g_{33}}(g^{21}g_{32}+g^{11}g_{13}).$$

Also, as a result of the condition of orthogonality, we have, on the surface,

$$g^{ij} = \frac{(-1)^{i+j}g_{3-i\ 3-j}}{\overline{g}}, \quad i,j=1,2,$$

with i, j fixed, and thus

$$P^1_2 = \frac{1}{J}\frac{\partial}{\partial \xi^3}(Jg^{13}) = -\frac{1}{g_{33}}\left(g^{21}\frac{\partial}{\partial \xi^3}g_{23} + g^{11}\frac{\partial}{\partial \xi^3}g_{13}\right) \tag{6.44}$$

on $\xi^3=\xi^3_0$. For the term P^i_1 on the surface $\xi^3=\xi^3_0$, we find, using (6.42) and (6.43),

$$\begin{aligned} P^i_1 &= \frac{1}{\sqrt{g_{33}\overline{g}}}\frac{\partial}{\partial \xi^j}(\sqrt{g_{33}\overline{g}}g^{ij}) \\ &= \frac{1}{\sqrt{\overline{g}}}\frac{\partial}{\partial \xi^j}(\sqrt{\overline{g}}g^{ij}) + \frac{g^{ij}}{2g_{33}}\frac{\partial}{\partial \xi^j}g_{33}, \quad i,j=1,2. \end{aligned} \tag{6.45}$$

The relations (6.44) and (6.45) yield

$$\begin{aligned} P^1 &= \frac{1}{\sqrt{\overline{g}}}\frac{\partial}{\partial \xi^j}(\sqrt{\overline{g}}g^{1j}) \\ &\quad + \frac{1}{g_{33}}\left[g^{11}\left(\frac{1}{2}\frac{\partial}{\partial \xi^1}g_{33} - \frac{\partial}{\partial \xi^3}g^{13}\right) + g^{12}\left(\frac{1}{2}\frac{\partial}{\partial \xi^2}g_{33} - \frac{\partial}{\partial \xi^3}g^{23}\right)\right] \\ &= \frac{1}{\sqrt{\overline{g}}}\frac{\partial}{\partial \xi^j}(\sqrt{\overline{g}}g^{1j}) + \frac{1}{g_{33}}\left(g^{11}\frac{\partial x^k}{\partial \xi^1}\frac{\partial^2 x^k}{\partial \xi^3 \partial \xi^3} + g^{12}\frac{\partial x^k}{\partial \xi^2}\frac{\partial^2 x^k}{\partial \xi^3 \partial \xi^3}\right). \end{aligned}$$

Analogously,

$$P^2 = \frac{1}{\sqrt{g}}\frac{\partial}{\partial \xi^j}(\sqrt{g}g^{2j}) + \frac{1}{g_{33}}\left(g^{21}\frac{\partial x^k}{\partial \xi^1}\frac{\partial^2 x^k}{\partial \xi^3 \partial \xi^3} + g^{22}\frac{\partial x^k}{\partial \xi^2}\frac{\partial^2 x^k}{\partial \xi^3 \partial \xi^3}\right).$$

So a general formula for P^i, $i = 1, 2$, on the coordinate surface $\xi^3 = \xi^3_0$ is

$$P^i = \frac{1}{\sqrt{g}}\frac{\partial}{\partial \xi^j}(\sqrt{g}g^{ij}) + \frac{1}{g_{33}}\left(g^{ij}\frac{\partial x^k}{\partial \xi^j}\frac{\partial^2 x^k}{\partial \xi^3 \partial \xi^3}\right), \tag{6.46}$$

where $i, j = 1, 2,\ k = 1, 2, 3$. Equations (6.46) can be used in the same manner as (6.33) and (6.34) to define the values of the control functions at the boundary segments for the purpose of providing orthogonality at the boundary with a specified normal spacing, through an iterative procedure.

If the vector $\boldsymbol{r}_{\xi^3\xi^3}$, $\boldsymbol{r} = (x^1, x^2, x^3)$, is parallel to the vector $\boldsymbol{r}_{\xi^3}$, for example when the curvature of the coordinate lines ξ^3 vanishes on the surface $\xi^3 = \xi^3_0$, then, from the condition of orthogonality, the second sum of (6.46) vanishes, which implies

$$P^i = \frac{1}{\sqrt{g}}\frac{\partial}{\partial \xi^j}(\sqrt{g}g^{ij}), \quad i = 1, 2, \tag{6.47}$$

on the surface $\xi^3 = \xi^3_0$.

As was mentioned, the covariant and contravariant elements of the surface $\xi^3 = \xi^3_0$ in the coordinates ξ^1, ξ^2 coincide with the elements g_{ij} and g^{ij}, respectively, for $i,\ j = 1, 2$. So the expression (6.47) for P^i is the value obtained by applying the Beltrami operator Δ_{B},

$$\Delta_{\mathrm{B}} = \frac{1}{\sqrt{g}}\frac{\partial}{\partial \xi^j}\left(\sqrt{g}g^{kj}\frac{\partial}{\partial \xi^k}\right), \quad j, k = 1, 2, \tag{6.48}$$

to the function $\xi^i(\boldsymbol{x})$, i.e.

$$\Delta_{\mathrm{B}}\xi^i = P^i, \quad i = 1, 2.$$

Analogous equations for P^i are valid for the coordinate surfaces $\xi^i = \xi^i_0$, $i = 1$ or $i = 2$.

6.3.4.3 Projection of the Poisson System on the Boundary Curve

Now we use the three-dimensional system (6.26) to find an expression for P^i, $i = 1, 2, 3$ on a coordinate line ξ^i.

Let the coordinate lines ξ^1 and ξ^2 be orthogonal between themselves and to the curve ξ^3 at all its points. Then on ξ^3

$$\begin{aligned} g^{ij} &= g_{ij} = 0, \quad i \neq j, \\ g^{ii} &= 1/g_{ii}, \quad \text{for each fixed } i = 1, 2, 3, \\ J^2 &= g_{11}g_{22}g_{33}. \end{aligned} \tag{6.49}$$

From the system (6.26) we obtain the formula for the control term P^3:

$$P^3 = \frac{1}{J}\frac{\partial}{\partial \xi^j}(Jg^{3j}) = \frac{1}{J}\left(\frac{\partial}{\partial \xi^1}(Jg^{31}) + \frac{\partial}{\partial \xi^2}(Jg^{32}) + \frac{\partial}{\partial \xi^3}(Jg^{33})\right). \tag{6.50}$$

The condition of orthogonality is obeyed along the whole coordinate curve ξ^3, and therefore we can carry out the differentiation with respect to ξ^3 of the quantities in (6.50) with the relations (6.49) substituted. Thus for the third sum of (6.50) we obtain

$$\begin{aligned}\frac{1}{J}\frac{\partial}{\partial \xi^3}(Jg^{33}) &= \frac{1}{J}\frac{\partial}{\partial \xi^3}\sqrt{\frac{g_{11}g_{22}}{g_{33}}} \\ &= \frac{1}{2J^2}\frac{\partial}{\partial \xi^3}(g_{11}g_{22}) - \frac{1}{2(g_{33})^2}\frac{\partial}{\partial \xi^3}g_{33}.\end{aligned} \tag{6.51}$$

From the formula (2.21),

$$\begin{aligned} g^{31} &= (g_{12}g_{23} - g_{13}g_{22})/J^2, \\ g^{32} &= (g_{13}g_{21} - g_{11}g_{23})/J^2. \end{aligned}$$

Using these relations and (6.50) in (6.51), we obtain

$$\begin{aligned} P^3 &= -\frac{1}{2(g_{33})^2}\frac{\partial}{\partial \xi^3}g_{33} + \frac{1}{J^2}\left(\frac{1}{2}\frac{\partial}{\partial \xi^3}(g_{11}g_{22}) - g_{22}\frac{\partial}{\partial \xi^3}g_{13} - g_{11}\frac{\partial}{\partial \xi^3}g_{23}\right) \\ &= -\frac{1}{2(g_{33})^2}\frac{\partial}{\partial \xi^3}g_{33} - \frac{1}{J^2}\left(g_{22}\frac{\partial^2 x^k}{\partial \xi^1 \partial \xi^1}\frac{\partial x^k}{\partial \xi^3} + g_{11}\frac{\partial^2 x^k}{\partial \xi^2 \partial \xi^2}\frac{\partial x^k}{\partial \xi^3}\right) \\ &= -\frac{1}{2(g_{33})^2}\frac{\partial}{\partial \xi^3}g_{33} - \frac{1}{J^2}(g_{22}[11,3] + g_{11}[22,3]) \quad \text{on } \xi^3. \end{aligned}$$

If the vectors $\boldsymbol{x}_{\xi^1\xi^1}$ and $\boldsymbol{x}_{\xi^2\xi^2}$ lie in the plane orthogonal to the coordinate ξ^3 (for instance, when the coordinate lines ξ^1 and ξ^2 are locally straight at the points intersecting the coordinate ξ^3) then the term in brackets in the above expression for P^3 is zero, and therefore we obtain

$$P^3 = -\frac{1}{2(g_{33})^2}\frac{\partial}{\partial \xi^3}g_{33}$$

on the curve ξ^3.

Analogous equations are obtained for the forcing terms P^i, $i = 1, 2$, along the coordinate curves ξ^i:

$$P^i = -\frac{1}{2(g_{ii})^2}\frac{\partial}{\partial \xi^i}g_{ii}, \quad i = 1, 2, \tag{6.52}$$

provided similar restrictions are applied. Note that the index i is fixed in (6.52).

In the case where the orthogonality condition for the coordinate lines is obeyed at every point of the domain X^3, we also readily obtain

$$P^i = \frac{1}{J}\frac{\partial}{\partial \xi^i}\sqrt{\frac{g_{jj}g_{kk}}{g_{ii}}}, \quad (i, j, k) \text{ cyclic with } i \text{ fixed.}$$

6.3.5 Control of the Angle of Intersection

Now, for two dimensions, we find how to use the source terms P^i to control the angle at which each grid line transverse to the boundary intersects it.

First, we note that the maximum principle for the Laplace operator guarantees that if

$$\nabla^2 f \geq \nabla^2 g$$

and

$$f|_{\partial X^n} = g|_{\partial X^n},$$

then

$$f(x) \leq g(x), \quad x \in X^n.$$

Therefore in the two-dimensional case a decrease of the values of P^i causes an increase of the values of ξ^i, and, correspondingly, a reduction of the intersection angle of the boundary line with the opposite family of coordinate curves. For example if P^i is negative at a grid point obtained through the inverted Laplace system, the point is moved towards the side where ξ^i is less. The opposite effect is produced when the values of P^i are increased. This observation allows one to influence the angle of intersection by choosing larger or smaller values of P^i.

A more sophisticated procedure to control the angle of intersection relies on a study of the dependence of the source terms P^i on the boundary distribution and the angle of intersection.

In place of the orthogonality condition $g_{12} = 0$ we use, therefore, the condition

$$\boldsymbol{x}_{\xi^1} \cdot \boldsymbol{x}_{\xi^2} = g_{12} = |\boldsymbol{x}_{\xi^1}||\boldsymbol{x}_{\xi^2}| \cos\theta = \sqrt{g_{11}g_{22}} \cos\theta, \tag{6.53}$$

where θ denotes the angle of intersection between a coordinate line, say $\xi^2 = \xi^2_0$, and the corresponding transverse coordinate curves $\xi^1 = \text{const}$. So θ is a function depending on ξ^1. A more convenient representation of this condition is

$$g_{12} = J \cot\theta, \tag{6.54}$$

which follows from (6.53) and the equation

$$J = |\boldsymbol{x}_{\xi^1}||\boldsymbol{x}_{\xi^2}| \sin\theta = \sqrt{g_{11}g_{22}} \sin\theta.$$

Since

$$J = \sqrt{g_{11}g_{22} - (g_{12})^2},$$

we obtain from (6.53) and (6.54)

$$\begin{aligned} g^{11} &= \frac{1}{g_{11}\sin^2\theta}, \\ g^{22} &= \frac{1}{g_{22}\sin^2\theta}, \\ g^{12} &= -\frac{\cos\theta}{\sqrt{g_{11}g_{22}}\sin^2\theta}. \end{aligned} \tag{6.55}$$

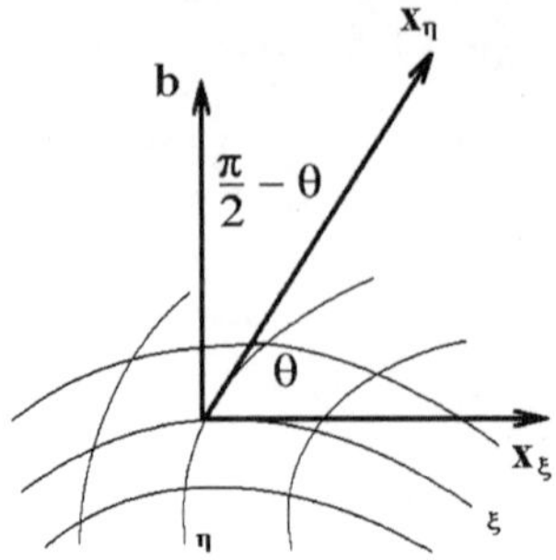

Fig. 6.3. Angle between the normal and tangential vectors

The angle between the tangent vectors $\boldsymbol{x}_{\xi^2}$ and $\boldsymbol{x}_{\xi^1}$ is θ, and therefore the vector $\boldsymbol{x}_{\xi^2}$ intersects the vector

$$\boldsymbol{b} = \left(-\frac{\partial x^2}{\partial \xi^1}, \frac{\partial x^1}{\partial \xi^1}\right),$$

which is orthogonal to the vector $\boldsymbol{x}_{\xi^1}$, at an angle of $\pi/2 - \theta$ (Fig. 6.3). The vectors $\boldsymbol{x}_{\xi^1}$ and $\boldsymbol{b}$ are orthogonal and have the same length: therefore

$$\boldsymbol{x}_{\xi^2} = F(\cos\theta \boldsymbol{x}_{\xi^1} + \sin\theta \boldsymbol{b}),$$

where $F = \sqrt{g_{22}/g_{11}}$. Hence

$$\begin{aligned}\frac{\partial x^1}{\partial \xi^2} &= F\left(\cos\theta\frac{\partial x^1}{\partial \xi^1} - \sin\theta\frac{\partial x^2}{\partial \xi^1}\right),\\ \frac{\partial x^2}{\partial \xi^2} &= F\left(\cos\theta\frac{\partial x^2}{\partial \xi^1} + \sin\theta\frac{\partial x^1}{\partial \xi^1}\right)\end{aligned} \tag{6.56}$$

on the coordinate curve $\xi^2 = \xi^2_0$. It is clear that (6.56) is a generalization of (6.36).

Now we compute the forcing term P^1 on the boundary curves $\xi^2 = 0$ and $\xi^2 = 1$ required to provide control of the angle θ at these segments. For this purpose we consider the representation of the Poisson system (6.13) in the form (6.30).

For $i = 1,\ n = 2$ the system (6.30) implies

$$\begin{aligned}P^1 &= -g^{11}\Gamma^1_{11} - 2g^{12}\Gamma^1_{12} - g^{22}\Gamma^1_{22}\\ &= -g^{11}\left(\frac{\partial^2 x^1}{\partial \xi^1\partial \xi^1}\frac{\partial \xi^1}{\partial x^1} + \frac{\partial^2 x^2}{\partial \xi^1\partial \xi^1}\frac{\partial \xi^1}{\partial x^2}\right)\\ &\quad - 2g^{12}\left(\frac{\partial^2 x^1}{\partial \xi^1\partial \xi^2}\frac{\partial \xi^1}{\partial x^1} + \frac{\partial^2 x^2}{\partial \xi^1\partial \xi^2}\frac{\partial \xi^1}{\partial x^2}\right)\\ &\quad - g^{22}\left(\frac{\partial^2 x^1}{\partial \xi^2\partial \xi^2}\frac{\partial \xi^1}{\partial x^1} + \frac{\partial^2 x^2}{\partial \xi^2\partial \xi^1}\frac{\partial \xi^1}{\partial x^2}\right).\end{aligned} \tag{6.57}$$

In the two-dimensional case we have

$$\begin{aligned}\frac{\partial \xi^1}{\partial x^1} &= \frac{\partial x^2}{\partial \xi^2}\Big/ J,\\ \frac{\partial \xi^1}{\partial x^2} &= -\frac{\partial x^2}{\partial \xi^1}\Big/ J,\end{aligned} \tag{6.58}$$

so, using the relation (6.56) valid along the coordinate $\xi^2 = \xi_0^2$, we obtain

$$\begin{aligned}\frac{\partial \xi^1}{\partial x^1} &= \frac{1}{g_{11}}\left(\cot\theta \frac{\partial x^2}{\partial \xi^1} + \frac{\partial x^1}{\partial \xi^1}\right),\\ \frac{\partial \xi^1}{\partial x^2} &= \frac{1}{g_{11}}\left(\frac{\partial x^2}{\partial \xi^1} - \cot\theta \frac{\partial x^1}{\partial \xi^1}\right).\end{aligned} \tag{6.59}$$

Let the quantities $\partial x^1/\partial \xi^1$ and $\partial x^2/\partial \xi^2$ not vanish at the point under consideration. Then the (6.57) is given by

$$\begin{aligned} g^{11}\Gamma^1_{11} &= g^{11}\left(\frac{\partial^2 x^1}{\partial \xi^1 \partial \xi^1}\frac{\partial \xi^1}{\partial x^1} + \frac{\partial^2 x^2}{\partial \xi^1 \partial \xi^1}\frac{\partial \xi^1}{\partial x^2}\right)\\ &= \frac{1}{(g_{11})^2 \sin^2\theta}\left[\left(\cot\theta \frac{\partial x^2}{\partial \xi^1} + \frac{\partial x^1}{\partial \xi^1}\right)\frac{\partial^2 x^1}{\partial \xi^1 \partial \xi^1}\right.\\ &\quad \left. + \left(-\cot\theta \frac{\partial x^1}{\partial \xi^1} + \frac{\partial x^2}{\partial \xi^1}\right)\frac{\partial^2 x^2}{\partial \xi^1 \partial \xi^1}\right]\\ &= \frac{1}{(g_{11})^2 \sin^2\theta}\left[\frac{1}{2}\frac{\partial}{\partial \xi^1} g_{11} - \cot\theta \left(\frac{\partial x^1}{\partial \xi^1}\right)^2 \frac{\partial}{\partial \xi^1}\left(\frac{\partial x^2}{\partial \xi^1}\Big/\frac{\partial x^1}{\partial \xi^1}\right)\right].\end{aligned} \tag{6.60}$$

In order to compute the second term of (6.57), we note first that from (6.56)

$$\frac{\partial x^1}{\partial \xi^2}\Big/\frac{\partial x^2}{\partial \xi^2} = \frac{\cos\theta \frac{\partial x^1}{\partial \xi^1} - \sin\theta \frac{\partial x^2}{\partial \xi^1}}{\sin\theta \frac{\partial x^1}{\partial \xi^1} + \cos\theta \frac{\partial x^2}{\partial \xi^1}}.$$

Therefore

$$\frac{\partial}{\partial \xi^1}\left(\frac{\partial x^1}{\partial \xi^2}\Big/\frac{\partial x^2}{\partial \xi^2}\right) = F^2\left(\frac{\partial x^2}{\partial \xi^2}\right)^2 [f_1(\xi^1) + f_2(\xi^1)], \tag{6.61}$$

where

$$\begin{aligned} f_1(\xi_1) &= -\left[\left(\sin\theta \frac{\partial x^1}{\partial \xi^1} + \cos\theta \frac{\partial x^2}{\partial \xi^1}\right)^2 + \left(\cos\theta \frac{\partial x^1}{\partial \xi^1} - \sin\theta \frac{\partial x^2}{\partial \xi^1}\right)^2\right]\theta_{\xi^1}\\ &= -g_{11}\theta_{\xi^1},\end{aligned}$$

$$
\begin{aligned}
f_2(\xi_1) &= \left(\cos\theta\frac{\partial^2 x^1}{\partial\xi^1\partial\xi^1} - \sin\theta\frac{\partial^2 x^2}{\partial\xi^1\partial\xi^1}\right)\left(\sin\theta\frac{\partial x^1}{\partial\xi^1} + \cos\theta\frac{\partial x^2}{\partial\xi^1}\right)\\
&\quad - \left(\sin\theta\frac{\partial^2 x^1}{\partial\xi^1\partial\xi^1} + \cos\theta\frac{\partial^2 x^2}{\partial\xi^1\partial\xi^1}\right)\left(\cos\theta\frac{\partial x^1}{\partial\xi^1} - \sin\theta\frac{\partial x^2}{\partial\xi^1}\right)\\
&= \frac{\partial^2 x^1}{\partial\xi^1\partial\xi^1}\frac{\partial x^2}{\partial\xi^1} - \frac{\partial^2 x^2}{\partial\xi^1\partial\xi^1}\frac{\partial x^1}{\partial\xi^1}\\
&= -\left(\frac{\partial x^1}{\partial\xi^1}\right)^2\frac{\partial}{\partial\xi^1}\left(\frac{\partial x^2}{\partial\xi^1}\Big/\frac{\partial x^1}{\partial\xi^1}\right).
\end{aligned}
$$

So, using the relations (6.57) and (6.61), we obtain

$$
\begin{aligned}
g^{12}\Gamma^1_{12} &= g^{12}\left(\frac{\partial^2 x^1}{\partial\xi^1\partial\xi^2}\frac{\partial\xi^1}{\partial x^1} + \frac{\partial^2 x^2}{\partial\xi^1\partial\xi^2}\frac{\partial\xi^1}{\partial x^2}\right)\\
&= \frac{g^{12}}{J}\left(\frac{\partial^2 x^1}{\partial\xi^1\partial\xi^2}\frac{\partial x^2}{\partial\xi^2} - \frac{\partial^2 x^2}{\partial\xi^1\partial\xi^2}\frac{\partial x^1}{\partial\xi^2}\right)\\
&= \frac{1}{J}g^{12}\left(\frac{\partial x^2}{\partial\xi^2}\right)^2\frac{\partial}{\partial\xi^1}\left(\frac{\partial x^1}{\partial\xi^2}\Big/\frac{\partial x^2}{\partial\xi^2}\right)\\
&= \frac{\cot\theta}{(g_{11})^2\sin^2\theta}\left[g_{11}\theta_{\xi^1} + \left(\frac{\partial x^1}{\partial\xi^1}\right)\frac{\partial}{\partial\xi^1}\left(\frac{\partial x^2}{\partial\xi^1}\Big/\frac{\partial x^1}{\partial\xi^1}\right)\right].
\end{aligned}
\tag{6.62}
$$

Analogously, the third term of (6.57) is given by

$$
\begin{aligned}
g^{22}\Gamma^1_{22} &= \frac{\partial^2 x^1}{\partial\xi^2\partial\xi^2}\frac{\partial\xi^1}{\partial x^1} + \frac{\partial^2 x^2}{\partial\xi^2\partial\xi^1}\frac{\partial\xi^1}{\partial x^2}\\
&= \frac{1}{J}\left(\frac{\partial^2 x^1}{\partial\xi^2\partial\xi^2}\frac{\partial x^2}{\partial\xi^2} - \frac{\partial^2 x^2}{\partial\xi^2\partial\xi^2}\frac{\partial x^1}{\partial\xi^2}\right)\\
&= \frac{1}{J}\left(\frac{\partial x^2}{\partial\xi^2}\right)^2\frac{\partial}{\partial\xi^2}\left(\frac{\partial x^1}{\partial\xi^2}\Big/\frac{\partial x^2}{\partial\xi^2}\right).
\end{aligned}
\tag{6.63}
$$

Now, using the relations (6.60), (6.62), and (6.63) in (6.57), we obtain

$$
\begin{aligned}
P^1 &= -\frac{1}{(g_{11})^2\sin^2\theta}\\
&\quad\times\left[\frac{1}{2}\frac{\partial}{\partial\xi^1}g_{11} + \cot\theta\left(\frac{\partial x^1}{\partial\xi^1}\right)^2\frac{\partial}{\partial\xi^1}\left(\frac{\partial x^2}{\partial\xi^1}\Big/\frac{\partial x^1}{\partial\xi^1}\right) + 2g_{11}\cot\theta\;\theta_{\xi^1}\right]\\
&\quad - \frac{1}{Jg_{22}\sin^2\theta}\left(\frac{\partial x^2}{\partial\xi^2}\right)^2\frac{\partial}{\partial\xi^2}\left(\frac{\partial x^1}{\partial\xi^2}\Big/\frac{\partial x^2}{\partial\xi^2}\right)
\end{aligned}
\tag{6.64}
$$

along the coordinate curve $\xi^2 = \xi^2_0$.

Analogously, for the second source term P^2 along the coordinate line $\xi^1 = \xi^1_0$,

$$P^2 = -\frac{1}{(g_{22})^2 \sin^2\theta} \\ \times \left[\frac{1}{2}\frac{\partial}{\partial \xi^2} g_{22} + \cot\theta \left(\frac{\partial x^2}{\partial \xi^2}\right)^2 \frac{\partial}{\partial \xi^2}\left(\frac{\partial x^1}{\partial \xi^2} \Big/ \frac{\partial x^2}{\partial \xi^2}\right) + 2 g_{11} \cot\theta \; \theta_{\xi^2}\right] \\ - \frac{1}{J g_{11} \sin^2\theta}\left(\frac{\partial x^1}{\partial \xi^1}\right)^2 \frac{\partial}{\partial \xi^1}\left(\frac{\partial x^2}{\partial \xi^1} \Big/ \frac{\partial x^1}{\partial \xi^1}\right). \tag{6.65}$$

If the coordinate curves $\xi^i = \text{const}$ are locally straight at the points of their intersection with the opposite coordinate lines, then the last line in (6.64) and (6.65) will vanish. Using the conventional notation x, y and ξ, η instead of x^1, x^2 and ξ^1, ξ^2, we have for the source terms P^1 and P^2 in this case

$$P^1 = -\frac{1}{(g_{11})^2 \sin^2\theta}\left[\frac{1}{2}\frac{\partial}{\partial \xi} g_{22} + \cot\theta \left(\frac{\partial x}{\partial \xi}\right)^2 \frac{\partial}{\partial \xi}\left(\frac{\partial y}{\partial \xi} \Big/ \frac{\partial x}{\partial \xi}\right) \\ + 2 g_{11} \cot\theta \; \theta_\xi\right], \tag{6.66}$$

$$P^2 = -\frac{1}{(g_{22})^2 \sin^2\theta}\left[\frac{1}{2}\frac{\partial}{\partial \eta} g_{22} + \cot\theta \left(\frac{\partial y}{\partial \eta}\right)^2 \frac{\partial}{\partial \eta}\left(\frac{\partial x}{\partial \eta} \Big/ \frac{\partial y}{\partial \eta}\right) \\ + 2 g_{11} \cot\theta \; \theta_\eta\right]. \tag{6.67}$$

6.4 Biharmonic Equations

The main drawback of a grid generation method based on a second-order elliptic differential equation is the limitation in controlling the boundary grid distribution and the direction of the coordinate lines emanating from the boundary. This results in considerable numerical difficulties in the solution of problems involving boundary conditions in the normal direction, for example problems of heat transfer and inviscid aerodynamics. Thus the technique described above which utilizes the forcing terms of the Poisson system to control the directions of the grid lines is not always acceptable.

A more reliable approach to this problem is the use of differential equations of increased order, in particular biharmonic equations. A system of biharmonic equations provides an efficient opportunity to simultaneously satisfy both Dirichlet and Neumann conditions on the boundaries. This provides the flexibility necessary to smoothly patch together the subgrids and control the locations of grid points.

6.4.1 Formulation of the Approach

As for the elliptic system of second order, the most acceptable system of biharmonic equations to produce suitable grids is formulated in terms of the coordinates x^i of the physical domain, used as independent variables, through a composition of Laplace operators:

$$\nabla^2(\nabla^2\xi^i) = 0, \quad i = 1, \dots, n. \tag{6.68}$$

This system is extended to the mixed-boundary-value problem by imposing the boundary conditions

$$\xi^i(\boldsymbol{x}) = f^i(\boldsymbol{x}), \qquad \frac{\partial \xi^i}{\partial \boldsymbol{n}}(\boldsymbol{x}) = 0, \quad \boldsymbol{x} \in \partial X^n. \tag{6.69}$$

The derivative $\partial/\partial \boldsymbol{n}$ is taken in the outward normal direction on the boundary of X^n. Applying the coupled approach, the problem of (6.68) and (6.69) yields the system

$$\begin{aligned} \nabla^2\xi^i &= p^i, \\ \nabla^2 p^i &= 0, \quad i = 1, \dots, n, \end{aligned} \tag{6.70}$$

and the boundary conditions for ξ^i and p^i

$$\begin{aligned} \boldsymbol{\xi}(\boldsymbol{x}) &= \boldsymbol{f}(\boldsymbol{x}), \quad \boldsymbol{x} \in \partial X^n, \\ \boldsymbol{p}(\boldsymbol{x}) &= \nabla^2\boldsymbol{\xi}(\boldsymbol{x}) - c\frac{\partial \boldsymbol{\xi}}{\partial \boldsymbol{n}}(\boldsymbol{x}), \quad \boldsymbol{x} \in \partial X^n, \end{aligned} \tag{6.71}$$

where c is an arbitrary nonzero constant, and

$$\boldsymbol{f} = (f^1, \dots, f^n), \qquad \boldsymbol{p} = (p^1, \dots, p^n).$$

6.4.2 Transformed Equations

In the computational domain Ξ^n with the dependent and independent variables interchanged, the original equations (6.70) become

$$\begin{aligned} g^{ij}\boldsymbol{x}_{\xi^i\xi^j} + p^i\boldsymbol{x}_{\xi^i} &= 0, \\ g^{ij}\boldsymbol{p}_{\xi^i\xi^j} &= 0, \quad i, j = 1, \dots, n. \end{aligned} \tag{6.72}$$

The boundary conditions for (6.72) are, in accordance with (6.71) and (6.29),

$$\begin{aligned} \boldsymbol{x}\big|_{\partial\Xi^n} &= \boldsymbol{f}^{-1}(\boldsymbol{\xi}), \\ p^k\big|_{\partial\Xi^n} &= g^{ij}\Gamma^k_{ij} - c\frac{\partial \xi^k}{\partial \boldsymbol{n}}, \quad i, j, k = 1, \dots, n. \end{aligned} \tag{6.73}$$

6.5 Orthogonal Systems

A system of equations suitable for generating orthogonal grids is commonly obtained in two ways. In the first approach, the system is derived from the following equations representing the condition of orthogonality:

$$g_{ij} \equiv \boldsymbol{x}_{\xi^i} \cdot \boldsymbol{x}_{\xi^j} = 0, \quad i \neq j.$$

The second approach is based on any differential identity which can be derived for a suitable system by eliminating the terms g_{ij}, $i \neq j$. The first approach is considered in Sect. 6.5.1 section for the generation of two-dimensional grids, while the second one is described in Sect. 6.5.2 for two and three-dimensional domains.

6.5.1 Derivation from the Condition of Orthogonality

One example of a differential system, considered by Haussling and Coleman (1981) to generate two-dimensional orthogonal and nearly orthogonal grids, is

$$\begin{aligned} \frac{\partial}{\partial \xi^1} g_{12} &= 0, \\ \frac{\partial}{\partial \xi^2} g_{12} &= 0. \end{aligned} \tag{6.74}$$

The constant solution $g_{12} = \text{const}$ to (6.74) exists only if it is consistent with the boundary data. Only at the corners of the computational region Ξ^2 can g_{12} be specified in advance. Thus the system (6.74) is suitable for obtaining an orthogonal grid when the region X^2 has right angles at the corners. Now we change to the customary notations x, y for x^1, x^2 and ξ, η for ξ^1, ξ^2.

Expanding (6.74) yields

$$x_\xi x_{\xi\eta} + x_{\xi\xi} x_\eta + y_\xi y_{\xi\eta} + y_{\xi\xi} y_\eta = 0, \tag{6.75}$$

$$x_\xi x_{\eta\eta} + x_{\xi\eta} x_\eta + y_\xi y_{\eta\eta} + y_{\xi\eta} y_\eta = 0. \tag{6.76}$$

To compute the transformation $\boldsymbol{r}(\xi, \eta) : \Xi^2 \to X^2$, $\boldsymbol{r} = (x, y)$, these equations are combined as follows. The product of (6.75) and x_η is added to the product of (6.76) and x_ξ, giving

$$\begin{aligned} &(x_\eta)^2 x_{\xi\xi} + (x_\xi)^2 x_{\eta\eta} + 2x_\xi x_\eta x_{\xi\eta} + x_\eta y_\eta y_{\xi\xi} + x_\xi y_\xi y_{\eta\eta} \\ &\quad + (x_\eta y_\xi + x_\xi y_\eta) y_{\xi\eta} = 0. \end{aligned} \tag{6.77}$$

The product of (6.75) and y_η is added to the product of (6.76) and y_ξ, yielding

$$\begin{aligned} &(y_\eta)^2 y_{\xi\xi} + (y_\xi)^2 y_{\eta\eta} + 2y_\xi y_\eta y_{\xi\eta} + x_\eta y_\eta x_{\xi\xi} + x_\xi y_\xi x_{\eta\eta} \\ &\quad + (x_\eta y_\xi + x_\xi y_\eta) x_{\xi\eta} = 0. \end{aligned} \tag{6.78}$$

The systems (6.77), and (6.78) are approximated by central differences and the resulting algebraic systems are solved iteratively using successive overrelaxation. The reason for replacing (6.75), and (6.76) with (6.77), and (6.78) is to obtain a nonzero coefficient for x_{ij} and y_{ij} in the finite-difference forms of (6.77) and (6.78). This eliminates the possibility of dividing by zero in the iteration process.

6.5.2 Multidimensional Equations

A multidimensional differential system for generating orthogonal and nearly orthogonal grids is usually obtained by the second approach, using some differential identities and then eliminating the terms g_{ij}, $i \neq j$. One example gives the identity (2.47),

$$\frac{\partial}{\partial \xi^j}\left(J\frac{\partial \xi^j}{\partial x^k}\right) \equiv 0, \quad j,k=1,\dots,n.$$

In accordance with (2.23),

$$\frac{\partial \xi^j}{\partial x^k} = g^{ij}\frac{\partial x^k}{\partial \xi^i}, \quad i,j,k=1,\dots,n.$$

Using this relation in the above equation, another form of the identity is obtained:

$$\frac{\partial}{\partial \xi^j}\left(Jg^{ij}\frac{\partial x^k}{\partial \xi^i}\right) \equiv 0, \quad i,j,k=1,\dots,n, \tag{6.79}$$

which also follows from the Beltrami equations

$$\frac{\partial^2 x^k}{\partial x^j \partial x^j} \equiv \frac{1}{J}\frac{\partial}{\partial \xi^j}\left(Jg^{ij}\frac{\partial x^k}{\partial \xi^i}\right) \equiv 0, \quad i,j,k=1,\dots,n. \tag{6.80}$$

Substituting the condition of orthogonality

$$g^{ij}=0, \quad i \neq j, \; i,j=1,\dots,n,$$

in the equations (6.79) for g^{ij}, $i \neq j$, we obtain the system of elliptic equations required to generate orthogonal coordinates:

$$\frac{\partial}{\partial \xi^j}\left(Jg^{jj}\frac{\partial x^k}{\partial \xi^j}\right) = 0, \quad j,k=1,\dots,n, \tag{6.81}$$

where

$$J = \sqrt{\prod_{i=1}^{n} g_{ii}},$$

$$g^{jj} = 1/g_{jj}, \quad \text{for each fixed } j=1,\dots,n.$$

In two dimensions, using the common notations x,y for the dependent variables and ξ,η for the independent variables, the system (6.81) is expressed as

$$\begin{aligned} \frac{\partial}{\partial \xi}\left(F\frac{\partial x}{\partial \xi}\right) + \frac{\partial}{\partial \eta}\left(\frac{1}{F}\frac{\partial x}{\partial \eta}\right) &= 0, \\ \frac{\partial}{\partial \xi}\left(F\frac{\partial y}{\partial \xi}\right) + \frac{\partial}{\partial \eta}\left(\frac{1}{F}\frac{\partial y}{\partial \eta}\right) &= 0, \end{aligned} \tag{6.82}$$

with $F=\sqrt{g_{22}/g_{11}}$.

Analogously, for the three-dimensional system, we obtain from (6.81)

$$\frac{\partial}{\partial \xi^i}\left(F_i \frac{\partial \boldsymbol{x}}{\partial \xi^i}\right) = 0, \quad i = 1, 2, 3, \ i \text{ fixed}, \tag{6.83}$$

where

$$F_i = g_{kk} g_{ll} / g_{ii}, \quad (i, k, l) \text{ cyclic and fixed},$$

i.e.

$$\begin{aligned} F_1 &= \sqrt{g_{22} g_{33} / g_{11}}, \\ F_2 &= \sqrt{g_{33} g_{11} / g_{22}}, \\ F_3 &= \sqrt{g_{11} g_{22} / g_{33}}. \end{aligned}$$

6.6 Hyperbolic and Parabolic Systems

The generation of a grid through elliptic systems can consume a large amount of computational time. This disadvantage gave rise to the development of grid techniques based on hyperbolic-type and parabolic-type partial differential equations.

Hyperbolic grid generation relies on the numerical solution of hyperbolic systems of equations. The hyperbolic equations allow one to use a marching numerical solution without any iteration or initial guess, which makes their use very simple and inexpensive.

Hyperbolic methods are efficient for generating grids in domains around bodies. The solution marches from the inner boundary toward the outer field generating loops of grids one by one, so the computational time is almost equal to that of one iteration of solving elliptic grid generation equations by an iterative scheme. So the computational time required to generate the grid by the marching algorithm is only a very small fraction of that for the elliptic grid generation equations and the fast-memory space required during grid generation can be substantially reduced from that required by the elliptic grid generation method. Furthermore, hyperbolic equations are very suitable for providing grid orthogonality and grid node clustering.

However, hyperbolic grid systems also have their inherent undesirable properties:

(1) since the hyperbolic methods are essentially a marching procedure, the specification of the entire boundary is not allowed, and therefore the methods are not appropriate for the computation of internal and closed systems;
(2) the techniques propagate singularities of the boundary into the interior of the domain;
(3) grid oscillation or even overlapping of grid lines is often encountered in hyperbolic grid generation unless artificial damping terms for stability are appropriately added to the equations.

There are two major approaches in formulating hyperbolic systems. In the first approach the Jacobian of the transformation is specified. The second imposes a specification of the cell aspect ratio.

Parabolic methods possess some of the advantages of both elliptic and hyperbolic techniques. The advantages of using parabolic partial differential equations to generate structured grids are as follows:

(1) parabolic equations allow for formulating initial-value problems, so grids are generated by a marching algorithm as in the hyperbolic grid generation method;
(2) the parabolic equations have most of the properties of the elliptic equations, in particular, the diffusion effect which smooths out any singularity of the inner boundary condition, and prescribed outer boundary conditions may be satisfied.

6.6.1 Specification of Aspect Ratio

The condition of orthogonality $g^{ij} = \boldsymbol{x}_{\xi^1} \cdot \boldsymbol{x}_{\xi^2} = 0$ alone is not sufficient for obtaining the coordinate transformation $\boldsymbol{x}(\boldsymbol{\xi}) : \Xi^2 \to X^2$. Two equations are needed, since both the x^1 and x^2 coordinates of the transformed grid points are to be found.

6.6.1.1 Initial-Value Problems

Here a method presented by Starius (1977) for determining orthogonal grids, based on nonlinear hyperbolic initial-value problems which are formally related to the Cauchy–Riemann equations, is considered. For convenience the ordinary notations x, y for x^1, x^2 and ξ, η for ξ^1, ξ^2 are utilized in this subsubsection.

The orthogonality requirement $g_{12} = 0$ yields the initial-value problem

$$\begin{aligned} x_\eta &= -y_\xi F, \quad x(\xi, 0) = x(\xi), \\ y_\eta &= x_\xi F, \quad y(\xi, 0) = y(\xi), \end{aligned} \tag{6.84}$$

where F is a positive function which is selected by meeting the following set of requirements:

(1) $F = F(\xi, \eta, x, y, x_\xi, y_\xi, x_\eta, y_\eta)$;
(2) a condition of invariance;
(3) a condition on the hyperbolic type of the system (6.84);
(4) geometrical conditions depending on the region X^2;
(5) sufficient conditions for well-posedness of the nonlinear hyperbolic initial-value problems.

The invariance conditions are simply invariance under transitions and rotations. Let (x, y) be a solution of the equations (6.84); then $(\overline{x}, \overline{y})$, defined by either

$$\begin{pmatrix} \overline{x} \\ \overline{y} \end{pmatrix} = \begin{pmatrix} x \\ y \end{pmatrix} + \begin{pmatrix} a \\ b \end{pmatrix}$$

or

$$\begin{pmatrix} \overline{x} \\ \overline{y} \end{pmatrix} = Q \begin{pmatrix} x \\ y \end{pmatrix}$$

where

$$Q = \begin{pmatrix} \cos\theta & -\sin\theta \\ \sin\theta & \cos\theta \end{pmatrix}$$

is the matrix of rotation, and a, b, θ are arbitrary constants, is also a solution. The first equation implies that F does not depend on (x, y), and the second implies the following partial differential equation for F:

$$-y_\xi F_{x_\xi} + x_\xi F_{y_\xi} - y_\eta F_{x_\eta} + x_\eta F_{y_\eta} = 0.$$

The solution of this equation is given by

$$F = F(\xi, \eta, g_{11}, g_{22}, g_{12}).$$

This is, in fact, the general solution. The last argument of this function F is zero because of the orthogonality requirement. Futher, by squaring (6.84) and adding we find that g_{22} is connected with g_{11} by the relation

$$g_{22} = F^2 g_{11},$$

i.e. F is the aspect ratio. Therefore only positive functions F depending on ξ, η and g_{11} have to be considered, i.e.

$$F = F(\xi, \eta, z), \quad z = \sqrt{g_{11}}.$$

The quantity z has an evident geometrical interpretation, namely, it is the length of the tangential vector $\boldsymbol{x}_\xi$.

In order the initial-value problem (6.84) is well-posed it is necessary that the system is hyperbolic. The type of the system (6.84) is defined by the eigenvalues of the matrix

$$M = \begin{pmatrix} -\frac{x_\xi y_\xi}{z} F_z & -F - \frac{y_\xi^2}{z} F_z \\ F + \frac{x_\xi^2}{z} F_z & \frac{x_\xi y_\xi}{z} F_z \end{pmatrix}$$

obtained through the linearization of the system (6.84). The system is hyperbolic if the matrix M has real eigenvalues. The characteristic equation for M is given by

$$\lambda^2 = -F(F + zF_z) = -\frac{f f_z}{z},$$

where

$$f = Fz.$$

Thus $f_z < 0$ if $F > 0$, and so the function $f = zF$ is strictly decreasing in z. The inequality $f_z < 0$ implies that M has distinct real eigenvalues. For $f_z = 0$ we have a multiple real eigenvalue but only one eigenvector.

Before a specific f can be chosen, the quantity z must be normalized in some sense. Let ξ be the arc length of ∂X^2; then $z = 1$ there. This means an equidistant grid spacing on the boundary $\eta = \eta_0$. A graded grid in the tangential direction is obtained with a suitable choice of the function $\xi(t)$. Grading in the other direction can be achieved in the same way, which implies that f contains an a factor depending on the direction of η. When different gradings in η are required for different values of ξ, the factor must depend on ξ as well.

In order to specify the function f, two cases are considered:

(1) the spacing of the mesh in the η direction is about the same along the whole boundary;
(2) the spacing is variable.

The spacing of the mesh along the η curve is expressed by

$$\int_0^{\eta_0} \sqrt{g_{22}}\mathrm{d}\eta = \int_0^{\eta_0} f\mathrm{d}\eta$$

for all ξ. From this equation it is seen that in the first case f does not need to depend explicitly on ξ, i.e. it will depend only on z.

When solving nonlinear hyperbolic problems, discontinuities and shocks generally appear in the solution or in its derivatives; f must be chosen so that this cannot happen as long as $z > 0$. It was proved by Starius that a suitable f is

$$f(z) = \frac{A + Bz^2}{C + z^2}, \tag{6.85}$$

where A, B, and C are constants such that $A > BC$, and $A + B = C + 1$. Since $f(0) = A/C$, the divergence factor in the η direction from the outer to the inner boundary of the curvilinear mesh can never exceed this quantity.

For the second case, it was assumed by Starius that

$$f(\xi, z) = a(\xi) f_0(\xi, z),$$

where $a(\xi)$ is a periodic function in ξ such that $0 < a \leq 1$. By using this f in (6.84) we can see that $a(\xi)$ is a relative measure of the grid spacing in the η direction.

6.6.2 Specification of Jacobian

6.6.2.1 Orthogonal Grids in Two Dimensions

In the two-dimensional case the hyperbolic coordinate equations obtained by specifying the Jacobian and a measure of orthogonality have the form

$$\begin{aligned} \boldsymbol{x}_{\xi^1} \cdot \boldsymbol{x}_{\xi^2} &= 0, \\ |\boldsymbol{x}_{\xi^1} \times \boldsymbol{x}_{\xi^2}| &= J, \end{aligned} \tag{6.86}$$

that is,

$$\begin{aligned} \frac{\partial x^1}{\partial \xi^1}\frac{\partial x^1}{\partial \xi^2} + \frac{\partial x^2}{\partial \xi^1}\frac{\partial x^2}{\partial \xi^2} &= 0, \\ \frac{\partial x^1}{\partial \xi^1}\frac{\partial x^2}{\partial \xi^2} - \frac{\partial x^1}{\partial \xi^2}\frac{\partial x^2}{\partial \xi^1} &= J, \end{aligned} \tag{6.87}$$

where J is a specified area source term. These equations form a system of nonlinear partial differential equations whose solution is based upon solving the system

$$A\boldsymbol{x}_{\xi^1} + B\boldsymbol{x}_{\xi^2} = \boldsymbol{f}, \tag{6.88}$$

where

$$A = \begin{pmatrix} \frac{1}{\sqrt{g^0_{11}g^0_{22}}}\frac{\partial x^1_0}{\partial \xi^2} - \frac{1}{g^0_{11}}\frac{\partial x^1_0}{\partial \xi^1} & \frac{1}{\sqrt{g^0_{11}g^0_{22}}}\frac{\partial x^2_0}{\partial \xi^2} - \frac{1}{g^0_{11}}\frac{\partial x^2_0}{\partial \xi^1} \\ \frac{\partial x^2_0}{\partial \xi^2} & -\frac{\partial x^1_0}{\partial \xi^2} \end{pmatrix},$$

$$B = \begin{pmatrix} \frac{1}{\sqrt{g^0_{11}g^0_{22}}}\frac{\partial x^1_0}{\partial \xi^1} - \frac{1}{g^0_{22}}\frac{\partial x^1_0}{\partial \xi^2} & \frac{1}{\sqrt{g^0_{11}g^0_{22}}}\frac{\partial x^2_0}{\partial \xi^1} - \frac{1}{g^0_{22}}\frac{\partial x^2_0}{\partial \xi^2} \\ -\frac{\partial x^2_0}{\partial \xi^1} & \frac{\partial x^1_0}{\partial \xi^1} \end{pmatrix},$$

$$\boldsymbol{x} = (x_1, x_2)^T, \qquad \boldsymbol{f} = (0,\ J + J_0)^T.$$

Equation (6.88) represents the linearization of (6.87) about the state $(\boldsymbol{x}_0)$. Taking ξ^2 as a marching direction, we obtain from (6.88)

$$\boldsymbol{x}_{\xi^2} + B^{-1}A\boldsymbol{x}_{\xi^1} = B^{-1}\boldsymbol{f}. \tag{6.89}$$

For the eigenvalues $\lambda_1,\ \lambda_2$ of the two-dimensional metric $B^{-1}A$ we have

$$\begin{aligned} \lambda_1\lambda_2 &= \det(B^{-1}A), \\ \lambda_1 + \lambda_2 &= \mathrm{Tr}(B^{-1}A). \end{aligned}$$

As

$$\begin{aligned} &\det(B^{-1}A) = F = |\boldsymbol{x}_{\xi^2}|/|\boldsymbol{x}_{\xi^1}| = \sqrt{g_{22}/g_{11}}, \\ &\mathrm{Tr}(B^{-1}A) = 0, \end{aligned}$$

we obtain $\lambda_1 = F,\ \lambda_2 = -F$. Hence the system (6.89) is hyperbolic and the local solution consists of a left- and a right-running wave. Equation (6.89) is typically modified by adding an artificial term $\epsilon\boldsymbol{x}_{\xi^1\xi^1}$ to stabilize the numerical scheme:

$$\boldsymbol{x}_{\xi^2} + B^{-1}A\boldsymbol{x}_{\xi^1} + \epsilon\boldsymbol{x}_{\xi^1\xi^1} = B^{-1}\boldsymbol{f}. \tag{6.90}$$

6.6.2.2 Two-Dimensional Nonorthogonal Grids

A more general hyperbolic system which does not include any constraints on the angle θ between the tangential vectors $\boldsymbol{x}_{\xi^1}$ and $\boldsymbol{x}_{\xi^2}$ is obtained from the identities

$$g_{12} = \sqrt{g_{11}g_{22}}\cos\theta, \qquad \sqrt{g_{11}g_{22}}\sin\theta = J, \tag{6.91}$$

where θ and J can be user-specified. Choosing the ξ^2 direction to be the marching direction and solving the system (6.90) for $\partial x^1/\partial\xi^2$, $\partial x^2/\partial\xi^2$, we obtain

$$\begin{aligned}\frac{\partial x^1}{\partial\xi^2} &= \sqrt{\frac{g_{22}}{g_{11}}}\left(\frac{\partial x^1}{\partial\xi^1}\cos\theta - \frac{\partial x^2}{\partial\xi^1}\sin\theta\right),\\ \frac{\partial x^2}{\partial\xi^2} &= \sqrt{\frac{g_{22}}{g_{11}}}\left(\frac{\partial x^2}{\partial\xi^1}\cos\theta + \frac{\partial x^1}{\partial\xi^1}\sin\theta\right).\end{aligned} \tag{6.92}$$

The linearization of these equations produces the system (6.88) with

$$A = \begin{pmatrix} \frac{1}{\sqrt{g^0_{11}g^0_{22}}}\frac{\partial x^1_0}{\partial\xi^2} - \frac{1}{g^0_{11}}\cos\theta\frac{\partial x^1_0}{\partial\xi^1} & \frac{1}{\sqrt{g^0_{11}g^0_{22}}}\frac{\partial x^2_0}{\partial\xi^2} - \frac{1}{g^0_{11}}\cos\theta_0\frac{\partial x^2_0}{\partial\xi^1} \\ \frac{\partial x^2_0}{\partial\xi^2} & -\frac{\partial x^1_0}{\partial\xi^2} \end{pmatrix},$$

$$B = \begin{pmatrix} \frac{1}{\sqrt{g^0_{11}g^0_{22}}}\frac{\partial x^1_0}{\partial\xi^1} - \frac{1}{g^0_{22}}\cos\theta\frac{\partial x^1_0}{\partial\xi^2} & \frac{1}{\sqrt{g^0_{11}g^0_{22}}}\frac{\partial x^2_0}{\partial\xi^1} - \frac{1}{g^0_{22}}\cos\theta_0\frac{\partial x^2_0}{\partial\xi^2} \\ -\frac{\partial x^2_0}{\partial\xi^1} & \frac{\partial x^1_0}{\partial\xi^1} \end{pmatrix},$$

$$\boldsymbol{f} = (\cos\theta + \cos\theta_0,\ J + J_0)^T.$$

The matrix B^{-1} exists when $\sin\theta \neq 0$, and

$$\lambda = \pm\sqrt{\frac{g_{22}}{g_{11}}}.$$

Hence the system (6.92) in this case is also hyperbolic.

The introduction of the angle θ into the system (6.92) allows one to solve the initial-value problem, i.e. to specify grid data on the initial boundary $\xi^2 = 0$ and the side boundaries $\xi^1 = 0$, $\xi^1 = 1$. For (6.92) the boundary curves $\xi^2 \to \boldsymbol{x}(\xi^1_0, \xi^2)$, $\xi^1_0 = 0$ or $\xi^1_0 = 1$, need not intersect the initial curve $\xi^2 = 0$ orthogonally and so the initial-value problem is typically ill-posed. Equations (6.92), however, give an opportunity to choose the angle terms near the boundary so that a consistent problem results.

6.6.2.3 Three-Dimensional Version

The three-dimensional hyperbolic grid generation approach, where the marching direction is say, ξ^3, is based on two orthogonality relations and an additional equation to control the Jacobian as follows:

$$\begin{aligned}\boldsymbol{x}_{\xi^1} \cdot \boldsymbol{x}_{\xi^3} &= 0,\\ \boldsymbol{x}_{\xi^2} \cdot \boldsymbol{x}_{\xi^3} &= 0,\\ \det\left(\frac{\partial x^i}{\partial \xi^j}\right) &= J, \quad i,j = 1,2,3.\end{aligned} \tag{6.93}$$

The local linearization of the system (6.93) with respect to $(\boldsymbol{x} - \boldsymbol{x}_0)$, where $\boldsymbol{x}_0(\xi)$ is a known state, neglecting products of small quantities that are second order in $(\boldsymbol{x} - \boldsymbol{x}_0)$, yields

$$A_0(\boldsymbol{x} - \boldsymbol{x}_0)_{\xi^1} + B_0(\boldsymbol{x} - \boldsymbol{x}_0)_{\xi^2} + C_0(\boldsymbol{x} - \boldsymbol{x}_0)_{\xi^3} = \boldsymbol{f}, \tag{6.94}$$

where $\boldsymbol{x} = (x^1, x^2, x^3)^T$, $\boldsymbol{x}_0 = (x_0^1, x_0^2, x_0^3)^T$, A_0, B_0, and C_0 are coefficient matrices that are evaluated from $\boldsymbol{x}_0(\boldsymbol{\xi})$, and the subscripts s^i, $i = 1,2,3$, denote partial derivatives.

6.6.3 Parabolic Equations

The parabolic grid approach lies between the elliptic and hyperbolic ones.

The two-dimensional parabolic grid generation equation where the marching direction is ξ^2 may be written in the following form:

$$\boldsymbol{x}_{\xi^2} = A_1 \boldsymbol{x}_{\xi^1\xi^1} - B_1 \boldsymbol{x} + \boldsymbol{S}, \tag{6.95}$$

where A_1, B_1, and B are matrix coefficients, and $\boldsymbol{S}$ is a source vector that contains the information about the outer boundary configuration. Analogously, the three-dimensional parabolic equations may be written as follows:

$$\boldsymbol{x}_{\xi^3} = A_i \boldsymbol{x}_{\xi^i\xi^i} - B_1 \boldsymbol{x} + \boldsymbol{S}, \quad i = 1,2. \tag{6.96}$$

6.6.4 Hybrid Grid Generation Scheme

The combination of the hyperbolic and parabolic schemes into a single scheme is attractive because it can use the advantages of both schemes. These advantages are; first, it is a noniterative scheme; second, the orthogonality of the grid near the initial boundary is well controlled; and third, the outer boundary can be prescribed.

A hybrid grid generation scheme in two dimensions for the particular marching direction ξ^2 can be derived by combining (6.89) and (6.95), in particular, as the sum of equations (6.89) and (6.95) multiplied by weights α and $1 - \alpha$, respectively:

$$\begin{aligned}&\alpha(B^{-1}A\boldsymbol{x}_{\xi^1} + \boldsymbol{x}_{\xi^2}) + (1-\alpha)(\boldsymbol{x}_{\xi^2} - A_1\boldsymbol{x}_{\xi^1\xi^1} + B_1\boldsymbol{x})\\ &\quad = \alpha B^{-1}\boldsymbol{f} + (1-\alpha)\boldsymbol{S}.\end{aligned} \tag{6.97}$$

The parameter α can be changed as desired to control the proportions of the two methods. If α approaches 1, (6.97) becomes the hyperbolic grid generation equation, while if α approaches zero it becomes the parabolic grid generation equation. In practical applications α is set to 1, when the grid generation starts from the initial boundary curve $\xi^2 = 0$, but it gradually decreases and approaches zero when the grid reaches the outer boundary.

An analogous combination of (6.94) and (6.96) can be used to generate three-dimensional grids through a hybrid of parabolic and hyperbolic equations.

6.7 Comments

A two-dimensional Laplace system (6.3) which implied the physical coordinates to be solutions in the logical domain Ξ^2 was introduced by Godunov and Prokopov (1967), Barfield (1970), and Amsden and Hirt (1973). A general two-dimensional elliptic system of the type (6.2) for generating structured grids was considered by Chu (1971).

A two-dimensional Laplace system (6.4) using the logical coordinates ξ^i as dependent variables was proposed by Crowley (1962) and Winslow (1967). The technique presented in this chapter to analyze the qualitative behavior near boundary segments of the coordinate lines obtained through the inverted Laplace equations was introduced by the author of this book. A rather geometric approach was described for this purpose in the monograph by Thompson, Warsi, and Mastin (1985). A more detailed analysis of grid behavior near the boundary of both surfaces and two- and three-dimensional domains was performed by Liseikin (2004, 2007).

Godunov and Prokopov (1972) obtained a system of the Poisson type (6.13) assuming that its solution is a composition of conformal and stretching transformations. The general Poisson system presented in the current book was justified by Thompson, Thames, and Mastin (1974) and Thompson, Warsi, and Mastin (1985) in their monograph.

The algorithm aimed at grid clustering at a boundary and forcing grid lines to intersect the boundary in a nearly normal fashion through the source terms of the Poisson system was developed by Steger and Sorenson (1979), Visbal and Knight (1982), and White (1990). Thomas and Middlecoff (1980) described a procedure to control the local angle of intersection between transverse grid lines and the boundary through the specification of the control functions. Control of grid spacing and orthogonality was performed by Tamamidis and Assanis (1991) by introducing a distortion function (the ratio of the diagonal metric elements) into the system of Poisson equations. Warsi (1982) replaced the source terms P^i in (6.13) by $g^{ii}P^i$ (i fixed) to improve the numerical behavior of the generator. As a result the modified system acquired the property of satisfying the maximum principle.

The technique based on setting to zero the off-diagonal elements of the elliptic system was proposed by Lin and Shaw (1991) to generate nearly orthogonal grids, while Soni et al. (1993) used a specification of the control functions for this purpose.

A composition of Poisson's and Laplace's equations in the computational domain to derive biharmonic equations of fourth order was used to generate smooth block-structured grids via the specification of grid line slopes and boundary point distributions by Bell, Shubin, and Stephens (1982). Schwarz (1986) used for this purpose equations of sixth order, which were composed of Poisson and Laplace systems with respect to the dependent physical coordinates x^i. An alternative method, based on the solution of biharmonic equations in the physical domain, was introduced by Sparis (1985). A recent implementation of the biharmonic equations to provide boundary orthogonality and off-boundary spacing as boundary conditions was presented by Sparis and Karkanis (1992).

A combination of elliptic and algebraic techniques was applied by Spekreijse (1995) to generate two- and three-dimensional grids. An approach to formulating an orthogonal system by differentiating nondiagonal metric elements was developed by Haussling and Coleman (1981). Ryskin and Leal (1983), in two dimensions, and Theodoropoulos and Bergeles (1989) in three dimensions, have developed elliptic methods for nearly orthogonal grid generation.

The first systematic analysis of the use of two-dimensional hyperbolic equations to generate orthogonal grids was made by Starius (1977) and Steger and Chaussee (1980), although hyperbolic grid generation can be traced back to McNally (1972). This system was generalized by Cordova and Barth (1988). They developed a two-dimensional hyperbolic system with an angle-control source term which allows one to constrain a grid with more than one boundary. A combination of grids using the hyperbolic technique of Steger and Chaussee (1980), which starts from each boundary segment was generated by Jeng and Shu (1995). The extension to three dimensions was performed by Steger and Rizk (1985), Chan and Steger (1992), and Tai, Chiang, and Su (1996), who introduced grid smoothing as well.

The generation of grids based on a parabolic scheme approximating the inverted Poisson equations was first proposed for two-dimensional grids by Nakamura (1982). A variation of the method of Nakamura was developed by Noack (1985) to use in space-marching solutions to the Euler equations. Extensions of this parabolic technique to generate solution adaptive grids were performed by Edwards (1985) and Noack and Anderson (1990).

A combination of hyperbolic and parabolic schemes that uses the advantages of the two but eliminates the drawbacks of each was proposed by Nakamura and Suzuki (1987).

7

Dynamic Adaptation

7.1 Introduction

Some basic differential methods for generating structured grids have been discussed in Chap. 6. However, a very important aspect of structured grid generation concerned with adaptation of grids to the numerical solution of partial differential equations has not been covered fully so far in the discussion. The goal of the current chapter is to partly eliminate this drawback by giving an elementary introduction to the subject of adaptive grid generation.

Adaptive grids are commonly considered as meshes which adjust to physical quantities in such a way that the quantities can be represented with the greatest possible accuracy and at optimal cost through a discrete solution on a given number of grid points or grid cells. For particular problems the adjustment can be accomplished on the basis of a theoretical or computational analysis, conducted beforehand, of those qualitative and/or quantitative properties of the solution which have the greatest influence on the accuracy of the numerical computation. However, it is more preferable for the development of automated grid codes to have a dynamic adaptation in which the process of grid generation is coupled with the numerical analysis of the physical problem, and consequently the grid points are built in response to the evolving solution without reliance on a priori knowledge of the properties of the solution.

In many practical problems there may exist narrow regions in the physical domains where the dependent quantities undergo large variations. These regions include shock waves in compressible flows, shear layers in laminar and turbulent flows, expansion fans, contact surfaces, slipstreams, phase-change interfaces, and boundary and interior layers, which, when interacting, can present significant difficulties in the numerical treatment. The need for a detailed description of the physical solutions to such problems requires the development of adaptive methods whose adaptivity is judged by their ability to provide a suitable concentration of grid nodes in these regions in comparison with the distribution of the nodes in the rest of the domain. One approach to generating a clustered grid in the appropriate zones was consid-

V.D. Liseikin, *Grid Generation Methods*, Scientific Computation,
DOI 10.1007/978-90-481-2912-6_7, © Springer Science+Business Media B.V. 2010

ered in Chap. 5. However, in the case of multidimensional nonlinear systems of equations, the locations of the zones of large solution variation are, as a rule, not known beforehand, and hence the explicit distribution of the grid points by the stretching method described in Chap. 5 cannot be successfully accomplished. One of the most efficient tools, which provides a real opportunity to enhance the efficiency of the numerical solution of these problems, is the adaptive grid technique, aimed at the dynamic distribution of the grid nodes with clustering in the zones of rapid change of the solution variables.

Analytical and numerical investigations have demonstrated that the adaptive grid technique has a significant potential to enhance the accuracy and efficiency of computational algorithms. This is especially true for the calculation of multidimensional and unstable problems with boundary and interior layers where the derivatives of the solution are large. Adaptivity can eliminate oscillations associated with inadequate resolution of large gradients more effectively, reduce the undesirable numerical viscosity, damp out instabilities, and considerably curtail the number of grid nodes needed to yield an acceptable solution of a problem relative to the number of nodes of a uniform grid. The interpolation of functions by discrete values is also more accurately performed over the whole region when the grid nodes are clustered in the zones of large derivatives of the functions.

Adaptive grids are, therefore, an important subject to study because of their potential for improving the accuracy and efficiency of the numerical solution of boundary value problems modeling various complex physical phenomena, and serious efforts have been undertaken to develop and enhance the methods of adaptive grid generation and to incorporate these methods into the numerical algorithms for solving field problems.

However, the problem of the development of robust adaptive grid methods is a serious challenge since conflicting demands are imposed on the methods; in particular, they should provide adequate resolution of the solution quantities in regions of high gradients, while also limiting the total number of points and excessive deformation of grid cells.

This chapter is concerned with the description of certain approaches to dynamic adaptation techniques. The emphasis is placed on the equidistribution techniques, realized by differential equations, some of which were discussed in Chap. 6. One more realization of the equidistribution techniques and realizations of other concepts through variational techniques will be considered in Chaps. 8 and 10.

7.2 One-Dimensional Equidistribution

One of the demands imposed on grid adaptation techniques is that of obtaining a numerical solution of a partial differential equation with optimal accuracy for a given number of grid nodes. A rather reasonable idea to realize this demand relies on a placement of the grid nodes in such a way that the error of the numerical solution is uniformly distributed throughout the domain.

Generally, in one dimension the solution error e can be expressed as the product of the local grid spacing $\triangle s$ raised to some power p and a combination of the local solution derivatives, i.e. $e = (\triangle s)^p Q[\mathrm{d}^i u/(\mathrm{d}x)^i]$. Therefore the most natural approach to realizing the idea of the uniform error distribution is one in which some function w, connected with the error by the relation $w \sim c|Q^{1/p}|$, is computed and the mesh size $\triangle s$ is specified so that $w \triangle s$ and consequently the error e is nearly the same at all points of the domain. As a result the grid nodes generated in accordance with this approach are clustered in the regions of large solution derivatives, where the error is expected to be higher.

In one dimension, the equidistribution techniques can be formulated in the most clear and theoretically justified way. These one-dimensional techniques have been the foundation for the demonstration and development of multi-dimensional adaptive methods. And numerous theoretical and computational investigations of the numerical algorithms for interpolation of functions and for solving boundary value problems for ordinary differential equations have shown that the equidistribution approaches have a high potential to provide an effective reduction in the numerical error of the algorithms and an improvement of the resolution of the dependent variables.

7.2.1 Example of an Equidistributed Grid

As an example demonstrating the efficiency of the equidistribution principle, we consider a numerical solution of the initial boundary value problem

$$\begin{aligned} \frac{\mathrm{d}u}{\mathrm{d}x} &= f(u,x), \quad x > a, \\ u(a) &= l \end{aligned} \tag{7.1}$$

approximated at the grid points x_i, $i = 0, 1, \dots, N$, by a Runge–Kutta scheme of the second order of accuracy:

$$\begin{aligned} \frac{u_{i+1} - u_i}{x_{i+1} - x_i} &= f(u_i, x_i) + \frac{1}{4}\{f[u_i + (x_{i+1} - x_i)f_i, x_{i+1}] \\ &\quad - f[u_i + (x_{i-1} - x_i)f_i, x_{i-1}]\}, \quad i > 0, \\ u_0 &= l, \end{aligned} \tag{7.2}$$

where $f_i = f(u_i, x_i)$, and the nonuniform grid nodes x_i, $i = 0, \dots, N$, with $x_0 = a$ are represented by a smooth one-to-one transformation $x(\xi)$, i.e. $x_i = x(ih)$, where h is the step size of a uniform grid in ξ space.

The error R of the approximation of (7.1) by the scheme (7.2) is estimated by the equation

$$R = \frac{h^2}{2}\left[\frac{1}{2}\frac{\mathrm{d}^2 u}{\mathrm{d}x^2}\frac{\mathrm{d}^2 x}{\mathrm{d}\xi^2} + \frac{1}{3}\frac{\mathrm{d}^3 u}{\mathrm{d}x^2}\left(\frac{\mathrm{d}x}{\mathrm{d}\xi}\right)^2\right] + O(h^3).$$

The optimal transformation, eliminating the major error term in R and thus increasing the global order of the approximation (7.2) and correspondingly providing an asymptotic optimization of the solution error, satisfies the equation

$$\frac{1}{2}\frac{\mathrm{d}^2 u}{\mathrm{d}x^2}\frac{\mathrm{d}^2 x}{\mathrm{d}\xi^2} + \frac{1}{3}\frac{\mathrm{d}^3 u}{\mathrm{d}x^2}\left(\frac{\mathrm{d}x}{\mathrm{d}\xi}\right)^2 = 0.$$

This equation is readily solved with respect to the dependent quantity $\mathrm{d}x/\mathrm{d}\xi$, resulting in $x_\xi = c(u_{xx})^{-2/3}$. Thus we can formulate an initial-value problem for the computation of the transformation $x(\xi)$ producing an optimal grid in the following form:

$$\frac{\mathrm{d}x}{\mathrm{d}\xi} = c\left|\frac{\mathrm{d}^2}{\mathrm{d}x^2}u(x)\right|^{-2/3}, \quad \xi > 0, \qquad (7.3)$$
$$x(0) = a.$$

If the quantity $\mathrm{d}^2u/\mathrm{d}x^2$ does not vanish then the solution of (7.3) is a smooth function $x(\xi)$ generating the optimal grid $x_i = x(ih)$, $i = 0, 1, \ldots$. From (7.3) we obtain

$$h_i \approx h\frac{\mathrm{d}x}{\mathrm{d}\xi}(ih) \approx ch\left|\frac{\mathrm{d}^2 u}{\mathrm{d}x^2}(x_i)\right|^{-2/3}$$

and consequently find that the grid steps h_i, $i = 0, 1, \ldots$, are generated in accordance with the equidistribution principle:

$$h_i\left|\frac{\mathrm{d}^2 u}{\mathrm{d}x^2}\right|^{2/3} \approx c.$$

From (7.1),

$$\frac{\mathrm{d}^2 u}{\mathrm{d}x^2} = ff_u + f_x,$$

and therefore the steps h_i of the optimal grid for the scheme (7.2) can be obtained from the relation

$$h_i w_i = c, \quad w_i = w(x_i), \qquad (7.4)$$

with the weight function

$$w = \sqrt[3]{(ff_u + f_x)^2}.$$

Thus this magnitude $w(x)$ is equally distributed over the steps of the optimal grid derived from a solution to the problem (7.3), and as a result we find that the step size h_i is inversly proportional to $w_i = w(x_i)$.

Now, in order to demonstrate the efficiency of the optimal grid obtained from (7.3), we analyze, as an example, the numerical solution to the following initial-value problem of the type (7.1):

$$\begin{aligned}\frac{\mathrm{d}u}{\mathrm{d}x} &= x^{-2}, \quad x > 1,\\ u(1) &= -1.\end{aligned} \tag{7.5}$$

The exact solution of the problem (7.5) is

$$u(x) = -1/x, \quad x \geq 1.$$

In accordance with (7.3), we find that $\mathrm{d}x/\mathrm{d}\xi = cx^2$. The solution of this equation with the initial boundary condition $x(0) = 1$ is the transformation $x(\xi)$ expressed as follows:

$$x(\xi) = \frac{1}{1 - c\xi}, \quad 0 \leq \xi < \frac{1}{c}.$$

Consequently, we find that the optimal grid points x_i are determined by the relations

$$x_i = 1/(1 - cih), \quad i = 0, \dots, N, \ chN < 1.$$

The approximation of the initial problem (7.5) by the scheme (7.2) on the grid nodes x_i has the form

$$\begin{aligned}&\frac{(u_{i+1} - u_i)[1 - c(i+1)h](1 - cih)}{ch}\\ &\quad = (1 - cih)^2 + \frac{1}{4}\{[1 - c(i+1)h]^2 - [1 - c(i-1)h]^2\}, \quad i \geq 1,\\ &u_0 = -1\end{aligned}$$

and its solution

$$u_i = cih - 1 = -\frac{1}{x_i}$$

is the exact solution $u(x)$ to the initial-value problem (7.5) at the grid nodes x_i, i.e. $u_i = u(x_i)$, $i = 1, \dots, N$. Thus we find that the numerical solution of (7.5) coincides with the exact solution at the grid nodes derived by the equidistribution principle.

7.2.2 Original Formulation

The original one-dimensional formulation of the equidistribution principle for generating grid steps was proposed by Boor (1974) for the purpose of obtaining more accurate interpolation of functions by splines. The principle was formulated as a rule for determining the grid nodes x_i, $i = 1, \dots, N$, in the interval $[a, b]$ in accordance with the relation

$$\begin{aligned}&\int_{x_i}^{x_{i+1}} w(x)\mathrm{d}x = c, \quad i = 0, 1, \dots, N-1,\\ &x_0 = a, \qquad x_N = b,\end{aligned} \tag{7.6}$$

where $w(x)$ is a certain positive quantity called either a monitor function or a weight function. A discrete form of (7.6) may be represented as

$$\begin{aligned} &h_i w_i = c, \quad i = 0, \ldots, N-1, \\ &x_0 = a, \qquad x_N = b, \qquad h_i = x_{i+1} - x_i, \end{aligned} \tag{7.7}$$

where $w_i = w(x_i'),\ x_i' \in h_i,\ i = 0, \ldots, N-1$. The constant c in (7.7) is determined from the condition $x_N = b$, i.e. after summing (7.6):

$$c = \frac{1}{N} \int_a^b w(x) \mathrm{d}x.$$

Equation (7.7) corresponds to (7.4), found for the generation of the optimal grid to the solution of the problem (7.1). The grid steps satisfying (7.7) will be small where the weight w is large, and vice versa. Thus the weight function provides information about the desired grid clustering.

This formulation of the equidistribution principle for generating one-dimensional grids has been used with success to generate one-dimensional adaptive grids for the numerical solution of both stationary and nonstationary problems with singularities. Commonly, the solution to the problem of interest is first found on an initial background grid, then the weight function w is computed at the points of this background grid and interpolated over the entire interval $[a, b]$, and afterwards grid points are either moved or added to satisfy (7.6). The process is then repeated until a desired convergence tolerance is achieved.

7.2.3 Differential Formulation

If the grid points in the interval $[a, b]$ are determined through a coordinate transformation $x(\xi)$ from the unit interval $[0, 1]$ onto the interval $[a, b]$ in the form $x_i = x(ih),\ i = 0, 1, \ldots, N,\ h = 1/N$, then (7.7) can be interpreted as a discrete approximation of the problem

$$\begin{aligned} &\frac{\mathrm{d}x}{\mathrm{d}\xi} w[x(\xi)] = c, \quad 0 < \xi < 1, \\ &x(0) = a, \qquad x(1) = b. \end{aligned} \tag{7.8}$$

Taking into account

$$\frac{\mathrm{d}x}{\mathrm{d}\xi} = 1 \Big/ \frac{\mathrm{d}\xi}{\mathrm{d}x},$$

the problem (7.8) is equivalent to the following one:

$$\begin{aligned} &\frac{\mathrm{d}\xi}{\mathrm{d}x} = cw(x), \quad a < x < b, \\ &\xi(a) = 0, \qquad \xi(1) = b. \end{aligned} \tag{7.9}$$

The integration of (7.9) over the interval $[a, b]$ gives

$$c = 1 \Big/ \int_a^b w(x)\mathrm{d}x,$$

and, consequently, by integrating (7.9) over the interval $[a, x]$, we obtain

$$\xi(x) = \int_a^x w(x) \Big/ \int_a^b w(x)\mathrm{d}x.$$

The inverse of $\xi(x)$ yields the transformation $x(\xi)$ satisfying (7.8), thus producing the equidistributed grid.

By differentiating (7.8) with respect to ξ we can eliminate the constant c from our consideration and obtain a two-point boundary value problem for the definition of $x(\xi)$:

$$\begin{aligned}
&\frac{\mathrm{d}}{\mathrm{d}\xi}\left(w[x(\xi)]\frac{\mathrm{d}x}{\mathrm{d}\xi}\right) = 0, \quad 0 < \xi < 1, \\
&x(0) = a, \qquad x(1) = b.
\end{aligned} \tag{7.10}$$

This problem is solved, as a rule, by a method based on an explicit or implicit iterative computation of the following parabolic boundary value problem:

$$\begin{aligned}
&\frac{\partial}{\partial t} = \frac{\partial}{\partial \xi}\left(w(x)\frac{\partial x}{\partial \xi}\right), \quad 0 < \xi < 1, \; t > 0, \\
&x(\xi, 0) = x_0(\xi), \qquad x(0, t) = a, \qquad x(1, t) = b.
\end{aligned} \tag{7.11}$$

One-dimensional adaptive numerical grids with the nodes $x_i = x(ih)$, $i = 0, 1, \ldots, N$, $h = 1/N$, constructed through the solution of either the initial-value problem (7.8) or the two-point boundary value problem (7.10) on the uniform grid $\xi_i = ih$, $i = 0, 1, \ldots, N$, have been used to study many practical problems with singularities.

It its apparent that any smooth, univariate, one-to-one coordinate transformation $x(\xi)$ can be found through the solution of the problems (7.8)–(7.10) with the monitor function

$$w(x) = \frac{\mathrm{d}\xi}{\mathrm{d}x}.$$

So, the problems (7.8)–(7.10) provide a general formulation for building univariate adaptive grids. The major task in this formulation is to define the appropriate weight functions $w(x)$.

7.2.4 Specification of Weight Functions

Commonly, the weight function $w(x)$ in (7.8)–(7.10) is defined in the form $w(x) = f[\boldsymbol{u}^{(k)}(x)]$, where f is some positive function, $\boldsymbol{u}(x)$ is either a solution to the physical problem or a function monitoring some features of the quality of the solution, and $\boldsymbol{u}^{(k)}(x)$ is the kth-order derivative of $\boldsymbol{u}(x)$ with respect to x. For example, three possible weight functions are

$$w(x) = \begin{cases} \alpha + \|\boldsymbol{u}(x)\|, \\ \sqrt{\alpha + \|\boldsymbol{u}_x\|^2}, \\ \sqrt{\alpha + \beta\|\boldsymbol{u}_x\|^2 + \gamma\|\boldsymbol{u}_{xx}\|^2}, \end{cases} \tag{7.12}$$

where $\alpha,\ \beta,\ \gamma$ are positive constants and $\|\boldsymbol{v}\|$ is some norm of the vector $\boldsymbol{v} = (v^1, \dots, v^n)$ at the point x, commonly a least-squares norm or $\max_i |v^i|$, $i = 1, \dots, n$. It has been shown analytically and by numerical experiments that adaptive grids built in accordance with the equidistribution principle (7.7) with weight functions of the form (7.12) may often eliminate oscillations and give a more accurate description of the solution in zones of large variations.

7.2.4.1 Optimally Distributed Grid

The idea of the equidistribution principle is to equidistribute the solution error by placing more grid nodes where the error is large, so as to gain high accuracy overall with a fixed number of grid points. The validity of this idea was demonstrated by the above example. The grid which minimizes the error of the numerical solution to a differential problem is an optimally distributed grid with the nodes optimally refined in the areas of large solution error. Thus, if a measure of the error $\boldsymbol{e}(x)$ is estimated for the grid interval h_i by the relation

$$\|\boldsymbol{e}(x)\| = (h_i)^p Q(x), \quad \boldsymbol{e}(x) = \boldsymbol{u}(x) - \boldsymbol{u}^h(x),$$

where $\boldsymbol{u}(x)$ is the exact solution to the physical problem and $\boldsymbol{u}^h(x)$ is the numerical solution, it is quite natural, for obtaining the optimally distributed grid, to define the weight function through a measure of the error by means of the following relation:

$$w(x) = [Q(x)]^{1/p}.$$

The error $\boldsymbol{e}(x)$ may be found with high accuracy as a solution to the boundary value problem

$$L(\boldsymbol{e}) = \boldsymbol{T}, \quad \mathbf{e}|_{\partial X^n} = 0, \tag{7.13}$$

where L is the operator of linearization of the governing equations for the physical boundary value problem, while $\boldsymbol{T}$ is the approximation error of the problem. The derivation of (7.13) is carried out analogously to that of stability equations.

Although an asymptotically accurate solution error $\boldsymbol{e}(x)$ can in principle be obtained from the equation for variation (7.13), in practice this task is very difficult and expensive, even for a one-dimensional problem, to which (7.13) is applicable.

A more promising way seems to lie in the generation of the grid in accordance with a uniform distribution of some norm of $\boldsymbol{T}(x)$, i.e. by defining the monitor function $w(x)$ through a measure of the approximation error $\boldsymbol{T}(x)$. In this case

$$w(x) = [Q_1(x)]^{1/p}$$

if

$$\|\boldsymbol{T}(x)\| = (h_i)^p Q_1(x),$$

and $\boldsymbol{T}(x)$ can be readily expressed through the solution derivatives. This assumption has some meaning since the relation

$$\|\boldsymbol{e}(x)\| \sim c(x)\|\boldsymbol{T}(x)\|$$

is generally valid for the numerical solution of boundary value problems. However, the utilization of a weight function defined through the error of approximation $\boldsymbol{T}(x)$ to generate adaptive grids for the numerical solution of singularly perturbed equations may result in too large a grid spacing in the areas which lie outside the boundary and interior layers.

As an example, we consider the following two-point boundary value problem with a boundary layer:

$$\begin{aligned} &\epsilon u'' + u' - u = f(x), \quad 0 < x < 1, \\ &u(0) = a, \qquad u(1) = b, \end{aligned} \tag{7.14}$$

where $1 \geq \epsilon > 0$ is a small parameter. An approximation of (7.14) using the upwind differencing

$$\begin{aligned} &\frac{2\epsilon}{h_i + h_{i-1}}\left(\frac{u_{i+1} - u_i}{h_i} - \frac{u_i - u_{i-1}}{h_{i-1}}\right) + \frac{u_{i+1} - u_i}{h_i} - u_i = f(x_i), \\ &0 < i < N, \ u_0 = a, \ u_N = b, \end{aligned} \tag{7.15}$$

on a nonuniform grid x_i, $i = 1, \ldots, N$, results in the approximation error

$$T \sim c\left\{\epsilon\left[h_i \frac{\mathrm{d}^3 u}{\mathrm{d}x^3}(x_i) + h_{i-1}\frac{\mathrm{d}^3 u}{\mathrm{d}x^3}(x_{i-1})\right] + h_i \frac{\mathrm{d}^2 u}{\mathrm{d}x^2}(x_i)\right\}. \tag{7.16}$$

Let $v(x)$ be a solution to the following initial-value problem associated with (7.14):

$$\begin{aligned} v' - v &= f(x), \quad x < 1, \\ v(1) &= b. \end{aligned}$$

If $|v(0) - a| > m$, where m is a positive constant independent of ϵ, then the solution $u(x)$ of (7.14) is a function of the exponential boundary-layer type (see Fig. 7.1) satisfying the inequality

$$|u(x) - v(x)| \leq M[\exp(-x/\epsilon) + \epsilon], \quad 0 \leq x \leq 1,$$

and its derivatives in the interval $[0, 1]$ are estimated by

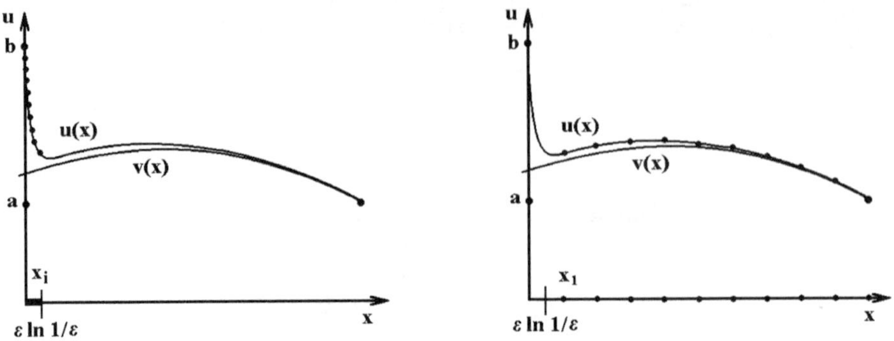

Fig. 7.1. Node placement generated by equidistribution of the approximation error (*left*) and of the solution error computed at the grid points (*right*)

$$\left|u^{(k)}(x)\right| \le M\left[\epsilon^{-k}\exp(-x/\epsilon)+1\right],$$

where M is a constant independent of ϵ. It is important to note that in the boundary layer interval $[0, k\epsilon\ln(1/\epsilon)]$ the derivatives of $u(x)$ are bounded from both sides by the estimate

$$M_1\left[\epsilon^{-k}\exp(-x/\epsilon)+1\right] \le \left|u^{(k)}(x)\right| \le M_2\left[\epsilon^{-k}\exp(-x/\epsilon)+1\right], \qquad (7.17)$$

where M_1, M_2 are positive constants which do not depend on ϵ. Thus, in accordance with (7.17), we obtain from (7.16) the following estimate for $T(x)$ in the boundary layer interval:

$$M_3h_i\left[\epsilon^{-2}\exp(-x/\epsilon)+1\right] \le T(x) \le M_4h_i\left[\epsilon^{-2}\exp(-x/\epsilon)+1\right],$$

with $M_i > 0$ and independent of ϵ. From this inequality it is natural to choose the weight $w(x)$, for the purpose of obtaining a uniform distribution of $T(x)$ with respect to the parameter ϵ, as

$$w(x) = \epsilon^{-2}\exp(-x/\epsilon)+1, \quad 0 \le x \le 1.$$

However, by computing (7.9) with this weight function $w(x)$, we obtain

$$\xi(x) = \frac{1-\exp(-x/\epsilon)+\epsilon x}{1-\exp(-1/\epsilon)+\epsilon}. \qquad (7.18)$$

If $x_0 = \epsilon\ln(1/\epsilon)$, (7.18) yields

$$\xi(x_0) = \frac{1-\epsilon+\epsilon^2\ln(1/\epsilon)}{1-\exp(-1/\epsilon)+\epsilon} \sim 1, \quad \text{if } 0<\epsilon\ll 1,$$

i.e. nearly all of the computational interval [0,1] (and consequently nearly the entire set of grid points) is mapped by $x(\xi)$ into the boundary layer. As a result the area outside the boundary layer is not provided with a sufficient number of grid nodes (Fig. 7.1, left).

Note that this phenomenon does not occur if the scale of the layer is less then 1, i.e. when the first derivative of $u(x)$ satisfies the estimate

$$|u'(x)| \leq M\epsilon^{-\alpha}, \quad 0 \leq x \leq 1, \ \alpha < 1,$$

with M independent of ϵ. This case occurs, for example, when the first derivative u' in (7.14) is either deleted or multiplied by a positive function vanishing at the point $x = 0$.

If the generation of the grid for the solution of (7.14) is determined by the condition of a uniform distribution of the solution error $e = u(x_i) - u_i(x_i)$, then both the boundary layer and the rest of the interval $[0, 1]$ are provided with a sufficient number of grid points. Indeed, the following estimate of e is valid:

$$|e_i| \leq Mh_i\left[\epsilon^{-1}\exp(-x/\epsilon) + 1\right],$$

and, consequently, assuming

$$w(x) = \epsilon^{-1}\exp(-x/\epsilon) + 1, \quad 0 \leq x \leq 1,$$

we obtain, solving (7.9),

$$\xi(x) = \frac{1 + x - \exp(-x/\epsilon)}{2 - \exp(-1/\epsilon)}, \quad 0 \leq x \leq 1.$$

This expression for $\xi(x)$ yields, for $x_0 = \epsilon\ln(1/\epsilon)$,

$$\xi(x_0) = \frac{1 + x_0 - \epsilon}{2 - \exp(-1/\epsilon)} \sim \frac{1}{2}, \quad \text{if } 0 < \epsilon \ll 1.$$

Thus, in this case, unlike to the previous one, nearly $N/2$ grid points will be placed in the boundary layer and the remaining $N/2$ nodes will be distributed outside the layer. The proportion of grid points placed in the boundary layer can be controlled by a constant c if we propose

$$w(x) = \epsilon^{-1}\exp(-x/\epsilon) + c,$$

which leads to the placement of approximately $cN/(1+c)$ nodes in the boundary layer.

Let us consider another phenomenon connected with the solution of (7.14) by the scheme (7.15). Since $|u^i| \leq M$, where u^i is the solution of (7.15) and M is independent of ϵ, we have from (7.15)

$$|e_i| \leq M\max_{i\geq 2}\left\{\frac{\epsilon}{(h_1)^2}, h_i[\exp(-x_i/\epsilon) + 1]\right\}, \quad i = 0, 1, \ldots, N-1.$$

Therefore, if $\epsilon \leq h(h_1)^2$, we obtain a uniform estimate of the solution error

$$|e_i| \leq Mh, \quad i = 0, \ldots, N-1.$$

In this case (Fig. 7.1, right), i.e. when all grid nodes lie outside the boundary layer, we find that the solution of the associated initial-value problem is solved more accurately than in the case where a part of the nodes is put in the boundary layer. As a result the solution to a problem of the type (7.14) may be more accurate at the points of a uniform grid than at the points of a grid with node clustering in the boundary layer when the parameter ϵ is sufficiently small, namely $\epsilon \leq 1/N^3$. However, this solution is not interpolated uniformly over the entire interval $[0, 1]$. This consideration shows that a measure of the solution error computed only at the grid points cannot be always used as a successful driving mechanism for the adaptive distribution of grid points.

One more disadvantage of generating grids in accordance with a uniform distribution of the error $\boldsymbol{T}(x)$ is the fact that for highly accurate approximations of the governing equations the expression for $\boldsymbol{T}(x)$ includes terms dependent on high-order derivatives of the solution, which may cause much numerical noise and instability.

Thus, though it is quite natural to incorporate directly the error measurements $\|\boldsymbol{e}(x)\|$ or $\|\boldsymbol{T}(x)\|$ into a formulation of the monitor functions, the computation of the optimally distributed grids defined by these measures may be an expensive and unsuccessful procedure and, in fact, relies on exact knowledge of the physical solution. So the requirement for efficiency of the algorithms leads the practitioners to specify the weight functions in more simple forms, applying for this purpose only lower-order derivatives of the solution. Generally, the largest numerical errors are found in regions of rapid variation of the lower derivatives of the solution, in particular, the first derivative. Therefore even the first derivatives of the solution can often be used to derive the weight functions. The following paragraphs give some examples of such weight functions composed of the first and/or second derivatives of the physical quantities.

7.2.4.2 Equidistant Mesh

A readily constructed grid, which can in some cases reduce the solution error, uses the measure of distance along the solution curve $\boldsymbol{u}(x)$ as the equally divided computational coordinate. This grid is called the equidistant mesh.

The second expression for $w(x)$ in (7.12) with $\alpha = 1$ provides the arc length monitor function $\boldsymbol{u}(x)$, whose curve is divided by the method into the equal intervals $\triangle s_i = c$, thus producing the equidistant mesh (Fig. 7.2). Equation (7.8), with this monitor function $w(x) = \sqrt{1 + \|\boldsymbol{u}_x\|^2}$ and the assumption for the norm $\|\boldsymbol{u}_x\|$

$$\|\boldsymbol{u}_x\|^2 = \sum_{i=1}^{k} \left(\frac{\mathrm{d}u^i}{\mathrm{d}x} \right)^2,$$

implies the following equation:

$$\frac{\mathrm{d}s}{\mathrm{d}\xi} = c$$

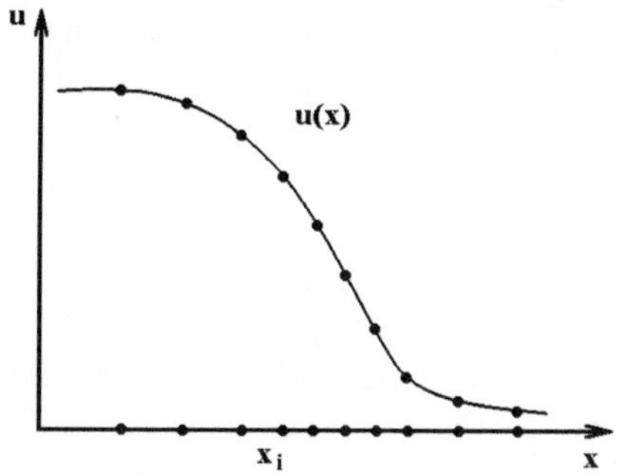

Fig. 7.2. Equidistant mesh

with respect to the dependent variable

$$s(x) = \int_a^x \sqrt{1 + \|\boldsymbol{u}_x\|^2} \mathrm{d}x.$$

The quantity $s(x)$ represents length of the curve

$$[x, \ \boldsymbol{u}(x)], \quad \boldsymbol{u}(x) = \left[u^1(x), \dots, u^k(x)\right]$$

in the space R^{k+1}.

The equidistant method is efficient for generating grids for the numerical solution of the singularly perturbed problems discussed in Chap. 4. A partial justification of this statement is based on the estimates for the solution derivatives, described in that chapter, which show that the area of large first derivatives of the solution to a singularly perturbed problem practically coincides with the area of large higher derivatives. In this case the monitor function

$$w(x) = \sqrt{1 + \|\boldsymbol{u}_x\|^2}$$

yields a coordinate transformation $x(\xi)$ such that the high-order derivatives with respect to ξ of the function $\boldsymbol{u}[x(\xi)]$ are nearly uniformly bounded. Thus the use of the arc length monitor may put an emphasis on error control in the vicinity of the narrow layers of rapid variation of the solution function.

The major factor influencing the use of the equidistant monitor is its simplicity. The utilization of the truncation error monitor for any difference scheme requires a computation of $\mathrm{d}^p/(\mathrm{d}x)^p\boldsymbol{u}(x)$ for some $p \geq 1$. In general, the complexity and expense of this calculation increase vary rapidly with p. So the equidistant method is optimal in a very real sense.

A generalization of the equidistant approach is commonly provided by the following weight functions:

$$w = \begin{cases} \alpha\|\boldsymbol{u}_x\| + \alpha_1, \\ \alpha_0 + \sum_i \alpha_i |u^i_x|^{\beta_i}, \\ \alpha_0 + \sum_i \alpha_i |u^i_\xi|^{\beta_i}, \\ (\alpha + \|\boldsymbol{u}^{(k)}\|^2)^{1/(2k)}. \end{cases} \tag{7.19}$$

Here u^i is the ith component of the vector-valued dependent variable $\boldsymbol{u}$ and α, α_i, β_i, l, k are positive constants. In the last case with $\alpha = 1$ and $k = 1$, the grid nodes are determined from the condition that the lengths of the segments of the curve $[x, \boldsymbol{u}(x)]$ are equal.

7.2.4.3 Utilization of the Second Derivative and Curvature

Weight functions are also often defined by utilizing the second derivative of the solution in addition to the first derivative, for example in the simplest form

$$w = 1 + \alpha\|\boldsymbol{u}_x\| + \beta\|\boldsymbol{u}_{xx}\|. \tag{7.20}$$

The second derivative is used to increase the concentration of grid points near the extrema of the function $\boldsymbol{u}(x)$. Weight functions which include the first and second derivatives of the solution in the form

$$w = 1 + a\|\boldsymbol{u}_\xi\|^\alpha + b\|\boldsymbol{u}_{\xi\xi}\|^\beta$$

or

$$w = 1 + a\|\boldsymbol{u}_x\|^\alpha + b\|\boldsymbol{u}_{xx}\|^\beta,$$

generalizing (7.20), are used to calculate one-dimensional waves and to construct multidimensional adaptive grids.

A more suitable grid may be obtained if the weight function $w(x)$ is constructed by combining the first derivative of the solution with the values of the curvature, for example in the forms

$$w = \begin{cases} (1 + \alpha|k|)\sqrt{1 + \|\boldsymbol{u}_x\|^2}, \\ 1 + \alpha\|\boldsymbol{u}_\xi\| + \frac{\beta}{|k|}\|\boldsymbol{u}_{\xi\xi}\|, \end{cases} \tag{7.21}$$

where k is the curvature of the curve $[x,\ \boldsymbol{u}(x)]$. In fact, the transformation $x(\xi)$ computed through the first representation of $w(x)$ in (7.21) is a composition of two transformations: one obtained by the equidistant approach and another serving to concentrate the grid points at the regions of large solution curvature. Note that in the case of a scalar function $u(x)$ the curvature is represented by the equation

$$k = u_{xx}/\sqrt{[1 + (u_x)^2]^3}.$$

In one more sophisticated approach, the weight function $w(x)$ is determined by solving the diffusion equation

$$-w_{xx} + \sigma w = \alpha\|\boldsymbol{u}_x\| + \beta|k|, \tag{7.22}$$

where σ, α, β are positive parameters. The term w_{xx} in (7.22) is included to provide a smoothing effect. This equation becomes a singularly perturbed one if σ is large, and as a result the weight function w in this case has a localized density, thus causing a localized adaptation.

7.3 Equidistribution in Multidimensional Space

The one-dimensional equidistribution approach discussed in Sect. 7.2 has led to the development of multidimensional equidistribution methods, in which the grid is generated in such a manner that the grid spacing in any direction is defined in accordance with the equidistribution of some prescribed quantity w, called either a weight function or a monitor function, along the coordinate line in this direction. It fact, most adaptive grid methods have much in common and can be interpreted as techniques based on an equal distribution of weight functions which represent some measures of the solution error.

7.3.1 One-Directional Equidistribution

An adaptation along a single family of coordinate lines is commonly carried out in accordance with the equidistribution principle $\triangle s w \sim$ const, where $\triangle s$ is the step size of a selected coordinate line ξ^i. Since $\triangle s \sim \triangle \xi^i \sqrt{g_{ii}}$ for a fixed i, where $g_{ij} = \boldsymbol{x}_{\xi^i} \cdot \boldsymbol{x}_{\xi^j}$, $i, j = 1, \ldots, n$, the principle can operate by specification of the ith diagonal element of the covariant metric tensor by assuming it to be inversely proportional to a weight w squared. Typical specifications of the monitor function $w(\boldsymbol{x})$ are performed through gradient magnitudes, curvatures, and cell quality measures. The distance s_i along any coordinate curve ξ^i satisfies the equation

$$\frac{\mathrm{d}s_i}{\mathrm{d}\xi^i} = \sqrt{g_{ii}}, \quad i \text{ fixed}.$$

On the other hand, in accordance with the equidistribution principle, we have for each fixed i

$$w_i \mathrm{d}s_i = c_i \mathrm{d}\xi^i.$$

The cancellation of $\mathrm{d}\xi^i$ from these two equations gives

$$g_{ii} = (c_i/w_i)^2$$

and thus

$$g_{ii}(w_{ii})^2 = (c_i)^2.$$

Differention with respect to ξ_i removes the constant c_i, yielding the partial differential equation

$$\frac{\partial g_{ii}}{\partial \xi_i} + \frac{2}{w_i}\frac{\partial w_i}{\partial \xi_i} g_{ii} = 0,$$

i.e.

$$\boldsymbol{x}_{\xi^i}\boldsymbol{x}_{\xi^i\xi^i} + \frac{1}{w_i}\frac{\partial w_i}{\partial \xi_i} g_{ii} = 0, \quad i \text{ fixed}. \tag{7.23}$$

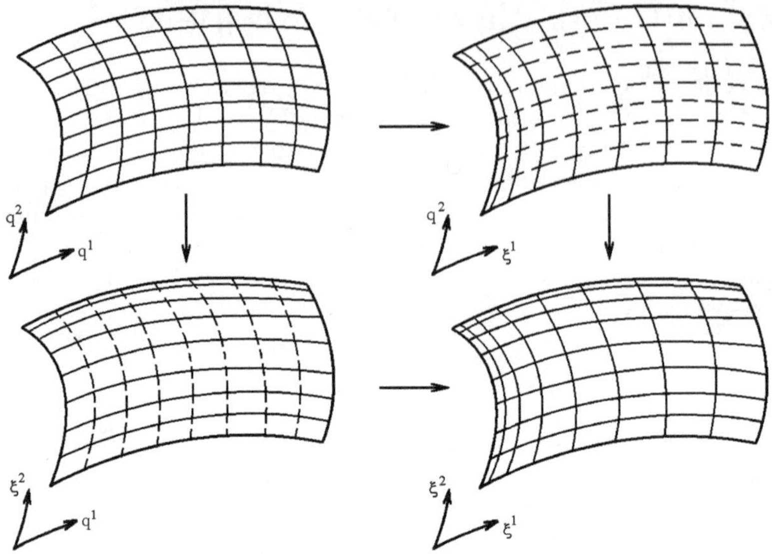

Fig. 7.3. Coordinate system derived by overlaying two independently deformed families (*solid curves*) of coordinate lines

7.3.2 Multidirectional Equidistribution

The generation of grids in multidimensional domains commonly requires adaptation in several directions. One way to perform such adaptation is to extend the methods of univariate equidistribution. A generalization of the univariate equidistribution approach to generate grids in a multidimensional domain can be accomplished by using either a combination or a composition of univariate equidistributions along fixed families of coordinate lines which are specified or computed beforehand in the physical domain.

7.3.2.1 Combination of One-Dimensional Equidistributions

In the combination approach the grid is derived by overlaying a series of grid lines obtained separately with univariate equidistributions in each direction (Fig. 7.3).

Thus, let the physical region X^n have a coordinate system $q^1, \ldots, q^n$ along which a univariate adaptation is supposed to be performed. Let the coordinate system q^i be determined by a coordinate transformation

$$\boldsymbol{x}(\boldsymbol{q}) : Q^n \to X^n .$$

Here the parametric (intermediate) domain Q^n can be considered as the unit cube. The equidistribution of the grid points along any fixed family of coordinate lines is carried out by the formulas of univariate equidistribution. This

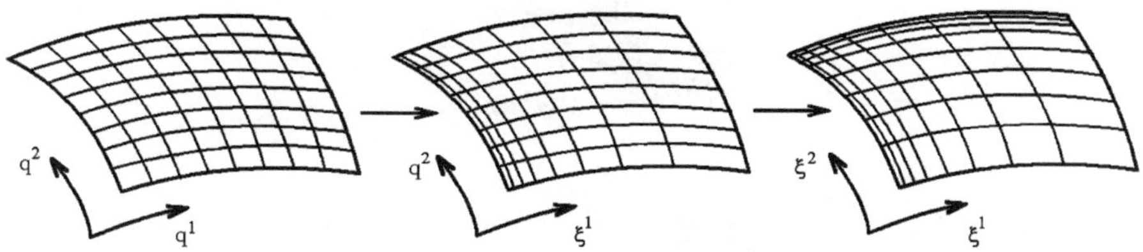

Fig. 7.4. Coordinate system derived by the composition of two successive deformations

process proceeds separately for each fixed coordinate family q^i by determining the respective weight function $w_i(\boldsymbol{q})$ and finding the coordinate transformation $q^i(\xi^i)$ for each fixed value $(q^1, \ldots, q^{i-1}, q^{i+1}, q^n)$ through the following equation, of the form (7.10):

$$\begin{aligned}&\frac{\partial}{\partial \xi^i}\left(\frac{\partial q^i}{\partial \xi^i} w_i(\boldsymbol{q})\right) = 0, \quad 0 < \xi^i < 1, \ i \text{ fixed}, \\ &q^i(0) = 0, \qquad q^i(1) = 1,\end{aligned} \tag{7.24}$$

and then forming the final grid by overlaying the coordinate curves obtained. In fact the grid is built by solving the system (7.24) for $i = 1, \ldots, n$. Of course, unlike the one-dimensional case, the solution to this system may not be a one-to-one transformation.

7.3.2.2 Composition of Univariate Equidistributions

In the composition approach, the process is split into a sequence of one-directional adaptations, in which the grid is obtained by successive application of one-dimensional equidistribution techniques (Fig. 7.4). In two dimensions this succession can be represented as

$$\boldsymbol{x}(q^1, q^2) \to \boldsymbol{x}\left[q^1(\xi^1, q^2), q^2\right] \to \boldsymbol{x}\{q^1\left[\xi^1, q^2(\xi^1, \xi^2)\right], q^2(\xi^1, \xi^2)\}.$$

This approach produces one-to-one coordinate transformations.

7.3.3 Control of Grid Quality

Unlike the case in one dimension, the multivariable equidistribution method, even in its simplest form, may produce mesh tangling or grids of poor quality in terms of smoothness, skewness, and aspect ratio. A lack of control of the grid quality may result in highly deformed cells, leading to a large truncation error. This section outlines one device aimed at performing the control and regularization of grid smoothness and skewness in the frame of the multivariable equidistribution method. Regularization is very important for the equations for moving the grid points in order to avoid excessive grid skewness. In fact

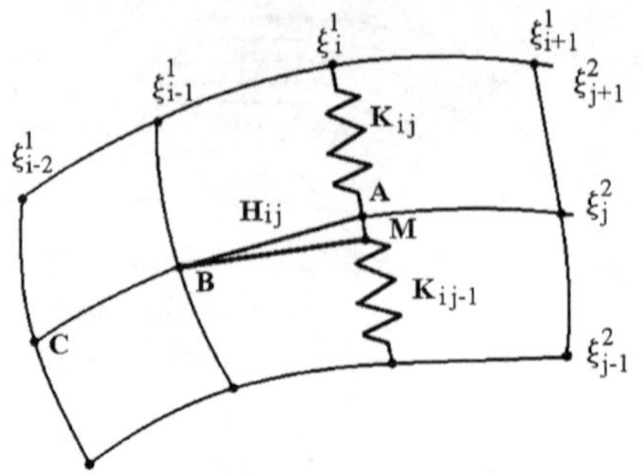

Fig. 7.5. Tension–torsion spring model

there are many approaches for improving the equidistribution techniques. For example, if the grid is adapted in one direction, then an orthogonality constraint is used in the other directions to control the grid distribution. A more practical scheme, however, is based on a tension and torsion spring analogy.

A model using tension and torsion springs for a multidirectional adaptation is a spring system in which each grid node is moved under the influence of tension and torsion springs. For simplicity we consider a two-dimensional formulation of the spring system.

The effect of the tension springs which connect adjacent grid points to each other along one particular coordinate line, say $\xi^1 = \xi^1_i$, results in the distribution of the grid points in accordance with a balance equation for the complete spring system

$$K_{ij} \triangle s_{ij} - K_{ij-1} \triangle s_{ij-1} = 0, \quad j = 1, \dots, N, \tag{7.25}$$

where K_{ij} is the spring constant of the spring connecting the point $\boldsymbol{\xi}_{ij+1}$ with $\boldsymbol{\xi}_{ij}$, while $\triangle s_{ij}$ is the arc length of the curve $\xi^1 = \xi^1_i$ between these two points (Fig. 7.5). Here and below we use $\boldsymbol{\xi}_{kl}$ to denote the point in X^2 which has the curvelinear coordinate values $(\xi^1_k,\ \xi^2_l)$ in the computational domain Ξ^2. Equation (7.25) is in fact the equation for the equidistribution principle (7.7) along the coordinate line $\xi^1 = \xi^1_i$ with the piecewise constant weight function

$$w(\xi^1_i, \xi^2) = K_{ij}, \quad \xi^2_j \le \xi^2 < \xi^2_{j+1}.$$

To distribute the grid lines along the $\xi^1 = \xi^1_i$ coordinate line in proportion to the dependent scalar variable f, the tension spring constant K_{ij} between the points $\boldsymbol{\xi}_{ij+1}$ and $\boldsymbol{\xi}_{ij}$ may be chosen according to

$$K_{ij} = F_{ij} + C_1 |f_{ij+1} - f_{ij}| / \triangle s_{ij}, \tag{7.26}$$

where C_1 and F_{ij} are constants controlling grid smoothness and adaptation, respectively.

A force to control the inclination of the other coordinate can be provided by torsion springs attached to nodes along the $\xi^2 = \xi^2_j$ line. The torsion spring attempts to change the inclination of the line $\xi^1 = \xi_i^1$ to that of some reference line by means of the force

$$F_{\text{torsion}} = -H_{i-1j}(\theta_{i-1j} - \phi_{i-1j}),$$

where H_{i-1j} is the spring constant of the torsion spring connecting the points $\boldsymbol{\xi}_{i-1j}$ and $\boldsymbol{\xi}_{ij}$ (B and A in Fig. 7.5), θ_{ij} is the inclination of the line connecting these points, and ϕ is the inclination of the reference line (BM in Fig. 7.5). The reference line for the line BA is prescribed in accordance with the user's choice, for example

(1) as an extension of the line CB to maintain smoothness,
(2) as a line normal to the $\xi^1 = \xi_i^1$ coordinate to keep the grid nearly orthogonal;
(3) as a preferred solution direction so that the grid follows it.

Typically, a combination of these reference lines is used.

The balance equation for the system with tension springs along the $\xi^2 = \xi_i^2$ coordinate and torsions springs connecting the $\xi^1 = \xi_i^1$ and $\xi^1 = \xi_{i-1}^1$ coordinate curves has the form

$$K_{ij} \bigtriangleup s_{ij} - K_{ij-1} \bigtriangleup s_{ij-1} - H_{i-1j}(\theta_{i-1j} - \phi_{i-1j}) = 0. \tag{7.27}$$

We can write

$$H_{i-1j}(\theta_{i-1j} - \phi_{i-1j}) = \overline{H}_{i-1j}(s_{ij} - \overline{s}_{ij}),$$

where s_{ij} is the arc length calculated from point $(i, 1)$ to point (i, j) along the $\xi^1 = \xi_i^1$ coordinate, $\overline{s}_{ij}$ is the projection of the reference line BM on the line $\xi^1 = \xi_i^1$, and the $\overline{H}_{i-1j}$ term is set equal to H_{i-1j} divided by the length of BM. Thus (7.27) is transformed to a tridiagonal system of equations for s_{ij}:

$$K_{ij-1}s_{ij-1} - (K_{ij} + K_{ij-1} + \overline{H}_{i-1j})s_{ij} + K_{ij}s_{ij+1} = -\overline{H}_{i-1j}\overline{s}_{ij}. \tag{7.28}$$

This model can be generalized to three dimensions with three-dimensional adaptation by introducing a spring system in which each grid point is suspended by six tension springs and twelve torsion springs. In this case the balance system analogous to (7.28) is rather cumbersome and requires significant computational effort. To facilitate the computation the procedure is commonly split into a sequence of one-directional adaptations.

7.3.4 Equidistribution over Cell Volume

The equidistribution principle in the multidimensional space R^n can also be formulated as a principle of equidistributing some grid measure $w(\boldsymbol{x})$ over each cell volume V_h:

$$w(\boldsymbol{x})V_h = \text{const}. \tag{7.29}$$

Since $V_h \approx \text{const}\det(\partial x^i/\partial \xi^j)$, $i, j = 1, \dots, n$, for the coordinate grids, we find that the equidistribution principle (7.29) can be reformulated as

$$Jw[\boldsymbol{x}(\boldsymbol{\xi})] = \text{const}, \tag{7.30}$$

where $J = \det(\partial x^i/\partial \xi^j)$, $i,j = 1,\dots,n$. If $w(x)$ is positive, then $J \neq 0$, and as a result the transformation $\boldsymbol{x}(\boldsymbol{\xi})$ is a one-to-one mapping and thus the grid cells are not folded.

The equidistribution principle (7.30) can be realized for arbitrary dimensions if the physical domain X^n coincides with the computational domain represented by the unit n-dimensional cube Ξ^n. For this purpose we consider a smooth, positive weight function $w(\boldsymbol{x})$ satisfying the normalization property

$$\int_{\Xi^n} \left(\frac{1}{w(\boldsymbol{x})} - 1\right) \mathrm{d}\boldsymbol{x} = 0. \tag{7.31}$$

It is evident that by means of a continuous deformation, the identity mapping $\mathrm{id} : \Xi^n \to \Xi^n$ can be transformed into a one-to-one mapping $\boldsymbol{x}(\boldsymbol{\xi}) : \Xi^n \to \Xi^n$ such that

$$\det\left(\frac{\partial x^i}{\partial \xi^j}\right) = \frac{1}{w[\boldsymbol{x}(\boldsymbol{\xi})]}, \quad i,j = 1,\dots,n, \text{ in } \Xi^n. \tag{7.32}$$

Thus the cells of the coordinate grid constructed using $\boldsymbol{x}(\boldsymbol{\xi})$ contract where $w(\boldsymbol{x})$ is large and expand where $w(\boldsymbol{x})$ is small.

Let this continuous deformation be designated by $\boldsymbol{x}(\boldsymbol{\xi}, t)$, where t is a parameter varying in the interval $[0, 1]$. In accordance with (2.87), we have for $\boldsymbol{x}(\boldsymbol{\xi}, t)$ the identity

$$\frac{\partial J}{\partial t} = J \frac{\partial}{\partial x^j} \frac{\partial x^j}{\partial t}(\boldsymbol{\xi}, t) = J \mathrm{div}_x \frac{\partial \boldsymbol{x}}{\partial t}, \quad j = 1,\dots,n, \tag{7.33}$$

where J is the determinant of $[\partial x^i(\boldsymbol{\xi}, t)/\partial \xi^j]$, $i,j = 1,\dots,n$. The necessary deformation $\boldsymbol{x}(\boldsymbol{\xi}, t)$ is defined as a solution to the initial-value problem at every point $\boldsymbol{\xi} \in \Xi^n$:

$$\begin{aligned} \frac{\mathrm{d}}{\mathrm{d}t}\boldsymbol{x}(\boldsymbol{\xi}, t) &= \frac{\boldsymbol{v}[\boldsymbol{x}(\boldsymbol{\xi}, t)]}{1 - t + tw[\boldsymbol{x}(\boldsymbol{\xi}, t)]}, \quad 0 < t \leq 1, \\ \boldsymbol{x}(\boldsymbol{\xi}, 0) &= \boldsymbol{\xi}, \end{aligned} \tag{7.34}$$

where the vector-valued function $\boldsymbol{v}(\boldsymbol{x}) = [v^1(\boldsymbol{x}), \dots, v^n(\boldsymbol{x})]$ is a solution to the following problem:

$$\begin{aligned} \mathrm{div}_x \boldsymbol{v}(\boldsymbol{x}) &= 1 - w(\boldsymbol{x}) \quad \text{in } \Xi^n, \\ \boldsymbol{v}(\boldsymbol{x}) &= 0 \quad \text{on } \partial\Xi^n. \end{aligned} \tag{7.35}$$

As $\boldsymbol{v}(\boldsymbol{x}) = 0$ at the boundary points of the domain Ξ^n, we find from (7.34) and (7.35) that the speed of the movement of the boundary grid points is zero. Therefore the deformation function $\boldsymbol{x}(\boldsymbol{\xi}, t)$ obtained as a solution to (7.34) with $\boldsymbol{v}(\boldsymbol{x})$ satisfying (7.35) does not move the boundary points, i.e.

$$\boldsymbol{x}(\boldsymbol{\xi}, t) = \boldsymbol{\xi} \quad \text{if } \boldsymbol{\xi} \in \partial\Xi^n,$$

for all $0 \leq t \leq 1$.

Using this function $\boldsymbol{x}(\boldsymbol{\xi}, t)$, the required coordinate transformation $\boldsymbol{x}(\boldsymbol{\xi})$ satisfying (7.32) is defined as $\boldsymbol{x}(\boldsymbol{\xi}) = \boldsymbol{x}(\boldsymbol{\xi}, 1)$. To prove (7.32), we consider the scalar function $H(\boldsymbol{\xi}, t)$ defined by

$$H(\boldsymbol{\xi}, t) = J\{1 - t + tw[\boldsymbol{x}(\boldsymbol{\xi}, t)]\}.$$

We have

$$\frac{\partial H}{\partial t} = \frac{\partial J}{\partial t}(1 - t + tw) + J\left(w - 1 + t\frac{\partial w}{\partial x^i}\frac{\partial x^i}{\partial t}\right),$$

and, taking advantage of (7.33),

$$\frac{\partial H}{\partial t} = J\left((1 - t + tw)\mathrm{div}_x \frac{\partial \boldsymbol{x}}{\partial t} + w - 1 + t\frac{\partial w}{\partial x^i}\frac{\partial x^i}{\partial t}\right). \tag{7.36}$$

Using (7.34) and (7.35) we obtain

$$\begin{aligned}
\mathrm{div}_x \frac{\partial \boldsymbol{x}}{\partial t} &= \mathrm{div}_x\left(\frac{\boldsymbol{v}}{1 - t + tw}\right) = \frac{\mathrm{div}_x \boldsymbol{v}}{1 - t + tw} - \frac{t\frac{\partial w}{\partial x^i}v^i}{(1 - t + tw)^2} \\
&= \frac{1 - w(\boldsymbol{x})}{1 - t + tw} - \frac{t\frac{\partial w}{\partial x^i}v^i}{(1 - t + tw)^2}.
\end{aligned}$$

Substituting this expression for $\mathrm{div}_x(\partial \boldsymbol{x}/\partial t)$ in (7.36), we obtain

$$\frac{\partial H}{\partial t} = J\left(1 - w(\boldsymbol{x}) - t\frac{\partial w}{\partial x^i}\frac{\partial x^i}{\partial t} + w(\boldsymbol{x}) - 1 + t\frac{\partial w}{\partial x^i}\frac{\partial x^i}{\partial t}\right) = 0.$$

Thus the function $H(\boldsymbol{\xi}, t)$ is independent of t, and consequently $H(\xi, 0) = H(\xi, 1)$. Since

$$H(\xi, 0) = \det\left(\frac{\partial x^i(\boldsymbol{\xi}, 0)}{\partial \xi^j}\right) = 1, \quad i, j = 1, \ldots, n,$$

we obtain

$$H(\xi, 1) = \det\left(\frac{\partial x^i(\boldsymbol{\xi}, 1)}{\partial \xi^j}\right) w[\boldsymbol{x}(\boldsymbol{\xi}, 1)] = 1, \quad i, j = 1, \ldots, n.$$

Thus the coordinate transformation $\boldsymbol{x}(\boldsymbol{\xi}) = \boldsymbol{x}(\boldsymbol{\xi}, 1)$ satisfies the equidistribution requirement (7.30).

Note that the computation of the vector field $\boldsymbol{v}(\boldsymbol{x})$ satisfying (7.35) can be done by using a least-squares method or by solving the following Poisson problem:

$$\begin{aligned}
-\triangle f &= w(\boldsymbol{x}) - 1 \quad \text{in } \Xi^n, \\
\frac{\partial f}{\partial \boldsymbol{n}} &= 0 \quad \text{on } \partial \Xi^n, \\
\boldsymbol{v}(x) &= \boldsymbol{\nabla} f.
\end{aligned} \tag{7.37}$$

The relation (7.31) guarantees a solution to (7.37). In fact, the deformation method described above can be extended to an arbitrary bounded domain with a Lipschitz continuous boundary.

Though the deformation technique was developed to construct coordinate transformations in the same domain, the use of the intermediate approach considered in Chap. 4 allows one to generate an adaptive grid in an arbitrary domain provided at least one nondegenerate transformation between the computational domain and the domain of interest exists.

7.4 Adaptation Through Control Functions

The differential approaches reviewed in Chap. 6 enable one to provide dynamic adaptation by means of the coefficients of the equations. These coefficients are referred to as the control functions.

7.4.1 Specification of the Control Functions in Elliptic Systems

7.4.1.1 Poisson System

In the elliptic method based on Poisson equations, considered in Chap. 6, the grid adaptation is realized through the choice of the control functions P^i in the inverted system of equations

$$g^{ij}\frac{\partial^2 x^k}{\partial \xi^i \partial \xi^j} = -P^i \frac{\partial x^k}{\partial \xi^i}, \quad i, j, k = 1, \dots, n. \tag{7.38}$$

In one dimension this system is reduced to

$$\left(\frac{\mathrm{d}\xi}{\mathrm{d}x}\right)^2 \frac{\mathrm{d}^2 x}{\mathrm{d}\xi^2} = -P\frac{\mathrm{d}x}{\mathrm{d}\xi}. \tag{7.39}$$

One way of determining the source function P in (7.39) so as to provide an opportunity to control the attraction or repulsion of grid points is to use an analogy between the one-dimensional equation (7.10), implementing the one-dimensional equidistribution principle, and (7.39). In accordance with the univariate equidistribution approach, the control function P in (7.39) realizing the equidistribution of the monitor function $w(x)$ should have the form

$$P = \left(\frac{\mathrm{d}\xi}{\mathrm{d}x}\right)^2 \frac{\mathrm{d}w}{\mathrm{d}\xi} \Big/ w, \tag{7.40}$$

since, with this expression for P, the one-dimensional inverted Poisson equation (7.39) is transformed to the equation

$$w\frac{\mathrm{d}}{\mathrm{d}\xi}\left(w\frac{\mathrm{d}x}{\mathrm{d}\xi}\right) = 0,$$

which is equivalent to (7.10).

The simplest approach to the specification of the control functions P^i in (7.38) lies in the generalization of (7.40). Commonly, this is realized by substituting g^{ii} for $(\mathrm{d}\xi/\mathrm{d}x)^2$ and $\partial w/\partial\xi^i$ for $\mathrm{d}w/\mathrm{d}\xi$, by the following representation of the control functions P^i, for example:

$$P^i = g^{ii}\frac{\partial w}{\partial \xi_i} \Big/ w, \quad \text{for } i \text{ fixed.} \tag{7.41}$$

A more general form also suggests

$$P^i = \sum_j g^{ij}\frac{\partial w^i}{\partial \xi^j} \Big/ w^i, \quad \text{for } i \text{ fixed.} \tag{7.42}$$

The functions w, w^i in (7.41) and (7.42) are defined in terms of the components of the solution of the physical problem and their derivatives. Thus in problems of the motion of a liquid in a reservoir, w is defined in terms of the depth function H:

$$w = 1 + H.$$

In gas-dynamics investigations, P^i and w are defined through a number of salient physical quantities. For flows with shock waves, use is made of the gas density, the Mach number, and the internal energy and pressure, while in boundary-layer calculations, the velocity or vorticity is used to formulate the control functions P^i. For instance, in supersonic compressible flows, the pressure p is commonly identified as such a quantity, since shock waves are detected by its rapid variation. To study these flows on a grid controlled by source terms P^i determined through (7.41), the weight function w is usually specified by the equation

$$w = 1 + \|\boldsymbol{\nabla} p\|.$$

Another example of the specification of the control functions P^i applied to the study of two-dimensional flows can be presented through the density ρ:

$$\begin{aligned} P^1 &= g^{11}\left(\frac{\partial w}{\partial \xi^1} - c_1\frac{\partial w}{\partial \xi^2}\right)w^{-1}, \\ P^2 &= g^{22}\left(-c_2\frac{\partial w}{\partial \xi^1} + \frac{\partial w}{\partial \xi^2}\right)w^{-1}, \\ w &= 1 + f(\rho). \end{aligned}$$

7.4.1.2 Other Equations

Elliptic equations of the form

$$\sum_{i,j=1}^{n} g^{ij}\frac{\partial}{\partial \xi^i}\left(b_{ij}\frac{\partial x^\alpha}{\partial \xi^j}\right) = 0, \quad \alpha = 1, 2, \ldots, n, \tag{7.43}$$

in which adaptation is carried out by choosing the coefficients b_{ij}, defined as a rule in terms of the gradients of the solution, are also used. Thus, in calculations of two-dimensional gas flows adaptive grids are constructed using (7.43) with

$$b_{11} = f_1, \qquad b_{22} = f_2, \qquad b_{12} = b_{21} = \sqrt{f_1 f_2},$$
$$f_i = f_i(p, g_{11}, g_{22}), \quad i = 1, 2.$$

The simplest form of (7.43) is obtained for

$$g^{ij} = \triangle^i_j, \qquad b_{ij} = f(\rho)\triangle^j_i,$$

resulting in

$$\frac{\partial}{\partial \xi^i}\left(f(\rho)\frac{\partial \boldsymbol{x}}{\partial \xi^i}\right) = 0, \quad i = 1, \ldots, n.$$

7.4.2 Hyperbolic Equations

Adaptation in the hyperbolic equations considered in Sect. 6.6 is accomplished by specifying in the equations the cell aspect ratio or the Jacobian of the transformation.

7.5 Grids for Nonstationary Problems

In the finite-difference techniques developed for the numerical solution of nonstationary problems the time variable t must be discretized in order to provide computation of all physical variables at each time slice t^n. As the physical solution is dependent on the variable t it is reasonable to adjust the grid to the solution to follow the trajectories of severe variations in the physical quantities. The goal is to compute as accurately as desired all physical quantities of interest at each grid location in space and time. As a result the placement of the grid points obtained with such an adjustment depends on the time t. The motion of the grid points is demonstrated in the transformed equations by the grid velocity $\boldsymbol{x}_t$ appearing in the transformed time derivative. In the process of numerical solution the grid speed is either found using the differences of the values of the function $\boldsymbol{x}(\boldsymbol{\xi}, t)$ on the $(n+1)$th and nth layers or specified in advance.

This section reviews some techniques aimed at the generation of structured nonstationary grids for time-dependent partial differential equations which, in addition to spatial derivatives, also contain derivatives with respect to the time variable t. An example is the following form:

$$\frac{\partial \boldsymbol{u}}{\partial t} = L(\boldsymbol{u}, \boldsymbol{x}, t), \quad \boldsymbol{x} \in X^n, \ t > 0, \tag{7.44}$$

where L is a differential operator involving spatial derivatives.

The accuracy of the numerical solution of time-dependent partial differential equations is significantly dependent on the time-step size as well as the spatial mesh size. However, high resolution in local regions of large solution variations allows larger time steps to be taken.

There are two basic approaches to solving nonstationary problems: the method of lines and the methods of moving grids.

7.5.1 Method of Lines

The idea of the method of lines consists in converting nonstationary equations into a system of ordinary differential equations with time as the independent variable. This is carried out by first approximating only the space derivatives on a specified space grid chosen a priori. This operation yields a system of ordinary differential equations with respect to the independent variable t. The system obtained is then approximated on a one-dimensional grid, generally nonuniform, discretizing the variable t. In this approach the spatial grid is fixed, and therefore it is applicable only to the numerical solution of problems with stationary solution singularities. However, many solutions to time-dependent problems have large variations which changed with time. Moreover, the solutions may have narrow layers of rapid change with respect to both space and time, for example a solution of the simple running-wave form

$$u(x,t) = f(x - ct),$$

where $f(y)$ is a scalar function, of boundary or interior-layer type.

7.5.2 Moving-Grid Techniques

There are two major approaches to controlling the movement of the points of the numerical grid with time. With the first, stationary grids are constructed by the same method on each time layer but the grid points move over time because of changes either of the control functions in the equation or of the boundary conditions for the coordinate transformations applying on the boundary of the computation region Ξ^n. With the second, an equation for the velocity $\boldsymbol{x}_t(\boldsymbol{\xi}, t)$ of grid point movement is determined and the boundary value problem is then solved for these equations.

7.5.2.1 Specification of Spatial Grid Distribution

The simplest way to generate grids for nonstationary problems is to use the equidistribution approach with a time-dependent weight function $w(x,t)$, for instance in the form

$$\int_{x_i}^{x_{i+1}} w(x,t)\mathrm{d}x = c(t) = \frac{1}{N}\int_a^b w(x,t)\mathrm{d}x, \quad i = 0, \ldots, N-1,$$

which is analogous to (7.6). This form can be readily reformulated as a boundary value problem of one of the types (7.8)–(7.10). In particular, in the case of one-dimensional nonstationary problems it may be realized by the replacement of the variables x, t by new variables ξ, t, where ξ is an arc-length-like coordinate, i.e. the result is the equidistant mesh. The advantage of this procedure is that in the variables ξ, t the solution $u(\xi, t)$ cannot develop large spatial gradients $\partial u/\partial \xi$.

7.5.2.2 Grid Movement Induced by Boundary Movement

Equations of the type (6.14) in which the control functions P^i depend on t are often used at each time step of the solution of gas-dynamics problems, the movement of grid nodes on the boundary of the region being determined by the motion of the medium; in particular, the velocity of the grid and the velocity of the medium are the same on the boundary. The movement of interior points in this approach depends solely upon the boundary motion. The boundaries of the regions may be shock waves, interior boundaries which separate the different regions of flow, or free surfaces.

Algebraic methods of constructing moving grids by interpolation from a moving boundary can also be considered in the same fashion.

7.5.2.3 Specification of Grid Speed

In multidimensional nonstationary problems, it is often necessary to generate adaptive grids which are adjusted to a moving solution of a problem. As a result the grid distribution is also nonstationary, that is, the derivative $\boldsymbol{x}_t(\boldsymbol{\xi}, t)$, which represents the speed of the grid movement, is nonzero in the general case. The expression $\boldsymbol{x}_t(\boldsymbol{\xi}, t)$ appears in the equations for nonstationary physical problems rewritten in the independent variables $\boldsymbol{\xi}$, t.

This subsection delineates some typical approaches for formulating grid equations by specification of the velocity of grid point movement to treat nonstationary problems. In our considerations we discuss a simplified form of the transformation $\boldsymbol{x}(\boldsymbol{\xi}, \tau)$, assuming $t = \tau$. In this case the first temporal derivatives $\partial/\partial t$ and $\partial/\partial \tau$ are subject to the relation

$$\frac{\partial}{\partial \tau} = \frac{\partial}{\partial t} + \frac{\partial}{\partial x^i}\frac{\partial x^i}{\partial \tau}.$$

Thus, assuming the identification $t = \tau$, (7.44) is transformed to

$$\frac{\partial \boldsymbol{u}}{\partial t} = -\frac{\partial \boldsymbol{u}}{\partial x^i}\frac{\partial x^i}{\partial t} + L[\boldsymbol{u}, \boldsymbol{x}(\boldsymbol{\xi}, t), t], \quad \boldsymbol{\xi} \in \Xi^n, \ t > 0, \tag{7.45}$$

with $\boldsymbol{u} = \boldsymbol{u}[\boldsymbol{x}(\boldsymbol{\xi}, t), t]$.

The simplest way to obtain grid equations which include the speed of grid movement is to add the term $\boldsymbol{x}_t(\boldsymbol{\xi}, t)$ to the equations developed for the generation of fixed grids, in particular, to the inverted Poisson equations (6.14).

The equations for the grid node velocities $\boldsymbol{x}_t(\boldsymbol{\xi}, t)$ can be determined, for example, from the conditions for minimizing functionals of measure of the deviation from Lagrangian properties. These functionals will be discussed in Chap. 8.

It is often proposed to solve nonstationary problems by determining equations for $\boldsymbol{x}_t(\boldsymbol{\xi}, t)$ from the condition that the solution to the physical problem in the new variables $(\boldsymbol{\xi}, t)$ is stationary. Also, the grid velocity can be formulated on the basis of providing conditions of stability in the difference scheme in nonstationary problems.

In one more approach, for the equations for multidimensional problems of the mechanics of a continuous medium, the equations for the grid velocities are determined from the condition that the convective terms are either a minimum or zero. For example, if the operator $L(\boldsymbol{u}, \boldsymbol{x}, t)$ in (7.44) has the form

$$L(\boldsymbol{u}, \boldsymbol{x}, t) = a^i \frac{\partial \boldsymbol{u}}{\partial x^i} + L_1(\boldsymbol{u}, \boldsymbol{x}, t), \quad i = 1, \dots, n,$$

where L_1 is an operator without first derivatives, then (7.45) in the new coordinates $t, \xi^1, \dots, \xi^n$ determined by the transformation $\boldsymbol{x}(\boldsymbol{\xi}, t)$ is expressed as follows:

$$\frac{\partial \boldsymbol{u}}{\partial t} = \left(a^i - \frac{\partial x^i}{\partial t}\right) + L_1[\boldsymbol{u}, \boldsymbol{x}(\boldsymbol{\xi}, t), t], \quad i = 1, \dots, n. \tag{7.46}$$

Thus the condition for zero convective terms in (7.46) gives the following equations for the components of the grid velocity $\boldsymbol{x}_t(\boldsymbol{\xi}, t)$:

$$\frac{\partial x^i}{\partial t} = a^i, \quad i = 1, \dots, n.$$

In such a way, for example, the Lagrangian coordinates are generated, so the approach aimed at the elimination of convective terms in the transformed equations by nonstationary coordinate mappings is referred to as the Lagrangian method.

The Poisson equations (6.14), formally differentiated with respect to t, are also used to obtain equations for $\boldsymbol{x}_t(\boldsymbol{\xi}, t)$. Also, differentation with respect to time of the equations modeling the equidistribution principle gives differential equations for the grid node velocity. This operation provides the necessary grid velocity equations, which are then integrated to obtain the grid motion as a function of time.

7.5.3 Time-Dependent Deformation Method

The deformation method considered in Sect. 7.3.4 for generating multidimensional structured grids in an n-dimensional domain X^n coinciding with the computational domain Ξ^n can be extended to produce time-dependent grids with nonstationary weight functions. This subsection describes such an extension, using for this purpose weight functions $w(\boldsymbol{x},t) > 0$ that satisfy the following normalization properties:

$$\int_{\Xi^n}\left(\frac{1}{w} - 1\right)\mathrm{d}\boldsymbol{\xi} = 0, \tag{7.47}$$

and $w(\boldsymbol{x},t_0) = 1$ for all $\boldsymbol{x} \in \Xi^n$. The nonstationary transformation $\boldsymbol{x}(\boldsymbol{\xi},t)$ satisfying the equidistribution condition (7.32) is found from the following initial-value problem for ordinary differential equations, formulated for each fixed point $\boldsymbol{\xi}$ in Ξ^n:

$$\begin{aligned} \frac{\partial}{\partial t}\boldsymbol{x}(\boldsymbol{\xi},t) &= \boldsymbol{v}[\boldsymbol{x}(\boldsymbol{\xi},t),t]w[\boldsymbol{x}(\boldsymbol{\xi},t),t], \quad t > t_0, \\ \boldsymbol{x}(\boldsymbol{\xi},t_0) &= \xi, \quad t = t_0, \end{aligned} \tag{7.48}$$

where the vector field $\boldsymbol{v}(\boldsymbol{x},t)$ is computed from the following system:

$$\begin{aligned} \mathrm{div}_x\boldsymbol{v}(\boldsymbol{x},t) &= -\frac{\partial}{\partial t}\left(\frac{1}{w(\boldsymbol{x},t)}\right) \quad \text{in } \Xi^n, \\ \mathrm{curl}\,\boldsymbol{v}(\boldsymbol{x},t) &= 0 \quad \text{in } \Xi^n, \\ \boldsymbol{v}\cdot\boldsymbol{n} &= 0 \quad \text{on } \partial\Xi^n. \end{aligned} \tag{7.49}$$

As in Sect. 7.3.4, we consider the derivative with respect to t of the function

$$H(\boldsymbol{\xi},t) = Jw[\boldsymbol{x}(\boldsymbol{\xi},t),t],$$

where $J = \det(\partial x^i/\partial\xi^j)$. Since the relation (7.33) is valid for the transformation $\boldsymbol{x}(\boldsymbol{\xi},t)$, we obtain

$$\frac{\partial H}{\partial t} = J\left(w\,\mathrm{div}_x\frac{\partial\boldsymbol{x}}{\partial t} + \frac{\partial w}{\partial x^i}\frac{\partial x^i}{\partial t} + \frac{\partial w}{\partial t}\right).$$

Application of (7.47) and (7.48) to this equation yields

$$\begin{aligned} \frac{\partial H}{\partial t} &= J\left[w\,\mathrm{div}_x\left(\frac{\boldsymbol{v}}{w}\right) + \frac{\partial w}{\partial x^i}\frac{v^i}{w} + \frac{\partial w}{\partial t}\right] \\ &= J\left(-\frac{\partial w}{\partial t} - \frac{v^i}{w}\frac{\partial w}{\partial x^i} + \frac{\partial w}{\partial x^i}\frac{v^i}{w} + \frac{\partial w}{\partial t}\right) = 0. \end{aligned}$$

Since $H(\boldsymbol{\xi},t_0) = 1$, we obtain

$$H(\boldsymbol{\xi},t) = Jw[\boldsymbol{x}(\boldsymbol{\xi},t),t] = 1,$$

i.e. the multidimensional equidistribution principle (7.32) is obeyed by the transformation $\boldsymbol{x}(\boldsymbol{\xi},t)$.

7.6 Comments

Reviews of adaptive methods for the generation of structured grids have been published by Anderson (1983), Thompson (1985), Eiseman (1987), Hawken, Gottlieb, and Hansen (1991), and Liseikin (1996b).

Equidistribution approaches of various kinds have been reported by a number of researchers. The original one-dimensional integral formulation of the equidistribution principle was proposed by Boor (1974), while the differential versions were presented by Yanenko, Danaev, and Liseikin (1977), Tolstykh (1978), and Dwyer, Kee, and Sanders (1980).

The most popular and general forms of the weight functions were reviewed by Thompson (1985). These functions were proposed by Russell and Christiansen (1978), Ablow and Schechter (1978), White (1979), Dwyer, Kee, and Sanders (1980), Nakamura (1983), and Anderson and Steinbrenner (1986). Acharya and Moukalled (1990) used a normalized second derivative of the solution as the weighting function. A linear combination of the first and second derivatives of the solution was used as a measure of the weighting function by Dwyer, Kee, and Sanders (1980), who successfully applied equidistribution along one family of grid lines within two-dimensional problems. A combination of first and second derivatives and the curvature of the solution variables to specify weight functions was applied by Ablow and Schechter (1978) and Noack and Anderson (1990). A selection of monitor functions in the form of weighted Boolean sums of various solution characteristics was defined by Weatherill and Soni (1991) and Soni and Yang (1992).

In a one-dimensional application, Gnoffo (1983) used a tension spring analogy in which the adapted grid point spacing along a family of coordinate lines resulted from the minimization of the spring system's potential energy. The idea of the spring analogy, represented in the equation for the weighting function, was extended by Nakahashi and Deiwert (1985) to demonstrate the feasibility and versatility of the equidistribution method. They also utilized in their considerations the notion of a torsion spring attached to each node in order to control the inclination of the grid lines.

Rai and Anderson (1981, 1982) and Eiseman (1985) have each used the idea of moving grid points under the influence of forcing or weighting functions that either attract or repel grid points relative to each other. Thus points with forcing (or weighting) functions greater than a specified average value attract each other, and those with values less than the average value repel each other.

Examples of the numerical solution of singularly perturbed problems on an equidistant mesh were studied by Andrew and Whrite (1979). The equidistant method was also advocated by Ablow (1982) and Catheral (1991), who applied it to some gas-dynamic calculations.

Dorfi and Drury (1987) used a very effective technique for incorporating smoothness into the univariate equidistribution principle. Their one-dimensional technique ensures that the ratio of adjacent grid intervals is restricted, thus controlling clustering and grid expansion. The power of this smoothing capability was clearly demonstrated in the valuable comparative studies by Furzland, Verwer, and Zegeling (1990) and Zegeling (1993). A multidimensional generalization of the approach of Dorfi and Drury (1987) was presented by Huang and Sloan (1994), who introduced control of concentration, scaling, and smoothness.

Morrison (1962) was apparently the first who managed to show analytically the efficiency of the error equidistribution principle for the generation of grids for the numerical solution of ordinary differential equations. Babuŝka and Rheinboldt (1978) proposed an error estimator based on the solution of a local variational problem. A truncation error measure for generating optimal grids was applied by Denny and Landis (1972), Liseikin and Yanenko (1977), White (1979, 1982), Ablow and Schechter (1978), Miller (1981), Miller and Miller (1981), Davis and Flaherty (1982), Adjerid and Flaherty (1986), and Petzold (1987). The approaches based on the equidistribution of the truncation error were developed by Pereyra and Sewell (1975) and Davis and Flaherty (1982), while the equidistribution of the residual was developed by Carey (1979), Pierson and Kutler (1980), and Rheinboldt (1981). An analysis of the strategies based on a uniform error distribution was also undertaken by Chen (1994). In particular, he proposed as optimal the following monitor functions:

$$M = \begin{cases} [(u')^2 + T^2]^{1/3}, \\ [(u'')^2 + T^2]^{1/3}, \\ [(u''')^2 + T^2]^{1/7}, \\ [(u''')^2 + (u'''')^2]^{1/7}[1 + (u')^2]^{1/2}. \end{cases}$$

However, the numerical experiments by Blom and Verwer (1989) show that the mesh generated from the error measurement monitor may be of poor quality.

Some methods which control the movement of the grid nodes in accordance with the equidistribution of the residuals of equations were developed by Miller (1981). Several versions of the moving-mesh method were also studied by Huang, Ren, and Russel (1994) to demonstrate their ability to accurately tracking rapid spatial and temporal transitions.

A curve-by-curve grid line equidistribution approach in the computational space was described by Eiseman (1987). The combination and composition versions of the multidimensional equidistribution principle were proposed by Darmaev and Liseikin (1987).

An orthogonalization technique of Potter and Tuttle (1973) for two-dimensional grid control, whereby the grids are adapted in one direction and orthogonality is imposed on the second, was proposed by Anderson and Rajendran (1984) and Dwyer and Onyejekwe (1985).

Certain anomalies which may arise in the process of the numerical solution of the nonlinear equations modeling the equidistribution principle and ways for surmounting them were discussed by Steinberg and Roache (1990) and Knupp (1991, 1992). Also, some adverse effects of dynamic grid adaptation on the numerical solution of physical problems were noted by Sweby and Yee (1990).

The idea of defining the control functions of the Poisson system through weight functions was formulated by Anderson (1983, 1987) and extended by Eiseman (1987). Some versions of the specification of the control functions through sums of derivatives of physical quantities and quality measures of the domain geometry were presented by Dannenhoffer (1990), Kim and Thompson (1990), Tu and Thompson (1991), Soni (1991), and Hall and Zingg (1995), while Hodge, Leone, and McCarry (1987) applied analytical expressions for this purpose.

A deformation method for generating multidimensional unfolded grids has been developed by Liao and Anderson (1992) and Semper and Liao (1995) on the basis of a deformation scheme originally introduced by Moser (1965). This deformation method was improved and applied to practical problems by Wan and Turek (2006a, 2006b, 2007), and Liao et al. (2008).

A large number of important moving-grid methods for the numerical solution of unsteady equations can be found in the survey by Hawken, Gottlieb, and Hansen (1991) and in the monograph by Zegeling (1993).

The movement of the nonstationary grid considered by Godunov and Prokopov (1972) was caused by the boundary point speeds, while in the interior of the domain the grid nodes were found by solving the inverted Poisson system. Hindman, Kutler, and Anderson (1981) formulated the grid motion in time by taking the time derivative of the inverted Poisson equations.

One more approach involving grid speed equations, but based on time differentiation of a more complicated set of Euler–Lagrange equations derived from the minimization of a Brackbill–Saltzman-type functional, was presented by Slater, Liou, and Hindman (1995).

A strategy for automatic time step selection based on equidistributing the local truncation error in both the time and the space discretization was proposed by Chen, Baines, and Sweby (1993).

White (1979) suggested a technique for numerically integrating systems of time-dependent first-order partial differential equations in one space variable x. His technique replaces the variables x, t by the new variables s, t, where s is an arc-length-like coordinate.

Some questions arising from coupling upwinding schemes with moving equidistributed meshes were discussed by Li and Petzold (1997). The stability problems related to an equidistributed mesh of the systems of differential equations for the grid velocities were studied by Coyle, Flaherty, and Ludwig (1986).

One approach to generating one-dimensional time-dependent grids was proposed by Dar'in, Mazhukin, and Samarskii (1988) with the help of the system of evolutionary equations

$$\frac{\partial x(\xi, t)}{\partial \xi} = \Psi,$$
$$\frac{\partial \Psi}{\partial t} = -\frac{\partial P}{\partial \xi}.$$

In order to concentrate the grid nodes in the high-gradient zones, various expressions for P containing the derivatives with respect to ξ of the physical quantities were considered. This approach was extended to the construction of two-dimensional adaptive grids by Dar'in and Mazhukin (1989).

The method of equidistribution and minimization of the heuristically determined error at each time step was used for calculations of nonstationary problems by Dorfi and Drury (1987), Dwyer, Kee, and Sanders (1980), Klopfer and McRae (1981), Miller (1983), Wathen (1990), and White (1982).

Formal addition of the velocity function $\boldsymbol{x}_t(\boldsymbol{\xi}, t)$ to (6.14), to the equations obtained from variational methods based on the minimization of grid quality functionals, or to expressions for the errors determined heuristically in terms of the spatial derivatives was analyzed by Rai and Anderson (1981), Bell, Shubin, and Stephens (1982), Harten and Hyman (1983), and Greenberg (1985).

Various physical analogies, such as those of springs (Bell and Shubin 1983, Rai and Anderson 1982), chemical reactions (Greenberg 1985) and concepts from continuum mechanics (Jacquotte 1987, Knupp 1995) have also been used to construct moving adaptive grids.

8

Variational Methods

8.1 Introduction

The calculus of variations provides an excellent opportunity to create new techniques for generation of structured grids by utilizing the idea of optimization of grid characteristics modeled through appropriate functionals. The grid characteristics include grid smoothness, departure from orthogonality or conformality, cell skewness, and cell volume. The minimization of a combination of the functionals representing the desired grid features generates the equations for those coordinate transformations which yield a grid with optimally balanced grid quality measures. The relative contributions of the functionals are determined by the user-prescribed weights.

The major task of the variational approach to grid generation is to describe all basic measures of the desired grid features in an appropriate functional form and to formulate a combined functional that provides a well-posed minimization problem. This chapter describes some basic functionals representing the grid quality properties and measures of grid features. These functionals can provide mathematical feedback in an automatic grid procedure.

8.2 Calculus of Variations

The goal of the calculus of variations is to find the functions which are optimal in terms of specified functionals. The optimal functions also are referred to as critical or stationary points of the respective functionals. The theory of the calculus of variations has been developed to formulate and describe the laws and relations concerned with the critical points of functionals. One of the most important achievements of this theory is the discovery that the optimal functions satisfy some easily formulated equations called the Euler–Lagrange equations. Thus the problem of computing the optimal functions is related to the problem of the solution of these equations. This section presents the

V.D. Liseikin, *Grid Generation Methods*, Scientific Computation,
DOI 10.1007/978-90-481-2912-6_8,

Euler–Lagrange equations derived by the minimization of functionals suitable for the purpose of grid generation.

The condition of the convexity of the functionals is of paramount importance to the well-posedness of both the minimization problem for the functionals and the boundary value problems for the resulting Euler–Lagrange equations. Therefore this section also discusses questions concerned with the convexity of functionals.

8.2.1 General Formulation

Commonly, in the calculus of variations, any functional over some admissible set of functions $\boldsymbol{f} : D^n \to R^m$ is defined by the integral

$$I(\boldsymbol{f}) = \int_{D^n} G(\boldsymbol{f})\mathrm{d}V, \tag{8.1}$$

where D^n is a bounded n-dimensional domain, and $G(\boldsymbol{f})$ is some operator specifying, for each vector-valued function $\boldsymbol{f} : D^n \to R^m$, a scalar function $G(\boldsymbol{f}) : D^n \to R$. The admissible set is composed of those functions $\boldsymbol{f}$ which satisfy a prescribed boundary condition

$$\boldsymbol{f}|_{\partial D^n} = \boldsymbol{\phi}$$

and for which the integral (8.1) is limited.

In the application of the calculus of variations to grid generation this set of admissible functions is a set of sufficiently smooth invertible coordinate transformations

$$\boldsymbol{\xi}(\boldsymbol{x}) : X^n \to \Xi^n$$

between the physical domain X^n and the computational domain Ξ^n or, vice versa, a set of sufficiently smooth invertible coordinate transformations from the computational domain Ξ^n onto the physical region X^n:

$$\boldsymbol{x}(\boldsymbol{\xi}) : Q^n \to X^n.$$

The integral (8.1) is defined over the domain X^n or Ξ^n, respectively.

In grid generation applications the operator G is commonly chosen as a combination of weighted local grid characteristics which are to be optimized. The choice depends, of course, on what is expected from the grid. Some forms of the weight functions were described in Sect. 7.2, while the local grid characteristics were formulated in Chap. 3 through the coordinate transformations and their first and second derivatives. Therefore, for the purpose of grid generation, it can be supposed that the most widely acceptable formula for the operator G in (8.1) is one which is derived from some expressions containing the first and second derivatives of the coordinate transformations. Thus we can assume that the functional (8.1), depending on the coordinate transformation $\boldsymbol{\xi}(\boldsymbol{x})$, is of the form

$$I(\boldsymbol{\xi}) = \int_{X^n} G(\boldsymbol{x}, \boldsymbol{\xi}, \boldsymbol{\xi}_{x^i}, \boldsymbol{\xi}_{x^i x^j}) \mathrm{d}\boldsymbol{x}, \tag{8.2}$$

where G is a smooth function of its variables

$$\boldsymbol{x}, \quad \boldsymbol{\xi}, \quad \boldsymbol{\xi}_{x^i} = \frac{\partial \boldsymbol{\xi}(\boldsymbol{x})}{\partial x^i}, \quad \text{and} \quad \boldsymbol{\xi}_{x^i x^j} = \frac{\partial^2 \boldsymbol{\xi}(\boldsymbol{x})}{\partial x^i \partial x^j}.$$

The admissible set for this functional is a set of the invertible vector-valued functions $\boldsymbol{\xi}(\boldsymbol{x}) : X^n \to \Xi^n$ satisfying the condition of smoothness up to the fourth order, i.e.

$$\xi^i(\boldsymbol{x}) \in C^4(X^n), \quad i = 1, \dots, n.$$

Analogously, the functional (8.1) formulated over a set of invertible coordinate transformations $\boldsymbol{x}(\boldsymbol{\xi})$ from $C^4(\Xi^n)$, has the form

$$I(\boldsymbol{x}) = \int_{\Xi^n} G(\boldsymbol{\xi}, \boldsymbol{x}, \boldsymbol{x}_{\xi^i}, \boldsymbol{x}_{\xi^i \xi^j}) \mathrm{d}\boldsymbol{\xi}. \tag{8.3}$$

In accordance with the assumption that the admissible set of functions for the functional (8.2) or (8.3) is composed of the corresponding invertible coordinate transformations, we can reformulate either of these two functionals in terms of the other by the following transition formulas:

$$\begin{aligned} \int_{X^n} f \mathrm{d}\boldsymbol{x} &= \int_{\Xi^n} (Jf) \mathrm{d}\boldsymbol{\xi}, \\ \int_{\Xi^n} f \mathrm{d}\boldsymbol{\xi} &= \int_{X^n} (f/J) \mathrm{d}\boldsymbol{x}. \end{aligned} \tag{8.4}$$

Thus for the functional (8.3) we obtain

$$\begin{aligned} I(\boldsymbol{x}) &= \int_{X^n} \Big(\frac{1}{J} G(\boldsymbol{\xi}, \boldsymbol{x}, \boldsymbol{x}_{\xi^i}, \boldsymbol{x}_{\xi^i \xi^j}) \Big) \mathrm{d}\boldsymbol{x} \\ &= \int_{X^n} G_1(\boldsymbol{x}, \boldsymbol{\xi}, \boldsymbol{\xi}_{x^i}, \boldsymbol{\xi}_{x^i x^j}) \mathrm{d}\boldsymbol{x} = I_1(\boldsymbol{\xi}) \end{aligned}$$

with an implied transition from $\boldsymbol{x}_{\xi^i}$ and $\boldsymbol{x}_{\xi^i \xi^j}$ to $\boldsymbol{\xi}_{x^l}$ and $\boldsymbol{\xi}_{x^l x^k}$.

8.2.2 Euler–Lagrange Equations

To be definite, we consider here the variational principle for grid generation in the form of the functional (8.2) over the set of invertible smooth coordinate transformations $\boldsymbol{\xi}(\boldsymbol{x})$ from the physical domain X^n onto the computational domain Ξ^n. In general, the functionals are formulated in the physical space X^n rather than in the parametric space Ξ^n. This is preferred because the physical-space formulation can be used more simply to obtain grid generation techniques that provide the necessary grid properties.

If the transformation $\boldsymbol{\xi}(\boldsymbol{x})$ is optimal for the functional (8.2) then it satisfies a system of Euler–Lagrange equations in the interior points of the domain X^n:

$$G_{\xi^i} - \frac{\partial}{\partial x^j} G_{\frac{\partial \xi^i}{\partial x^j}} + \frac{\partial^2}{\partial x^j \partial x^k} G_{\frac{\partial^2 \xi^i}{\partial x^j \partial x^k}} = 0, \quad i, j, k, = 1, \dots, n, \tag{8.5}$$

where the subscripts ξ^i, $\partial \xi^i / \partial x^j$, and $\partial^2 \xi^i / \partial x^j \partial x^k$ mean the corresponding partial derivatives of G. We remind the reader that the repeated indices here and below imply summation over them unless otherwise noted.

In many applications the integrand G is dependent only on $\boldsymbol{x}$, the function $\boldsymbol{\xi}(\boldsymbol{x})$, and its first derivatives, i.e. $G = G(\boldsymbol{x}, \boldsymbol{\xi}, \boldsymbol{\xi}_{x^i})$. In this case the admissible set of functions $\boldsymbol{\xi}^i(\boldsymbol{x})$ can be from the class $C^2(X^n)$, and the system of Euler–Lagrange equations (8.5) is reduced to

$$G_{\xi^i} - \frac{\partial}{\partial x^j} G_{\frac{\partial \xi^i}{\partial x^j}} = 0, \quad i, j = 1, \dots, n. \tag{8.6}$$

We give a schematic deduction of (8.6). Equations (8.5) are obtained in a similar manner.

Let the transformation $\boldsymbol{\xi}(\boldsymbol{x})$ be a critical point of the functional (8.2) with $G = G(\boldsymbol{x}, \boldsymbol{\xi}, \boldsymbol{\xi}_{x^j})$. In order to prove that $\boldsymbol{\xi}(\boldsymbol{x})$ satisfies (8.6), we first choose a scalar smooth function from $C^2(X^n)$ which equals zero on the boundary of the domain X^n. Let this function be denoted by $\phi(\boldsymbol{x})$. Now, using $\phi(\boldsymbol{x})$, we define for a fixed index i a vector-valued function $\boldsymbol{\psi}(\boldsymbol{x}) = [\psi^1(\boldsymbol{x}), \dots, \psi^n(\boldsymbol{x})]$ dependent on $\boldsymbol{x}$ as follows:

$$\begin{aligned} \psi^j(\boldsymbol{x}) &= 0, \quad j \neq i, \\ \psi^i(\boldsymbol{x}) &= \phi(\boldsymbol{x}), \quad j = i, \end{aligned}$$

i.e.

$$\boldsymbol{\psi}(\boldsymbol{x}) = \phi(\boldsymbol{x}) \boldsymbol{e}_i,$$

where $\boldsymbol{e}_i$ is the ith basic Cartesian vector. As was assumed, the transformation $\boldsymbol{\xi}(\boldsymbol{x})$ is critical for the functional (8.2), and therefore the following scalar smooth function

$$y(\epsilon) = I(\boldsymbol{\xi} + \epsilon \boldsymbol{\psi}) = \int_{X^n} G(\boldsymbol{x}, \boldsymbol{\xi} + \epsilon \boldsymbol{\psi}, \boldsymbol{\xi}_{x^i} + \epsilon \boldsymbol{\psi}_{x^i}) \mathrm{d}\boldsymbol{x}, \tag{8.7}$$

where ϵ is a real variable, has an extremum at the point $\epsilon = 0$. This results in the relation $y'(0) = 0$. In accordance with the rule of differentiation of integrals we obtain

$$y'(0) = \int_{X^n} \left(\phi(\boldsymbol{x}) G_{\xi^i} + \frac{\partial \phi}{\partial x^j} G_{\frac{\partial \xi^i}{\partial x^j}} \right) \mathrm{d}\boldsymbol{x}, \quad j = 1, \dots, n.$$

Taking into account that

$$\frac{\partial \phi}{\partial x^j} G_{\frac{\partial \xi^i}{\partial x^j}} = \frac{\partial}{\partial x^j}(\phi G_{\frac{\partial \xi^i}{\partial x^j}}) - \phi \frac{\partial}{\partial x^j} G_{\frac{\partial \xi^i}{\partial x^j}}, \quad j = 1, \ldots, n,$$

we have

$$y'(0) = \int_{X^n} \left[\phi \left(G_{\xi^i} - \frac{\partial}{\partial x^j} G_{\frac{\partial \xi^i}{\partial x^j}} \right) + \frac{\partial}{\partial x^j} \left(\phi G_{\frac{\partial \xi^i}{\partial x^j}} \right) \right] \mathrm{d}\boldsymbol{x} = 0. \tag{8.8}$$

Using the divergence theorem

$$\int_{X^n} \left(\sum_{i=1}^{n} \frac{\partial A^i}{\partial x^i} \right) \mathrm{d}\boldsymbol{x} = \int_{\partial X^n} (\boldsymbol{A} \cdot \boldsymbol{n}) \mathrm{d}S, \tag{8.9}$$

valid for a smooth arbitrary vector function $\boldsymbol{A}(\boldsymbol{x}) = [A^1(\boldsymbol{x}), \ldots, A^n(\boldsymbol{x})]$, we conclude that

$$\int_{X^n} \frac{\partial}{\partial x^j} (\phi G_{\frac{\partial \xi^i}{\partial x^j}}) \mathrm{d}\boldsymbol{x} = \int_{\partial X^n} \phi (\boldsymbol{G}^i \cdot \boldsymbol{n}) \mathrm{d}S = 0,$$

where

$$\boldsymbol{G}^i = (G_{\frac{\partial \xi^i}{\partial x^1}}, \ldots, G_{\frac{\partial \xi^i}{\partial x^n}}),$$

since the selected function ϕ equals zero at all points of ∂X^n. Thus we find that the second summation term in the integral (8.8) can be omitted, and consequently we find that

$$y'(0) = \int_{X^n} \phi \left(G_{\xi^i} - \frac{\partial}{\partial x^j} G_{\frac{\partial \xi^i}{\partial x^j}} \right) \mathrm{d}\boldsymbol{x} = 0 \tag{8.10}$$

for every smooth function $\phi(\boldsymbol{x})$ satisfying the condition proposed above, that $\phi(\boldsymbol{x}) = 0$ if $\boldsymbol{x} \in \partial X^n$. From this relation we readily find that the optimal coordinate transformation $\boldsymbol{\xi}(\boldsymbol{x})$ obeys (8.6) at every interior point of the domain X^n. If this were not so, then there would exist an interior point $\boldsymbol{x}_0$ such that the function

$$f(\boldsymbol{x}) = \left(G_{\xi^i} - \frac{\partial}{\partial x^j} G_{\frac{\partial \xi^i}{\partial x^j}} \right)(\boldsymbol{x})$$

does not vanish at this point, i.e. $f(\boldsymbol{x}_0) \neq 0$, say $f(\boldsymbol{x}_0) > 0$. The function $f(\boldsymbol{x})$ is continuous, so there exists a positive number $r > 0$ such that $f(\boldsymbol{x})$ does not change its sign for all $\boldsymbol{x}$ satisfying $|\boldsymbol{x} - \boldsymbol{x}_0| \leq r$, i.e. $f(\boldsymbol{x}) > 0$ at these points. Now we use for the function $\phi(\boldsymbol{x})$ in (8.10) a nonnegative mapping which equals zero at all points $\boldsymbol{x}$ that are outside the subdomain $|\boldsymbol{x} - \boldsymbol{x}_0| \leq r$, and $\phi(\boldsymbol{x}_0) > 0$. For example, one such function $\phi(\boldsymbol{x})$ is expressed as follows:

$$\phi(\boldsymbol{x}) = \begin{cases} \exp(\frac{1}{(\boldsymbol{x} - \boldsymbol{x}_0)^2 - r^2}), & |\boldsymbol{x} - \boldsymbol{x}_0| < r, \\ 0, & |\boldsymbol{x} - \boldsymbol{x}_0| \geq r. \end{cases}$$

This function is from the class $C^\infty(X^n)$. In accordance with the assumed property of $f(\boldsymbol{x})$, we find that (8.10) is not satisfied for the function $\phi(\boldsymbol{x})$

specified above, namely $y'(0) > 0$, i.e. the function $\boldsymbol{\xi}(\boldsymbol{x})$ is not critical for the functional (8.2), which is contrary to the initial assumption about the extremum of the functional at the point $\boldsymbol{\xi}(\boldsymbol{x})$. From this contradiction we conclude that the optimal transformation $\boldsymbol{\xi}(\boldsymbol{x})$ is a solution to (8.6) at every interior point of the domain X^n.

Analogously, there can be obtained a system of Euler–Lagrange equations for the functional (8.3),

$$G_{x^i} - \frac{\partial}{\partial \xi^j} G_{\frac{\partial x^i}{\partial \xi^j}} + \frac{\partial^2}{\partial \xi^j \partial \xi^k} G_{\frac{\partial^2 x^i}{\partial \xi^j \partial \xi^k}} = 0, \quad i, j = 1, \ldots, n, \tag{8.11}$$

which is satisfied by the optimal coordinate transformation $\boldsymbol{x}(\boldsymbol{\xi})$.

8.2.3 Functionals Dependent on Metric Elements

The most common interior characteristics of grid cells were defined in Chap. 3 by the elements of the metric tensors. As the covariant elements can be derived from the contravariant ones, we can assume that the functional (8.2) representing an integral measure of some grid feature is defined by the integrand G, depending on $\boldsymbol{x}$ and the elements $g^{ij} = \nabla \xi^i \cdot \nabla \xi^j = \frac{\partial \xi^i}{\partial x^k} \frac{\partial \xi^j}{\partial x^k}$ only, i.e.

$$I(\boldsymbol{\xi}) = \int_{X^n} G(\boldsymbol{x}, g^{ij}) \mathrm{d}\boldsymbol{x}.$$

In this case the corresponding system of the Euler–Lagrange equations (8.6) has the following divergent form

$$\frac{\partial}{\partial x^j} \left(\frac{\partial G}{\partial g^{lk}} \frac{\partial g^{lk}}{\partial (\partial \xi^i / \partial x^j)} \right) = 0, \quad i, j, k, l = 1, \ldots, n. \tag{8.12}$$

As

$$\frac{\partial g^{lk}}{\partial (\partial \xi^i / \partial x^j)} = \delta^i_l \frac{\partial \xi^k}{\partial x^j} + \delta^i_k \frac{\partial \xi^l}{\partial x^j}, \quad i, j, k, l = 1, \ldots, n, \tag{8.13}$$

we have the result that

$$\frac{\partial G}{\partial g^{lk}} \frac{\partial g^{lk}}{\partial (\partial \xi^i / \partial x^j)} = \left(\frac{\partial G}{\partial g^{ik}} + \frac{\partial G}{\partial g^{ki}} \right) \frac{\partial \xi^k}{\partial x^j}, \quad i, j, k, l = 1, \ldots, n.$$

Therefore the system of the Euler–Lagrange equations (8.12) is equivalent to

$$\frac{\partial}{\partial x^j} \left[\left(\frac{\partial G}{\partial g^{ik}} + \frac{\partial G}{\partial g^{ki}} \right) \frac{\partial \xi^k}{\partial x^j} \right] = 0, \quad i, j, k = 1, \ldots, n, \tag{8.14}$$

with summation over the repeated indices j and k. Taking advantage of the identity (2.56), the system (8.14) can be converted to

$$\frac{\partial}{\partial \xi^j} \left[J g^{kj} \left(\frac{\partial G}{\partial g^{ik}} + \frac{\partial G}{\partial g^{kj}} \right) \right] = 0, \quad i, j, k = 1, \ldots, n, \tag{8.15}$$

written with respect to the independent variables ξ^i. In particular, when the integrand G is defined by the diagonal elements g^{ii} of the contravariant metric tensor (g^{ij}) then (8.14) and (8.15) are as follows:

$$\frac{\partial}{\partial x^j}\left(\frac{\partial G}{\partial g^{ii}}\frac{\partial \xi^i}{\partial x^j}\right) = 0,$$
$$\frac{\partial}{\partial \xi^j}\left(Jg^{ij}\frac{\partial G}{\partial g^{ii}}\right) = 0, \quad i,j = 1,\dots,n$$

with fixed index i.

8.2.4 Functionals Dependent on Tensor Invariants

Chapter 3 presents a description of some local grid quality properties which are defined by the invariants $I_1,\dots,I_n$ of the metric tensor $g_{ij} = \boldsymbol{x}_{\xi^i}\cdot\boldsymbol{x}_{\xi^j}$. The integration of these properties over the physical or the computational domain represents the global grid properties in the form of functionals depending on the invariants. Taking into account the general identity

$$\int_{\Xi^n} G\mathrm{d}\boldsymbol{\xi} = \int_{X^n} [G/(I_n^{1/2})]\mathrm{d}\boldsymbol{x},$$

where

$$I_n = g = \det(g_{ij}) = \overset{2}{\det}\left(\frac{\partial x^i}{\partial \xi^j}\right),$$

we can consider all these functionals as integrals over the domain X^n in the form

$$I(\boldsymbol{\xi}) = \int_{X^n} G(\boldsymbol{x}, I_1,\dots,I_n)\mathrm{d}\boldsymbol{x}. \tag{8.16}$$

The Euler–Lagrange equations (8.6) in this case are represented as follows:

$$\frac{\partial}{\partial x^j}\left(G_{I_k}\frac{\partial I_k}{\partial(\partial \xi^i/\partial x^j)}\right) = 0, \quad i,j,k = 1,\dots,n. \tag{8.17}$$

8.2.4.1 Two-Dimensional Tensor

For two-dimensional coordinate transformations $\boldsymbol{x}(\boldsymbol{\xi}) : \Xi^2 \to X^2$, the invariants are defined by (3.23). Since we consider the functionals depending on the invariants as integrals over the domain X^n, we need to rewrite the invariants through the terms of the contravariant metric tensor (g^{ij}). This is readily accomplished by applying (2.20). Thus we have in two dimensions

$$I_1 = g(g^{11} + g^{22}),$$
$$I_2 = g = 1/\det(g^{ij}) = 1\Big/\left[\overset{2}{\det}\left(\frac{\partial \xi^i}{\partial x^j}\right)\right].$$

From these relations we obtain

$$\begin{aligned}
\frac{\partial I_2}{\partial(\partial\xi^i/\partial x^j)} &= -2J^3(-1)^{i+j}\frac{\partial\xi^{3-i}}{\partial x^{3-j}} = -2I_2\frac{\partial x^j}{\partial\xi^i},\\
\frac{\partial I_1}{\partial(\partial\xi^i/\partial x^j)} &= 2g\frac{\partial\xi^i}{\partial x^j} - 2J^3(g^{11}+g^{22})(-1)^{i+j}\frac{\partial\xi^{3-i}}{\partial x^{3-j}}\\
&= 2\left(I_2\frac{\partial\xi^i}{\partial x^j} - I_1\frac{\partial x^j}{\partial\xi^i}\right), \quad i,j=1,2,
\end{aligned} \tag{8.18}$$

without summation over the repeated indices i and j. Thus we find that in two dimensions the Euler–Lagrange system (8.17) can be expressed as follows:

$$\frac{\partial}{\partial x^j}\left[G_{I_1}\left(I_2\frac{\partial\xi^i}{\partial x^j} - I_1\frac{\partial x^j}{\partial\xi^i}\right) - G_{I_2}I_2\frac{\partial x^j}{\partial\xi^i}\right] = 0, \quad i,j=1,2. \tag{8.19}$$

The application of the identity (2.56) to each equation of (8.19) leads to the following system:

$$\frac{\partial}{\partial\xi^j}\{J[g^{ij}G_{I_1}I_2 - \delta^i_j(G_{I_1}I_1 + G_{I_2}I_2)]\} = 0, \quad i,j=1,2. \tag{8.20}$$

In particular, for the integral measure of the two-dimensional grid density expressed locally by (3.66) with $n=2$, we assume

$$I_{\mathrm{cn}} = \int_{X^2}(I_1/I_2)\mathrm{d}\boldsymbol{x} = \int_{X^2}(g^{11}+g^{22})\mathrm{d}\boldsymbol{x},$$

and therefore the system of Euler–Lagrange equations (8.19) for I_{cn} is the system of Laplace equations

$$\frac{\partial}{\partial x^j}\frac{\partial\xi^i}{\partial x^j} = 0, \quad i,j=1,2.$$

8.2.4.2 Three-Dimensional Tensor

As in two dimensions, the invariants (3.24) of the three-dimensional tensor (g_{ij}) can be expressed through the elements of the contravariant tensor (g^{ij}). Using for this purpose (2.21), we obtain

$$\begin{aligned}
I_1 &= g(g^{11}g^{22} + g^{11}g^{33} + g^{22}g^{33} - g^{12}g^{21} - g^{13}g^{31} - g^{23}g^{32}),\\
I_2 &= (g^{11}+g^{22}+g^{33})/\det(g^{ij}),\\
I_3 &= g = 1/\det(g^{ij}).
\end{aligned}$$

Therefore we have for $i,j=1,2,3$

$$\begin{aligned}
\frac{\partial I_3}{\partial g^{ij}} &= -g^2 \quad \text{cofactor of } g^{ij} = -gg_{ij} = -g_{ij}I_3,\\
\frac{\partial I_2}{\partial g^{ij}} &= \delta^i_j g - gg_{ij}(g^{11}+g^{22}+g^{33}) = \delta^i_j I_3 - g_{ij}I_2,\\
\frac{\partial I_1}{\partial g^{ij}} &= \delta^i_j g(g^{11}+g^{22}++g^{33}) - g^{ij} - g^{ij}I_3 = \delta^i_j I_2 - g_{ij}I_3 - g^{ij}.
\end{aligned} \tag{8.21}$$

Taking into account (8.13), we obtain

$$\frac{\partial I_k}{\partial(\partial\xi^i/\partial x^j)} = \left(\frac{\partial I_k}{\partial g^{im}} + \frac{\partial I_k}{\partial g^{mi}}\right)\frac{\partial \xi^m}{\partial x^j}, \quad i,j,k,m = 1,2,3.$$

Therefore the use of (8.21) leads to

$$\begin{aligned}
\frac{\partial I_3}{\partial(\partial\xi^i/\partial x^j)} &= -2I_3\frac{\partial x^j}{\partial \xi^i}, \quad i,j = 1,2,3,\\
\frac{\partial I_2}{\partial(\partial\xi^i/\partial x^j)} &= -2I_3\frac{\partial \xi^i}{\partial x^j} - 2I_2\frac{\partial x^j}{\partial \xi^i}, \quad i,j = 1,2,3,\\
\frac{\partial I_1}{\partial(\partial\xi^i/\partial x^j)} &= 2\left(I_2\frac{\partial \xi^i}{\partial x^j} - I_3\frac{\partial x^j}{\partial \xi^i} - g^{im}\frac{\partial \xi^m}{\partial x^j}\right), \quad i,j,m = 1,2,3.
\end{aligned}$$

Using these relations in (8.17) we find that the three-dimensional Euler–Lagrange equations for the functional (8.16) can be written as

$$\begin{aligned}
&\frac{\partial}{\partial x^j}\left[G_{I_1}\left(I_2\frac{\partial \xi^i}{\partial x^j} - I_3\frac{\partial x^j}{\partial \xi^i} - g^{im}\frac{\partial \xi^m}{\partial x^j}\right)\right.\\
&\quad \left. - G_{I_2}\left(I_3\frac{\partial \xi^i}{\partial x^j} + I_2\frac{\partial x^j}{\partial \xi^i}\right) - G_{I_3}I_3\frac{\partial x^j}{\partial \xi^i}\right] = 0,\\
&i,j,m = 1,2,3.
\end{aligned} \tag{8.22}$$

This system, after application of (2.56) to every equation, is transformed to the

$$\begin{aligned}
&\frac{\partial}{\partial \xi^j}\{J[g^{ij}(G_{I_1}I_2 - G_{I_2}I_3) - g^{im}g^{mj} - \delta^i_j(G_{I_1}I_3 + G_{I_2}I_2 + G_{I_3}I_3)]\}\\
&= 0, \quad i,j,m = 1,2,3,
\end{aligned} \tag{8.23}$$

written with respect to the independent variables ξ^i.

8.2.5 Convexity Condition

Convexity is a very important property imposed on functionals in the calculus of variations. In the case of the functional (8.2) it is formulated by the condition of positiveness of the tensors G^i, $i = 1,\dots,n$:

$$G^i = (G_{\frac{\partial \xi^i}{\partial x^j}\frac{\partial \xi^i}{\partial x^k}}), \quad \text{with } i \text{ fixed.}$$

Namely, every tensor G^i, $i = 1,\dots,n$, must be strongly positive. Recall that a matrix is strongly positive if every principal minor is larger then zero. In this case there exists a constant $c_i > 0$ for every fixed index $i = 1,\dots,n$, such that

$$G_{\frac{\partial \xi^i}{\partial x^j}\frac{\partial \xi^i}{\partial x^k}} b^j b^k \geq c_i b^l b^l, \quad j,k,l = 1,\dots,n, \tag{8.24}$$

for an arbitrary vector $\boldsymbol{b} = (b^1, \ldots, b^n)$. The inequality (8.24) means that the system of Euler–Lagrange equations (8.6) is elliptic.

Convex functionals generate well-posed problems for their minimization and for the solution of the Dirichlet boundary value problem for the corresponding Euler–Lagrange equations (8.5) or (8.6). In particular, the relation (8.24) guarantees that there exists a unique, isolated optimal transformation which satisfies the system of Euler–Lagrange equations with Dirichlet boundary conditions.

8.3 Integral Grid Characteristics

Grid properties can play an extremely important role in influencing the accuracy and efficiency of the numerical solutions of partial differential equations. In particular, the truncation error is affected by the grid skewness, grid size, grid size ratio, angle between the grid lines, grid nonuniformity, and consistency of the grid with the features of the physical solution. Thus, by controlling these grid quantities one can control the efficiency of the numerical solution of boundary value problems. The calculus of variations allows one to formulate, through appropriate functionals, natural techniques which can serve as tools to control various grid properties.

A description of some local grid characteristics was given in Chap. 3 through the elements of the metric tensors. The procedure of integration of these characteristics defines functionals which reflect global properties of the grid. In this section some basic functionals modeling global grid characteristics are formulated. It needs to be emphasized that some local characteristics are dimensionally homogeneous. Therefore, in order to preserve this quality globally, the integration of the corresponding quantities should be carried out over a scaled region. If we assume that the logical domain Ξ^n is the unit cube, we can utilize it as such a normalized domain.

8.3.1 Dimensionless Functionals

Dimensionless functionals are formed by integrating the dimensionless grid characteristics reviewed in Chap. 3 over the computational domain Ξ^n.

8.3.1.1 Grid Skewness

The integrated measures of three-dimensional grid skewness are obtained from the formulas (3.58)–(3.60), derived by means of the cosines and cotangents of the angles between the tangential and normal vectors. These characteristics are dimensionally homogeneous and, in accordance with the argument about the integration of dimensionless quantities over the domain Ξ^n, we formulate the global grid skewness measures as

$$
\begin{aligned}
I_{\mathrm{sk},1} &= \int_{\Xi^3} \left(\frac{(g_{12})^2}{g_{11}g_{22}} + \frac{(g_{13})^2}{g_{11}g_{33}} + \frac{(g_{23})^2}{g_{22}g_{33}} \right) \mathrm{d}\boldsymbol{\xi}, \\
I_{\mathrm{sk},2} &= \int_{\Xi^3} \frac{1}{g} \left(\frac{(g_{12})^2}{g^{33}} + \frac{(g_{13})^2}{g^{22}} + \frac{(g_{23})^2}{g^{11}} \right) \mathrm{d}\boldsymbol{\xi} \\
&= \int_{\Xi^3} \left(\frac{(g_{12})^2}{g_{11}g_{22} - (g_{12})^2} + \frac{(g_{13})^2}{g_{11}g_{33} - (g_{13})^2} \right. \\
&\quad \left. + \frac{(g_{23})^2}{g_{22}g_{33} - (g_{23})^2} \right) \mathrm{d}\boldsymbol{\xi}, \\
I_{\mathrm{sk},3} &= \int_{\Xi^3} \left(\frac{(g^{12})^2}{g^{11}g^{22}} + \frac{(g^{13})^2}{g^{11}g^{33}} + \frac{(g^{23})^2}{g^{22}g^{33}} \right) \mathrm{d}\boldsymbol{\xi}, \\
I_{\mathrm{sk},4} &= \int_{\Xi^3} g \left(\frac{(g^{12})^2}{g_{33}} + \frac{(g^{13})^2}{g_{22}} + \frac{(g^{23})^2}{g_{11}} \right) \mathrm{d}\boldsymbol{\xi} \\
&= \int_{\Xi^3} \left(\frac{(g^{12})^2}{g^{11}g^{22} - (g^{12})^2} + \frac{(g^{13})^2}{g^{11}g^{33} - (g^{13})^2} \right. \\
&\quad \left. + \frac{(g^{23})^2}{g^{22}g^{33} - (g^{23})^2} \right) \mathrm{d}\boldsymbol{\xi}.
\end{aligned}
\tag{8.25}
$$

Since the elements g^{ij} of the contravariant metric tensor are expressed directly through the derivatives of the functions $\partial\xi^i/\partial x^j$, we see that the Euler–Lagrange equations for the functionals $I_{\mathrm{sk},3}$ and $I_{\mathrm{sk},4}$ can be obtained more easily if these functionals are reformulated over the domain X^3. This can be accomplished by using the relation (8.4). For example, we have, for the functional $I_{\mathrm{sk},3}$,

$$
I_{\mathrm{sk},3} = \int_{X^3} \sqrt{\det(g^{ij})} \left(\frac{(g^{12})^2}{g^{11}g^{22}} + \frac{(g^{13})^2}{g^{11}g^{33}} + \frac{(g^{23})^2}{g^{22}g^{33}} \right) \mathrm{d}\boldsymbol{x}.
$$

The functionals $I_{\mathrm{sk},1}$ and $I_{\mathrm{sk},2}$ can be transformed to functionals dependent on $\boldsymbol{\xi}(\boldsymbol{x})$ by the rule of transition (2.21) from the elements (g_{ij}) in the integrand to the elements g^{ij}.

In two dimensions we obtain from (8.25) only two functionals of dimensionally homogeneous skewness:

$$
\begin{aligned}
I_{\mathrm{sk},1} &= \int_{\Xi^2} \frac{(g_{12})^2}{g_{11}g_{22}} \mathrm{d}\boldsymbol{\xi} = \int_{X^2} \sqrt{\det(g^{ij})} \frac{(g^{12})^2}{g^{11}g^{22}} \mathrm{d}\boldsymbol{x}, \\
I_{\mathrm{sk},2} &= \int_{\Xi^2} g (g^{12})^2 \mathrm{d}\boldsymbol{\xi} = \int_{X^2} \frac{1}{\sqrt{\det(g^{ij})}} (g^{12})^2 \mathrm{d}\boldsymbol{x}.
\end{aligned}
\tag{8.26}
$$

The first functional is defined through the cosines of the angles while the second is determined by the cotangents of the angles.

8.3.1.2 Deviation from Orthogonality

Dimensionally homogeneous functionals indicating the global grid nonorthogonality in three dimensions can be derived from the local nonorthogonality measures (3.61). As a result, we have in three dimensions

$$
\begin{aligned}
I_{o,1} &= \int_{\Xi^3} \frac{g_{11}g_{22}g_{33}}{g} \mathrm{d}\boldsymbol{\xi} \\
&= \int_{X^3} \frac{[g^{22}g^{33} - (g^{23})^2][g^{11}g^{33} - (g^{13})^2][g^{11}g^{22} - (g^{12})^2]}{\sqrt{\det^5(g^{ij})}} \mathrm{d}\boldsymbol{x}, \quad (8.27) \\
I_{o,2} &= \int_{\Xi^3} g(g^{11}g^{22}g^{33})\mathrm{d}\boldsymbol{\xi} = \int_{X^3} \frac{1}{\sqrt{\det(g^{ij})}}(g^{11}g^{22}g^{33})\mathrm{d}\boldsymbol{x}.
\end{aligned}
$$

In two dimensions (8.27) yields only one functional of departure from orthogonality:

$$
I_{o,1} = \int_{\Xi^2} \frac{g_{11}g_{22}}{g} \mathrm{d}\boldsymbol{\xi} = \int_{X^2} \frac{g^{11}g^{22}}{\sqrt{\det(g^{ij})}} \mathrm{d}\boldsymbol{x}. \qquad (8.28)
$$

8.3.1.3 Deviation from Conformality

Integration of the dimensionless characteristics (3.67), (3.76), and (3.79) over Ξ^2 or Ξ^3 generates a quantity which reflects an integral departure of the grid from a conformal grid. Thus we obtain in two dimensions,

$$
I_{cf,1} = \int_{\Xi^2} (I_1/\sqrt{I_2})\mathrm{d}\boldsymbol{\xi} = \int_{X^2} (I_1/I_2)\mathrm{d}\boldsymbol{x}. \qquad (8.29)
$$

An analogous consideration of (3.77) yields, in three dimensions,

$$
I_{cf,1} = \int_{\Xi^3} [I_2/\sqrt[3]{(I_3)^2}]\mathrm{d}\boldsymbol{\xi} = \int_{X^3} [I_2/(I_3)^{7/6}]\mathrm{d}\boldsymbol{x}. \qquad (8.30)
$$

The quantity (3.80) for $n = 3$ defines one more three-dimensional form of the nonconformality functional:

$$
\begin{aligned}
I_{cf,2} &= \int_{\Xi^3} [I_1/(I_3)^{1/3}]\mathrm{d}\boldsymbol{\xi} \\
&= \int_{\Xi^3} \frac{g_{11} + g_{22} + g_{33}}{\sqrt[3]{\det(g_{ij})}} \mathrm{d}\boldsymbol{\xi}. \qquad (8.31)
\end{aligned}
$$

Reformulation of $I_{cf,2}$ over the domain X^3 yields

$$
\begin{aligned}
I_{cf,2} = \int_{X^3} \frac{1}{\sqrt[6]{\det^5[g_{ij})}} \{ & g^{11}g^{22} + g^{11}g^{33} + g^{22}g^{33} \\
& - [(g^{12})^2 + (g^{13})^2 + (g^{23})^2]\}\mathrm{d}\boldsymbol{x}, \qquad (8.32)
\end{aligned}
$$

using (8.4) and (2.21).

We can use as the integrand the dimensionally homogeneous quantity:

$$Q_{\mathrm{cf},3} = \left[I_2/\sqrt[3]{(I_3)^2}\right]^\alpha = (Q_{\mathrm{cf},1})^\alpha, \quad \alpha > 0, \tag{8.33}$$

which also reaches its minimum value when the three-dimensional coordinate transformation $\boldsymbol{x}(\boldsymbol{\xi})$ is conformal. In this case we can control the form of the Euler–Lagrange equations with the parameter α. The corresponding functional with the quantitative characteristic (8.33) is as follows:

$$\begin{aligned} I_{\mathrm{cf},3} &= \int_{\Xi^3} \left[I_2/\sqrt[3]{(I_3)^2}\right]^\alpha \mathrm{d}\boldsymbol{\xi} \\ &= \int_{X^3} (I_2)^\alpha/(I_3)^{2\alpha/3+1/2} \mathrm{d}\boldsymbol{x}. \end{aligned}$$

If $\alpha = 3/2$ we obtain

$$\begin{aligned} I_{\mathrm{cf},3} &= \int_{X^3} (I_2/I_3)^{3/2} \mathrm{d}\boldsymbol{x} \\ &= \int_{X^3} (g^{11} + g^{22} + g^{33})^{3/2} \mathrm{d}\boldsymbol{x}, \end{aligned}$$

taking into account (3.24). The system of Euler–Lagrange equations (8.14) or (8.19) for this functional has the form

$$\frac{\partial}{\partial x^j}\left(\sqrt{g^{11} + g^{22} + g^{33}}\frac{\partial \xi^i}{\partial x^j}\right) = \frac{\partial}{\partial x^j}\left(\sqrt{\frac{I_2}{I_3}}\frac{\partial \xi^i}{\partial x^j}\right) = 0,$$
$$i, j = 1, 2, 3. \tag{8.34}$$

Taking advantage of (2.56) or (8.23), we obtain for the inverted system of (8.34)

$$\frac{\partial}{\partial \xi^j}(\sqrt{I_2} g^{ij}) = 0, \quad i, j = 1, 2, 3. \tag{8.35}$$

Also, multiplication of (8.34) by $\partial x^k/\partial \xi^i$ and summation over i yields one more inverted system of (8.34):

$$g^{ij}\frac{\partial^2 x^k}{\partial \xi^i \partial \xi^j} = \sqrt{\frac{I_3}{I_2}}\frac{\partial}{\partial x^k}\sqrt{\frac{I_2}{I_3}}, \quad i, j = 1, 2, 3. \tag{8.36}$$

An analogously simple system of Euler–Lagrange equations is derived for n-dimensional functionals of nonconformality by replacing the local measure (3.79) with

$$Q_{\mathrm{cf},3} = \left[I_{n-1}/(I_n)^{1-1/n}\right]^\alpha, \quad \alpha > 0.$$

For the functional $I_{\mathrm{cf},3}$ with $\alpha = n/2$, we obtain

$$\begin{aligned} I_{\mathrm{cf},3} &= \int_{\Xi^n} \left[I_{n-1}/(I_n)^{1-1/n}\right]^{n/2} \mathrm{d}\boldsymbol{\xi} = \int_{X^n} (I_{n-1}/I_n)^{n/2} \mathrm{d}\boldsymbol{x} \\ &= \int_{X^n} (g^{11} + \cdots + g^{nn})^{n/2} \mathrm{d}\boldsymbol{x}. \end{aligned} \tag{8.37}$$

The system of Euler–Lagrange equations for this functional is

$$\frac{\partial}{\partial x^j}\left((g^{11}+\cdots+g^{nn})^{n/2-1}\frac{\partial \xi^i}{\partial x^j}\right)=0, \quad i,j=1,\ldots,n. \tag{8.38}$$

Multiplying (8.38) by $\partial x^k/\partial \xi^i$ and summing over i, we obtain in analogy with (8.36), the system with respect to the dependent variables x^i and independent variables ξ^i:

$$g^{ij}\frac{\partial^2 x^k}{\partial \xi^i \partial \xi^j}=H^{-1}\frac{\partial}{\partial x^k}H, \quad i,j,k=1,\ldots,n, \tag{8.39}$$

where

$$H=(g^{11}+\cdots+g^{nn})^{n/2-1}=(I_{n-1}/I_n)^{n/2-1}.$$

8.3.2 Dimensionally Heterogeneous Functionals

8.3.2.1 Smoothness Functionals

The characteristic of local grid concentration is expressed through the invariants by (3.66). In general, this quantity is not dimensionless, and therefore its integration is carried out over the physical domain X^n. The resulting functional,

$$I_s=\int_{X^n}(I_{n-1}/I_n)\mathrm{d}\boldsymbol{x}=\int_{X^n}(g^{11}+\cdots+g^{nn})\mathrm{d}\boldsymbol{x}, \tag{8.40}$$

formulated for an arbitrary n-dimensional domain X^n, is called the functional of smoothness. We see that the functional of smoothness (8.40) for $n=2$ coincides with the functional of conformality (8.29). However, in the three-dimensional case the functionals (8.30) and (8.40) are different. The Euler–Lagrange equations for the smoothness functional (8.40) comprise a simple system of Laplace equations:

$$\frac{\partial}{\partial x^j}\left(\frac{\partial \xi^i}{\partial x^j}\right)=0, \quad i,j=1,\ldots,n.$$

The inverted system with respect to the dependent variable $\boldsymbol{x}$ is obtained in the ordinary manner by multiplying (8.40) by $\partial x^k/\partial \xi^i$ and summing over i. As a result we obtain the n-dimensional inverted Laplace system

$$g^{ij}\frac{\partial \boldsymbol{x}}{\partial \xi^i \partial \xi^j}=0, \quad i,j=1,\ldots,n. \tag{8.41}$$

8.3.2.2 Functionals of Orthogonality

The characteristics (3.62) and (3.63) of the local deviation of the three-dimensional cells from orthogonal cells define two functionals of orthogonality. For the purpose of simplicity of the resulting Euler–Lagrange equations, it is more suitable to integrate (3.62) over Ξ^3 and (3.63) over X^3. So we obtain the following functionals, which represent some measures of grid nonorthogonality and also can be interpreted as measures of grid skewness:

$$\begin{aligned} I_{o,3} &= \int_{\Xi^3} ((g_{12})^2 + (g_{13})^2 + (g_{23})^2) \mathrm{d}\boldsymbol{\xi}, \\ I_{o,4} &= \int_{X^3} ((g^{12})^2 + (g^{13})^2 + (g^{23})^2) \mathrm{d}\boldsymbol{x}. \end{aligned} \tag{8.42}$$

The corresponding Euler–Lagrange equations (8.11) and (8.6) have the form

$$\begin{aligned} g_{ik} \frac{\partial}{\partial \xi^j} \left(\frac{\partial x^k}{\partial \xi^j} \right) &= 0, \\ g^{ik} \frac{\partial}{\partial x^j} \left(\frac{\partial \xi^k}{\partial x^j} \right) &= 0, \quad i, j, k = 1, 2, 3, \ i \neq k. \end{aligned} \tag{8.43}$$

By applying (2.56) to every equation of the second system of (8.43), a converted system is obtained:

$$g^{ik} \frac{\partial}{\partial \xi^k} J = 0, \quad k = 1, 2, 3, \ i \neq k. \tag{8.44}$$

The systems (8.43) and (8.44) derive ill-posed boundary value problems, and therefore the functionals of orthogonality are commonly combined with the functional of smoothness (8.40) to yield well-posed problems of grid generation. In two dimensions, the orthogonality functionals (8.42) are

$$\begin{aligned} I_{o,3} &= \int_{\Xi^2} (g_{12})^2 \mathrm{d}\boldsymbol{\xi}, \\ I_{o,4} &= \int_{X^2} (g^{12})^2 \mathrm{d}\boldsymbol{x}. \end{aligned} \tag{8.45}$$

The local departure of a two-dimensional grid from an orthogonal one with a prescribed cell aspect ratio F may be estimated by the measure

$$Q_{o,5} = \left(\frac{1}{\sqrt{F}} \frac{\partial x^1}{\partial \xi^1} - \sqrt{F} \frac{\partial x^2}{\partial \xi^2} \right)^2 + \left(\frac{1}{\sqrt{F}} \frac{\partial x^2}{\partial \xi^1} + \sqrt{F} \frac{\partial x^1}{\partial \xi^2} \right)^2, \tag{8.46}$$

since $Q_{o,5} = 0$ if and only if the grid is orthogonal and $g_{11} = F^2 g_{22}$. From (8.46), we obtain

$$Q_{o,5} = \frac{1}{F} g_{11} + F g_{22} - 2J.$$

This quantity $Q_{o,5}$ defines one more functional of departure from orthogonality:

$$\begin{aligned} I_{o,5} &= \int_{\Xi^2} \left(\frac{1}{F} g_{11} + F g_{22} - 2J \right) \mathrm{d}\boldsymbol{\xi} \\ &= \int_{\Xi^2} \left(\frac{1}{F} g_{11} + F g_{22} \right) \mathrm{d}\boldsymbol{\xi} - 2S, \end{aligned} \tag{8.47}$$

where S is the area of the domain X^2, with the following Euler–Lagrange equations:

$$\frac{\partial}{\partial \xi^1}\left(\frac{1}{F}\frac{\partial x^i}{\partial \xi^1} + F\frac{\partial x^i}{\partial \xi^2}\right) = 0, \quad i = 1, 2. \tag{8.48}$$

Analogously, the functional of departure from orthogonality is defined through the elements g^{ii} as

$$I_{o,6} = \int_{X^2} \left(\frac{1}{F} g^{11} + F g^{22}\right) \mathrm{d}\boldsymbol{x} - 2. \tag{8.49}$$

8.3.3 Functionals Dependent on Second Derivatives

This subsection reviews a formulation of the functionals in the form (8.2) or (8.3), where the integrands include terms dependent on second derivatives of coordinate transformations.

8.3.3.1 Functionals of Eccentricity

The eccentricity functionals are derived from the local grid eccentricity measures (3.81) and (3.82). Since $Q_{\epsilon,1}$, from (3.81), is expressed through the first and second derivatives of $\boldsymbol{x}(\boldsymbol{\xi})$ with respect to ξ^i, we will integrate this quantity over Ξ^n. For a similar reason, the relation (3.82) is integrated over X^n. As a result, we obtain the integral characteristics of grid eccentricity in the form

$$\begin{aligned} I_{\epsilon,1} &= \int_{\Xi^n} \sum_{i=1}^{n} \left(\frac{\partial}{\partial \xi^i} \ln \sqrt{g_{ii}}\right)^2 \mathrm{d}\boldsymbol{\xi}, \\ I_{\epsilon,2} &= \int_{X^n} \sum_{i=1}^{n} \left(\frac{\partial}{\partial x^i} \ln \sqrt{g^{ii}}\right)^2 \mathrm{d}\boldsymbol{x}. \end{aligned} \tag{8.50}$$

Unlike the functionals determined by the first derivatives of the varied functions $\boldsymbol{\xi}(\boldsymbol{x})$ or $\boldsymbol{x}(\boldsymbol{\xi})$, the functionals (8.50) include second derivatives. The system of Euler–Lagrange equations (8.11) for grid generation is therefore of fourth order. This makes it possible not only to specify the boundary nodal distribution when generating a grid by solving such a system, but also to specify the directions of the coordinate lines emerging from the boundaries, which is important when one needs to construct smoothly abutting grids in complicated regions, as in the case of applying block grid techniques. Some questions related to the formulation of the boundary conditions and the correctness of the boundary value problems, and also to the numerical justification for the systems of Euler–Lagrange equations for constructing grids are still to be resolved, however.

8.3.3.2 Functionals of Grid Warping and Grid Torsion

The functionals of grid warping and grid torsion are formulated analogously through the respective local measures (3.83) and (3.84). Like the functionals of eccentricity (8.50), these functionals are dependent on second derivatives, thus generating Euler–Lagrange equations of the fourth order.

8.4 Adaptation Functionals

Numerical grids can significantly influence various characteristics of the efficiency of the numerical solution of partial differential equations. One of the most important characteristics is the accuracy of the numerical solution, which is formulated through the error of the numerical calculation. In this matter, the theory of the calculus of variations provides an excellent opportunity to formulate the requirement of a minimal error for a given number of grid points in a straightforward form through the functional of error. The minimization of this functional generates an optimal grid in the sense of accuracy. Thus the variational approach is a natural tool for generating grids adapted to the physical solution.

The simplest and most logical way of defining the error functional I_{er} seems to be through the integral measure of the local numerical error $\boldsymbol{r} = \boldsymbol{u} - \boldsymbol{u}^h$,

$$I_{\mathrm{er},1} = \int_{\Xi^n} \|\boldsymbol{r}\| \mathrm{d}\boldsymbol{\xi}, \tag{8.51}$$

or through the integral of the measure of the approximation error $\boldsymbol{T}$,

$$I_{\mathrm{er},2} = \int_{\Xi^n} \|\boldsymbol{T}\| \mathrm{d}\boldsymbol{\xi}, \tag{8.52}$$

with the notation $\boldsymbol{r}$ and $\boldsymbol{T}$ discussed in Sect. 7.2. However, this logical formulation results in a very cumbersome and high-order system of Euler–Lagrange equations; namely, its order is twice the order of the derivatives in $\boldsymbol{r}$ or $\boldsymbol{T}$. The numerical solution of these Euler–Lagrange equations is a very difficult task, especially in the case of multidimensional space. Thus the optimal grid can be obtained only at the expense of the efficiency of the grid generation process. Evidently, a more optimal approach to generating adaptive grids through the variational technique lies in formulating simpler error functionals in order to balance the accuracy of the solution against the cost of obtaining the grid.

A common approach aimed at the minimization of the numerical error relies on concentration of grid nodes in the subregions of high truncation error. One version of this approach, reviewed in Chap. 7, was formulated through the equidistribution principle. In fact this principle is universal, since all adaptive methods aimed at the concentration of grid nodes in the regions of large solution variations are related to the one-dimensional equidistribution principle, which requires the grid spacing to be inversely proportional to a weight function. The equidistribution principle can be formulated in a number of different ways. In this section a variational version of the equidistribution approach is discussed.

8.4.1 One-Dimensional Functionals

The basic one-dimensional differential model for the equidistribution principle with the weight function w was formulated as the two-point boundary value problem (7.10). In the case where the weight function w is defined in the interval ξ and thus does not vary when $x(\xi)$ changes, the problem (7.10) is a boundary value problem for the Euler–Lagrange equation obtained by optimizing the functional

$$I_{\mathrm{eq}} = \int_0^1 w\left(\frac{\mathrm{d}x}{\mathrm{d}\xi}\right)^2 \mathrm{d}\xi. \tag{8.53}$$

The functional (8.53) physically models the energy which arises in a system of nodes x_i connected by springs with stiffness $2w_i$. The equilibrium condition of this system also determines the positions of the grid points x_i defined by a coordinate transformation $x(\xi)$ satisfying (7.10).

There also is a geometric interpretation of the following numerical approximation of the integral (8.53) on a uniform grid $\xi_i = ih, \ h = 1/N$:

$$I^h_{\mathrm{eq}} = N \sum_{i=0}^{N-1} w_{i+1/2}(h_{i+1/2})^2, \tag{8.54}$$

where

$$\begin{aligned} w_{i+1/2} &= [w(\xi_i) + w(\xi_{i+1})]/2, \\ h_{i+1/2} &= x(\xi_{i+1}) - x(\xi_i). \end{aligned}$$

The expression (8.54) describes a hyperellipsoid for each value of I^h_{eq}, if $h_{i+1/2}$ is considered as the ith coordinate in the $(N-1)$-dimensional Euclidean space, and its minimization means that the hyperplane

$$\sum_{i=1}^{N-1} h_{i+1/2} = b - a,$$

where $b-a$ is the length of the segment x, is an $(N-2)$-dimensional tangent plane for this hyperellipsoid.

Since, in general, the error of the solution in the interval x is described by an expression of the form $\boldsymbol{r}_i = \boldsymbol{C}(h_i)^k \approx \boldsymbol{C}h^k(\partial x/\partial\xi)^k$, the functional (8.53) can be interpreted as the integral error of the second-order approximation of a one-dimensional differential problem. The error functional of the approximation of order k can be represented by the integral

$$I_{\text{eq}} = \int_0^1 w \left(\frac{\mathrm{d}x}{\mathrm{d}\xi}\right)^k \mathrm{d}\xi, \quad k > 0. \tag{8.55}$$

A geometric interpretation of the functional (8.55) for $k = 4$ is possible if u_x^2 is taken as the weight function. In this case the value of the functional (8.55) is proportional to the sum of the squares of the areas of the rectangles which border the curve $u = u(x)$ in the (u, x) plane.

Commonly the weight function w is defined in the physical region, and therefore the variational formulation of the equidistribution method typically utilizes a functional with respect to transformations $\xi(x)$ with specified boundary conditions:

$$I_{\text{eq}} = \int_X w_1(x) \left(\frac{\mathrm{d}\xi}{\mathrm{d}x}\right)^k \mathrm{d}x, \quad k > 0, \tag{8.56}$$

whose Euler–Lagrange equation is obtained in accordance with (8.6). Thus the optimal transformation $\xi(x)$ for this functional is the solution to the boundary value problem

$$\begin{aligned} &\frac{\mathrm{d}}{\mathrm{d}x}\left[w_1(x)\left(\frac{\mathrm{d}\xi}{\mathrm{d}x}\right)^{k-1}\right] = 0, \quad a < x < b, \\ &\xi(a) = 0, \qquad \xi(b) = 1. \end{aligned} \tag{8.57}$$

From (8.57), the following relation follows directly:

$$w_1^{1/(k-1)} \frac{\mathrm{d}\xi}{\mathrm{d}x} = \text{const} \quad \Longrightarrow \quad w_1^{1/(1-k)} \frac{\mathrm{d}x}{\mathrm{d}\xi} = \text{const},$$

which results in small values of $\mathrm{d}x/\mathrm{d}\xi$ when $w_1^{1/(1-k)}$ is large and vice versa. A problem equivalent to (7.10) is obtained from (8.57) when $k = 2,\ w_1(x) = w^{-1}(x),\ w_1 > 0$.

Thus the equidistribution method considered in Sect. 7.2 can be interpreted as a variational method for constructing grids by minimizing the functionals (8.55) or (8.56). An analytic justification for using functionals of these forms when constructing adaptive grids is provided by the example in Sect. 7.2: the formula (7.3) for an optimal grids can be obtained from the condition of a minimum of the functional (8.56) with $k = 2$ and

$$w_1(x) = \left(\left\|\frac{\mathrm{d}u}{\mathrm{d}x}\right\|\right)^{-p}, \quad p = 2/3,$$

by integrating (8.57).

8.4.2 Multidimensional Approaches

In this subsection the variational formulations (8.55) and (8.56) of the one-dimensional equidistribution approach are taken as a starting point for the extension to multiple dimensions.

The basic elements of the functionals (8.55) and (8.56) are the weight functions and the first derivative of the transformation $x(\xi)$ or $\xi(x)$. When generalizing to an n-dimensional region X^n, this derivative can be interpreted as the Jacobian of transformation $\boldsymbol{x}(\boldsymbol{\xi})$ or $\boldsymbol{\xi}(\boldsymbol{x})$, or as the square roots of the values of the diagonal elements of the covariant metric tensor (g_{ij}) or the contravariant metric tensor (g^{ij}). Thus, in many of the generalizations of the functionals (8.55) and (8.56) that have been proposed for constructing adaptive grids in an n-dimensional domain X^n, the expression $\mathrm{d}x/\mathrm{d}\xi$ is replaced by $J = \det(\partial x^i/\partial \xi^j)$, and $\mathrm{d}\xi/\mathrm{d}x$ by $\det(\partial \xi^i/\partial x^j) = 1/J$, or combinations of the diagonal elements of the covariant or contravariant metric tensor (g_{ij}) and (g^{ij}) are used. Since $J = \sqrt{g}$, all these functionals can be formulated through the metric tensors (g_{ij}) or (g^{ij}). Thus the Euler–Lagrange equations for these functionals are readily obtained by using (8.14) or (8.15).

8.4.2.1 Volume-Weighted Functional

For example, the functional defined through the Jacobian $J = \sqrt{g}$, called the volume-weighted functional, has the form

$$I_{\mathrm{vw}} = \int_{X^n} w(\boldsymbol{x}) g^k \mathrm{d}\boldsymbol{x}, \quad k > 0. \tag{8.58}$$

The expected result of the minimization of this functional is small values of the Jacobian when $w(x)$ is large and vice versa.

In analogy with the first line of (8.21), we have for arbitrary dimensions

$$\frac{\partial g}{\partial g^{lk}} = -g g_{lk}, \quad l, k = 1, \ldots, n.$$

Therefore, using (8.14), we obtain a system of Euler–Lagrange equations for the functional (8.58):

$$\frac{\partial}{\partial x^j}\left(w g^k g_{im} \frac{\partial \xi^m}{\partial x^j}\right) = \frac{\partial}{\partial x^j}\left(w g^k \frac{\partial x^j}{\partial \xi^i}\right) = 0, \quad i, j, k, m = 1, \ldots, n. \tag{8.59}$$

In order to obtain compact equations which include only the derivatives with respect to ξ^i we use the identity

$$\frac{\partial}{\partial x^j}\left(g^{-1/2} \frac{\partial x^j}{\partial \xi^i}\right) \equiv 0, \quad i, j = 1, \ldots, n,$$

which is a mere reformulation of (2.47). Therefore, from (8.59), we obtain

$$\begin{aligned} \frac{\partial}{\partial x^j}\left(w g^k \frac{\partial x^j}{\partial \xi^i}\right) &= g^{-1/2} \frac{\partial}{\partial x^j}(w g^{k+1/2}) \frac{\partial x^j}{\partial \xi^i} \\ &= g^{-1/2} \frac{\partial}{\partial \xi^i}(w g^{k+1/2}) = 0. \end{aligned} \tag{8.60}$$

8.4.2.2 Tangent-Length-Weigthed Functionals

An adaptation functional which use the diagonal elements of the metric tensor (g_{ij}) can be expressed as follows:

$$I_{\mathrm{tw},1} = \int_{\Xi^n} \Big(w(\boldsymbol{\xi}) \sum_i g_{ii} \Big) \mathrm{d}\boldsymbol{\xi} = \int_{\Xi^n} w(\boldsymbol{\xi}) I_1 \mathrm{d}\boldsymbol{\xi}. \tag{8.61}$$

A functional aimed at providing an individual grid concentration in each grid direction ξ^i can be formulated through a combination of the edge length characteristics g_{ii} with individually specified weights:

$$I_{\mathrm{tw},2} = \int_{\Xi^n} \Big(\sum_i w^i(\boldsymbol{\xi}) g_{ii} \Big) \mathrm{d}\boldsymbol{\xi}. \tag{8.62}$$

The weight functions w^i control the grid spacing along each coordinate independently.

The system of Euler–Lagrange equations for the functional (8.62) is of simple elliptic type,

$$\frac{\partial}{\partial \xi^j} \Big(w^i \frac{\partial x^i}{\partial \xi^j} \Big) = 0, \quad i, j = 1, \dots, n,$$

with the index i fixed. The functionals (8.61) and (8.62) influence the grid node distribution in the direction of the coordinate lines.

8.4.2.3 Normal-Length-Weighted Functionals

Analogous adaptation functionals determined by weighted diagonal elements g^{ii} of the contravariant metric tensor (g^{ij}) have the form

$$I_{\mathrm{nw},1} = \int_{X^n} \Big(w(\boldsymbol{x}) \sum_i g^{ii} \Big) \mathrm{d}\boldsymbol{x} \tag{8.63}$$

and

$$I_{\mathrm{nw},2} = \int_{X^n} \Big(\sum_i w_i(\boldsymbol{x}) g^{ii} \Big) \mathrm{d}\boldsymbol{x}. \tag{8.64}$$

The functionals (8.63) and (8.64) are also referred to as diffusion functionals. They are formulated for the purpose of distributing the nodes in the direction of the normals to the coordinate surfaces $\xi^i = c$, with clustering of the grid points in the neighborhoods of large values of the weighting functions and rarefying of the nodes in the vicinity of small values of the weights.

The resulting Euler–Lagrange equations for critical functions $\boldsymbol{\xi}(\boldsymbol{x})$ of the functionals (8.63) and (8.64) have the form

$$\frac{\partial}{\partial x^j}\left(w_i \frac{\partial \xi^i}{\partial x^j}\right) = 0, \quad i,j = 1,\ldots,n, \tag{8.65}$$

with the index i fixed. If $w^i > 0$ then this system is elliptic. The relation $w^i = w,\ i = 1,\ldots,n$, in this system corresponds to the functional (8.63). In this case the transformed equations with the dependent and independent variables interchanged are readily obtained by multiplying the system of Euler–Lagrange equations (8.65) by $\partial x^k/\partial \xi^i$ and summing over i. As a result, we obtain

$$g^{ij}\frac{\partial^2 x^k}{\partial \xi^i \partial \xi^j} - \frac{1}{w}\frac{\partial w}{\partial x^k} = 0, \quad i,j,k = 1,\ldots,n. \tag{8.66}$$

Using the relation (2.23), we have

$$\frac{\partial w}{\partial x^k} = \frac{\partial w}{\partial \xi^i}\frac{\partial \xi^i}{\partial x^k} = g^{ij}\frac{\partial w}{\partial \xi^i}\frac{\partial x^k}{\partial \xi^j}.$$

Therefore we obtain, from the inverted system of Euler–Lagrange equations (8.66),

$$g^{ij}\left(\frac{\partial^2 x^k}{\partial \xi^i \partial \xi^j} - \frac{1}{w}\frac{\partial w}{\partial \xi^i}\frac{\partial x^k}{\partial \xi^j}\right) = w g^{ij}\frac{\partial}{\partial \xi^i}\left(\frac{1}{w}\frac{\partial x^k}{\partial \xi^j}\right) = 0.$$

Thus we obtain another compact form of the Euler–Lagrange equations for the functional (8.63):

$$g^{ij}\frac{\partial}{\partial \xi^i}\left(\frac{1}{w}\frac{\partial x^k}{\partial \xi^j}\right) = 0, \quad i,j,k = 1,\ldots,n. \tag{8.67}$$

8.4.2.4 Metric-Weighted Functionals

The length-weighted functionals permit a natural generalization in the form of metric-weighted functionals:

$$I_{\mathrm{mw},1} = \int_{\Xi^n} w^{ij}(\boldsymbol{\xi}) g_{ij} \mathrm{d}\boldsymbol{\xi} \tag{8.68}$$

and

$$I_{\mathrm{mw},2} = \int_{X^n} w_{ij}(\boldsymbol{x}) g^{ij} \mathrm{d}\boldsymbol{x}. \tag{8.69}$$

The condition of convexity (8.24) will be satisfied for these functionals if the matricies (w_{ij}) and (w^{ij}), respectively, are positive. Without loss of generality, we can assume in (8.68) and (8.69) that $w_{ij} = w_{ji}$, $w^{ij} = w^{ji}$. The corresponding systems of Euler–Lagrange equations for (8.68) and (8.69) have then the form

$$\begin{aligned}
&\frac{\partial}{\partial \xi^j}\left(w^{ik}\frac{\partial x^k}{\partial \xi^j}\right) = 0, \\
&\frac{\partial}{\partial x^j}\left(w_{ik}\frac{\partial \xi^k}{\partial x^j}\right) = 0, \quad i,j,k = 1,\ldots,n.
\end{aligned} \tag{8.70}$$

8.4.2.5 General Approach

A more general formulation of the adaptation functionals utilizes the weighted elements of the matrix $[(\partial x^i/\partial \xi^j)^2]$ or the matrix $[(\partial \xi^i/\partial x^j)^2]$. For example,

$$\begin{aligned} I_{\mathrm{ad},6} &= \int_{\Xi^n} \left[\sum_{i,j} w^{ij}(\boldsymbol{\xi}) \left(\frac{\partial x^i}{\partial \xi^j} \right)^2 \right] \mathrm{d}\boldsymbol{\xi}, \\ I_{\mathrm{ad},7} &= \int_{X^n} \left[\sum_{i,j} w^{ij}(\boldsymbol{x}) \left(\frac{\partial \xi^i}{\partial x^j} \right)^2 \right] \mathrm{d}\boldsymbol{x}. \end{aligned} \tag{8.71}$$

The corresponding Euler–Lagrange equations are

$$\begin{aligned} &\frac{\partial}{\partial \xi^j} \left(w^{ij} \frac{\partial x^i}{\partial \xi^j} \right) = 0, \\ &\frac{\partial}{\partial x^j} \left(w^{ij} \frac{\partial \xi^i}{\partial x^j} \right) = 0, \quad i, j = 1, \dots, n, \end{aligned} \tag{8.72}$$

where the summation is carried out only over j and the index i is fixed.

8.4.2.6 Nonstationary Functionals

In the construction of adaptive grids for spatial nonstationary elastoplastic and gas-dynamics problems, the adaptation functional characterizing the concentration of grid nodes in the high-gradient region of the flow velocity $\boldsymbol{u} = (u^1, u^2, u^3)$ is defined in terms of the velocity components of the grid nodes $\partial x^i/\partial t$:

$$I_{\mathrm{ad},6} = \int_{X^3} (k \operatorname{div} \overline{\boldsymbol{w}} - \operatorname{div} \overline{\boldsymbol{u}})^2 \mathrm{d}\boldsymbol{x}, \tag{8.73}$$

where

$$\begin{aligned} \overline{\boldsymbol{w}} &= (\overline{w}^1, \dots, \overline{w}^n), \quad \overline{w}^i = \sum_{j=1}^{3} \frac{\partial x^j}{\partial t} \frac{\partial \xi^i}{\partial x^j}, \\ \overline{\boldsymbol{u}} &= (\overline{u}^1, \dots, \overline{u}^n), \quad \overline{u}^i = \sum_{j=1}^{3} \frac{\partial \xi^i}{\partial x^j} u^j, \quad i = 1, 2, 3. \end{aligned}$$

Minimization of the functional (8.73) involves equidistribution of the weight function $w(x)$ with respect to the values, to degree k, of the grid cell volumes, generating a grid with small cell volumes in the neighborhood of large values of $w(x)$.

8.4.2.7 Weight Functions

The weight functions w, w_i, w^i, $w_{i,j}$, and $w^{i,j}$ in the formulations of adaptation functionals considered above are usually taken as combinations of the moduli of the derivatives of those components u^i of the solution of the physical problem for which these derivatives can take large values. For instance, for the flow of a viscous heat-conducting gas, the function $w(\boldsymbol{x})$ has the form

$$w = \left(\epsilon + \sum_i |\mathrm{grad}\, u^i|^{\alpha_j}\right)^{\beta} + \phi_0(\boldsymbol{x}),$$

where $\phi_0(\boldsymbol{x})$ is a positive function and ϵ, α_i, and β are positive constants. The weight function is of this form because of the need to construct a grid which is invariant under Galilean transformations and provides node clustering in the region of high gradients of $\boldsymbol{u}$. Other weight functions were outlined in Sect. 7.2.

8.5 Functionals of Attraction

For some multidimensional problems, there are natural families of lines or vector fields which should be aligned with the grid lines or basic vector fields for reasons of computational efficiency. In gas dynamics, for instance, these are the streamlines (or lines of potential), lines of predominant direction of flow, and a family of the Lagrange coordinates; in plasma theory they are the preferred vector directions defined by the magnetic field. The solution to viscous transonic flow problems usually contains shock structures which should be aligned with one coordinate direction, while boundary layers should be aligned with the coordinates from the other family; namely they need to be parallel to a streamwise coordinate. Some problems also have an underlying symmetry which should be matched with the coordinate system.

The alignment of the coordinate lines with natural families of curves of this kind leads to efficiency in the numerical modeling. For example, the use of Lagrange coordinates in problems of fluid motion simplifies the representation of the equations, and makes it possible to localize the moving region and to follow the motion of the fluid particles during the numerical solution. The requirement to generate aligned coordinates can be readily realized by variational techniques through suitable functionals of departure. This section presents a formulation of certain functionals of this type.

8.5.1 Lagrangian Coordinates

The condition for a coordinate ξ^i to be Lagrangian in a three-dimensional fluid flow is given by the equation

$$\frac{\partial \xi^i}{\partial t} + \overline{u}^i = 0, \quad i = 1, 2, 3, \tag{8.74}$$

where $\overline{u}^i$ is the ith component of the velocity vector in the moving system of coordinates (t, ξ^1, ξ^2, ξ^3), i.e.

$$\overline{u}^i = \sum_{i=1}^{3} u^j \frac{\partial \xi^i}{\partial x^j},$$

with u^j, $j = 1, 2, 3$, representing the jth velocity component in the Cartesian system (t, x^1, x^2, x^3). Since

$$\frac{\partial \xi^i}{\partial t} = -\sum_{j=1}^{3} \frac{\partial \xi^i}{\partial x^j} \frac{\partial x^j}{\partial t},$$

(8.6) is equivalent to

$$\overline{w}^i - \overline{u}^i = 0, \tag{8.75}$$

where $\overline{w}^i$ is the ith component of the grid velocity vector expanded in the tangential vectors $\boldsymbol{x}_{\xi^j}$, $j = 1, \ldots, n$.

Equation (8.75) can be used to determine the functional of deviation from a Lagrangian coordinate grid:

$$\begin{aligned} I_{\mathrm{L},1} &= \int_{X^3 \times I} w \sum_{i=1}^{3} (\overline{w}^i - \overline{u}^i)^2 \mathrm{d}\boldsymbol{x}\, \mathrm{d}t \\ &= \int_{X^3 \times I} w \sum_{i=1}^{m} \left(\frac{\partial \xi^i}{\partial t} + \overline{u}^i \right)^2 \mathrm{d}\boldsymbol{x}\, \mathrm{d}t, \end{aligned} \tag{8.76}$$

where I is the range of the variable t. The functional (8.76) is formulated so as to provide attraction of the grid lines to the Lagrangian coordinates.

The Euler–Lagrange equations (8.6) derived from this functional are as follows:

$$\frac{\partial}{\partial t}[w(\overline{w}^i - \overline{u}^i)] - \frac{\partial}{\partial x^j}[w(\overline{w}^i - \overline{u}^i)u^j] = 0, \quad i, j = 1, 2, 3. \tag{8.77}$$

By applying (2.95), they are transformed to the system

$$\frac{\partial}{\partial t}[Jw(\overline{w}^i - \overline{u}^i)] + \frac{\partial}{\partial \xi^j}[Jw(\overline{w}^i - \overline{u}^i)(\overline{w}^j - \overline{u}^j)] = 0, \quad i, j = 1, 2, 3, \tag{8.78}$$

with respect to the dependent variables t, ξ^1, ξ^2, ξ^3.

When all of the coordinates are Lagrangian, the conditions (8.75) with $i = 1, 2, 3$ are equivalent to the system of equations

$$x^i_t(t, \xi^1, \xi^2, \xi^3) - u^i = 0, \quad i = 1, 2, 3.$$

This relation is used to define a functional that controls the attraction of the generated grid to the Lagrangian grid in the form

$$I_{\mathrm{L},2} = \int_{\Xi^3 \times I} w \sum_{i=1}^{3} \left(u^i - \frac{\partial x^i}{\partial t} \right)^2 \mathrm{d}\boldsymbol{\xi}\, \mathrm{d}t. \tag{8.79}$$

8.5.2 Attraction to a Vector Field

Alignment can be very useful when there is a natural anisotropy in the physical problem, for example a dominant flow direction which is expressed by a vector field. The variational approach can be helpful in generating techniques to obtain such alignment. Functionals which take into account the direction of the prescribed vector fields $\boldsymbol{A}_i(x)$, $i = 1, 2, 3$, for constructing three-dimensional coordinate transformations are introduced in the form

$$I_{\mathrm{vf},1} = \int_{X^3} \left(w \sum_{i=1}^{3} (\boldsymbol{A}_i \times \boldsymbol{\nabla}\xi^i)^2 \right) \mathrm{d}\boldsymbol{x}, \tag{8.80}$$

where w is the weight function, and $\boldsymbol{\nabla}\xi^i = \operatorname{grad} \xi^i$. In the process of the minimization of this functional the normals to the surface $\xi^i = c$ tend to become parallel to $\boldsymbol{A}^i$. From (2.28), we have for the integrand of the functional (8.80)

$$\begin{aligned} w \sum_{i=1}^{3} (\boldsymbol{A}_i \times \boldsymbol{\nabla}\xi^i)^2 &= w(|\boldsymbol{A}_i|^2 g^{ii} - (\boldsymbol{A}_i \cdot \boldsymbol{\nabla}\xi^i)^2) \\ &= w\left(|\boldsymbol{A}_i|^2 g^{ii} - A_i^k A_i^p \frac{\partial \xi^i}{\partial x^k} \frac{\partial \xi^i}{\partial x^p} \right). \end{aligned}$$

From this relation we readily obtain the Euler–Lagrange equations for the functional (8.80):

$$\frac{\partial}{\partial x^j} \left[w \left(|\boldsymbol{A}_i|^2 \frac{\partial \xi^i}{\partial x^j} - A_i^j A_i^p \frac{\partial \xi^i}{\partial x^p} \right) \right], \quad i, j, p = 1, 2, 3, \tag{8.81}$$

with the index i fixed.

Analogously, a functional of alignment of the tangent vectors $\boldsymbol{x}_{\xi^i}$ with the prescribed vector fields $\boldsymbol{A}_i$, $i = 1, 2, 3$, can be defined:

$$I_{\mathrm{vf},2} = \int_{X^3} \left(w \sum_{i=1}^{3} (\boldsymbol{A}_i \times \boldsymbol{x}_{\xi^i})^2 \right) \mathrm{d}\boldsymbol{x}. \tag{8.82}$$

This functional can serve to attract the coordinate lines to the streamlines of the vector fields $\boldsymbol{A}_i$, $i = 1, 2, 3$.

8.5.3 Jacobian-Weighted Functional

A Jacobian-weighted functional represents the deviation of the Jacobian matrix $(\partial\xi^i/\partial x^j)$ of the transformation $\boldsymbol{\xi}(x)$ from the prescribed matrix $S(\boldsymbol{x}) = [S^{ij}(\boldsymbol{x})]$, $i, j = 1, \ldots, n$, via a least-squares fit. In particular, the functional can have the following form:

$$I_{\mathrm{jw},1}(\boldsymbol{\xi}) = \int_{X^n} G(\boldsymbol{x}, \boldsymbol{\nabla}\xi^i)\mathrm{d}x, \tag{8.83}$$

with

$$G(\boldsymbol{x}, \boldsymbol{\nabla}\xi^i) = \sum_{i,j=1}^{n} \left(\frac{\partial\xi^i}{\partial x^j} - S^{ij}(\boldsymbol{x})\right)^2.$$

In fact, the integrand $G(\boldsymbol{x}, \boldsymbol{\nabla}\xi^i)$ is the square of the Frobenius norm of the matrix

$$M = \left(\frac{\partial\xi^i}{\partial x^j} - S^{ij}\right), \quad i, j = 1, \ldots, n,$$

i.e.

$$G(\boldsymbol{x}, \boldsymbol{\nabla}\xi^i) = \mathrm{tr}(M^T M).$$

The Euler–Lagrange equations derived from the minimization of the functional I_{jw} have the form

$$\frac{\partial}{\partial x^j}\left(\frac{\partial\xi^i}{\partial x^j} - S^{ij}\right) = 0, \quad i, j = 1, \ldots, n. \tag{8.84}$$

These equations are elliptic and are in fact, a variant of the Poisson system (6.13) with

$$P^i = \frac{\partial}{\partial x^j} S^{ij}, \quad i, j = 1, \ldots, n.$$

Multiplying this system by $\partial x^k/\partial\xi^i$ and summing the result over i, we obtain the following transformed equation for the transformed dependent variable $\boldsymbol{x}(\boldsymbol{\xi})$, written in vector form:

$$g^{ij}\frac{\partial^2 \boldsymbol{x}}{\partial\xi^i\xi^j} + \frac{\partial}{\partial x^j}(S^{ij})\boldsymbol{x}_{\xi^i} = 0. \tag{8.85}$$

In accordance with (2.70),

$$\frac{\partial}{\partial x^j} S^{ij} = \frac{1}{J}\frac{\partial}{\partial\xi^k}(J\overline{S}^{ik}), \quad i, j, k = 1, \ldots, n,$$

where

$$\overline{S}^{ik} = S^{ij}\frac{\partial\xi^k}{\partial x^j}, \quad i, j, k = 1, \ldots, n.$$

Hence (8.85) can be written as follows:

$$g^{ij}\frac{\partial^2 \boldsymbol{x}}{\partial \xi^i \xi^j} + \frac{1}{J}\frac{\partial}{\partial \xi^j}(J\overline{S}^{ij})\boldsymbol{x}_{\xi^i} = 0. \tag{8.86}$$

Analogously, a Jacobian-weighted functional $I_{\mathrm{jw},2}$, which measures the squared departure of the Jacobi matrix $(\partial x^i/\partial \xi^j)$ of the coordinate transformation $\boldsymbol{x}(\boldsymbol{\xi})$ from the prescribed matrix $[S^{ij}(\boldsymbol{\xi})]$ can be defined as

$$I_{\mathrm{jw},2} = \int_{\Xi^n} \sum_{i,j=1}^{n} \left(\frac{\partial x^i}{\partial \xi^j} - S^{ij}\right)^2 \mathrm{d}\boldsymbol{\xi}. \tag{8.87}$$

The Jacobian-weighted functionals $I_{\mathrm{jw},1}$ and $I_{\mathrm{jw},2}$ also can be interpreted as one more form for the functionals of alignment for vector fields. Let a vector field be given by n vectors

$$\boldsymbol{v}_i(\boldsymbol{x}), \quad i = 1, \ldots, n, \qquad \boldsymbol{v}_i = (v_i^1, \ldots, v_i^n).$$

Then the following form of a functional of grid attraction to the given vector field can be defined:

$$I_{\mathrm{vf},3} = \int_{X^n} \sum_{i=1}^{n} |\boldsymbol{\nabla}\xi^i - \boldsymbol{v}_i|^2 \mathrm{d}\boldsymbol{x}. \tag{8.88}$$

This functional is in fact the Jacobian-weighted functional (8.84) with

$$S^{ij} = v_i^j, \quad i, j = 1, \ldots, n.$$

Note that the functionals of the form (8.83) and (8.87) are more efficient for attracting grid lines to the corresponding vector fields than the functionals of the type (8.88). This is because the former are concerned with attraction to the specified directions only, while with the latter an attraction to both the directions and the specified lengths is required.

The form of the Jacobian-weighted functional gives a clear guideline for producing nonfolded grids. This guideline is based upon the following global univalence theorem.

Theorem 1. *Let $F : U \to R^n$ be a differentiable mapping, where U is the rectangular region of R^n : $U = \{\boldsymbol{x} : \boldsymbol{x} \in R^n | a_i \le x_i \le b_i\}$. If the Jacobian matrix of F at $\boldsymbol{x}$ is positive for every $\boldsymbol{x} \in U$, then F is globally one-to-one in U.*

Recall that an $n \times n$ real matrix A is positive if every principale minor of A is positive. Thus, in order to obtain one-to-one coordinate transformations this theorem suggests that one should use only positive matrices as the matrix S.

The minimization of the functional $I_{\mathrm{jw},1}$ generates a transformation $\boldsymbol{\xi}(\boldsymbol{x})$ whose Jacobian matrix may be so close to the matrix S that the matrix $(\partial \xi^i/\partial x^j)$ is also positive. Thus the matrix $(\partial x^i/\partial \xi^j)$ is positive as well, and in accordance with the above theorem, the transformation $\boldsymbol{x}(\boldsymbol{\xi}) : \Xi^n \to X^n$ is a one-to-one mapping.

8.6 Energy Functionals of Harmonic Function Theory

The theory of harmonic maps is useful for formulating variational grid generation techniques which provide well-posed problems of grid generation.

8.6.1 General Formulation of Harmonic Maps

First, we consider the definition of a harmonic map between two general n-dimensional Riemannian manifolds X^n and Z^n with covariant metric tensors d_{ij} and D_{ij} in some local coordinates x^i, $i = 1, \dots, n$, and z^i, $i = 1, \dots, n$, respectively.

Every $C^1(X^n)$ map $\boldsymbol{z}(\boldsymbol{x}) : X^n \to Z^n$ defines an energy density by the following formula:

$$e(\boldsymbol{z}) = \frac{1}{2} d^{ij}(\boldsymbol{x}) D_{kl}(\boldsymbol{z}) \frac{\partial z^k}{\partial x^i} \frac{\partial z^l}{\partial x^j}, \quad i, j, k, l = 1, \dots, n, \tag{8.89}$$

where (d^{ij}) is the contravariant metric tensor of X^n, i.e. $d_{ij} d^{jk} = \delta_i^k$. The total energy associated with the mapping $\boldsymbol{z}(\boldsymbol{x})$ is then defined as the integral of (8.89) over the manifold X^n:

$$E(\boldsymbol{z}) = \int_{X^n} e(\boldsymbol{z}) \mathrm{d}X^n. \tag{8.90}$$

A transformation $\boldsymbol{z}(\boldsymbol{x})$ of class $C^2(X^n)$ is referred to as a harmonic mapping if it is a critical point of the functional of the total energy (8.90). The Euler–Lagrange equations whose solution minimizes the energy functional (8.90) are given by

$$\frac{1}{\sqrt{d}} \frac{\partial}{\partial x^k} \left(\sqrt{d} d^{kj} \frac{\partial z^l}{\partial x^j} \right) + d^{kj} \Gamma^l_{mp} \frac{\partial z^m}{\partial x^k} \frac{\partial z^p}{\partial x^j} = 0, \tag{8.91}$$

where $d = \det(d_{ij})$ and Γ^l_{mp} are Christoffel symbols of the second kind on the manifold Z^n:

$$\Gamma^l_{mp} = \frac{1}{2} D^{lj} \left(\frac{\partial D_{jm}}{\partial z^p} + \frac{\partial D_{jp}}{\partial z^m} - \frac{\partial D_{mm}}{\partial z^j} \right). \tag{8.92}$$

The following theorem guarantees the uniqueness of the harmonic mapping.

Theorem 2. *Let X^n, with metric d_{ij}, and Z^n, with metric D_{ij}, be two Riemannian manifolds with boundaries ∂X^n and ∂Z^n and let $\boldsymbol{\phi} : X^n \to Z^n$ be a diffeomorphism. If the curvature of Z^n is nonpositive and ∂Z^n is convex (with respect to the metric D_{ij}) then there exists a unique harmonic map $\boldsymbol{z}(\boldsymbol{x}) : X^n \to Z^n$ such that $\boldsymbol{z}(\boldsymbol{x})$ is a homotopy equivalent to $\boldsymbol{\phi}$. In other words, one can deform $\boldsymbol{z}$ to $\boldsymbol{\phi}$ by constructing a continuous family of maps $\boldsymbol{g}_t : X^n \to Z^n$, $t \in [0, 1]$, such that $\boldsymbol{g}_0(\boldsymbol{x}) = \boldsymbol{\phi}(\boldsymbol{x})$, $\boldsymbol{g}_1(\boldsymbol{x}) = \boldsymbol{z}(\boldsymbol{x})$, and $\boldsymbol{g}_t(\boldsymbol{x}) = \boldsymbol{z}(\boldsymbol{x})$ for all $\boldsymbol{x} \in \partial X^n$.*

8.6.2 Application to Grid Generation

In application of the harmonic theory to grid generation, the manifold Z^n is assumed to correspond to the computational domain Ξ^n, with a Euclidean metric $D_{ij} = \delta^i_j$. Since the Euclidean space Ξ^n is flat, i.e. it has zero curvature, and the domain Ξ^n is constructed by the user, both requirements of the above theorem can be satisfied. For the manifold X^n, one uses a set of the points of a physical domain X^n with an introduced Riemannian metric d_{ij}. The functional of the total energy (8.90) has then the form

$$E(\boldsymbol{\xi}) = \frac{1}{2} \int_{X^n} \left(\sqrt{d} d^{kl} \frac{\partial \xi^i}{\partial x^k} \frac{\partial \xi^i}{\partial x^l} \right) \mathrm{d}\boldsymbol{x}, \quad i, k, l = 1, \dots, n. \tag{8.93}$$

And for the Euler–Lagrange equations (8.91), we have

$$\frac{1}{\sqrt{d}} \frac{\partial}{\partial x^k} \left(\sqrt{d} d^{kj} \frac{\partial \xi^i}{\partial x^j} \right) = 0, \quad i, j, k = 1, \dots, n, \tag{8.94}$$

since, from (8.92), $\Gamma^l_{mp} = 0$. The left-hand part of (8.94) is the Beltrami operator Δ_B, so (8.94) is equivalent to

$$\Delta_B(\boldsymbol{\xi}) = 0. \tag{8.95}$$

Equations (8.94), in contrast to (8.91), are linear and of elliptic type, and have a conservative form. Therefore they satisfy the maximum principle, and the Dirichlet boundary value problem is a well-posed problem for this system of equations, i.e. the above theorem is proved very easily for the functional (8.93).

Equations (8.94) can be reformulated with interchanged dependent and independent variables in the typical manner, by multiplying the system by $\partial x^l / \partial \xi^i$ and summing over i. As a result, we obtain

$$\overline{d}^{km} \frac{\partial^2 x^l}{\partial \xi^k \partial \xi^m} - \frac{1}{\sqrt{d}} \frac{\partial}{\partial x^k} (\sqrt{d} d^{kl}) = 0, \quad k, l, m = 1, \dots, n, \tag{8.96}$$

where

$$\overline{d}^{km} = d^{ij} \frac{\partial \xi^k}{\partial x^i} \frac{\partial \xi^m}{\partial x^j}, \quad i, j, k, m = 1, \dots, n,$$

are the elements of the contravariant metric tensor of the Riemannian manifold X^n in the coordinates ξ^i.

8.6.3 Relation to Other Functionals

Some of the functionals given earlier are identical to the functionals of energy (8.90) and (8.93). For example, the smoothness functional (8.40) is the functional of the form (8.93) with the Euclidean metric $d_{ij} = \delta^i_j$ in X^n.

Analogously, the diffusion functional (8.63) can be interpreted as the functional (8.93), with the contravariant metric tensor d_{ij} in X^n satisfying the condition

$$\sqrt{d}d^{ij} = w\delta^i_j, \quad i,j = 1,\dots,n.$$

From this relation we readily obtain

$$d_{ij} = w^{2/(n-2)}\delta^i_j, \quad i,j = 1,\dots,n.$$

Thus $d_{ij} = w^2\delta^i_j$ for $n = 3$. The above formula fails to define the corresponding metric in the two-dimensional domain X^2. However, the formulation of the diffusion functional (8.63) as the energy functional (8.93) can be accomplished by using the Euclidean metric in X^n and the Riemannian metric in Ξ^n for arbitrary n, by setting

$$D_{ij} = w\delta^i_j, \quad i,j = 1,\dots,n.$$

The functional of inhomogeneous diffusion (8.64) can be interpreted in a similar manner by taking

$$d_{ij} = w\delta^i_j, \qquad D_{ij} = w_i\delta^i_j, \quad i,j = 1,\dots,n.$$

Similarly, the functionals (8.61), (8.62), and (8.72) can be identified with the functional of energy (8.93).

Other applications of the functionals of energy to generate surface and hypersurface grids will be discussed in Chaps. 9 and 10.

8.7 Combinations of Functionals

The functionals described in Sects. 8.2–8.6 are used to control and realize various grid properties. This is carried out by combining these functionals with weights in the form

$$I = \sum_i \lambda_i I_i, \quad i = 1,\dots,k. \tag{8.97}$$

Here λ_i, $i = 1,\dots,k$, are specified parameters which determine the individual contribution of each functional I_i to I. The ranges of the parameters λ_i controlling the relative contributions of the functionals can be defined readily when the functionals I_i are dimensionally homogeneous. However, if they are dimensionally inhomogeneous, then the selection of a suitable value for λ_i presents some difficulties. A common rule for selecting the parameters λ_i involves making each component $\lambda_i I_i$ in (8.97) of a similar scale by using a dimensional analysis.

The most common practice in forming the combination (8.97) uses both the functionals of adaptation to the physical solution and the functionals of grid regularization. The first reason for using such a strategy is connected with the fact that the process of adaptation can excessively distort the form of the grid cells. The distortion can be prevented by functionals which impede cell deformation. These functionals are ones which control grid skewness, smoothness, and conformality. The second reason for using the regularization functionals is connected with the natural requirement for the well-posedness of the grid generation process. This requirement is achieved by the utilization of convex functionals in variational grid generators. The convex functionals are represented by energy-type functionals, producing harmonic maps, and by the functionals of conformality.

The various functionals described above provide broad opportunities to control and realize the required grid properties, though problems still remain; these require more detailed studies of all properties of the functionals. The knowledge of these properties will allow one to utilize the functionals as efficient tools to generate high-quality grids.

8.7.1 Natural Boundary Conditions

In order to achieve the desired result more efficiently when generating grids by the variational approach, one needs to adjust the boundary conditions to the resulting Euler–Lagrange equations. As an illustration, we can consider the process of generating two-dimensional conformal grids with the Laplace equations derived from the functional of conformality (8.29). A conformal grid is not obtained with arbitrary boundary conditions, but only with strictly specified ones.

The natural boundary conditions for the Euler–Lagrange equations yielded by functionals are those for which the boundary contribution to the variation is zero. The natural boundary conditions are derived in the typical way, by writing out the first variation of the functional.

8.8 Comments

A detailed description of the fundamentals and theoretical results of the calculus of variations can be found in the monographs by Gelfand and Fomin (1963) and by Ladygenskaya and Uraltseva (1973).

Liseikin and Yanenko (1977), Danaev, Liseikin, and Yanenko (1978), Ghia, Ghia, and Shin (1983), Brackbill and Saltzman (1982) and Bell and Shubin (1983) have each used the variational principle for grid adaptation.

The diffusive form of the adaptive functional (8.63) was formulated originally by Danaev, Liseikin, and Yanenko (1980) and Winslow (1981). A generalization of this functional to (8.64) with an individual weight function for each direction was realized by Eiseman (1987) and Reed, Hsu, and Shiau (1988). The most general variational formulation of the modified anisotropic diffusion approach was presented by Hagmeijer (1994). A variational principle for the Jacobian-weighted functional was formulated, studied, and developed by Knupp (1995, 1996) and Knupp, Margolin, and Shashkov (2002). A considarable amount of work in determining and studying the conditions required to guarantee invertibility of coordinate transformations was published by Pathasarathy (1983) and Clement, Hagmeijer, and Sweers (1996).

The functional measuring the alignment of the two-dimensional grid with a specified vector field was formulated by Giannakopoulos and Engel (1988). The extension of this approach to three dimensions was discussed by Brackbill (1993).

A variational method optimizing cell aspect ratios was presented and analyzed by Mastin (1992). A dimensionally homogeneous functional of two-dimensional grid skewness was proposed by Steinberg and Roache (1986).

The property of eccentricity for the univariate transformation $\boldsymbol{x}(\boldsymbol{\xi})$ was introduced by Sidorov (1966), while a three-dimensional extension was performed by Serezhnikova, Sidorov, and Ushakova (1989). A form of smoothness based on the eccentricity term was developed by Winkler, Mihalas, and Norman (1985).

The variational formulation of grid properties was described by Warsi and Thompson (1990).

The geometric interpretation of the approximation (8.51) was given by Steinberg and Roache (1986).

The introduction of the volume-weighted functional was originally proposed in two dimensions by Yanenko, Danaev, and Liseikin (1977).

The approach of determining functionals which depend on invariants of orthogonal transformations of the metric tensor (g_{ij}), to ensure that the problems are well-posed and to obtain more compact formulas for the Euler–Lagrange equations, was proposed by Jacquotte (1987). In his paper, the grids were constructed through functionals obtained by modeling different elastic and plastic properties of a deformed body.

The metric-weighted functional was formulated by Belinsky et al. (1975) for the purpose of generating quasiconformal grids.

The possibility of using harmonic function theory to provide a general framework for developing multidimensional mesh generators was discussed by Dvinsky (1991). The interpretation of the functional of diffusion as a version of the energy functional was presented by Brackbill (1993). A detailed survey of the theory of harmonic mappings was published by Eells and Lenaire (1988).

9

Curve and Surface Grid Methods

9.1 Introduction

Curvilinear lines and surfaces are common geometrical objects in both structured and unstructured grid generation techniques. Curves appear in grid considerations as the boundary segments of two-dimensional domains or surface patches and as the edges of three-dimensional blocks. Surfaces arise as the boundary segments and/or faces of three-dimensional domains or blocks.

The main goal of grid generation on a curve is to provide boundary data for boundary-fitted grid generators for two-dimensional planar domains and surfaces. Analogously, surface grid generation is needed chiefly to build grids on the boundaries of three-dimensional domains or blocks in order to provide boundary data for volume grid techniques.

In the structured concept, grid generation on a surface follows the construction of a set of surface patches, specification of a parametrization for every patch, and generation of one-dimensional curve grids on the edges of the surface patches to provide the boundary conditions for the surface grid generator. In fact, for the purpose of simplicity and for maintaining adherence of the surface grid techniques to the physical geometry, the grid is commonly generated in a parametric two-dimensional domain and then mapped onto the original patch of the surface.

Thus the process of surface grid generation may by divided into three steps: forward mapping, grid generation, and backward mapping. The forward mapping is a representation of the background surface patch from a three-dimensional physical domain to a two-dimensional parameter area. Once the forward mapping is complete, the grid is generated in the parameter space and then mapped back into the physical space (backward mapping). A surface patch is formed as a curvilinear triangle or a quadrilateral, with three or four boundary segments, respectively. The corresponding parametric domain may also have the shape of a triangle or a quadrilateral with curved boundary edges. The backward transformation from the parametric domain to the patch

V.D. Liseikin, *Grid Generation Methods*, Scientific Computation,
DOI 10.1007/978-90-481-2912-6_9, © Springer Science+Business Media B.V. 2010

is defined by specifically adjusted interpolations. The specification depends on variations in surface features.

The generation of the grid on the parametric domain is derived by the same types of approach—algebraic, differential, and variational—as those which were described for planar domains. However, these approaches are adjusted by including the necessary surface characteristics, expressed in terms of the surface quadratic forms (see Sect. 3.3), to satisfy the required grid properties on the surface.

This chapter gives a review of some advanced techniques of curve and surface grid generation.

9.2 Grids on Curves

Methods for generation of grids on curves are the simplest to formulate and analyze. These methods provide the background for the development of surface grid techniques. Some common structured approaches to grid generation on curves are discussed in this section.

9.2.1 Formulation of Grids on Curves

A curve in n-dimensional space is represented parametrically by a smooth, nonsingular, vector-valued function from a normalized interval $[0,1]$:

$$\boldsymbol{x}(\varphi) : [0,1] \to R^n, \quad \boldsymbol{x}(\varphi) = [x^1(\varphi), \dots, x^n(\varphi)]. \tag{9.1}$$

Let the curve with the parametrization $\boldsymbol{x}(\varphi)$ be designated by S^{x1}. The transformation (9.1) provides a discrete grid on the curve S^{x1} by mapping the nodes of a uniform grid in the interval $[0,1]$ into S^{x1} with $\boldsymbol{r}(\varphi)$, i.e. the grid points $\boldsymbol{x}_i$, $i = 0, 1, \dots, N$, are defined as

$$\boldsymbol{x}_i = \boldsymbol{x}(ih), \quad h = 1/N.$$

However, the need to produce a grid with particular desirable properties requires the introduction of a control tool. Such control of the generation of a curve grid is carried out with strongly monotonic and smooth intermediate transformations

$$\varphi(\xi) : [0,1] \to [0,1], \tag{9.2}$$

which generate the grid nodes φ_i on the interval $[0,1]$, where

$$\varphi_i = \varphi(ih), \quad i = 0, 1, \dots, N, \ h = 1/N.$$

The transformation $\varphi(\xi)$ is chosen is such a way that the composition

$$\boldsymbol{x}[\varphi(\xi)] : [0,1] \to R^n, \tag{9.3}$$

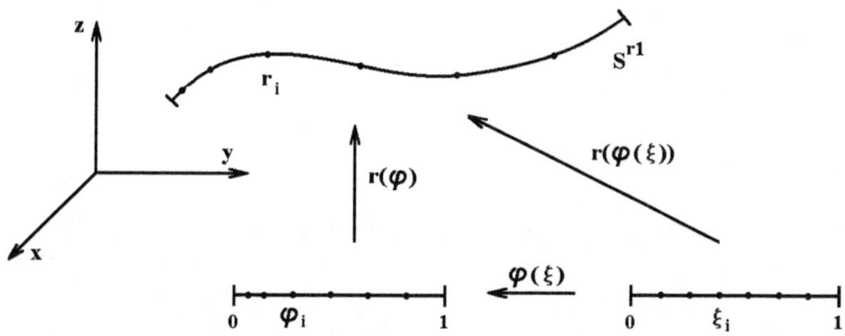

Fig. 9.1. Scheme of generation of a curve grid

which represents a new parametrization of S^{x1}, generates the grid nodes

$$\boldsymbol{x}_i = \boldsymbol{x}[\varphi(ih)] = \boldsymbol{x}(\varphi_i), \quad i = 0, 1, \ldots, N, \ h = 1/N, \tag{9.4}$$

with the desired properties. Figure 9.1 demonstrates the scheme of the generation of a curve grid. Thus, the process of grid generation on a curve is turned into the definition of an intermediate transformation $\varphi(\xi)$ so as to provide a suitable parametrization of the curve. One such natural transformation is connected with the scaled arc length parameter ξ which, in analogy with (3.1), is defined by

$$\xi(\varphi) = \frac{1}{c}\int_0^{\varphi} \sqrt{g^{x\varphi}}\mathrm{d}x, \quad c = \int_0^1 \sqrt{g^{x\varphi}}\mathrm{d}x, \tag{9.5}$$

where

$$g^{x\varphi} = \boldsymbol{x}_\varphi \cdot \boldsymbol{x}_\varphi = |\boldsymbol{x}_\varphi|^2.$$

The function $\varphi(\xi)$, inverse to $\xi(\varphi)$, is subject to the condition

$$\frac{\mathrm{d}\varphi}{\mathrm{d}\xi} = \frac{c}{\sqrt{g^{x\varphi}}}. \tag{9.6}$$

Therefore we have, for the grid nodes $\varphi_i = \varphi(ih)$ in the interval $[0, 1]$, the relation

$$\frac{\varphi_{i+1} - \varphi_i}{h} \approx \frac{\mathrm{d}\varphi}{\mathrm{d}\xi} = \frac{c}{\sqrt{g^{x\varphi}}}, \quad i = 0, 1, \ldots, N, \ h = 1/N, \tag{9.7}$$

and consequently we obtain, for the grid nodes $\boldsymbol{x}_i$ on the curve S^{x1},

$$|\boldsymbol{x}_{i+1} - \boldsymbol{x}_i| \approx |\boldsymbol{x}_\varphi|(\varphi_{i+1} - \varphi_i) \approx ch, \quad i = 0, 1, \ldots, N, \ h = 1/N. \tag{9.8}$$

Equations (9.6)–(9.8) are examples of the equidistribution principle considered in Sect. 7.3, which is based on a specification of distances between the grid points in accordance with a rule of inverse proportionality to a weight function.

9.2.2 Grid Methods

The main approach to generating one-dimensional grids on curves is based on a specification of the grid step spacing. The approach is realized by direct reparametrization of the curve with suitable univariate intermediate transformations $\varphi(\xi)$, such as the ones considered in Chap. 4, or by the formulation of equations and functionals through the first derivative of the intermediate transformations.

9.2.2.1 Differential Approach

The simplest differential method for the definition of an intermediate transformation $\varphi(\xi)$ relies on the solution of the initial-value problem in the form of (9.6)

$$\begin{aligned} \frac{\mathrm{d}\varphi}{\mathrm{d}\xi} &= \frac{c}{F(\varphi)}, \quad 0 < \varphi \leq 1, \\ \varphi(0) &= 0, \qquad c = \int_0^1 F(\varphi)\mathrm{d}\varphi, \end{aligned} \tag{9.9}$$

where $F(\varphi)$ is a nonnegative function specified by the user. Differentiation of (9.9) with respect to ξ allows one to eliminate the constant c and obtain the two-point boundary value problem

$$\begin{aligned} &\frac{\mathrm{d}}{\mathrm{d}\xi}\left(\frac{\mathrm{d}\varphi}{\mathrm{d}\xi}F(\varphi)\right) = 0, \quad 0 < \varphi < 1, \\ &\varphi(0) = 0, \qquad \varphi(1) = 1. \end{aligned} \tag{9.10}$$

Equations (9.9) and (9.10) represent the formulation of the equidistribution principle of Chap. 7. Taking advantage of (9.6)–(9.8), we see that the solution of (9.9) or (9.10) produces a grid on the curve S^{x1} with a grid spacing inversely proportional to $\sqrt{g^{x\varphi}}F(\varphi)$. Thus, in the weight-concept formulation

$$\frac{|\boldsymbol{x}_{i+1} - \boldsymbol{x}_i|}{h} \approx \frac{c_1}{w(\varphi_i)}, \quad i = 0, \ldots, N-1,$$

we obtain

$$F(\varphi) = w(\varphi)\sqrt{g^{x\varphi}}, \tag{9.11}$$

which, in the case $w(\varphi) = 1$, corresponds to the scaled-arc-length parametrization (9.6).

The weight function $w(\varphi)$ is specified by the user in accordance with the requirement to cluster the grid points in the zones of particular interest. It can be defined through the derivatives of the physical quantities or through the measures of the curve features described in Sect. 3.5, in particular, through the metric tensor $g^{x\varphi}$, curvature k, or tension τ. The specification determines the concentration of the curve grid, which becomes larger in the areas of large values of the weight function.

9.2.2.2 Variational Approach

In accordance with the results of Chap. 8, the differential formulation (9.10) of the equidistribution principle is obtained from the minimization of the functional

$$I = \int_0^1 \frac{1}{F(\varphi)} \left(\frac{\mathrm{d}\xi}{\mathrm{d}\varphi} \right)^2 \mathrm{d}\varphi, \tag{9.12}$$

whose Euler–Lagrange equations

$$\frac{\partial}{\partial \varphi} \left(\frac{\mathrm{d}\xi}{\mathrm{d}\varphi} \frac{1}{F(\varphi)} \right) = 0$$

(see Sect. 8.2) are equivalent to (9.10). Taking advantage of (9.11), we obtain an equivalent form of (9.12) through the weight function $w(\varphi)$:

$$\begin{aligned} I = \int_0^1 \frac{1}{w(\varphi)\sqrt{g^{x\varphi}}} \left(\frac{\mathrm{d}\xi}{\mathrm{d}\varphi} \right)^2 \mathrm{d}\varphi &= \int_{S^{x1}} \frac{1}{g^{x\varphi} w(\varphi)} \left(\frac{\mathrm{d}\xi}{\mathrm{d}\varphi} \right)^2 \mathrm{d}S^{x1} \\ &= \int_{S^{x1}} \frac{1}{g^{x\xi} w(\varphi)} \mathrm{d}S^{x1}, \end{aligned} \tag{9.13}$$

where

$$g^{x\xi} = \frac{\mathrm{d}\boldsymbol{x}[\varphi(\xi)]}{\mathrm{d}\xi} \cdot \frac{\mathrm{d}\boldsymbol{x}[\varphi(\xi)]}{\mathrm{d}\xi} = \boldsymbol{x}_\varphi \cdot \boldsymbol{x}_\varphi \left(\frac{\mathrm{d}\varphi}{\mathrm{d}\xi} \right)^2 = g^{x\varphi} \left(\frac{\mathrm{d}\varphi}{\mathrm{d}\xi} \right)^2 .$$

In analogy with the differential approach, the weight function $w(\varphi)$ in (9.13) is defined by the values of the solution or its derivatives and/or by the curve quality measures.

9.2.2.3 Monitor Formulation

The monitor approach for controlling the grid steps on a curve S^{x1} relies on the introduction of a monitor curve which is defined by the values of some vector function $\boldsymbol{f} : X^n \to R^k$, $\boldsymbol{f}(\boldsymbol{x}) = [f^1(\boldsymbol{x}), \ldots, f^k(\boldsymbol{x})]$, over the curve, where X^n is a domain containing the curve. The parameter function (9.1) and $\boldsymbol{f}(\boldsymbol{x})$ define a parametrization $\boldsymbol{r}(\varphi)$ of the monitor curve designated by S^{r1}:

$$\boldsymbol{r}(\varphi) : [0,1] \to R^{n+k}, \quad \boldsymbol{r}(\varphi) = \left[r^1(\varphi), \ldots, r^{n+k}(\varphi) \right], \tag{9.14}$$

where

$$r^i(\varphi) = x^i(\varphi), \quad i = 1, \ldots, n, \qquad r^{n+j}(\varphi) = f^j[\boldsymbol{x}(\varphi)], \quad j = 1, \ldots, k.$$

We obtain

$$g^{r\varphi} = g^{x\varphi} + g^{f\varphi}, \tag{9.15}$$

where

$$g^{f\varphi} = \frac{\mathrm{d}\boldsymbol{f}[\boldsymbol{x}(\varphi)]}{\mathrm{d}\varphi} \cdot \frac{\mathrm{d}\boldsymbol{f}[\boldsymbol{x}(\varphi)]}{\mathrm{d}\varphi} = \frac{\partial f^j}{\partial x^l}\frac{\partial f^j}{\partial x^m}\frac{\mathrm{d}x^l}{\mathrm{d}\varphi}\frac{\mathrm{d}x^m}{\mathrm{d}\varphi},$$
$$j = 1, \dots, k, \; l, m = 1, \dots, n.$$

In the monitor approach the grid on the curve S^{x1} is obtained by mapping a uniform grid on S^{r1} with the projection function $P : R^{n+k} \to R^n$ $(P(x^1, \dots, x^{n+k}) = (x^1, \dots, x^n))$. The uniform grid on S^{r1} is derived by means of the arc-length approach, realized with the initial-value problem (9.6), with the two-point boundary value problem (9.10), or with the variational problem (9.12) for $F(\varphi) = \sqrt{g^{r\varphi}}$. As a result, we have for the intermediate transformation $\varphi(\xi)$

$$\frac{\mathrm{d}}{\mathrm{d}\xi}\left(\frac{\mathrm{d}\varphi}{\mathrm{d}\xi}\sqrt{g^{r\varphi}}\right) = 0, \quad 0 < \xi < 1,$$
$$\varphi(0) = 0, \qquad \varphi(1) = 1, \tag{9.16}$$

with $g^{r\varphi}$ specified by (9.15). The transformation $\boldsymbol{x}[\varphi(\xi)]$ defines the grid on the surface S^{x1} which coincides with the grid projected from S^{r1}. Since

$$\frac{|\boldsymbol{x}_{i+1} - \boldsymbol{x}_i|}{h} \approx \left|\frac{\mathrm{d}x}{\mathrm{d}\varphi}\right|\frac{\mathrm{d}\varphi}{\mathrm{d}\xi} = \frac{\sqrt{g^{x\varphi}}}{\sqrt{g^{r\varphi}}} = \frac{1}{\sqrt{1 + g^{f\varphi}/g^{x\varphi}}},$$

the monitor approach provides node clustering in the zones of large values of $g^{f\varphi}$ and, consequently, where the derivatives of the function $\boldsymbol{f}(\boldsymbol{x})$ are large.

Note that in accordance with (9.13), the variational formulation of the monitor approach is given by the functional

$$I = \int_{S^{r1}} \frac{1}{g^{r\varphi}} \mathrm{d}S^{r1}. \tag{9.17}$$

9.3 Formulation of Surface Grid Methods

It is assumed in this chapter that the surface under consideration lies in the Euclidean space R^3. Without loss of generality, we suggest that the surface, denoted as S^{x2}, is locally represented by a parametrization

$$\boldsymbol{x}(\boldsymbol{s}) : S^2 \to R^3, \quad \boldsymbol{x}(\boldsymbol{s}) = \left[x^1(\boldsymbol{s}), x^2(\boldsymbol{s}), x^3(\boldsymbol{s})\right], \; \boldsymbol{s} = (s^1, s^2), \tag{9.18}$$

where S^2 is a two-dimensional parametric domain with the Cartesian coordinates s^1, s^2, while $\boldsymbol{x}(\boldsymbol{s})$ is a smooth, nondegenerate function.

9.3.1 Mapping Approach

The generation of a grid on the surface S^{x2} is based on the introduction of a standard computational domain Ξ^2 with a reference grid and a one-to-one transformation

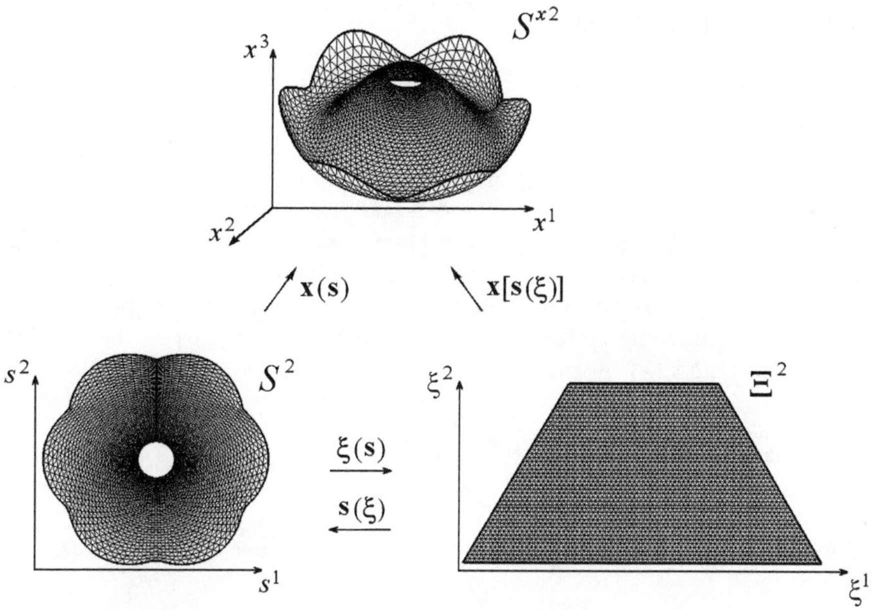

Fig. 9.2. Framework for generation of triangular surface grids

$$\boldsymbol{s}(\boldsymbol{\xi}) : \Xi^2 \to S^2, \quad \boldsymbol{s}(\boldsymbol{\xi}) = \left[s^1(\boldsymbol{\xi}), s^2(\boldsymbol{\xi})\right], \ \boldsymbol{\xi} \in \Xi^2. \tag{9.19}$$

This mapping (9.19) generates in fact a grid in the two-dimensional domain S^2. However, a required grid on the surface S^{x2} is defined by mapping some reference grid in Ξ^2 onto S^{x2} by the composite transformation

$$\boldsymbol{x}[\boldsymbol{s}(\boldsymbol{\xi})] : \Xi^2 \to S^{x2}, \tag{9.20}$$

or, equivalently, by mapping with $\boldsymbol{x}(\boldsymbol{s})$ the grid generated in S^2 by some suitable transformation $\boldsymbol{s}(\boldsymbol{\xi}) : \Xi^2 \to S^2$ (see Fig. 9.2).

Thus the problem of the generation of a surface grid is turned into the problem of choosing an appropriate computational domain Ξ^2 with a suitable reference grid and of constructing an adequate transformation between the computational and the parametric two-dimensional domains.

Some techniques for generating two-dimensional coordinate transformations of domains were considered in Chaps. 4–8. However, the direct application of the approaches discussed above to the generation of two-dimensional planar grids may not lead to satisfactory grids on the surface, since the grid on the surface, obtained by the backward mapping $\boldsymbol{x}(\boldsymbol{s})$, may become significantly distorted because of the mapping. The formulation of the proper methods should take into account the geometric features of the surface under consideration and properties of the parametrization $\boldsymbol{x}(\boldsymbol{s})$.

9.3.2 Associated Metric Relations

The features of the surface are described through the first and second fundamental forms (Sect. 3.3), in particular, by the metric elements which are derived from the dot products of the tangential vectors to the coordinate lines.

The scheme of grid generation on the surface S^{x2} implies, in fact, two parametrizations of the surface: the original parametrization $\boldsymbol{x}(\boldsymbol{s})$ with the parametric space S^2 and the final one $\boldsymbol{x}[\boldsymbol{s}(\boldsymbol{\xi})]$ with the parametric space Ξ^2. The original parametrization is considered as an input, while the final parametrization $\boldsymbol{x}[\boldsymbol{s}(\boldsymbol{\xi})]$ is an output of the surface grid generation process. The role of the intermediate transformation $\boldsymbol{s}(\boldsymbol{\xi})$ is to correct the drawbacks of the original mapping $\boldsymbol{x}(\boldsymbol{s})$ by transforming it to $\boldsymbol{x}[\boldsymbol{s}(\boldsymbol{\xi})]$, which should generate a grid with the properties required by the user.

The covariant metric tensor of the surface in the coordinates ξ^1, ξ^2, denoted by

$$G^{x\xi} = (g_{ij}^{x\xi}), \quad i,j = 1,2,$$

is defined by the dot product of the tangent vectors $\boldsymbol{x}_{\xi^i} = \partial\boldsymbol{x}(\boldsymbol{s})/\partial\xi^i(\boldsymbol{\xi})$, $i = 1,2$, i.e.

$$g_{ij}^{x\xi} = \boldsymbol{x}_{\xi^i} \cdot \boldsymbol{x}_{\xi^j}, \quad i,j = 1,2.$$

Analogously, the elements of the covariant metric tensor

$$G^{xs} = (g_{ij}^{xs}), \quad i,j = 1,2,$$

in the coordinates s^i, $i = 1,2$, are expressed in the following form:

$$g_{ij}^{xs} = \frac{\partial \boldsymbol{x}}{\partial s^i} \cdot \frac{\partial \boldsymbol{x}}{\partial s^j}, \quad i,j = 1,2.$$

It is clear that

$$\begin{aligned} g_{ij}^{x\xi} &= g_{mk}^{xs} \frac{\partial s^m}{\partial \xi^i} \frac{\partial s^k}{\partial \xi^j}, \\ g_{ij}^{xs} &= g_{mk}^{x\xi} \frac{\partial \xi^m}{\partial s^i} \frac{\partial \xi^k}{\partial s^j}, \quad i,j,k,m = 1,2, \end{aligned} \tag{9.21}$$

using the convention of summation over repeated indices.

The contravariant metric tensor of the surface in the coordinates ξ^i, $i = 1,2$, denoted by

$$G_{\xi x} = (g_{\xi x}^{ij}), \quad i,j = 1,2,$$

and in the coordinates s^i, $i = 1,2$, written as

$$G_{sx} = (g_{sx}^{ij}), \quad i,j = 1,2,$$

is the matrix inverse to $G^{x\xi}$ and G^{xs}, respectively. Thus

$$
\begin{aligned}
g^{ij}_{\xi x} &= (-1)^{i+j} g^{x\xi}_{3-i\ 3-j}/g^{x\xi}, \qquad & g^{x\xi}_{ij} &= (-1)^{i+j} g^{x\xi} g^{3-i,3-j}_{\xi x}, \\
g^{ij}_{sx} &= (-1)^{i+j} g^{xs}_{3-i\ 3-j}/g^{xs}, \qquad & g^{xs}_{ij} &= (-1)^{i+j} g^{xs} g^{3-i,3-j}_{sx},
\end{aligned}
\tag{9.22}
$$

where $i, j = 1, 2$, and

$$
g^{x\xi} = \det G^{x\xi}, \qquad g^{xs} = \det G^{xs}.
$$

Note that here, on the right-hand side of every relation in (9.22), the summation convention is not applied over i and j.

Similarly to (9.21), the elements of the contravariant metric tensor in the coordinates ξ^i and s^i, $i = 1, 2$, are connected by the relations

$$
\begin{aligned}
g^{ij}_{\xi x} &= g^{mk}_{sx} \frac{\partial \xi^i}{\partial s^m} \frac{\partial \xi^j}{\partial s^k}, \\
g^{ij}_{sx} &= g^{mk}_{\xi x} \frac{\partial s^i}{\partial \xi^m} \frac{\partial s^j}{\partial \xi^k}, \qquad i, j, k, m = 1, 2.
\end{aligned}
\tag{9.23}
$$

These relations and (9.21) readily yield

$$
\begin{aligned}
g^{ii}_{sr} &= g^{mk}_{\xi x} g^{s\xi}_{mk}, \qquad i, j, k, m = 1, 2, \\
g^{x\xi} &= g^{xs} g^{s\xi},
\end{aligned}
\tag{9.24}
$$

where $g^{s\xi}_{mk}$, $k, m = 1, 2$, are the elements of the covariant metric tensor of S^2 in the coordinates ξ^i, i.e. $g^{s\xi}_{mk} = \boldsymbol{s}_{\xi^m} \cdot \boldsymbol{s}_{\xi^k}$, while $g^{s\xi} = \det(g^{s\xi}_{ij}) = \det^2(\partial s^i / \partial \xi^j)$.

Now we proceed to the description of some advanced grid generation techniques for the generation of grids on the surface S^{x2}.

9.4 Beltramian System

It is desirable to develop methods of surface grid generation which are invariant of the parametrizations $\boldsymbol{x}(\boldsymbol{s}) : S^2 \to S^{x2}$. One such surface grid generation system is obtained from the Beltrami second-order differential operator.

9.4.1 Beltramian Operator

The Beltrami operator Δ_B is defined as

$$
\Delta_B[f] = \frac{1}{\sqrt{g^{xs}}} \frac{\partial}{\partial s^j} \left(\sqrt{g^{xs}} g^{mj}_{sr} \frac{\partial}{\partial s^m} f \right), \qquad j, m = 1, 2. \tag{9.25}
$$

When S^{x2} is a plane and the coordinate system s^1, s^2 is orthonormal, i.e. $g^{xs}_{ij} = g^{ij}_{sx} = \delta^i_j$, then (9.25) is the Laplace operator. Thus the operator (9.25) is a generalization of the Laplacian on a surface.

The Beltrami operator does not depend on the parametrization of the surface. For instance, let u^i, $i = 1, 2$, be another parametrization. Taking into account the general relation (2.56), we obtain, for arbitrary smooth functions A^1 and A^2,

$$\frac{\partial A^j}{\partial s^j} = \frac{1}{\sqrt{g^{su}}}\frac{\partial}{\partial u^k}\left(\sqrt{g^{su}}A^m\frac{\partial u^k}{\partial s^m}\right), \quad i,j,k,m = 1,2, \qquad (9.26)$$

where

$$\sqrt{g^{su}} = \det\left(\frac{\partial s^i}{\partial u^m}\right), \quad i,m = 1,2.$$

Assuming

$$A^j = \sqrt{g^{xs}}g^{mj}_{sr}\frac{\partial}{\partial s^m}f, \quad j = 1,2,$$

we have from (9.25)

$$\frac{1}{\sqrt{g^{xs}}}\frac{\partial A^j}{\partial s^j} = \Delta_B[f], \quad j = 1,2.$$

Therefore (9.26) yields

$$\Delta_B[f] = \frac{1}{\sqrt{g^{ru}}}\frac{\partial}{\partial u^j}\left(\sqrt{g^{ru}}g^{mp}_{sr}\frac{\partial u^j}{\partial s^p}\frac{\partial u^k}{\partial s^m}\frac{\partial f}{\partial u^k}\right), \quad j,k,m,p = 1,2,$$

where $g^{ru} = g^{xs}g^{su}$. Thus, using (9.23), we obtain

$$\frac{1}{\sqrt{g^{xs}}}\frac{\partial}{\partial s^j}\left(\sqrt{g^{xs}}g^{mj}_{sr}\frac{\partial}{\partial s^m}f\right) = \frac{1}{\sqrt{g^{ru}}}\frac{\partial}{\partial u^j}\left(\sqrt{g^{ru}}g^{jk}_{ur}\frac{\partial f}{\partial u^k}\right), \quad j,k,m = 1,2.$$

So the value of the operator Δ_B at a function f does not depend on any parametrization of the surface S^{x2}.

9.4.2 Surface Grid System

A surface grid system can be formed, in analogy with the Laplace system (6.4), by the Beltrami equations with respect to the components $\xi^i(\boldsymbol{s})$ of the inverse mapping of the intermediate transformation $\boldsymbol{s}(\boldsymbol{\xi})$:

$$\Delta_B[\xi^i] = 0, \quad i = 1,2,$$

i.e. taking advantage of (9.25),

$$\frac{1}{\sqrt{g^{xs}}}\frac{\partial}{\partial s^j}\left(\sqrt{g^{xs}}g^{mj}_{sr}\frac{\partial \xi^i}{\partial s^m}\right) = 0, \quad i,j,m = 1,2. \qquad (9.27)$$

The system (9.27) is a generalization of the two-dimensional Laplace system (6.4) applied to the generation of planar grids. As in the case of the Laplace equations, the solution to a Dirichlet boundary value problem for (9.27) satisfies the conditions necessary for efficient surface grid generation. In particular, if the computational domain Ξ^2 is chosen to be convex, then the values of the function $\boldsymbol{\xi}(\boldsymbol{s})$ satisfying (9.27) lie in Ξ^2 if $\boldsymbol{\xi}(\boldsymbol{s})$ maps the boundary of S^2 onto the boundary of Ξ^2. Moreover, the transformation $\boldsymbol{\xi}(\boldsymbol{s})$ is a one-to-one transformation if it is homeomorphic on the boundary. This is the main justification of the formulation of (9.27) with respect to the inverse transformation $\boldsymbol{\xi}(\boldsymbol{s})$.

In order to generate a grid on S^2, the system (9.27) is inverted to interchange its dependent and independent variables. This is done in the typical manner, by multiplying the system by $\partial s^l/\partial \xi^i$ and summing over i. Thus we obtain

$$\frac{1}{\sqrt{g^{xs}}}\frac{\partial}{\partial s^j}\left(\sqrt{g^{xs}}g_{sr}^{mj}\frac{\partial \xi^i}{\partial s^m}\right)\frac{\partial s^l}{\partial \xi^i}$$
$$= -g_{sr}^{mj}\frac{\partial \xi^i}{\partial s^m}\frac{\partial \xi^k}{\partial s^j}\frac{\partial^2 s^l}{\partial \xi^i \partial \xi^k} + \frac{1}{\sqrt{g^{xs}}}\frac{\partial}{\partial s^j}(\sqrt{g^{xs}}g_{sr}^{lj}) = 0.$$

So, taking into account (9.23), the system (9.27) is transformed to the following nonlinear system with respect to the dependent variables $s^i(\boldsymbol{\xi})$:

$$g_{\xi x}^{ij}\frac{\partial^2 s^l}{\partial \xi^i \partial \xi^j} = \frac{1}{\sqrt{g^{xs}}}\frac{\partial}{\partial s^i}(\sqrt{g^{xs}}g_{sr}^{il}), \quad i,j,l = 1,2. \tag{9.28}$$

We shall call these equations as the inverted Beltrami equations.

The right-hand part of (9.28) is in fact the value of the Beltrami operator applied to the function s^l; thus the system (9.28) can be written out as follows:

$$g_{\xi x}^{ij}\frac{\partial^2 s^l}{\partial \xi^i \partial \xi^j} = \Delta_B[s^l], \quad i,j,l = 1,2.$$

Taking advantage of (9.22), the system (9.28) is transformed to a system whose coefficients are determined by the elements of the metric tensors $(g_{ij}^{x\xi})$ and (g_{ij}^{xs}):

$$B_2^{x\xi}[s^l] = (-1)^l\frac{g^{x\xi}}{\sqrt{g^{xs}}}\left(\frac{\partial}{\partial s^2}\frac{g_{3-l\,1}^{xs}}{\sqrt{g^{xs}}} - \frac{\partial}{\partial s^1}\frac{g_{3-l\,2}^{xs}}{\sqrt{g^{xs}}}\right), \quad l = 1,2, \tag{9.29}$$

where $B_2^{x\xi}$ is an operator defined as follows:

$$B_2^{x\xi}[y] = g_{22}^{x\xi}\frac{\partial^2 y}{\partial \xi^1 \partial \xi^1} - 2g_{12}^{x\xi}\frac{\partial^2 y}{\partial \xi^1 \partial \xi^2} + g_{11}^{x\xi}\frac{\partial^2 y}{\partial \xi^2 \partial \xi^2}.$$

In particular, let the surface S^{x2} be a monitor surface formed by the values of a scalar height function $u(\boldsymbol{s})$ and consequently be represented by the parametrization

$$\boldsymbol{x}(\boldsymbol{s}) = [s^1, s^2, u(\boldsymbol{s})].$$

Then we obtain

$$g_{ij}^{xs} = \delta_j^i + u_{s^i} u_{s^j}, \qquad g_{sx}^{ij} = (-1)^{i+j} \frac{\delta_j^i + u_{s^{3-i}} u_{s^{3-j}}}{1 + (u_{s^1})^2 + (u_{s^2})^2}, \qquad i, j = 1, 2,$$

$$g^{xs} = 1 + (u_{s^1})^2 + (u_{s^2})^2, \qquad g^{x\xi} = J^2 g^{xs}, \quad J = \det(\partial s^i / \partial \xi^j)$$

without summing over i and j, and correspondingly,

$$\Delta_B[s^l] = \left(\frac{(-1)^{j+l}}{\sqrt{1 + (u_{s^1})^2 + (u_{s^2})^2}} \right) \frac{\partial}{\partial s^j} \left(\frac{\delta_l^j + u_{s^{3-j}} u_{s^{3-l}}}{\sqrt{1 + (u_{s^1})^2 + (u_{s^2})^2}} \right), \quad j, l = 1, 2,$$

with l fixed. Thus the system (9.29) has in this case the form

$$B_2^{x\xi}[s^l] = (-1)^{k+l} J^2 \sqrt{1 + (u_{s^1})^2 + (u_{s^2})^2} \frac{\partial}{\partial s^k} \left(\frac{\delta_l^k + u_{s^{3-k}} u_{s^{3-l}}}{\sqrt{1 + (u_{s^1})^2 + (u_{s^2})^2}} \right),$$
$$i, j, k, l = 1, 2, \ l \text{ fixed.}$$

One more equivalent form of the system (9.29) in this case is as follows:

$$B_2^{x\xi}[s^l] + B_2^{x\xi}[u_1] \frac{\partial u}{\partial s^l} = 0, \quad l = 1, 2,$$

where $u_1(\boldsymbol{\xi}) = u[\boldsymbol{s}(\boldsymbol{\xi})]$.

The right-hand sides of the systems (9.28) and (9.29) are defined through the metric elements of the original surface parametrization in the coordinates s^i, $i = 1, 2$, and do not depend on the transformation $\boldsymbol{s}(\boldsymbol{\xi}) : \Xi^2 \to S^2$; therefore they can be computed in advance or at a previous step of an iterative solution procedure.

The values of the numerical solution of a boundary value problem for (9.28) or (9.29) on a reference grid in Ξ^2 define a grid in the parametric domain S^2. The final grid on the surface S^{x2}, generated through the inverted Beltrami system (9.28), is obtained by mapping the above grid with the original parametric transformation $\boldsymbol{x}(\boldsymbol{s})$.

9.5 Interpretations of the Beltramian System

Equations (9.27) are a generalization of the two-dimensional Laplace system on a surface. In this section we give some interpretations and justifications of the systems (9.27)–(9.29) concerned with grid generation.

9.5.1 Variational Formulation

As was shown in Sect. 8.2, the Laplace system (6.4) for generating grids in domains can be obtained from the minimization of the functional of smoothness

$$I_s = \int_{X^n} \left(\sum_{i=1}^n g^{ii} \right) d\boldsymbol{x}, \quad g^{ij} = \sum_{m=1}^n \frac{\partial \xi^i}{\partial x^m} \frac{\partial \xi^j}{\partial x^m}. \tag{9.30}$$

A similar functional, whose Euler–Lagrange equations are equivalent to (9.27), can be formulated for the Beltrami equations as well. This functional has the form of (9.30) with g^{ii} and X^n replaced by $g^{ii}_{\xi x}$ and S^{x2}, respectively:

$$I_{\mathrm{s}} = \int_{S^{x2}} \left(\sum_{i=1}^{2} g^{ii}_{\xi r} \right) \mathrm{d}S^{x2} = \int_{S^{x2}} (\operatorname{tr} G_{\xi x}) \mathrm{d}S^{x2}. \tag{9.31}$$

In order to write out the Euler–Lagrange equations for this functional, we consider the integration over S^2. We obtain, using (9.23) and the relation $\mathrm{d}S^{x2} = \sqrt{g^{xs}}\mathrm{d}\boldsymbol{s}$,

$$I_{\mathrm{s}} = \int_{S^2} \sqrt{g^{xs}} g^{mk}_{sr} \frac{\partial \xi^i}{\partial s^m} \frac{\partial \xi^i}{\partial s^k} \mathrm{d}\boldsymbol{s}. \tag{9.32}$$

The functional (9.32) is formulated on a set of smooth functions $\xi^i(\boldsymbol{s})$, and the terms $\sqrt{g^{xs}} g^{mk}_{sr}$ of its integrand are defined through the metric elements of the original parametrization $\boldsymbol{x}(\boldsymbol{s})$ and therefore do not depend on these functions. So the Euler–Lagrange equations derived from the functional (9.32) have the form of (8.6), namely,

$$\frac{\partial}{\partial s^j} \left(\sqrt{g^{xs}} g^{mj}_{sr} \frac{\partial \xi^i}{\partial s^m} \right) = 0, \quad i, j, m = 1, 2, \tag{9.33}$$

and they are equivalent to (9.27). This variational formulation allows one to generate surface grids by a variational method. For this purpose, the functional (9.31) written out with respect to the integral over the domain Ξ^2,

$$I_{\mathrm{s}} = \int_{\Xi^2} \frac{1}{\sqrt{g^{x\xi}}} (g^{x\xi}_{11} + g^{x\xi}_{22}) \mathrm{d}\boldsymbol{\xi},$$

is used.

9.5.2 Harmonic-Mapping Interpretation

The integral (9.31), and consequently (9.32), according to the accepted terminology in differential geometry, is the total energy associated with the function $\boldsymbol{\xi}(\boldsymbol{s}) : S^2 \to \Xi^2$ which represents a mapping between the manifold S^{x2} with the metric tensor g^{xs}_{ij} and the computational domain Ξ^2 with the Cartesian coordinates ξ^i. A function which is a critical point of the energy functional is called a harmonic function. It follows from the theory of harmonic functions on manifolds that if there is a diffeomorphism $\boldsymbol{f}(\boldsymbol{s}) : S^2 \to \Xi^2$ and the boundary of the manifold Ξ^2 is convex, while its curvature is nonpositive, then a harmonic function coinciding with $\boldsymbol{f}$ on the boundary of the manifold S^{x2} is also a diffeomorphism between S^2 and Ξ^2. In the case under consideration the coordinates of the manifold Ξ^2 are Cartesian, and therefore its curvature is nonpositive. So if the boundary of the computational domain Ξ^2 is convex (e.g. Ξ^2 is a rectangle) and the surface S^{x2} is diffeomorphic to Ξ^2 then the mapping $\boldsymbol{\xi}(\boldsymbol{s})$ that minimizes the functional of smoothness (9.32) is a one-to-one transformation and the grid obtained by the proposed variational method is therefore nondegenerate.

9.5.3 Formulation Through Invariants

We note that, in accordance with (3.22), the trace $\mathrm{tr}\, G_{\xi x}$ of the contravariant 2×2 metric tensor can be expressed through the invariants I_1, I_2 of the orthogonal transforms of the covariant metric tensor $G^{x\xi}$, namely,

$$\mathrm{tr}\, G_{\xi x} = \frac{I_1}{I_2}.$$

Therefore the functional of smoothness (9.31) can also be expressed through these invariants:

$$I_{\mathrm{s}} = \int_{S^{x2}} \left(\frac{I_1}{I_2}\right) \mathrm{d}S^{x2}. \tag{9.34}$$

The invariant I_2 is the Jacobian of the matrix $G^{x\xi}$ and it equals the area squared of the parallelogram formed by the tangent vectors $\boldsymbol{x}_{\xi^1}$ and $\boldsymbol{x}_{\xi^2}$. The invariant I_1 is $g_{11}^{x\xi} + g_{22}^{x\xi}$ and means the sum of the lengths squared of the sides of the parallelogram. So

$$\frac{I_1}{I_2} = \frac{g_{11}^{x\xi} + g_{22}^{x\xi}}{g^{x\xi}}. \tag{9.35}$$

It is obvious that

$$g^{x\xi} = g_{11}^{x\xi}(d_2)^2, \qquad g^{x\xi} = g_{22}^{x\xi}(d_1)^2,$$

where d_i, $i = 1, 2$, is the distance between the vertex of the vector $\boldsymbol{x}_{\xi^i}$ and the other vector $\boldsymbol{x}_{\xi^j}$, $j = 3 - i$. Therefore, from (9.35),

$$\frac{I_1}{I_2} = \frac{1}{(d_1)^2} + \frac{1}{(d_2)^2}. \tag{9.36}$$

The quantity d_i is connected with the distance l_i between the grid lines $\xi^i = c$ and $\xi^i = c + h$ by the relation

$$l_i = d_i h + O(h)^2.$$

So, from (9.36),

$$\frac{I_1}{I_2} = \sum_{i=1}^{2} (h/l_i)^2 + O(h).$$

The quantity $(h/l_i)^2$ increases as the grid nodes cluster in the direction normal to the coordinate $\xi^i = \mathrm{const}$, and therefore it can be considered as some measure of the grid concentration in this direction and, consequently, the functional (9.34) defines an integral measure of the grid clustering in all directions.

Hence the problem of the minimization of the smoothness functional (9.31) can be interpreted as a problem of finding a grid with uniform clustering on the surface S^{x2}. So the Beltrami equations (9.27) tend to generate a uniform grid on the surface in the same manner as the system of the Laplace equations does in a domain.

9.5.4 Formulation Through the Surface Christoffel Symbols

Equivalent forms of the surface system of equations (9.27) and (9.28) can be obtained by a consideration of the formulas of Gauss. These formulas represent the derivatives of the tangential vectors $\boldsymbol{x}_{\xi^1}$ and $\boldsymbol{x}_{\xi^2}$ through the basis $(\boldsymbol{x}_{\xi^1}, \boldsymbol{x}_{\xi^2}, \boldsymbol{n})$, where $\boldsymbol{n}$ is a unit normal to the surface.

9.5.4.1 Surface Gauss Equations

In accordance with (2.5), we can write

$$\boldsymbol{x}_{\xi^i\xi^j} = a^{lm}(\boldsymbol{x}_{\xi^i\xi^j} \cdot \boldsymbol{a}_l)\boldsymbol{a}_m, \quad i,j = 1,2, \ l,m = 1,2,3, \tag{9.37}$$

where $\boldsymbol{a}_1 = \boldsymbol{x}_{\xi^1}$, $\boldsymbol{a}_2 = \boldsymbol{x}_{\xi^2}$, $\boldsymbol{a}_3 = \boldsymbol{n}$, and (a^{lm}) is the matrix inverse to $(a_{kp}) = (\boldsymbol{a}_k \cdot \boldsymbol{a}_p)$. We readily obtain, for the elements of the matrix (a_{ij}),

$$a_{ij} = g_{ij}^{x\xi}, \qquad a_{i3} = a_{3i} = 0, \quad i,j = 1,2, \qquad a_{33} = 1.$$

Therefore

$$a^{ij} = g_{\xi x}^{ij}, \qquad a^{i3} = a^{3i} = 0, \qquad a^{33} = 1, \quad i,j = 1,2. \tag{9.38}$$

Thus (9.37) results in

$$\boldsymbol{x}_{\xi^i\xi^j} = g_{\xi x}^{km}(\boldsymbol{x}_{\xi^i\xi^j} \cdot \boldsymbol{r}_{\xi m})\boldsymbol{x}_{\xi k} + (\boldsymbol{x}_{\xi^i\xi^j} \cdot \boldsymbol{n})\boldsymbol{n}, \quad i,j,k,m = 1,2. \tag{9.39}$$

The quantities $\boldsymbol{x}_{\xi^i\xi^j} \cdot \boldsymbol{x}_{\xi m}$ in (9.39) are the surface Christoffel symbols of the first kind, denoted by $[ij,m]$. These quantities coincide with the space Christoffel symbols (2.39) for the indices $i,j,k = 1,2$, and, in the same manner as (2.44), they are subject to the relations

$$[ij,k] = \frac{1}{2}\left(\frac{\partial g_{ik}^{x\xi}}{\partial \xi^j} + \frac{\partial g_{jk}^{x\xi}}{\partial \xi^i} - \frac{\partial g_{ij}^{x\xi}}{\partial \xi^k}\right), \quad i,j,k = 1,2.$$

The surface Christoffel symbols of the second kind, denoted by Υ_{ij}^k, are defined in analogy with (2.42) by

$$\Upsilon_{ij}^k = g_{\xi x}^{km}[ij,m] = g_{\xi x}^{km}\boldsymbol{x}_{\xi^i\xi^j} \cdot \boldsymbol{x}_{\xi^m}, \quad i,j,k,m = 1,2. \tag{9.40}$$

The quantities $\boldsymbol{x}_{\xi^i\xi^j} \cdot \boldsymbol{n}$ in (9.39), designated as b_{ij}, i.e.

$$b_{ij} = \boldsymbol{x}_{\xi^i \xi^j} \cdot \boldsymbol{n}, \quad i, j = 1, 2, \tag{9.41}$$

are called the coefficients of the second fundamental form. With these designations, we have for (9.39)

$$\boldsymbol{x}_{\xi^i \xi^j} = \Upsilon^k_{ij} \boldsymbol{x}_{\xi^k} + b_{ij} \boldsymbol{n}. \tag{9.42}$$

The relations (9.42) are referred to as the surface Gauss identities.

9.5.4.2 Weingarten Equation

Some other important relations are concerned with the first derivatives of the unit normal $\boldsymbol{n}$. Since $\boldsymbol{n}_{\xi^i} \cdot \boldsymbol{n} = 0$, $i = 1, 2$, the vectors $\boldsymbol{n}_{\xi^i}$, $i = 1, 2$, are orthogonal to $\boldsymbol{n}$ and hence can be expanded in the tangential vectors $\boldsymbol{x}_{\xi^i}$, $i = 1, 2$. Taking advantage of (2.5) and (9.38), we obtain

$$\boldsymbol{n}_{\xi^i} = g^{lm}_{\xi x} (\boldsymbol{n}_{\xi^i} \cdot \boldsymbol{x}_{\xi^l}) \boldsymbol{x}_{\xi^m}, \quad i, l, m = 1, 2. \tag{9.43}$$

Since

$$\boldsymbol{n}_{\xi^i} \cdot \boldsymbol{x}_{\xi^l} = (\boldsymbol{n} \cdot \boldsymbol{x}_{\xi^l})_{\xi^i} - \boldsymbol{n} \cdot \boldsymbol{r}_{\xi^l \xi^i} = -b_{li},$$

(9.43) has the form

$$\boldsymbol{n}_{\xi^i} = -g^{lm}_{\xi x} b_{li} \boldsymbol{x}_{\xi^m}, \quad i, l, m = 1, 2. \tag{9.44}$$

The relations (9.44) are called the Weingarten equations.

9.5.4.3 Mean Curvature

The value of the Beltrami operator over the position vector is connected with the mean curvature. The quantity known as the mean surface curvature is expressed in accordance with (3.19) by

$$K_{\mathrm{m}} = \frac{1}{2} g^{ij}_{\xi x} b_{ij}, \quad i, j = 1, 2. \tag{9.45}$$

The mean surface curvature does not depend on the surface parametrization. In fact, if s^i, $i = 1, 2$, are new surface coordinates, then

$$\boldsymbol{x}_{\xi^i \xi^j} = \frac{\partial s^k}{\partial \xi^i} \frac{\partial s^m}{\partial \xi^j} \boldsymbol{x}_{s^k s^m} + \frac{\partial^2 s^m}{\partial \xi^i \partial \xi^j} \boldsymbol{x}_{s^m},$$

and hence

$$b_{ij} = \frac{\partial s^k}{\partial \xi^i} \frac{\partial s^m}{\partial \xi^j} \boldsymbol{x}_{s^k s^m} \cdot \boldsymbol{n}, \quad i, j, k, m = 1, 2.$$

As the contravariant tensor $(g^{im}_{\xi x})$ is transformed to the coordinates s^i in accordance with the relations (9.23), we now readily see that the mean curvature is an invariant of the surface parametrizations. Therefore it can be defined from the original parametrization $\boldsymbol{x}(\boldsymbol{s})$.

Recall that in geometry the first and second fundamental forms in the coordinates ξ^1, ξ^2 are, respectively,

$$\mathrm{I} = g_{ij}^{x\xi} \mathrm{d}\xi^i \mathrm{d}\xi^j, \qquad \mathrm{II} = b_{ij} \mathrm{d}\xi^i \mathrm{d}\xi^j, \quad i,j = 1,2.$$

The first fundamental form, derived from the surface metric, describes the inner geometry of the surface, while the second form deals with the outer geometry, since it is derived from the specification of the surface immersion into R^3. Obviously the second fundamental form gives an expression for the local departure of the surface from its tangent plane.

9.5.4.4 Relation Between Beltrami's Equation and Christoffel Symbols

We assume that the basic surface vectors $(\boldsymbol{x}_{\xi^1}, \boldsymbol{x}_{\xi^2}, \boldsymbol{n})$ compose a right-handed triad. This leads to

$$\boldsymbol{n} = \frac{1}{\sqrt{g^{x\xi}}} (\boldsymbol{x}_{\xi^1} \times \boldsymbol{r}_{\xi^2}). \tag{9.46}$$

Note that the tangent vectors $\boldsymbol{x}_{\xi^i}$, $i = 1,2$, are neither normalized nor orthogonal to each other, while $\boldsymbol{n}$ is normalized and orthogonal to both $\boldsymbol{x}_{\xi^1}$ and $\boldsymbol{x}_{\xi^2}$. Using the general vector identity

$$\boldsymbol{u} \times (\boldsymbol{v} \times \boldsymbol{w}) = (\boldsymbol{u} \cdot \boldsymbol{w})\boldsymbol{v} - (\boldsymbol{u} \cdot \boldsymbol{v})\boldsymbol{w}$$

we obtain taking into account (9.46),

$$\boldsymbol{x}_{\xi^i} \times \boldsymbol{n} = \frac{1}{\sqrt{g^{x\xi}}} (g_{i2}^{x\xi} \boldsymbol{x}_{\xi^1} - g_{i1}^{x\xi} \boldsymbol{x}_{\xi^2}).$$

With (9.22), this results in the relation

$$(-1)^{i+1} \boldsymbol{x}_{\xi^{3-i}} \times \boldsymbol{n} = \sqrt{g^{x\xi}} g_{\xi x}^{ij} \boldsymbol{x}_{\xi^j}, \quad i,j = 1,2, \tag{9.47}$$

with i fixed. The application of (9.47) to the Beltramian operator yields

$$\begin{aligned}
\Delta_B[\boldsymbol{x}] &= \frac{1}{\sqrt{g^{x\xi}}} \frac{\partial}{\partial \xi^j} \left(\sqrt{g^{x\xi}} g_{\xi x}^{ij} \boldsymbol{x}_{\xi^i} \right) \\
&= (-1)^{j+1} \frac{1}{\sqrt{g^{x\xi}}} \frac{\partial}{\partial \xi^j} (\boldsymbol{x}_{\xi^{3-j}} \times \boldsymbol{n}) \\
&= (-1)^{j+1} \frac{1}{\sqrt{g^{x\xi}}} (\boldsymbol{x}_{\xi^{3-j}} \times \boldsymbol{n}_{\xi^j}), \quad i,j = 1,2.
\end{aligned} \tag{9.48}$$

Taking advantage of the Weingarten equations (9.44) and (9.46), we have

$$(-1)^{j+1}\frac{1}{\sqrt{g^{x\xi}}}(\boldsymbol{x}_{\xi^{3-j}}\times\boldsymbol{n}_{\xi^j}) = (-1)^j\frac{1}{\sqrt{g^{x\xi}}}g^{lm}_{\xi x}b_{lj}(\boldsymbol{x}_{\xi^{3-j}}\times\boldsymbol{x}_{\xi^m})$$
$$= g^{lj}_{\xi x}b_{lj}\boldsymbol{n}, \quad j,l,m=1,2.$$

Thus (9.48) also has the form

$$\Delta_B[\boldsymbol{x}] = b_{ij}g^{ij}_{\xi x}\boldsymbol{n} = 2K_{\mathrm{m}}\boldsymbol{n}, \tag{9.49}$$

where the quantity K_{m}, defined by (9.45), is the mean curvature of the surface. Equation (9.49) means that the surface position vector $\boldsymbol{x}(\boldsymbol{\xi})$ is transformed by the Beltrami operator to the vector $\Delta_B[\boldsymbol{x}]$, which is orthogonal to the surface. Now expanding the differentiation in $\Delta_B[\boldsymbol{x}]$ and using the expression for $\Delta_B[\xi^i]$, we obtain one more form of $\Delta_B[\boldsymbol{x}]$:

$$\Delta_B[\boldsymbol{x}] = g^{ij}_{\xi x}\boldsymbol{x}_{\xi^i\xi^j} + (\Delta_B[\xi^i])\boldsymbol{x}_{\xi^i}, \quad i,j=1,2. \tag{9.50}$$

Equating the right-hand sides of (9.49) and (9.50), we have the identity

$$g^{ij}_{\xi x}\boldsymbol{x}_{\xi^i\xi^j} + (\Delta_B[\xi^i])\boldsymbol{r}_{\xi^i} = 2K_{\mathrm{m}}\boldsymbol{n}, \quad i,j=1,2. \tag{9.51}$$

Thus, if the surface coordinate system ξ^1, ξ^2 is obtained by the solution of (9.27) then from (9.51) we obtain

$$g^{ij}_{\xi x}\boldsymbol{x}_{\xi^i\xi^j} = 2K_{\mathrm{m}}\boldsymbol{n}, \quad i,j=1,2.$$

We obtain one more identity by multiplying (9.42) by $g^{ij}_{\xi r}$:

$$g^{ij}_{\xi x}\boldsymbol{x}_{\xi^i\xi^j} = g^{ij}_{\xi x}\Upsilon^k_{ij}\boldsymbol{x}_{\xi^k} + 2K_{\mathrm{m}}\boldsymbol{n}.$$

Comparing this identity with (9.51), we have the identity

$$\Delta_B[\xi^i] = -g^{kl}_{\xi x}\Upsilon^i_{kl}, \quad i,k,l=1,2, \tag{9.52}$$

i.e. the value obtained by applying the Beltrami operator to the function $\xi^i(\boldsymbol{s})$ is defined through the surface Christoffel symbols and the surface metric elements. Note that the surface identity (9.52) is a reformulation of the identity (6.29), valid for domains.

Taking advantage of (9.52), the Beltrami system (9.27) can be written in the following equivalent form as

$$g^{kl}_{\xi x}\Upsilon^i_{kl} = 0, \quad i,k,l=1,2, \tag{9.53}$$

or, using (9.40), as

$$g^{kl}_{\xi x}g^{ij}_{\xi x}[kl,j] \equiv g^{kl}_{\xi x}g^{ij}_{\xi x}\boldsymbol{x}_{\xi^k\xi^l}\cdot\boldsymbol{x}_{\xi^j} = 0, \quad i,j,k,l=1,2. \tag{9.54}$$

Multiplying (9.54) by $g^{x\xi}_{im}$ and summing over i yields

$$g^{kl}_{\xi x}\boldsymbol{x}_{\xi^k\xi^l}\cdot\boldsymbol{x}_{\xi^m}=0,\quad k,l,m=1,2. \tag{9.55}$$

Recall that the first and second Christoffel symbols in (9.53) and (9.54) are defined in terms of the coordinates ξ^i.

In the particular case where the surface S^{x2} is a monitor surface defined by the values of a height function $u(\boldsymbol{s})$ over the domain S^2, i.e.

$$\boldsymbol{x}(\boldsymbol{s})=[s^1,s^2,u(\boldsymbol{s})],$$

multiplication of (9.55) by $g^{x\xi}(\partial\xi^m/\partial s^i)$ and summation over m produces the following system of inverted surface Beltrami equations:

$$B^{x\xi}_2[s^i]+B^{x\xi}_2[u]\frac{\partial u}{\partial s^i}=0,\quad i=1,2, \tag{9.56}$$

where

$$B^{x\xi}_2[y]\equiv g^{x\xi}g^{ij}_{\xi x}\frac{\partial^2 y}{\partial\xi^i\partial\xi^j}\equiv g^{x\xi}_{22}\frac{\partial^2 y}{\partial\xi^1\partial\xi^1}-2g^{x\xi}_{12}\frac{\partial^2 y}{\partial\xi^1\partial\xi^2}+g^{x\xi}_{11}\frac{\partial^2 y}{\partial\xi^2\partial\xi^2},\quad i,j=1,2.$$

Another form of the inverted Beltrami system can be obtained from the elliptic system (9.28) for generating surface grids. Namely, applying (9.52) to (9.28) with the identification $\xi^i=s^i,\ i=1,2$, we obtain

$$B^{x\xi}_2[s^l]+g^{x\xi}g^{km}_{sx}\Upsilon^l_{km}=0,\quad i,j,k,l,m=1,2, \tag{9.57}$$

where the Υ^l_{km} are the second surface Christoffel symbols in the coordinates s^i. An equivalent form of (9.56) is also obtained by utilizing (9.40) for $\xi^i=s^i,\ i=1,2$:

$$B^{x\xi}_2[s^l]+g^{x\xi}g^{km}_{sx}g^{lp}_{sx}[km,p]=0,\quad i,j,k,l,m,p=1,2,$$

where $[km,p]=\boldsymbol{x}_{s^ks^m}\cdot\boldsymbol{x}_{s^p}$.

In the particular case where the surface S^{x2} is a monitor surface defined by the values of a height function $u(\boldsymbol{s})$ over the domain S^2, i.e.

$$\boldsymbol{x}(\boldsymbol{s})=[s^1,s^2,u(\boldsymbol{s})],$$

the Beltrami equations (9.57) have the form

$$g^{ij}_{\xi x}\frac{\partial^2 s^l}{\partial\xi^i\partial\xi^j}+g^{km}_{sx}g^{lp}_{sx}\frac{\partial^2 u}{\partial s^k\partial s^m}\frac{\partial u}{\partial s^p}=0,\quad i,j,k,l,m,p=1,2,$$

where

$$g^{km}_{sx}=(-1)^{k+m}\frac{\delta^k_m+\frac{\partial u}{\partial s^{3-k}}\frac{\partial u}{\partial s^{3-m}}}{1+(\frac{\partial u}{\partial s^1})^2+(\frac{\partial u}{\partial s^2})^2},\quad k,m=1,2;$$

here the indices k and m are fixed.

9.5.5 Relation to Conformal Mappings

A two-dimensional grid on the two-dimensional surface S^{x2} can also be generated by the conformal mapping of the unit rectangle Ξ^2 onto the surface S^{x2}. Such an approach for the generation of grids on two-dimensional surfaces lying in the three-dimensional space R^3 was formulated by Khamayseh and Mastin (1996). This subsection gives a description of their approach.

The coordinate transformation $\boldsymbol{x}[\boldsymbol{s}(\boldsymbol{\xi})] : \Xi^2 \to S^{x2}$ on the surface S^{x2} represented by the parametrization $\boldsymbol{x}(\boldsymbol{s}) : S^2 \to R^3$ is a conformal mapping if it is orthogonal and has a constant aspect ratio. These conditions lead to the system of equations

$$g_{12}^{x\xi} = 0, \qquad d^2 g_{11}^{x\xi} = g_{22}^{x\xi}.$$

Using (9.21), we see that the above system of conditions for conformal mappings is equivalent to

$$g_{11}^{xs}\frac{\partial s^1}{\partial \xi^1}\frac{\partial s^1}{\partial \xi^2} + g_{12}^{xs}\left(\frac{\partial s^1}{\partial \xi^1}\frac{\partial s^2}{\partial \xi^2} + \frac{\partial s^1}{\partial \xi^2}\frac{\partial s^2}{\partial \xi^1}\right) + g_{22}^{xs}\frac{\partial s^2}{\partial \xi^1}\frac{\partial s^2}{\partial \xi^2} = 0,$$

$$d^2\left[g_{11}^{xs}\left(\frac{\partial s^1}{\partial \xi^1}\right)^2 + 2g_{12}^{xs}\frac{\partial s^1}{\partial \xi^1}\frac{\partial s^2}{\partial \xi^1} + g_{22}^{xs}\left(\frac{\partial s^2}{\partial \xi^1}\right)^2\right]$$
$$= g_{11}^{xs}\left(\frac{\partial s^1}{\partial \xi^2}\right)^2 + 2g_{12}^{xs}\frac{\partial s^1}{\partial \xi^2}\frac{\partial s^2}{\partial \xi^2} + g_{22}^{xs}\left(\frac{\partial s^2}{\partial \xi^2}\right)^2.$$

The combination of these equations results in the complex equation

$$g_{11}^{xs}r^2 + 2g_{12}^{xs}zw + g_{22}^{xs}w^2 = 0, \tag{9.58}$$

where

$$z = d\frac{\partial s^1}{\partial \xi^1} + \mathrm{i}\frac{\partial s^1}{\partial \xi^2}, \qquad w = d\frac{\partial s^2}{\partial \xi^1} + \mathrm{i}\frac{\partial s^2}{\partial \xi^2}.$$

The solution of the quadratic equation (9.58) for z gives

$$z = \frac{-g_{12}^{xs} \pm \mathrm{i}\sqrt{g^{xs}}}{g_{11}^{xs}}w, \quad g^{xs} = \det G^{xs}.$$

Equating the real and the imaginary parts of this expression gives a system of two real equations

$$\begin{aligned} d\frac{\partial s^1}{\partial \xi^1} &= \frac{1}{g_{11}^{xs}}\left(-d\, g_{12}^{xs}\frac{\partial s^2}{\partial \xi^1} \pm \sqrt{g^{xs}}\frac{\partial s^2}{\partial \xi^2}\right), \\ \frac{\partial s^1}{\partial \xi^2} &= \frac{1}{g_{11}^{xs}}\left(-g_{12}^{xs}\frac{\partial s^2}{\partial \xi^2} \pm d\sqrt{g^{xs}}\frac{\partial s^2}{\partial \xi^1}\right). \end{aligned} \tag{9.59}$$

This system generates the first-order elliptic system

$$d\frac{\partial s^1}{\partial \xi^1} = a\frac{\partial s^2}{\partial \xi^2} - b\frac{\partial s^1}{\partial \xi^2},$$
$$d\frac{\partial s^2}{\partial \xi^1} = b\frac{\partial s^2}{\partial \xi^2} - c\frac{\partial s^1}{\partial \xi^2}, \tag{9.60}$$

where

$$a = \frac{-g_{22}^{xs}}{\pm\sqrt{g^{xs}}}, \qquad b = \frac{g_{12}^{xs}}{\pm\sqrt{g^{xs}}}, \qquad c = \frac{-g_{11}^{xs}}{\pm\sqrt{g^{xs}}}. \tag{9.61}$$

The sign $\pm$ needs to be chosen such that the Jacobian of the coordinate transformation $s(\xi) : \Xi^2 \to S^2$ is strongly positive:

$$J = \frac{\partial s^1}{\partial \xi^1}\frac{\partial s^2}{\partial \xi^2} - \frac{\partial s^1}{\partial \xi^2}\frac{\partial s^2}{\partial \xi^1} > 0.$$

The difference between the upper equation of the system (9.60) multiplied by $\partial s^2/\partial \xi^2$ and the lower one multiplied by $\partial s^1/\partial \xi^2$ produces the equation for the Jacobian

$$\mathrm{d}J = a\left(\frac{\partial s^2}{\partial \xi^2}\right)^2 - 2b\frac{\partial s^1}{\partial \xi^2}\frac{\partial s^2}{\partial \xi^2} + c\left(\frac{\partial s^1}{\partial \xi^2}\right)^2.$$

The quantities g_{11}^{xs} and g_{22}^{xs} are positive by definition, so the negative sign in (9.61) will make $a > 0$ and $c > 0$. The equation $ac - b^2 = 1$ leads to the inequality $|b| < |(ac)^{1/2}|$ and, therefore,

$$\mathrm{d}J > \left(a^{1/2}\frac{\partial s^2}{\partial \xi^2} - c^{1/2}\frac{\partial s^1}{\partial \xi^2}\right)^2 \geq 0;$$

hence $J > 0$. Thus we take in (9.61)

$$a = \frac{1}{\sqrt{g^{xs}}}g_{22}^{xs}, \qquad b = -\frac{1}{\sqrt{g^{xs}}}g_{12}^{xs}, \qquad c = \frac{1}{\sqrt{g^{xs}}}g_{11}^{xs}. \tag{9.62}$$

Taking into account the relations

$$\frac{\partial \xi^i}{\partial s^j} = (-1)^{i+j}\frac{\partial s^{3-j}}{\partial \xi^{3-i}}/J, \quad i,j = 1,2,$$

(i and j fixed) in (9.60) one can derive either the system

$$d\frac{\partial \xi^2}{\partial s^2} = a\frac{\partial \xi^1}{\partial s^1} + b\frac{\partial \xi^1}{\partial s^2},$$
$$d\frac{\partial \xi^2}{\partial s^1} = -b\frac{\partial \xi^1}{\partial s^1} - c\frac{\partial \xi^1}{\partial s^2},$$

or

$$\frac{1}{d}\frac{\partial \xi^1}{\partial s^2} = -a\frac{\partial \xi^2}{\partial s^1} - b\frac{\partial \xi^2}{\partial s^2},$$
$$\frac{1}{d}\frac{\partial \xi^1}{\partial s^1} = c\frac{\partial \xi^2}{\partial s^2} + b\frac{\partial \xi^2}{\partial s^1}.$$

These systems are uncoupled by differentiating the first equation of each system with respect to s^1 and the second one with respect to s^2 and then subtracting the results. The final second-order elliptic system has then the following form:

$$\frac{\partial}{\partial s^1}\left(a\frac{\partial \xi^i}{\partial s^1}\right)+\frac{\partial}{\partial s^1}\left(b\frac{\partial \xi^i}{\partial s^2}\right)+\frac{\partial}{\partial s^2}\left(b\frac{\partial \xi^i}{\partial s^1}\right)+\frac{\partial}{\partial s^2}\left(c\frac{\partial \xi^i}{\partial s^2}\right)=0, \qquad (9.63)$$

for $i=1,2$. Note that the coefficients a,b,c satisfy the following formulas:

$$a=\sqrt{g^{xs}}g^{11}_{sx}, \qquad b=\sqrt{g^{xs}}g^{12}_{sx}, \qquad c=\sqrt{g^{xs}}g^{22}_{sx}, \qquad (9.64)$$

obtained from (9.22) and (9.62). The substitution of these expressions in the above second-order elliptic system (9.63) for the coefficients a,b,c leads to the system of equations

$$\frac{\partial}{\partial s^m}\left(\sqrt{g^{xs}}\,g^{ml}_{sx}\frac{\partial \xi^i}{\partial s^l}\right)=0, \quad i,l,m=1,2,$$

which is equivalent to the system of the Beltrami equations (9.27).

9.5.6 Projection of the Laplace System

In this subsection we demonstrate that if the surface S^{x2} is a coordinate surface of a three-dimensional coordinate system ξ^1, ξ^2, ξ^3 of a domain X^3, subject to the Laplace equations (6.4) for $n=3$, then the two-dimensional Beltrami equations can be obtained by projecting the three-dimensional Laplace system on S^{x2}, provided the coordinate lines emanating from S^{x2} are orthogonal to and locally straight at the surface.

Thus we consider a domain X^3 with the coordinate parametrization represented by a smooth one-to-one transformation

$$\boldsymbol{x}(\boldsymbol{\xi}): \Xi^3 \to X^3.$$

Let the surface S^{x2}, represented by the parametrization

$$\boldsymbol{x}(\boldsymbol{s}): S^2 \to R^3,$$

coincide with the coordinate surface, say $\xi^3=\xi^3_0$. Therefore this surface has one more parametrization, induced by $\boldsymbol{x}(\boldsymbol{\xi})$,

$$\boldsymbol{x}_1(\xi^1,\xi^2): \Xi^2 \to R^3, \quad \boldsymbol{x}_1(\xi^1,\xi^2)=\boldsymbol{x}(\xi^1,\xi^2,\xi^3_0),$$

which is connected with $\boldsymbol{x}(\boldsymbol{s})$ by some smooth one-to-one transformation

$$\boldsymbol{s}(\xi^1,\xi^2): \Xi^2 \to S^2, \quad \text{i.e. } \boldsymbol{x}_1(\xi^1,\xi^2)=\boldsymbol{x}[\boldsymbol{s}(\xi^1,\xi^2)].$$

In accordance with Sect. 6.3.4, the condition of orthogonality of the ξ^3 coordinate lines to the surface $\xi^3=\xi^3_0$ generates the system

$$\frac{1}{\sqrt{g^{x\xi}}}\frac{\partial}{\partial \xi^j}(\sqrt{g^{x\xi}}g^{ij}_{\xi x}) + \frac{1}{g_{33}}g^{ij}_{\xi r}(\boldsymbol{x}_{\xi^j}\cdot \boldsymbol{x}_{\xi^3\xi^3}) = 0, \quad i,j=1,2, \tag{9.65}$$

which is obtained from (6.46) for $P^i = 0$, $i = 1, 2, 3$. This system (9.65) is in fact a projection of the Laplace equations

$$\nabla \xi^i = 0, \quad i = 1, 2, 3,$$

on the coordinate surface $\xi^3 = \xi^3_0$. In the above equation

$$g^{x\xi} = \det(g_{ij}), \quad i,j = 1,2,$$
$$g^{ij}_{\xi x} = (-1)^{i+j} g_{3-i,3-j}/g^{x\xi} = g^{ij}, \quad i,j = 1,2.$$

The condition of local straightness of the ξ^3 coordinate curves implies

$$\boldsymbol{x}_{\xi^3\xi^3} = b(\boldsymbol{\xi})\boldsymbol{x}_{\xi^3}$$

and from the condition of orthogonality we find, that

$$\boldsymbol{x}_{\xi^j}\cdot \boldsymbol{x}_{\xi^3\xi^3} = 0, \quad j = 1,2.$$

Therefore if the ξ^3 coordinate curves are straight at the points of the surface $\xi^3 = \xi^3_0$, the system (9.65) is reduced to

$$\frac{1}{\sqrt{g^{x\xi}}}\frac{\partial}{\partial \xi^j}(\sqrt{g^{x\xi}}g^{ij}_{\xi x}) = 0, \quad i,j = 1,2. \tag{9.66}$$

As $g_{ij} = \boldsymbol{x}_{\xi^i}\cdot \boldsymbol{x}_{\xi^j}$, $\boldsymbol{x} = (x^1, x^2, x^3)$, the matrix (g_{ij}), $i,j = 1,2$, coincides with the covariant metric tensor

$$G^{x\xi} = (g^{x\xi}_{ij}), \quad i,j = 1,2,$$

of the surface $\xi^3 = \xi^3_0$ in the coordinates ξ^1, ξ^2. So (9.66) is merely $\Delta_B[\xi^i] = 0$ when the conditions specified above are observed. Since the value of the Beltrami operator is invariant of the parametrization of the surface, we obtain from (9.66), assuming $\xi^i = s^i$,

$$\frac{1}{\sqrt{g^{xs}}}\frac{\partial}{\partial s^j}\left(\sqrt{g^{xs}}g^{lj}_{sr}\frac{\partial \xi^i}{\partial s^l}\right) = 0, \quad i,j,l = 1,2,$$

i.e. the system (9.27).

9.6 Control of Surface Grids

9.6.1 Control Functions

One approach to controlling the generation of a surface grid is to add forcing terms to the Beltrami operator in analogy with the Poisson system, i.e. to extend the system (9.27) to the following one:

$$\Delta_B[\xi^i] = P^i, \quad i = 1, 2. \tag{9.67}$$

The system inverse to (9.67) with interchanged dependent and independent variables is obtained by the same procedure as was applied to produce (9.28) and has the form

$$g^{ij}_{\xi x} \frac{\partial^2 s^l}{\partial \xi^i \partial \xi^j} + \frac{\partial s^l}{\partial \xi^i} P^i = \Delta_B[s^l], \quad i, j, l = 1, 2, \tag{9.68}$$

or with the application of the operator $B^{x\xi}_2$ from (9.29)

$$B^{x\xi}_2[s^l] + g^{x\xi} \frac{\partial s^l}{\partial \xi^i} P^i = g^{x\xi} \Delta_B[s^l], \quad i, l = 1, 2.$$

The control functions P^i, $i = 1, 2$, can be chosen in the same manner as was done in Sects. 6.3 and 7.2. For example, for the generation of grids that are nearly orthogonal at the boundaries, we may apply the approach of Sect. 6.3, determining the values of P^i on the boundary curves and then propagating them into the interior.

For this purpose we have to find the values of $\Delta_B[\xi^i]$ on the boundary curves.

9.6.2 Projection on the Boundary Line

Let the family of the surface coordinate lines $\xi^1 = \text{const}$ be orthogonal to the boundary coordinate curve $\xi^2 = \xi^2_0$. Then, along this curve,

$$\begin{aligned} g^{x\xi}_{12} &= 0, & g^{x\xi} &= g^{x\xi}_{11}\, g^{x\xi}_{22}, \\ g^{11}_{\xi x} &= \frac{1}{g^{x\xi}_{11}}, & g^{22}_{\xi r} &= \frac{1}{g^{x\xi}_{22}}. \end{aligned} \tag{9.69}$$

Now we find the values of the quantity $\Delta_B[\xi^i]$ at the points of the boundary curve $\xi^2 = \xi^2_0$. For this purpose we use the representation of the Beltramian operator in the form (9.52):

$$\Delta_B[\xi^i] = -g^{kl}_{\xi x} \Upsilon^i_{kl}, \quad i, k, l = 1, 2.$$

Taking advantage of (9.40), we also have

$$\Delta_B[\xi^i] = -g^{kl}_{\xi x} g^{im}_{\xi x}[kl, m], \quad i, k, l, m = 1, 2. \tag{9.70}$$

Therefore

$$\begin{aligned} \Delta_B[\xi^1] &= -(g^{11}_{\xi x})^2[11, 1] - g^{11}_{\xi r} g^{22}_{\xi x}[22, 1] \\ &= -\frac{1}{2(g^{x\xi}_{11})^2} \frac{\partial g^{x\xi}_{11}}{\partial \xi^1} - g^{11}_{\xi x} g^{22}_{\xi x} \boldsymbol{x}_{\xi^2 \xi^2} \cdot \boldsymbol{x}_{\xi^1} \quad \text{on } \xi^2 = \xi^2_0. \end{aligned} \tag{9.71}$$

Since

$$[22,1] = \frac{\partial}{\partial \xi^2} g_{12} - \frac{1}{2}\frac{\partial}{\partial \xi^1} g_{22},$$

we also obtain applying (9.69),

$$\Delta_B[\xi^1] = \frac{1}{2g_{22}^{x\xi}}\frac{\partial}{\partial \xi^1}(g_{22}^{x\xi}/g_{11}^{x\xi}) - \frac{1}{g_{11}^{x\xi}g_{22}^{x\xi}}\frac{\partial}{\partial \xi^2}(g_{12}^{x\xi}) \quad \text{on } \xi^2 = \xi_0^2. \tag{9.72}$$

Equations (9.71) and (9.72) give an expression for the forcing term P^1 on the boundary curve $\xi^2 = \xi_0^2$ that is necessary for the system (9.67) to generate a grid with orthogonally emanating coordinate lines $\xi^1 = \text{const}$. Analogously, we have, using (9.69) on the boundary line $\xi^2 = \xi_0^2$,

$$\Delta_B[\xi^2] = \frac{1}{2g_{11}^{x\xi}}\frac{\partial}{\partial \xi^2}(g_{11}^{x\xi}/g_{22}^{x\xi}). \tag{9.73}$$

In the same way, if the family of the surface coordinate lines $\xi^2 = \text{const}$ is orthogonal to the coordinate curve $\xi^1 = \xi_0^1$, we obtain from (9.69) and (9.70)

$$\begin{aligned} \Delta_B[\xi^1] &= \frac{1}{2g_{22}^{x\xi}}\frac{\partial}{\partial \xi^1}(g_{22}^{x\xi}/g_{11}^{x\xi}), \\ \Delta_B[\xi^2] &= \frac{1}{2g_{11}^{x\xi}}\frac{\partial}{\partial \xi^2}(g_{11}^{x\xi}/g_{22}^{x\xi}) - \frac{1}{g_{11}^{x\xi}g_{22}^{x\xi}}\frac{\partial}{\partial \xi^1}(g_{12}^{x\xi}), \quad \text{on } \xi^1 = \xi_0^1. \end{aligned} \tag{9.74}$$

The values of $\Delta_B[\xi^i]$ on the boundary segments can be used as the boundary conditions for the control functions P^i. Utilizing the considerations discussed in Sect. 6.3, we can implement approaches similar to those of that section to generate surface grids which are nearly orthogonal at the boundary curves. With this scheme, a grid with coordinate lines nearly orthogonal at the boundary segments can be generated by the following iterative procedure. The values of P^i on the boundary curves are defined by (9.72)–(9.74) using the grid data from the previous step. Then the control functions are determined in the whole domain Ξ^2 by algebraic or elliptic techniques. Applying the system (9.68), the grid corresponding to the next step is built. The process continues until a specified convergence requirement is met.

9.6.3 Monitor Approach

Another approach to controlling the generation of a grid on a surface relies on the concept of the monitor surface defined by the values of some vector function $\boldsymbol{f} = [f^1(\boldsymbol{s}), \ldots, f^k(\boldsymbol{s})]$ over the surface S^{x2}. The monitor surface, denoted by S^{r2}, lies in R^{3+k} and can be represented through the parametric coordinates s^1, s^2 by the equation

$$\boldsymbol{r}(\boldsymbol{s}) = \left[x^1(\boldsymbol{s}), x^2(\boldsymbol{s}), x^3(\boldsymbol{s}), f^1(\boldsymbol{s}), \ldots, f^k(\boldsymbol{s})\right], \quad \boldsymbol{s} = (s^1, s^2).$$

The covariant (contravariant) metric elements of the monitor surface S^{r2} in the coordinates v^1, v^2 are denoted as g_{ij}^{rv} (g_{vr}^{ij}). In this approach, the Beltrami equations on the surface S^{r2} are used to generate the grid on the parametric domain S^2. Mapping this grid on S^{x2} with the transformation $\boldsymbol{x}(\boldsymbol{s}) : S^2 \to S^{x2}$ produces a grid on the surface S^{x2} dependent on the control function $\boldsymbol{f}$.

The equations for generating the grid on the parametric domain S^2 are obtained in the same manner as the equations (9.27) and have, with respect to the dependent coordinates s^i and the independent coordinates ξ^i, a form similar to (9.28):

$$g_{\xi r}^{ij} \frac{\partial^2 s^l}{\partial \xi^i \partial \xi^j} = \Delta_B[s^l], \quad i, j, l = 1, 2, \tag{9.75}$$

where $g_{\xi r}^{ij}$, $i, j = 1, 2$, are the elements of the contravariant tensor of the surface S^{r2} in the coordinates ξ^1, ξ^2, defined through the elements of the covariant tensor

$$g_{ij}^{r\xi} = \frac{\partial \boldsymbol{r}}{\partial \xi^i} \cdot \frac{\partial \boldsymbol{r}}{\partial \xi^j} = g_{ij}^{x\xi} + \frac{\partial \boldsymbol{f}[\boldsymbol{s}(\boldsymbol{\xi})]}{\partial \xi^i} \cdot \frac{\partial \boldsymbol{f}[\boldsymbol{s}(\boldsymbol{\xi})]}{\partial \xi^j}, \quad i, j = 1, 2,$$

by the formula

$$g_{\xi r}^{ij} = (-1)^{i+j} g_{3-i3-j}^{r\xi} / g^{r\xi}, \quad g^{r\xi} = \det(g_{ij}^{r\xi}),$$

with fixed i and j. The Beltrami operator in the right-hand part of (9.75) is expressed as

$$\Delta_B = \frac{1}{\sqrt{g^{rs}}} \frac{\partial}{\partial s^j} \left(\sqrt{g^{rs}} g_{sr}^{ij} \frac{\partial}{\partial s^i} \right)$$

through the metric elements in the coordinates s^1, s^2 of the monitor surface S^{r2}.

The same considerations as in Sect. 9.5 show that the grid on the monitor surface S^{r2} obtained through (9.75) tends to become uniform, and its projection onto the surface S^{x2} produces a grid with a node concentration in the regions of large variations of the control function $\boldsymbol{f}$. One example of such adaptive grid is illustrated by Fig. 9.3.

9.6.4 Control by Variational Methods

The variational approaches of Chap. 8 can be successfully applied to surface grid generation. What is needed is the formulation of the corresponding surface-grid quality measures. Chapter 3 gives a detailed description of domain-grid characteristics in terms of the space metric elements and the space Christoffel symbols. The quantities expressing the local grid properties are readily reformulated for grids on surfaces, using for this purpose the surface metric elements and surface Christoffel symbols. The integration of these

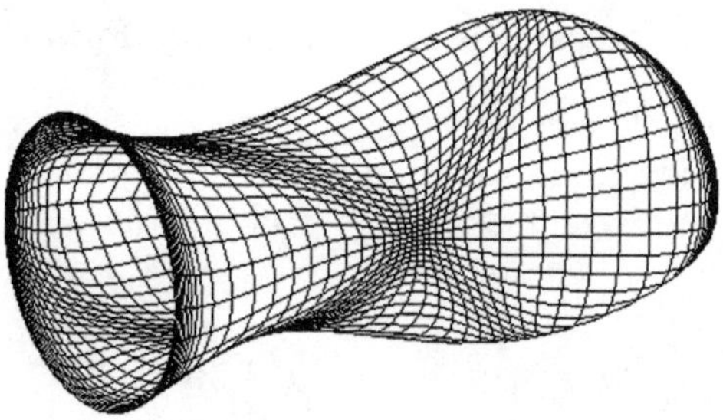

Fig. 9.3. Quadrilateral adaptive surface grid

quantities provides functionals which reflect the integral measures of the respective grid qualities. The grid on the surface is then generated by using the standard scheme of Chap. 8, by optimizing the grid characteristics with the minimization of a combination of functionals to obtain a grid with certain desired properties. In this subsection we describe some surface functionals which represent popular geometric quality measures of surface grids. For generality the functionals are formulated with weights which are determined by derivatives of the solution quantities or by measures of the quality features of the surface.

9.6.4.1 Functionals Dependent on Invariants

In analogy with Chap. 8 there are three basic surface functionals determined by the invariants I_1 and I_2 of the metric tensor $(g_{ij}^{x\xi})$ in the coordinates ξ^i. The first is the length-weighted functional

$$I_{\mathrm{lw}} = \int_{\Xi^2} w I_1 \mathrm{d}\boldsymbol{\xi} = \int_{\Xi^2} w(g_{11}^{x\xi} + g_{22}^{x\xi})\mathrm{d}\boldsymbol{\xi}. \tag{9.76}$$

The second is the area-weighted functional

$$I_{\mathrm{a}} = \int_{S^2} w(I_2)^m \mathrm{d}\boldsymbol{s} = \int_{S^2} w(g^{x\xi})^m \mathrm{d}\boldsymbol{s}. \tag{9.77}$$

And the third is the density-weighted functional

$$I_{\mathrm{d}} = \int_{\Xi^2} w(I_1/\sqrt{I_2})\mathrm{d}\boldsymbol{\xi} = \int_{S^2} w\sqrt{g^{xs}}(g_{\xi x}^{11} + g_{\xi x}^{22})\mathrm{d}\boldsymbol{s}. \tag{9.78}$$

The length-weighted functional controls the lengths of the cell edges, while the functional (9.77) regulates the areas of the surface grid cells. The control of the grid density may be carried out through the functional (9.78). Note that the functional (9.78) with $w = 1$ is in fact the smoothness functional (9.31).

9.6.4.2 Weight Skewness and Orthogonality Functionals

Analogously, functionals can be formulated which measure the surface grid skewness and deviation from orthogonality. Thus, in accordance with (8.26), the functional for the weighted surface grid skewness may be either of the following forms:

$$\begin{aligned} I_{\mathrm{sk},1} &= \int_{\Xi^2} w \frac{(g_{12}^{x\xi})^2}{g_{11}^{x\xi} g_{22}^{x\xi}} \mathrm{d}\boldsymbol{\xi}, \\ I_{\mathrm{sk},2} &= \int_{\Xi^2} w \frac{(g_{\xi x}^{12})^2}{g_{\xi x}^{11} g_{\xi x}^{22}} \mathrm{d}\boldsymbol{\xi}. \end{aligned} \tag{9.79}$$

Similarly to (8.28) and (8.45), the orthogonality functionals can be expressed as follows:

$$\begin{aligned} I_{\mathrm{o},1} &= \int_{\Xi^2} w \left(\frac{g_{11}^{x\xi} g_{22}^{x\xi}}{g^{x\xi}} \right) \mathrm{d}\boldsymbol{\xi}, \\ I_{\mathrm{o},2} &= \int_{\Xi^2} w (g_{12}^{x\xi})^2 \mathrm{d}\boldsymbol{\xi}, \\ I_{\mathrm{o},3} &= \int_{S^{x2}} w (g^{12})^2 \mathrm{d}S^{x2} = \int_{S^2} w \sqrt{g^{xs}} (g^{12})^2 \mathrm{d}\boldsymbol{s}. \end{aligned} \tag{9.80}$$

9.6.4.3 Weight Functions

Commonly, the weight functions for surface grid generation are formulated through the derivatives of the solution quantities and through the features of the surface. Section 7.2 presents some popular expressions for the weights for generating planar and volume grids which can also be utilized in surface grid generation. Surface grid generation also requires adjustments of the measures of the grid cells to the curvature of the surface. This can be carried out by applying the curvature measures as weight functions.

A surface S^{x2} has two curvatures: the mean curvature

$$K_{\mathrm{m}} = \frac{1}{2} g_{sx}^{ij} b_{ij}, \quad i,j = 1,2, \tag{9.81}$$

and the Gaussian curvature

$$K_{\mathrm{G}} = \frac{1}{g^{xs}} (b_{11} b_{22} - (b_{12})^2), \tag{9.82}$$

where

$$b_{ij} = \frac{1}{\sqrt{g^{xs}}} \boldsymbol{x}_{s^i s^j} \cdot (\boldsymbol{x}_{s^1} \times \boldsymbol{x}_{s^2}), \quad i,j = 1,2.$$

The corresponding weights can be formulated in the form

$$w = (K_{\mathrm{m}})^2 \quad \text{or} \quad w = (K_{\mathrm{G}})^2. \tag{9.83}$$

An expression for the quantity $b_{11}b_{22} - (b_{12})^2$ can be obtained through derivatives of the elements of the metric tensor and the coefficients of the second fundamental form of the surface. This is accomplished by using the expansion (9.39) with the substitution of ξ^i for s^i, which results in the following relation:

$$\boldsymbol{x}_{s^1s^1} \cdot \boldsymbol{x}_{s^2s^2} - \boldsymbol{x}_{s^1s^2} \cdot \boldsymbol{x}_{s^1s^2} = g^{xs}_{ij}(\Upsilon^i_{11}\Upsilon^j_{22} - \Upsilon^i_{12}\Upsilon^j_{12}) + b_{11}b_{22} - (b_{12})^2,$$
$$i,j = 1,2. \tag{9.84}$$

The left-hand part of (9.84) equals

$$\boldsymbol{x}_{s^1s^1} \cdot \boldsymbol{x}_{s^2s^2} - \boldsymbol{x}_{s^1s^2} \cdot \boldsymbol{x}_{s^1s^2} = \frac{\partial}{\partial s^1}(\boldsymbol{x}_{s^2s^2} \cdot \boldsymbol{x}_{s^1}) - \frac{\partial}{\partial s^2}(\boldsymbol{x}_{s^1s^2} \cdot \boldsymbol{x}_{s^1}).$$

Since

$$\boldsymbol{x}_{s^2s^2} \cdot \boldsymbol{x}_{s^1} = [22,1] = \frac{1}{2}\left(2\frac{\partial g^{xs}_{12}}{\partial s^2} - \frac{\partial g^{xs}_{22}}{\partial s^1}\right),$$
$$\boldsymbol{x}_{s^1s^2} \cdot \boldsymbol{x}_{s^1} = [12,1] = \frac{1}{2}\frac{\partial g^{xs}_{11}}{\partial s^2},$$

we obtain from (9.85)

$$\boldsymbol{x}_{s^1s^1} \cdot \boldsymbol{x}_{s^2s^2} - \boldsymbol{x}_{s^1s^2} \cdot \boldsymbol{x}_{s^1s^2} = \frac{1}{2}\left(2\frac{\partial^2 g^{xs}_{12}}{\partial s^1 \partial s^2} - \frac{\partial^2 g^{xs}_{11}}{\partial s^2 \partial s^2} - \frac{\partial^2 g^{xs}_{22}}{\partial s^1 \partial s^1}\right).$$

Therefore (9.82) results in

$$K_{\mathrm{G}} = \frac{1}{g^{xs}}\left[\frac{1}{2}\left(2\frac{\partial^2 g^{xs}_{12}}{\partial s^1 \partial s^2} - \frac{\partial^2 g^{xs}_{11}}{\partial s^2 \partial s^2} - \frac{\partial^2 g^{xs}_{22}}{\partial s^1 \partial s^1}\right) - g^{xs}_{ij}(\Upsilon^i_{11}\Upsilon^j_{22} - \Upsilon^i_{12}\Upsilon^j_{12})\right],$$
$$i,j = 1,2. \tag{9.85}$$

9.6.5 Orthogonal Grid Generation

Orthogonal elliptic coordinate systems on a surface are formulated in the standard way, as for domains. Namely, an appropriate identity is chosen, which is then transformed to the orthogonal system by substituting zero for the nondiagonal metric elements.

For example, using the Beltrami operator

$$\Delta_B \xi^i = \frac{1}{\sqrt{g^{x\xi}}}\frac{\partial}{\partial \xi^j}(\sqrt{g^{x\xi}}g^{ij}_{\xi r}), \quad i,j = 1,2,$$

we obtain

$$\Delta_B \xi^i \frac{\partial x^l}{\partial \xi^i} = -g^{ij}_{\xi x}\frac{\partial^2 s^l}{\partial \xi^i \partial \xi^j} + \Delta_B s^l, \quad i,j,l = 1,2.$$

Thus we have the identity

$$\frac{1}{\sqrt{g^{x\xi}}}\frac{\partial}{\partial\xi^j}(\sqrt{g^{x\xi}}g^{ij}_{\xi r})\frac{\partial s^l}{\partial\xi^i}+g^{ij}_{\xi r}\frac{\partial^2 s^l}{\partial\xi^i\partial\xi^j}=\Delta_B s^l,$$

which implies

$$\frac{1}{\sqrt{g^{x\xi}}}\left[\frac{\partial}{\partial\xi^j}\left(\sqrt{g^{x\xi}}g^{ij}_{\xi r}\frac{\partial s^l}{\partial\xi^i}\right)\right]=\Delta_B s^l,\quad i,j,l=1,2.$$

Substituting in these equations the condition of orthogonality $g^{12}_{\xi r}=0$ for the term $g^{12}_{\xi x}$, we obtain an elliptic system for generating orthogonal or nearly orthogonal grids on the surface S^{x2}:

$$\frac{1}{\sqrt{g^{x\xi}}}\frac{\partial}{\partial\xi^j}\left(\sqrt{g^{x\xi}}g^{jj}_{\xi x}\frac{\partial s^l}{\partial\xi^j}\right)=\Delta_B s^l, \qquad (9.86)$$

where

$$g^{x\xi}=g^{x\xi}_{11}\,g^{x\xi}_{22},\qquad g^{11}_{\xi x}=1/g^{x\xi}_{11},\qquad g^{22}_{\xi r}=1/g^{x\xi}_{22}.$$

Thus we have from (9.87)

$$\frac{1}{\sqrt{g^{x\xi}_{11}\,g^{x\xi}_{22}}}\left[\frac{\partial}{\partial\xi^1}\left(\sqrt{g^{x\xi}_{22}/g^{x\xi}_{11}}\frac{\partial s^l}{\partial\xi^1}\right)+\frac{\partial}{\partial\xi^2}\left(\sqrt{g^{x\xi}_{11}/g^{x\xi}_{22}}\frac{\partial s^l}{\partial\xi^2}\right)\right]=\Delta_B s^l,$$
$$l=1,2. \qquad (9.87)$$

The system (9.88) is in fact a generalization of the planar system (6.82).

9.7 Hyperbolic Method

Although the elliptic methods described above can provide satisfactory grids for most applications, there are situations when it is more convenient to use hyperbolic methods, in particular, when the four boundaries of the surface grid need not be specified and constructed prior to the generation of the interior grid. Such a situation, for example, occurs in the generation of grids for intersecting geometric components where the surface grid is generated hyperbolically by marching away from the intersection curve. For the overset grid approach, it is frequently the case that the location of some boundary components is not restricted. Also, domain decomposition is simplified under the overset grid approach, and the grid generation time with a hyperbolic technique is relatively fast since only one boundary needs to be specified.

The hyperbolic method of surface grid generation involves marching a grid away from an initial boundary curve by a user-specified distance. This is achieved by the numerical solution of a set of hyperbolic partial differential equations. Desirable grid attributes such as grid point clustering and orthogonality control are naturally achieved. The grid points obtained are projected onto the underlying surface after each marching step.

This section describes the hyperbolic method of surface grid generation proposed by Steger (1991) and Chan and Buning (1995). The grid generation procedure is formulated in physical space rather than parameter space. This is preferred since the physical-space formulation provides direct control of the grid spacing and orthogonality.

9.7.1 Hyperbolic Governing Equations

Let ξ^1 and ξ^2 be the coordinates of the surface, where ξ^1 runs along some initial boundary curve and ξ^2 is the marching direction away from the curve on the surface. Also, let $\boldsymbol{n} = (n^1, n^2, n^3)$ be the local unit normal, which is assumed to be computable anywhere on the surface. The constraints of orthogonality of the families of grid lines and specified mesh cell area are

$$\begin{aligned} \boldsymbol{x}_{\xi^1} \cdot \boldsymbol{x}_{\xi^2} &= 0, \\ \boldsymbol{n} \cdot (\boldsymbol{x}_{\xi^1} \times \boldsymbol{x}_{\xi^2}) &= \triangle S, \end{aligned} \tag{9.88}$$

where $\triangle S$ is a user-specified surface mesh cell area. A third equation, needed to close the system, is provided by requiring that the marching direction of the grid be orthogonal to the surface normal at the local grid point, i.e. the marching direction is along the tangent plane of the underlying surface at this point. This gives

$$\boldsymbol{n} \cdot \boldsymbol{x}_{\xi^l} = 0. \tag{9.89}$$

A unit vector in the marching direction ξ^2 can be obtained from the cross product of $\boldsymbol{n}$ with a unit vector in the initial curve direction ξ^1.

Equations (9.89) and (9.90), in the usual variables x, y, z and ξ, η, can be written as

$$\begin{aligned} & x_\xi x_\eta + y_\xi y_\eta + z_\xi z_\eta = 0, \\ & n^1(y_\xi z_\eta - z_\xi y_\eta) + n^2(z_\xi x_\eta - x_\xi z_\eta) + n^3(x_\xi y_\eta - y_\xi x_\eta) = \triangle S, \\ & n^1 x_\eta + n^2 y_\eta + n^3 z_\eta = 0. \end{aligned} \tag{9.90}$$

These equations form a hyperbolic system for marching in the η direction.

Equations (9.90) are written in terms of the physical coordinates instead of the parametric coordinates. In order to preserve the specified surface shape, the physical coordinates are repeatedly projected onto the surface in the course of the iteration.

9.8 Comments

A number of algorithms for generation of curve grids were discussed by Eiseman (1987) and Knupp and Steinberg (1993).

The use of Beltrami's equations to generate surface grids was proposed by Warsi (1982), in analogy with the widely utilized Laplace grid generator of Crowley (1962) and Winslow (1967). Warsi (1990) has also justified these equations by using some fundamental results of differential geometry. Theoretical analyses of the relation of Beltrami's equations to the equations of Gauss and Weingarten were given by Warsi (1982, 1990) and Garon and Camarero (1983). An implementation of the Beltrami operator to derive a fourth-order surface elliptic system was performed by Ronzheimer et al. (1994), while Spekreijse, Nijhuis, and Boerstoel (1995) applied this operator with algebraic techniques to generate surface grids with orthogonality at the edges.

A surface grid generation scheme that uses a quasi-two-dimensional elliptic system, obtained by projecting the inverted three-dimensional Laplace system, to generate grids on smooth surfaces analytically specified by the equation $z = f(x, y)$ was proposed by Thomas (1982). The method was extended and updated by Takagi et al. (1985) and Warsi (1986) for arbitrary curved surfaces using a parametric surface representation. An adaptive surface grid technique based on control functions and parametric specifications was also considered by Lee and Loellbach (1989).

Some robust blending functions for algebraic surface grid generation were proposed by Soni (1985) using the normalized arc lengths of the physical edges of the surface patches. These functions and various techniques for the surface patch parametrization were discussed by Samareh-Abolhassani and Stewart (1994) for the purpose of the development of a surface grid software system.

The hyperbolic approach based on grid orthogonality was extended to surfaces by Steger (1991). An analogous technique for generating overset surface grids was presented by Chan and Buning (1995).

Liseikin (1991a, 1992, 2004, 2007) used an elliptic system derived from a variational principle to produce n-dimensional harmonic coordinate transformations which, in a particular two-dimensional case, generate both uniform and adaptive grids on surfaces. Harmonic mapping was also used by Arina and Casella (1991) to derive a surface elliptic system. The conformal mapping technique for generating surface grids presented in this chapter was formulated by Khamayseh and Mastin (1996).

Variational approaches for generating grids on surfaces were described by Saltzman (1986), Liseikin (1991a), and Steinberg and Roache (1992). A variational adaptive technique for deriving a surface grid approaching orthogonality was developed by Desbois and Jacquotte (1991). A variational approach was also given by Castilio (1991) for the control of spacing, cell area, orthogonality, and quality measures. Several grid generation anomalies which appear while implementing some surface variational techniques were discovered by Steinberg and Roache (1986).

An optimization approach to surface grid generation which aimed to maximize grid smoothness and orthogonality was discussed by Pearce (1990). Some techniques for clustering the grid points in regions of larger curvature were considered by Weilmuenster, Smith, and Everton (1991).

10

Comprehensive Method

10.1 Introduction

Many physical phenomena involve the rapid formation, propagation, and disintegration of small-scale structures such as shock waves in compressible flows, shear layers in laminar and turbulent flows, phase boundaries in nonequilibrim and boundary and interior layers, tearing layers and magnetic reconnection regions in magnetically confined plasmas, etc. A promising tool to deal with the numerical problems related to these structures is adaptive grid generation technology. With the increasing complexity of the physical problem, there is an increased need for comprehensive methods which enable one to generate suitable adaptive meshes in a uniform "block box" mode, with or without human interaction.

This chapter describes one such comprehensive grid generation method that relies on variational and elliptic approaches to the generation of grids on hypersurfaces. The method enables one to generate in a unified manner both fixed and adaptive grids in domains and on surfaces with complicated geometry and/or complex physical solution.

The elliptic method of grid generation based on the numerical solution of the system of inverted Poisson equations, proposed by Godunov and Prokopov (1972) and further developed by Thompson, Thames, and Mastin (1974) and other researchers, is being used in a broad range of practical applications. The method allows the users to generate numerical grids in fairly complicated domains and on surfaces that arise while analyzing multidimensional field problems. Practically all big grid generation codes incorporate it as a basic tool for generating grids. However, this basic method has at least three serious drawbacks:

(1) uncertainties in the specification of the control functions required to provide the required types of adaptation;
(2) the cells of the grids produced by this method may be folded;

V.D. Liseikin, *Grid Generation Methods*, Scientific Computation,
DOI 10.1007/978-90-481-2912-6_10, © Springer Science+Business Media B.V. 2010

(3) poor opportunities to provide effective automation of the grid generation process.

The first drawback is due to the lack of clear understanding of connections between geometric characteristics and required properties of grids and the control functions in the system of Poisson equations. One of the reasons for the second drawback is that the system does not obey the maximum principle. The third one is due to the first and second drawbacks and also because of the lack of any inherent potential to provide unification of the method when applying it to domains and/or surfaces with different dimensions or while performing adaptation.

These drawbacks partly account for the rise of interactive grid generation as a substitute for automatic generation. However, the major driving factors of comprehensive grid codes must first be automation and then graphical interaction.

Much research effort has being spent on circumventing these drawbacks within the framework of the Poisson system, by segmenting the domain into simpler subdomains, implementing grid smoothing techniques, developing automatic control function specifications, and applying sophisticated interactive systems. However, the present codes, though in principle they enable one to generate grids in three-dimensional domains, are not yet efficient enough. In particular, they take too many man-hours to generate acceptable grids in complicated regions, and there are requirements from the users that this time must be reduced by one order of magnitude at least for adequate efficiency of these grid codes for the numerical solution of large problems.

A more promising way lies, apparently, in creating new methods that are free or suffer in less degree from the disadvantages pertinent to the technique based on the Poisson system. In this respect, considerable progress may be achieved by the development of the methods which apply coordinate transformations that are inverse to harmonic functions. The salient features of such transformations are that they are solutions to elliptic systems and guarantee one-to-one mappings for two dimensions, at least if the parametric field is convex and there is a diffeomorphism between it and the physical domain, which can be eather convex or concave. Therefore the grids obtained do not fold in this case.

The basic comprehensive method discussed in this chapter is one version of the harmonic-function approach for generating both adaptive and fixed grids. The method relies on a variational technique for the generation of grids on hypersurfaces with the help of an energy functional with respect to control metrics. One energy functional is the functional of smoothness which is defined for a general hypersurface through the invariants of its metric tensor in the grid coordinates and has a clear geometric interpretation of a measure of grid nonuniformity. In fact, the grid in this method is derived from a coordinate transformation that is inverse to the solution of a system of Beltrami equations which are the Euler–Lagrange equations of the functional.

The method of grid generation founded on the minimization of the functional of energy or the numerical solution of the corresponding inverted Beltrami equations allows code designers to merge the two tasks of surface grid generation and volume grid generation into one task when developing a comprehensive grid generation code. Since the grid generation is based on harmonic coordinate transformations that are able to generate unfolded grids in regions with complex geometry, the method can also relieve an array of bottlenecks of codes by reducing the number of blocks required for partitioning a complicated physical domain. The functions representing the monitor surface and control metrics are easily determined, thus providing efficient and straightforwardly controlled grid adaptation of various types. Thus, the method is free from the drawbacks of the elliptic method based on Poisson equations, and its numerical implementation should provide a uniform environment for the generation of fixed and adaptive grids in arbitrary regions. This gives grounds to expect that the method will be relevant to a large number of application areas.

This chapter describes some properties of the method and discusses its relation to other techniques, in particular, to the approaches using harmonic functions and Beltrami equations.

10.2 Hypersurface Geometry and Grid Formulation

For generality, we consider arbitrary n-dimensional hypersurfaces lying in the $(n+k)$-dimensional space R^{n+k}, $n \geq 1$, $k \geq 0$, though in practical applications the dimension n equals $1, 2$, and 3. Each hypersurface under consideration is supposed to be represented by a parametrization

$$\boldsymbol{x}(\boldsymbol{s}) : S^n \to R^{n+k}, \quad \boldsymbol{x} = (x^1(\boldsymbol{s}), \dots, x^{n+k}(\boldsymbol{s})), \ \boldsymbol{s} = (s^1, \dots, s^n), \tag{10.1}$$

where S^n is some n-dimensional parametric domain in R^n with Cartesian coordinates s^i, $i = 1, \dots, n$. The hypersurface represented by (10.1) is designated by S^{xn}. This section gives a natural multidimensional generalization of the notions and relations considered for n-dimensional domains, curves, and two-dimensional surfaces in Chaps. 2, 3, and 9.

10.2.1 Hypersurface Grid Formulation

In order to generate a numerical grid on an arbitrary hypersurface S^{xn}, represented by a parametrization $\boldsymbol{x}(\boldsymbol{s}) : S^n \to R^{n+k}$, both an n-dimensional computational domain $\Xi^n \subset R^n$ and an intermediate invertible, smooth transformation

$$\boldsymbol{s}(\boldsymbol{\xi}) : \Xi^n \to S^n, \quad \boldsymbol{s}(\boldsymbol{\xi}) = \left[s^1(\boldsymbol{\xi}), \dots, s^n(\boldsymbol{\xi})\right],$$
$$\boldsymbol{\xi} = (\xi^1, \dots, \xi^n), \ \boldsymbol{\xi} \in \Xi^n, \tag{10.2}$$

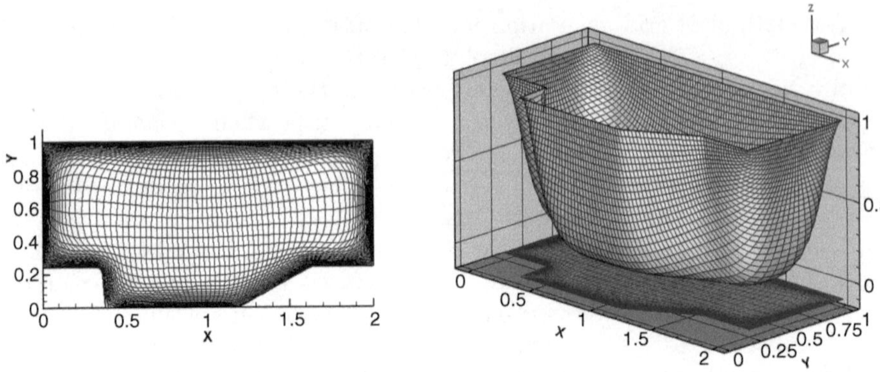

Fig. 10.1. Scheme of grid adaptation by the use of a monitor hypersurface

between Ξ^n and the parametrization domain S^n are determined. Then the numerical grid on the hypersurface S^{xn} is built by mapping some reference grid, specified in the computational domain Ξ^n, onto the hypersurface S^{xn} with the aid of the composition of the transformations $\boldsymbol{x}(\boldsymbol{s})$ and $\boldsymbol{s}(\boldsymbol{\xi})$, i.e. with

$$\boldsymbol{x}[\boldsymbol{s}(\boldsymbol{\xi})] : \Xi^n \to R^{n+k}.$$

This transformation defines a new coordinate system $\xi^1, \ldots, \xi^n$, hereafter referred to as the grid coordinate system, of the hypersurface S^{xn}.

The original parametrization $\boldsymbol{x}(\boldsymbol{s})$ also generates a grid on S^{xn} by mapping some reference grid in S^n. However, this grid may be unsatisfactory. The role of the intermediate transformation $\boldsymbol{s}(\boldsymbol{\xi})$ is to make the grid on S^{xn} satisfy the necessary properties.

The computational domain Ξ^n, along with the cells of its reference grid, may be rectangular or have another configuration, say, tetrahedral or prismatic, when $n = 3$, with the cells of its grid being similar or different in shape, in particular, the reference grid may be unstructured which results in an unstructured grid in S^{xn}.

10.2.2 Monitor Hypersurfaces

A hypersurface, for the purpose of adaptive grid generation, is commonly represented by a monitor hypersurface formed by values of some vector function over the physical domain or surface. This vector function can be a solution to the problem of interest, a combination of its components or derivatives, or any other variable vector quantity that suitably monitors the features of the physical solution or of the geometry of the physical domain or surface which significantly affect the accuracy of the calculations. The vector functions provide an efficient opportunity to control the grid quality, in particular, the concentration of grid nodes in the zones of large variations of a such function (see Fig. 10.1 for $n = 2$, $k = 1$). The parametrization of the monitor hypersurface is established very simply. For example, in one case, important for the generation of adaptive grids in a physical domain $X^n \subset R^n$, the monitor hypersurface designated further by S^{rn} is defined as an n-dimensional hypersurface formed by the values of some vector function

$$\boldsymbol{f}(\boldsymbol{x}) : X^n \to R^k, \quad \boldsymbol{x} = (x^1, \dots, x^n), \ \boldsymbol{f} = \big[f^1(\boldsymbol{x}), \dots, f^k(\boldsymbol{x})\big],$$

over X^n. This monitor hypersurface S^{rn} lies in the $(n+k)$-dimensional space R^{n+k} (see Fig. 10.1 for $n = 2$, $k = 1$). It is apparent that the parametric domain S^n can coincide with X^n and, consequently, the parametric mapping designated further by $\boldsymbol{r}(\boldsymbol{s}) : S^n \to R^{n+k}$ of S^{rn} can be defined as

$$\boldsymbol{r}(\boldsymbol{s}) = [\boldsymbol{s}, \boldsymbol{f}(\boldsymbol{s})] = \big[s^1, \dots, s^n, f^1(\boldsymbol{s}), \dots, f^k(\boldsymbol{s})\big], \quad \boldsymbol{s} = \boldsymbol{x}. \tag{10.3}$$

If the monitor hypersurface is formed by the values of the function $\boldsymbol{f}(\boldsymbol{x})$ over a two-dimensional surface S^{x2} lying in the space R^3 and represented by the parametrization $\boldsymbol{x}(\boldsymbol{s}) = [x^1(\boldsymbol{s}), x^2(\boldsymbol{s}), x^3(\boldsymbol{s})]$ from a two-dimensional domain S^2 in the space R^2, then the monitor hypersurface S^{r2} lies in the space R^{3+k} and it can be described by a parametrization from S^2 in the form

$$\boldsymbol{r}(\boldsymbol{s}) : S^2 \to R^{3+k}, \quad \boldsymbol{r}(\boldsymbol{s}) = \big\{\boldsymbol{x}(\boldsymbol{s}), \boldsymbol{f}[\boldsymbol{x}(\boldsymbol{s})]\big\}. \tag{10.4}$$

Analogously, a one-dimensional monitor surface S^{r1} (curve) over a curve S^{x1} lying in R^n, $n = 1, 2, 3$, and represented by

$$\boldsymbol{x}(\varphi) : [a, b] \to R^n,$$

can be defined by the parametrization

$$\boldsymbol{r}(s) : [a, b] \to R^{n+k}, \quad \boldsymbol{r}(s) = \big\{\boldsymbol{x}(s), \boldsymbol{f}[\boldsymbol{x}(s)]\big\}, \ s = \varphi. \tag{10.5}$$

10.2.3 Metric Tensors

As for ordinary two-dimensional surfaces lying in R^3, the interior features of the hypersurface S^{xn} are defined by the elements of the covariant metric tensor which in the coordinates v^i are denoted as g^{xv}_{ij} and computed by the dot products of the vectors $\boldsymbol{x}_{v^i} = \partial \boldsymbol{x}/\partial v^i$ that are tangent to the coordinate lines, i.e.

$$g^{xv}_{ij} = \boldsymbol{x}_{v^i} \cdot \boldsymbol{x}_{v^j}, \quad i, j = 1, \dots, n. \tag{10.6}$$

The determinant of (g^{xv}_{ij}) is denoted by g^{xv}.

The contravariant metric tensor of the hypersurface S^{xn} in the coordinates v^i is the inverse of (g^{xv}_{ij}) and is denoted by (g^{ij}_{vx}).

The covariant metric tensor $(g^{x\xi}_{ij})$ of the hypersurface S^{xn} in the grid coordinates $\xi^1, \dots, \xi^n$ has n invariants $I_1, \dots, I_n$, whose geometrical meaning is described through the geometrical measures of the edges, faces, etc. of the n-dimensional parallelepiped formed by the tangent vectors $\boldsymbol{x}_{\xi^i}$, $i = 1, \dots, n$. In analogy with Chap. 3, the invariants of the metric tensor of the hypersurface S^{xn} can be used to formulate some quality properties of the grid on S^{xn}.

Similar to (10.6) we shall designate by $g^{rv}_{ij}(g^{ij}_{vr})$ the covariant (contravariant) metric elements of the monitor hypersurface S^{rn} in the coordinates $v^1, \dots, v^n$, i.e.

$$g^{rv}_{ij} = \boldsymbol{r}_{v^i} \cdot \boldsymbol{r}_{v^j}, \quad i,j = 1, \dots, n. \tag{10.7}$$

Thus in the case of the parametrization (10.3) of the monitor hypersurface S^{rn} over the domain S^n, we obtain

$$g^{rs}_{ij} = \delta^i_j + \boldsymbol{f}_{s^i} \cdot \boldsymbol{f}_{s^j}, \quad i,j = 1, \dots, n. \tag{10.8}$$

If $\boldsymbol{f}$ is a scalar function f then the determinant g^{rs} of the tensor (g^{rs}_{ij}) is readily computed:

$$g^{rs} = 1 + \frac{\partial f}{\partial s^i}\frac{\partial f}{\partial s^i} = 1 + (\boldsymbol{\nabla} f)^2, \quad i = 1, \dots, n.$$

The parametrization (10.4) of the monitor surface S^{r2} determined by the values of $\boldsymbol{f}$ over the surface S^{x2} yields

$$g^{rs}_{ij} = g^{xs}_{ij} + \frac{\partial \boldsymbol{f}[\boldsymbol{x}(\boldsymbol{s})]}{\partial s^i} \cdot \frac{\partial \boldsymbol{f}[\boldsymbol{x}(\boldsymbol{s})]}{\partial s^j}, \quad i,j = 1,2, \tag{10.9}$$

where (g^{xs}_{ij}) is the covariant metric tensor of the surface S^{x2} in the coordinates s^i, $i = 1,2$. In the case of a scalar monitor function f, we obtain

$$\begin{aligned} g^{rs} &= g^{xs} + g^{xs}_{11}\left(\frac{\partial f}{\partial s^2}\right)^2 - 2g^{xs}_{12}\frac{\partial f}{\partial s^1}\frac{\partial f}{\partial s^2} + g^{xs}_{22}\left(\frac{\partial f}{\partial s^1}\right)^2 \\ &= g^{xs}\left(1 + g^{ij}_{sx}\frac{\partial f}{\partial s^i}\frac{\partial f}{\partial s^j}\right), \quad i,j = 1,2, \end{aligned} \tag{10.10}$$

where $g^{xs} = \det(g^{xs}_{ij})$, (g^{ij}_{sx}) is the inverse of (g^{xs}_{ij}), and by $\partial f/\partial s^i$ is meant $\partial f[\boldsymbol{x}(\boldsymbol{s})]/\partial s^i$ here.

10.2.4 Christoffel Symbols

The quantities

$$[ij, l] = \boldsymbol{x}_{s^i s^j} \cdot \boldsymbol{x}_{s^l}, \quad i,j,l = 1, \dots, n,$$

are the hypersurface Christoffel symbols of the first kind of the hypersurface S^{xn} in the parametric coordinates s^i. These symbols are, in fact, the space Christoffel symbols and, therefore, are subject to the relations (2.44).

In analogy with (9.40), the hypersurface Christoffel symbols of the second kind in the coordinates s^i are defined by the relation

$$\Upsilon^l_{ij} = g^{lm}_{sx}[ij, m], \quad i,j,l,m = 1, \dots, n. \tag{10.11}$$

From (2.44),

$$\Upsilon^j_{ji} = g^{jm}_{sx}[ji,m] = \frac{1}{2} g^{jm}_{sx} \left(\frac{\partial g^{xs}_{jm}}{\partial s^i} + \frac{\partial g^{xs}_{im}}{\partial s^j} - \frac{\partial g^{xs}_{ij}}{\partial s^m} \right)$$
$$= \frac{1}{2} g^{jm}_{sx} \frac{\partial g^{xs}_{jm}}{\partial s^i}, \quad i,j,m = 1,\dots,n,$$

and in accordance with the formula (2.46) for differentiation of the Jacobian we have an analog of the identity (2.45) in the form

$$\frac{\partial}{\partial s^i} \sqrt{g^{xs}} = \frac{1}{2} \sqrt{g^{xs}} g^{jm}_{sx} \frac{\partial g^{xs}_{jm}}{\partial s^i}$$
$$= \sqrt{g^{xs}} \Upsilon^j_{ji}, \quad i,j,m = 1,\dots,n. \tag{10.12}$$

Now we determine the role of the Christoffel symbols in the expansion of the derivatives of the tangent vectors $\boldsymbol{x}_{s^i}$. Let $\boldsymbol{d}_{ij}$ be the vector defined by the relation

$$\boldsymbol{d}_{ij} = \boldsymbol{x}_{s^i s^j} - \Upsilon^l_{ij} \boldsymbol{x}_{s^l}, \quad i,j,l = 1,\dots,n.$$

We have

$$\boldsymbol{d}_{ij} \cdot \boldsymbol{x}_{s^m} = [ij,m] - \Upsilon^l_{ij} g^{xs}_{lm} = 0, \quad i,j,l,m = 1,\dots,n,$$

from (10.11). Thus we find that in the following expansion of the vectors $\boldsymbol{x}_{s^i s^j}$,

$$\boldsymbol{x}_{s^i s^j} = \Upsilon^l_{ij} \boldsymbol{x}_{s^l} + \boldsymbol{d}_{ij}, \quad i,j,l = 1,\dots,n, \tag{10.13}$$

the vectors $\boldsymbol{d}_{ij}$ lie in the k-dimensional hypersurface which is orthogonal to the tangent n-dimensional hypersurface defined by the tangent vectors $\boldsymbol{x}_{s^i}$. Note that if some vectors $\boldsymbol{v}_1,\dots,\boldsymbol{v}_k$ from R^{n+k} comprise an orthonormal basis for this k-dimensional hypersurface, i.e.

$$\boldsymbol{v}_m \cdot \boldsymbol{x}_{s^j} = 0, \quad m = 1,\dots,k, \ j = 1,\dots,n,$$
$$\boldsymbol{v}_m \cdot \boldsymbol{v}_p = \delta^m_p, \quad m,p = 1,\dots,k,$$

then, in accordance with (2.5), we find that

$$\boldsymbol{x}_{s^i s^j} = \Upsilon^l_{ij} \boldsymbol{x}_{s^l} + (\boldsymbol{x}_{s^i s^j} \cdot \boldsymbol{v}_m) \boldsymbol{v}_m, \quad l,i,j = 1,\dots,n, \ m = 1,\dots,k.$$

Thus we obtain from (10.13):

$$\boldsymbol{d}_{ij} = (\boldsymbol{x}_{s^i s^j} \cdot \boldsymbol{v}_m) \boldsymbol{v}_m, \quad m = 1,\dots,k, \ i,j = 1,\dots,n,$$

and so (10.13) is a generalization of (2.35) and (9.42) with the identification $\xi^i = s^i$.

10.2.5 Relations Between Metric Elements

The mapping $\boldsymbol{x}[\boldsymbol{s}(\boldsymbol{\xi})]$ which generates a grid on the hypersurface S^{xn} determines a new coordinate system ξ^i, $i = 1, \dots, n$, on S^{xn} and it also defines the values of the covariant metric tensor $(g_{ij}^{x\xi})$ in the coordinates ξ^i, whose elements are the scalar products of the vectors $\boldsymbol{x}_i = \partial \boldsymbol{x}[\boldsymbol{s}(\boldsymbol{\xi})]/\partial \xi^i$, $i = 1, \dots, n$, i.e.

$$g_{ij}^{x\xi} = \boldsymbol{x}_i \cdot \boldsymbol{x}_j, \quad i, j = 1, \dots, n.$$

The elements of the covariant tensor of S^{xn} in the coordinates s^i and ξ^i are connected by the following relations:

$$\begin{aligned} g_{ij}^{x\xi} &= g_{ml}^{xs} \frac{\partial s^m}{\partial \xi^i} \frac{\partial s^l}{\partial \xi^j}, \\ g_{ij}^{xs} &= g_{ml}^{x\xi} \frac{\partial \xi^m}{\partial s^i} \frac{\partial \xi^l}{\partial s^j}, \quad i, j, l, m = 1, \dots, n, \end{aligned} \tag{10.14}$$

where $\partial \xi^m / \partial s^i$, $i, m = 1, \dots, n$, is the first derivative with respect to s^i of the mth component $\xi^m(\boldsymbol{s})$ of the mapping

$$\boldsymbol{\xi}(\boldsymbol{s}) : S^n \to \Xi^n, \quad \boldsymbol{\xi}(\boldsymbol{s}) = [\xi^1(\boldsymbol{s}), \dots, \xi^n(\boldsymbol{s})], \ \boldsymbol{s} = (s^1, \dots, s^n), \ \boldsymbol{s} \in S^n,$$

which is inverse to the intermediate mapping

$$\boldsymbol{s}(\boldsymbol{\xi}) : \Xi^n \to S^n, \quad \boldsymbol{s}(\boldsymbol{\xi}) = [s^1(\boldsymbol{\xi}), \dots, s^n(\boldsymbol{\xi})].$$

The latter mapping serves to generate a suitable grid on the hypersurface S^{xn}.

The contravariant metric tensor $(g_{\xi x}^{ij})$ of the hypersurface S^{xn} in the curvilinear coordinates ξ^i, is the inverse to the covariant tensor $(g_{ij}^{x\xi})$, that is,

$$g_{ij}^{x\xi} g_{\xi x}^{jl} = \delta_l^i, \quad i, j, l = 1, \dots, n.$$

In analogy with the relations (9.23) for two-dimensional surfaces, the elements of the contravariant tensor of S^{xn} in the coordinates s^i and ξ^i are connected as follows:

$$\begin{aligned} g_{\xi x}^{ij} &= g_{sx}^{ml} \frac{\partial \xi^i}{\partial s^m} \frac{\partial \xi^j}{\partial s^l}, \\ g_{sx}^{ij} &= g_{\xi x}^{ml} \frac{\partial s^i}{\partial \xi^m} \frac{\partial s^j}{\partial \xi^l}, \quad i, j, l, m = 1, \dots, n. \end{aligned} \tag{10.15}$$

Similarly to (9.24), we also have the relations

$$\begin{aligned} g_{sx}^{ii} &= g_{\xi x}^{mk} g_{mk}^{s\xi}, \\ g^{x\xi} &= J^2 g^{xs}, \quad i, k, m = 1, \dots, n, \end{aligned}$$

where

$$g^{x\xi} = \det(g_{ij}^{x\xi}), \qquad g_{ij}^{s\xi} = \frac{\partial \boldsymbol{s}}{\partial \xi^i} \cdot \frac{\partial \boldsymbol{s}}{\partial \xi^j}, \qquad J = \det\left(\frac{\partial s^i}{\partial \xi^j}\right), \quad i, j = 1, \dots, n.$$

10.3 Functional of Smoothness

One of the ways of finding the intermediate coordinate transformations $\boldsymbol{s}(\boldsymbol{\xi})$ between the computational and parametric domains required to generate grids on hypersurfaces is to use variational methods. In accordance with the grid methods for domains and two-dimensional surfaces considered above, the most appropriate functional for this purpose is the smoothness one, since it generates a system of Beltrami equations possessing the unique properties desired for grid generation.

10.3.1 Formulation of the Functional

Similarly to (9.31), the expression for the functional of grid smoothness on the hypersurface S^{xn} with the parametrization (10.1) is represented as

$$I_{\mathrm{s}} = \int_{S^{xn}} \left(\sum_{i=1}^{n} g^{ii}_{\xi x} \right) \mathrm{d}S^{xn} = \int_{S^{xn}} (\mathrm{tr}\, g^{ij}_{\xi x}) \mathrm{d}S^{xn}, \tag{10.16}$$

defined on the set of invertible functions $\boldsymbol{\xi}(\boldsymbol{s}) \in C^2(S^n)$. Since

$$\mathrm{d}S^{xn} = \sqrt{g^{xs}} \mathrm{d}\boldsymbol{s} = \sqrt{g^{x\xi}} \mathrm{d}\boldsymbol{\xi}$$

we obtain from (10.16)

$$I_{\mathrm{s}} = \int_{S^n} \sqrt{g^{xs}} (\mathrm{tr}\, g^{ij}_{\xi x}) \mathrm{d}\boldsymbol{s} = \int_{\Xi^n} \sqrt{g^{x\xi}} (\mathrm{tr}\, g^{ij}_{\xi x}) \mathrm{d}\boldsymbol{\xi}.$$

Thus for $n = 1, 2$, and 3 we have

$$I_{\mathrm{s}} = \begin{cases} \int_{S^1} \sqrt{g^{xs}} g^{11}_{\xi x} \mathrm{d}s, & n = 1, \\ \int_{S^2} \sqrt{g^{xs}} (g^{11}_{\xi x} + g^{22}_{\xi x}) \mathrm{d}s^1 \mathrm{d}s^2, & n = 2, \\ \int_{S^3} \sqrt{g^{xs}} (g^{11}_{\xi x} + g^{22}_{\xi x} + g^{33}_{\xi x}) \mathrm{d}s^1 \mathrm{d}s^2 \mathrm{d}s^3, & n = 3, \end{cases}$$

with the corresponding contravariant metric elements $g^{ij}_{\xi x}$ and determinants g^{xs} for each $n = 1, 2, 3$. Analogously, using suitable formulas for the elements of inverse matrices, we obtain the formulation of the inverted smoothness functional in terms of the covariant metric elements $g^{x\xi}_{ij}$:

$$I_{IS} = \begin{cases} \int_{\Xi^1} \frac{1}{\sqrt{g^{x\xi}}} \mathrm{d}\xi, & n = 1, \\ \int_{\Xi^2} \frac{1}{\sqrt{g^{x\xi}}} (g^{x\xi}_{11} + g^{x\xi}_{22}) \mathrm{d}\xi^1 \mathrm{d}\xi^2, & n = 2, \\ \int_{\Xi^3} \frac{1}{\sqrt{g^{x\xi}}} [g^{x\xi}_{11} g^{x\xi}_{22} + g^{x\xi}_{11} g^{x\xi}_{33} + g^{x\xi}_{22} g^{x\xi}_{33} & \\ \quad -(g^{x\xi}_{12})^2 - (g^{x\xi}_{13})^2 - (g^{x\xi}_{23})^2] \mathrm{d}\xi^1 \mathrm{d}\xi^2 \mathrm{d}\xi^3, & n = 3. \end{cases}$$

Note that the functional I_{IS}, in this formulation, is defined on the set of invertible transformations $\boldsymbol{s}(\boldsymbol{\xi}) \in C^2(\Xi^n)$.

When the hypersurface S^{xn} is a three-dimensional region X^3, the functional (10.16) is the very functional of grid smoothness on X^3,

$$I_s = \int_{X^3} \left(\sum_{i=1}^{3} g^{ii} \right) \mathrm{d}\boldsymbol{x}, \qquad g^{ij} = \sum_{m=1}^{3} \frac{\partial \xi^i}{\partial x^m} \frac{\partial \xi^j}{\partial x^m}, \tag{10.17}$$

described in Chap. 8. Therefore the functional (10.16), being the generalization of (10.17), is called the functional of grid smoothness on the hypersurface S^{xn}. Further, it will be shown that such a generalization of the functional (10.17) to n-dimensional hypersurfaces preserves all salient features of grids obtained by applying the smoothness functional on domains.

10.3.2 Geometric Interpretation

This subsection describes a geometric meaning of the smoothness functional which justifies to some extent its expression (10.16) for the generation of quasiuniform grids on hypersurfaces, in particular, on monitor hypersurfaces S^{rn} specified by the parameterizations (10.3) and (10.4) and, consequently, adaptive grids in domains and on surfaces. The explanation of the geometric interpretation of the functional follows in general the considerations presented in Sect. 3.7 for domain grid generation and in Sect. 9.5.3 for two-dimensional surface grid generation.

First, note that the trace of the contravariant n-dimensional tensor $(g^{ij}_{\xi x})$ can be expressed through the invariants I_{n-1} and I_n of the orthogonal transforms of the covariant tensor $(g^{x\xi}_{ij})$, namely

$$\mathrm{tr}(g^{ij}_{\xi x}) = \frac{I_{n-1}}{I_n}.$$

Therefore the functional of smoothness (10.16) can also be expressed through these invariants:

$$I_s = \int_{S^{xn}} \left(\frac{I_{n-1}}{I_n} \right) \mathrm{d}S^{xn}. \tag{10.18}$$

Now, for the purpose of simplicity, we restrict our consideration to three dimensions. The functional (10.18) then has the form

$$I_s = \int_{S^{x3}} \left(\frac{I_2}{I_3} \right) \mathrm{d}S^{r3}.$$

In three dimensions the invariant I_3 is the Jacobian of the matrix $(g^{x\xi}_{ij})$ and it represents the volume V^3 of the three-dimensional parallelepiped P^3 formed by the basic tangent vectors $\boldsymbol{x}_i$, $i = 1, 2, 3$. The invariant I_2 of the matrix $(g^{x\xi}_{ij})$ is the sum of its principal minors of order 2. Every principal minor of order 2 equals the Jacobian of the two-dimensional matrix A^2 obtained from $(g_{ij}^{x\xi})$ by crossing out a row and a column which intersect on the diagonal. Therefore each element of the matrix A^2 is a dot product of two tangential vectors of the basis $\boldsymbol{x}_i$, $i = 1, 2, 3$, and, consequently, the Jacobian of A^2 equals the square of the area of the parallelogram formed by these two vectors. So the invariants I_2, I_3 can be expressed as

$$I_2 = \sum_{m=1}^{3} (V_m^2)^2, \qquad I_3 = (V^3)^2,$$

where V_m^2 is the area of the boundary segment of the parallelepiped P^3 formed by the vectors $\boldsymbol{x}_i$, $i = 1, 2, 3$, except for $\boldsymbol{x}_m$, and V^3 is the volume of P^3. Therefore

$$\frac{I_2}{I_3} = \sum_{m=1}^{3} (V_m^2)^2/(V^3)^2. \tag{10.19}$$

It is obvious that

$$V^3 = d_m V_m^2, \quad m = 1, 2, 3,$$

where d_m is the distance between the vertex of the vector $\boldsymbol{x}_m$ and the plane spanned by the vectors $\boldsymbol{x}_i$, $i \neq m$. Hence, from (10.19),

$$\frac{I_2}{I_3} = \sum_{m=1}^{3} (1/d^m)^2. \tag{10.20}$$

Now let us consider two grid surfaces $\xi^m = c$ and $\xi^m = c + h$ obtained by mapping a uniform rectangular grid with a step size h in the computational domain Ξ^3 onto the hypersurface S^{x3}. The distance l_m between a node on the coordinate surface $\xi^m = c$ and the nearest node on the surface $\xi^m = c+h$ equals $d_m h + O(h)^2$. Therefore (10.20) is equivalent to

$$\frac{I_2}{I_3} = \sum_{m=1}^{3} (h/l_m)^2 + O(h).$$

The quantity $(h/l_m)^2$ increases as the grid nodes cluster in the direction normal to the surface $\xi^m = c$, and therefore it can be considered as some measure of the grid concentration in this direction; consequently, the functional (10.18) for $n = 3$ defines an integral measure of the grid clustering in all directions. Therefore, as in the case of two-dimensional surfaces considered in Chap. 9, the problem of minimizing the functional of smoothness (10.16) for $n = 3$ can be interpreted as a problem of finding a grid with a minimum of nonuniform clustering, namely a quasiuniform grid on the surface S^{x3}. Analogous interpretations are valid for arbitrary dimensions. The interpretation of the smoothness functional considered above justifies, to some extent, its potential to generate adaptive grids in a domain or surface by projecting onto the domain or surface quasiuniform grids built on monitor hypersurfaces (see Fig. 10.1) by the minimization of the functional.

10.3.3 Dimensionless Functionals

This section formulates some dimensionally homogeneous functionals through the invariants of the metric tensor $(g_{ij}^{x\xi})$ which, in analogy with the smoothness functional, measure global nonuniformity of hypersurface grids.

As was demonstrated in Sect. 10.3.2, the quantity I_{n-1}/I_n represents a measure of the local clustering of a hypersurface grid. Integration of this measure over the hypersurface S^{xn} derives the smoothness functional whose minimization tends to yield a uniform grid on S^{xn}. The smoothness functional possesses the spectacular properties reviewed in Sects. 10.3.2 and 10.3.3. In the particular case when $n = 2$, this functional is also dimensionless. However, it is not dimensionless in three-dimensions, which may by important for the generation of spatial adaptive grids. Nevertheless, using the invariants of the covariant metric tensor $(g_{ij}^{x\xi})$, we can formulate dimensionless functionals measuring grid nonuniformity for arbitrary $n > 1$.

For this purpose we, in analogy with (8.33), consider a dimensionless measure of the local departure of a hypersurfaces grid, in particular of a monitor surface grid, from a conformal one:

$$Q_{\mathrm{cf},3} = \left[I_{n-1}/(I_n)^{1-1/n}\right]^\alpha, \quad \alpha > 0,\ n > 1.$$

As was mentioned in Sect. 8.3.1, the dimensionless functionals are formulated by integrating local dimensionless measures over the unit cube Ξ^n. Thus we obtain one functional, using the quantity $Q_{\mathrm{cf},3}$,

$$\begin{aligned} I_{\mathrm{cf},3} &= \int_{\Xi^n} \left[I_{n-1}/(I_n)^{1-1/n}\right]^\alpha \mathrm{d}\boldsymbol{\xi} \\ &= \int_{S^{xn}} \frac{1}{\sqrt{I_n}} \left[I_{n-1}/(I_n)^{1-1/n}\right]^\alpha \mathrm{d}S^{xn}. \end{aligned} \tag{10.21}$$

Functional (10.21) can be interpreted as an integral measure of the departure from conformality. When $(1 - 1/n)\alpha = 1/2$ this functional is independent of the parametrization of S^{xn}.

Another dimensionless functional measuring the departure from conformality is defined, in analogy with (8.31), through the first and nth invariants:

$$\begin{aligned} I_{\mathrm{cf},2} &= \int_{\Xi^n} \left[I_1/(I_n)^{1/n}\right]^\alpha \mathrm{d}\boldsymbol{\xi} \\ &= \int_{S^{xn}} \frac{1}{\sqrt{I_n}} \left[I_1/(I_n)^{1/n}\right]^\alpha \mathrm{d}S^{xn}, \quad \alpha > 0,\ n > 1. \end{aligned} \tag{10.22}$$

This functional is independent on parameterizations of S^{xn} if $\alpha = n/2$. The parameter α in (10.21) and (10.22) can be used to control the form of the functionals. In particular, assuming $\alpha = n/2$ in (10.21), we obtain a dimensionless functional which is defined through the measure of the local grid clustering $I_{n-1}/I_n = g_{\xi x}^{ii}$, $i = 1, \ldots, n$:

$$I_{\mathrm{cf},3} = \int_{S^{xn}} (I_{n-1}/I_n)^{n/2} \mathrm{d}S^{xn}. \tag{10.23}$$

Note that functional (10.23) coincides with the smoothness functional for $n = 2$. For $n = 3$ we obtain, using (10.15),

$$\begin{aligned} I_{\mathrm{cf},3} &= \int_{S^{r3}} \left(\frac{I_2}{I_3}\right)^{3/2} \mathrm{d}S^{r3} \\ &= \int_{S^3} \sqrt{g^{xs}} \left(g_{sx}^{kl} \frac{\partial \xi^i}{\partial s^k} \frac{\partial \xi^i}{\partial s^l}\right)^{3/2} \mathrm{d}\boldsymbol{s}, \quad i,j,k,l = 1,2,3. \end{aligned} \tag{10.24}$$

Thus the Euler–Lagrange equations for this functional have the form

$$\frac{\partial}{\partial s^j}\left(\sqrt{g^{xs}} \sqrt{\frac{I_2}{I_3}} g_{sx}^{kj} \frac{\partial \xi^i}{\partial s^k}\right) = 0, \quad i,j,k = 1,2,3, \tag{10.25}$$

while the corresponding inverted equations are represented as

$$g_{\xi x}^{ij} \frac{\partial^2 s^l}{\partial \xi^i \partial \xi^j} = \frac{1}{\sqrt{g^{xs}}} \sqrt{\frac{I_3}{I_2}} \frac{\partial}{\partial s^k}\left(\sqrt{g^{xs}} \sqrt{\frac{I_2}{I_3}} g_{sx}^{lk}\right), \quad i,j,k,l = 1,2,3, \tag{10.26}$$

which are a generalization of (8.39) with $n = 3$ for three-dimensional hypersurfaces. In the same manner there can be written Euler–Lagrange equations for the functional $I_{\mathrm{cf},3}$ for arbitrary $n > 1$.

Analogously, by assuming $\alpha = n/[2(n-1)]$ we find a simpler form of the dimensionless functional (10.22):

$$I_{\mathrm{cf},2} = \int_{S^{xn}} (I_1/I_n)^{n/[2(n-1)]} \mathrm{d}S^{xn}. \tag{10.27}$$

This functional coincides with the smoothness functional for $n = 2$, while for $n = 3$ it has the form

$$I_{\mathrm{cf},2} = \int_{S^{r3}} (I_1/I_3)^{3/4} \mathrm{d}S^{r3}. \tag{10.28}$$

Substituting in formulas (10.21)–(10.28) S^{xn} for S^{rn} yields the corresponding dimensionless functionals for generating adaptive grids in S^{xn}.

10.3.4 Euler–Lagrange Equations

The substitution of the parametric domain S^n for the integration hypersurface S^{xn} in (10.16) yields the smoothness functional in the following equivalent form with integration over S^n:

$$I_{\mathrm{s}} = \int_{S^n} \sqrt{g^{xs}}\, g_{sx}^{ml} \frac{\partial \xi^i}{\partial s^m} \frac{\partial \xi^i}{\partial s^l} \mathrm{d}\boldsymbol{s}, \quad i,m,l = 1,\dots,n. \tag{10.29}$$

The quantities g^{xs} and g^{ml}_{sx} in (10.29) are defined through the specified parametrization $\boldsymbol{x}(\boldsymbol{s}) : S^n \to R^n$, and therefore they remain unchanged when the functions $\xi^i(\boldsymbol{s})$ are varied. So the Euler–Lagrange equations derived from the functional of smoothness are readily obtained and, in accordance with (8.6), have the following form:

$$\frac{\partial}{\partial s^m}\left(\sqrt{g^{xs}}\, g^{ml}_{sx}\frac{\partial \xi^i}{\partial s^l}\right) = 0, \quad i, m, l = 1, \dots, n. \tag{10.30}$$

If, in particular, S^{xn} is an n-dimensional domain S^n, then the system (10.30) is equivalent to the system of Laplace equations

$$\nabla^2 \xi^i \equiv \frac{\partial}{\partial s^j}\left(\frac{\partial \xi^i}{\partial s^j}\right) = 0, \quad i, j, = 1, \dots, n,$$

introduced by Crowley (1962) and Winslow (1967) for the generation of fixed grids in domains. Therefore the method for generating grids on hypersurfaces S^{xn} by solving a boundary value problem for (10.30) derived from the functional of smoothness can also be considered as an extension of the Crowley–Winslow approach.

We considered in Chap. 6 the technique for generating adaptive grids in S^n which is based on the numerical solution of the Poisson system

$$\frac{\partial}{\partial s^j}\left(\frac{\partial \xi^i}{\partial s^j}\right) = P^i, \quad i, j, = 1, \dots, n, \tag{10.31}$$

where the P^i are the control functions. In the case of a monitor hypersurface S^{rn} represented by (10.3) or (10.4) the system (10.30) with the identification $g^{xs} = g^{rs}$ and $g^{ml}_{sx} = g^{ml}_{sr}$ can also be interpreted as a system of elliptic equations with a control function. The control function is the monitor mapping $\boldsymbol{f}(\boldsymbol{s})$ whose values over the physical domain or surface form the monitor hypersurface S^{rn}. The influence of the control function $\boldsymbol{f}(\boldsymbol{s})$ is realized through the magnitudes $\boldsymbol{f}_{s^i} \cdot \boldsymbol{f}_{s^j}$ in the terms g^{rs} and g^{ml}_{sr}. These terms are determined by the tensor elements g^{rs}_{ij} in the form (10.8) or (10.9) which defines the covariant metric tensor of the hypersurface S^{rn} in the coordinates s^i represented by the parametrization (10.3) or (10.4). The system (10.30), in contrast to that of (10.31), has a divergent form and its solution is a harmonic function, as was mentioned above.

Notice that the Dirichlet problem for both the equations (10.30) with respect to the metric of a monitor hypersurface S^{rn} and (10.31) is well posed. Let us consider a twice differentiable function $\boldsymbol{\xi}_0(\boldsymbol{s}) : S^n \to \Xi^n$, $\boldsymbol{\xi}_0(\boldsymbol{s}) = (\xi^1_0(s), \dots, \xi^1_0(s))$ that is a one-to-one transformation at the boundary points but it transforms two different interior points of S^n into one point in Ξ^n, that is $\boldsymbol{\xi}_0(\boldsymbol{s})$ is not invertible. Then assuming in (10.31)

$$P^i = \frac{\partial}{\partial s^j}\left(\frac{\partial \xi^i_0}{\partial s^j}\right), \quad i, j, = 1, \dots, n,$$

we obtain that $\boldsymbol{\xi}_0(\boldsymbol{s})$ is the solution of (10.31), however, the inverted Poisson system with such P^i does not have a solution.

10.3.5 Equivalent Forms

In analogy with two-dimensional space, the left-hand part of (10.30) multiplied by $1/\sqrt{g^{xs}}$ is the value of the Beltrami operator

$$\Delta_B = \frac{1}{\sqrt{g^{xs}}}\frac{\partial}{\partial s^m}\left(\sqrt{g^{xs}}\, g^{ml}_{sx}\frac{\partial}{\partial s^l}\right), \quad l, m = 1, \ldots, n, \tag{10.32}$$

applied to the function $\xi^i(x)$. Thus we find that the system of Euler–Lagrange equations (10.30) for the generation of quasiuniform grids on hypersurfaces is equivalent to

$$\Delta_B[\xi^i] \equiv \frac{1}{\sqrt{g^{x\xi}}}\frac{\partial}{\partial \xi^j}(\sqrt{g^{x\xi}} g^{jl}_{\xi x}) = 0 \quad i, j, l = 1, \ldots, n. \tag{10.33}$$

Now we obtain other forms of (10.33). For this purpose we first compute the value of $\Delta_B[\boldsymbol{x}]$, where the operator Δ_B is defined by (10.32). Expanding the differentiation in $\Delta_B[\boldsymbol{x}]$, we have

$$\Delta_B[\boldsymbol{x}] = \frac{1}{\sqrt{g^{x\xi}}}\frac{\partial}{\partial \xi^j}(\sqrt{g^{x\xi}}\, g^{jl}_{\xi x}\boldsymbol{x}_{\xi^l}) = g^{jl}_{\xi x}\boldsymbol{x}_{\xi^j\xi^l} + \Delta_B[\xi^l]\boldsymbol{x}_{\xi^l},$$
$$j, l = 1, \ldots, n. \tag{10.34}$$

Using the expansion (10.13) with the assumption $\xi^i = s^i$, $i = 1, \ldots, n$, we obtain

$$\boldsymbol{x}_{\xi^i\xi^j} = \Upsilon^l_{ij}\boldsymbol{x}_{\xi^l} + \boldsymbol{d}_{ij}, \quad i, j, l = 1, \ldots, n,$$

where

$$\Upsilon^l_{ij} = g^{lm}_{\xi x}[ij, m], \quad [ij, m] = \boldsymbol{x}_{\xi^i\xi^j} \cdot \boldsymbol{x}_{\xi^m}, \; i, j, l, m = 1, \ldots, n, \tag{10.35}$$

and the $\boldsymbol{d}_{ij}$ are the vectors orthogonal to the tangent n-dimensional hyperplane defined by the vectors $\boldsymbol{x}_{\xi^i}$, $i = 1, \ldots, n$. From the above expansion of $\boldsymbol{x}_{\xi^i\xi^j}$, we have

$$g^{jl}_{\xi x}\boldsymbol{x}_{\xi^j\xi^l} = g^{jl}_{\xi x}\Upsilon^m_{jl}\boldsymbol{x}_{\xi^m} + g^{jl}_{\xi x}\boldsymbol{d}_{jl}, \quad j, l, m = 1, \ldots, n.$$

Substitution of these identities in (10.34) yields

$$\Delta_B[\boldsymbol{x}] = (g^{jl}_{\xi x}\Upsilon^m_{jl} + \Delta_B[\xi^m])\boldsymbol{x}_{\xi^m} + g^{jl}_{\xi x}\boldsymbol{d}_{jl}, \quad j, l, m = 1, \ldots, n. \tag{10.36}$$

Now we show that the vector $\Delta_B[\boldsymbol{x}]$, as well as the vectors $\boldsymbol{d}_{ij}$, lies in the k-dimensional hyperplane which is orthogonal to the tangent hyperplane, i.e.

$$\Delta_B[\boldsymbol{x}] \cdot \boldsymbol{x}_{\xi^i} = 0 \quad \text{for all } i = 1, \ldots, n.$$

Indeed, we have

$$\begin{aligned}\Delta_B[\boldsymbol{x}]\cdot\boldsymbol{x}_{\xi^i} &= \frac{1}{\sqrt{g^{x\xi}}}\left(\frac{\partial}{\partial\xi^j}\sqrt{g^{x\xi}}g^{jm}_{\xi x}\boldsymbol{x}_{\xi^m}\right)\cdot\boldsymbol{x}_{\xi^i}\\ &= \frac{1}{\sqrt{g^{x\xi}}}\frac{\partial}{\partial\xi^j}(\sqrt{g^{x\xi}}g^{jm}_{\xi x}\boldsymbol{x}_{\xi^m}\cdot\boldsymbol{x}_{\xi^i}) - g^{jm}_{\xi x}\boldsymbol{x}_{\xi^m}\cdot\boldsymbol{x}_{\xi^i\xi^j}\\ &= \frac{1}{\sqrt{g^{x\xi}}}\frac{\partial}{\partial\xi^i}\sqrt{g^{x\xi}} - \Upsilon^j_{ji}, \quad i,j,m=1,\dots,n. \end{aligned}\tag{10.37}$$

Now, using the identity (10.12) (valid for arbitrary parametrization) in the coordinates ξ^i, we obtain

$$\Delta_B[\boldsymbol{x}]\cdot\boldsymbol{x}_{\xi^i} = 0, \quad i=1,\dots,n,$$

from (10.37). Therefore the coefficients before $\boldsymbol{x}_{\xi^m}$ in (10.36) are equal to zero, i.e. we have the identity

$$\Delta_B[\xi^m] = -g^{jl}_{\xi x}\Upsilon^m_{jl}, \quad j,l,m=1,\dots,n. \tag{10.38}$$

Thus (10.36) becomes

$$\Delta_B[\boldsymbol{x}] = g^{jl}_{\xi x}\boldsymbol{d}^{jl}, \quad j,l=1,\dots,n. \tag{10.39}$$

Note that (10.39) is an extension of (6.27) and (9.49) to general hypersurfaces.

From (10.35), the identity (10.38) also has the form

$$\begin{aligned}\Delta_B[\xi^m] &= -g^{jl}_{\xi x}\, g^{mi}_{\xi x}[jl,i] = -g^{jl}_{\xi x}\, g^{mi}_{\xi x}(\boldsymbol{x}_{\xi^j\xi^l}\cdot\boldsymbol{x}_{\xi^i}),\\ & i,j,l,m=1,\dots,n. \end{aligned}\tag{10.40}$$

Analogously, we have, assuming $s^i=\xi^i$, $i=1,\dots,n$, in (10.40),

$$\Delta_B[s^m] = -g^{jl}_{sx}\, g^{mi}_{sx}(\boldsymbol{x}_{s^js^l}\cdot\boldsymbol{x}_{s^i}), \quad i,j,l,m=1,\dots,n. \tag{10.41}$$

The identities (10.38) and (10.40) represent an extension of the identities (9.52) valid for two-dimensional surfaces.

Since the Beltrami system (10.33) is equivalent to (10.30), we obtain from (10.40) one more system of equations:

$$g^{jl}_{\xi x}\, g^{mi}_{\xi x}[jl,i] = 0, \quad i,j,l,m=1,\dots,n,$$

which is equivalent to the Euler–Lagrange equations (10.30). Multiplication of this system by $g^{x\xi}_{mp}$ and summation over m yields one more equivalent system

$$g^{jl}_{\xi x}[jl,p] = 0, \quad j,l,p=1,\dots,n,$$

namely

$$g^{jl}_{\xi x}\frac{\partial^2 \boldsymbol{x}[\boldsymbol{s}(\boldsymbol{\xi})]}{\partial\xi^j\partial\xi^l}\cdot\frac{\partial\boldsymbol{x}[\boldsymbol{s}(\boldsymbol{\xi})]}{\partial\xi^p}=0,\quad j,l,p=1,\dots,n. \tag{10.42}$$

In particular, if S^{rn} is a monitor surface S^{xn} over a domain S^n, represented by (10.3), then the system (10.42) is

$$g^{jl}_{\xi r}\left\{\frac{\partial^2\boldsymbol{s}}{\partial\xi^j\partial\xi^l}\cdot\frac{\partial\boldsymbol{s}}{\partial\xi^p}+\frac{\partial^2\boldsymbol{f}[\boldsymbol{s}(\boldsymbol{\xi})]}{\partial\xi^j\partial\xi^l}\cdot\frac{\partial\boldsymbol{f}[\boldsymbol{s}(\boldsymbol{\xi})]}{\partial\xi^p}\right\}=0,\quad j,l,p=1,\dots,n. \tag{10.43}$$

This system of equations can be used to generate adaptive grids in S^{xn} and nearly orthogonal grids in the vicinity of boundary segments of a domain X^n by specifying the monitor hypersurface with a suitable choice of a monitor function $\boldsymbol{f}(\boldsymbol{x})$.

10.4 Hypersurface Grid Systems

The numerical grid on the hypersurface S^{xn} is built by mapping a reference grid in the computational domain Ξ^n with the coordinate transformation $\boldsymbol{x}[\boldsymbol{s}(\boldsymbol{\xi})] : \Xi^n \to S^{xn}$. The method based on the numerical solution of the inverted Beltrami equations in the metric of S^{xn} produces quasiuniform grids on S^{xn}. When the method is applied to a monitor surface S^{rn} over S^{xn} formed by the values of some control function $\boldsymbol{f}(\boldsymbol{s})$ the grid generated is in fact an adaptive grid on S^{xn}. Thus, in order to determine the nodes of the quasiuniform grid on the monitor hypersurface S^{rn} and the nodes of the adaptive grid on S^{xn}, it is sufficient to know the values of the function $\boldsymbol{s}(\boldsymbol{\xi}) : \Xi^n \to S^n$ at the points of the reference grid in the computational domain Ξ^n.

10.4.1 Inverted Beltrami Equations

The equations which the components $s^i(\boldsymbol{\xi})$ of the function $\boldsymbol{s}(\boldsymbol{\xi})$ satisfy are derived by inverting the system (10.30). The inverse system of (10.30) is obtained first by multiplying each ith component of this system by the derivative $\partial s^m/\partial\xi^i$ and then by summing the result over i. This operation produces the following system:

$$\begin{aligned}&\frac{\partial}{\partial s^p}\left(\sqrt{g^{xs}}\,g^{pl}_{sx}\frac{\partial\xi^i}{\partial s^l}\right)\frac{\partial s^m}{\partial\xi^i}\\&=\frac{\partial}{\partial s^p}\left(\sqrt{g^{xs}}\,g^{pl}_{sx}\frac{\partial\xi^i}{\partial s^l}\frac{\partial s^m}{\partial\xi^i}\right)-\sqrt{g^{xs}}\,g^{pl}_{sx}\frac{\partial\xi^i}{\partial s^l}\frac{\partial\xi^j}{\partial s^p}\frac{\partial^2 s^m}{\partial\xi^i\partial\xi^j}=0,\\&i,j,l,m,p=1,\dots,n.\end{aligned}$$

By multiplying this system of equations by $1/\sqrt{g^{xs}}$ and taking into account (10.15) and the relation

$$\frac{\partial\xi^i}{\partial s^l}\frac{\partial s^m}{\partial\xi^i}=\delta^l_m,\quad i,l,m=1,\dots,n,$$

the system of the inverse equations with s^i as dependent and ξ^i as independent variables has the form

$$g_{\xi x}^{ip}\frac{\partial^2 s^m}{\partial \xi^i \partial \xi^p} = \frac{1}{\sqrt{g^{xs}}}\frac{\partial}{\partial s^l}\left(\sqrt{g^{xs}}\, g_{sx}^{lm}\right), \quad i,l,m,p = 1,\dots,n. \tag{10.44}$$

This is a system of quasilinear equations and, which is important for the creation of iterative numerical algorithms, its right-hand part is defined only by the tensor elements g_{ij}^{xs} of the hypersurface S^{xn} and, therefore, remains unchanged when the function $\boldsymbol{s}(\boldsymbol{\xi})$ is varied. Moreover, each ith equation of the right-hand part of the system (10.44) is the value of the Beltrami operator applied to the function s^i. Thus the system (10.44) can be written in the following equivalent form:

$$g_{\xi x}^{ij}\frac{\partial^2 s^l}{\partial \xi^i \partial \xi^j} = \Delta_B[s^l], \quad i,j,l = 1,\dots,n. \tag{10.45}$$

As in (6.16) introducing an operator

$$B_n^{x\xi}[y] \equiv g^{x\xi} g_{\xi x}^{ij}\frac{\partial^2 y}{\partial \xi^i \partial \xi^j}, \quad i,j = 1,\dots,n, \tag{10.46}$$

gives the system (10.45) another form

$$B_n^{x\xi}[s^l] = g^{x\xi}\Delta_B[s^l], \quad l = 1,\dots,n. \tag{10.47}$$

Also taking advantage of (10.41), we obtain from (10.47) one more system of equations equivalent to (10.44):

$$B_n^{x\xi}[s^l] = -g^{x\xi} g_{sx}^{jm}\, g_{sx}^{li}[jm,i], \quad i,j,l,m = 1,\dots,n, \tag{10.48}$$

where $[jm,i] = \boldsymbol{x}_{s^j s^m} \cdot \boldsymbol{x}_{s^i}$. Thus (10.48) is a generalization of (9.57).

If S^{xn} is a monitor surface S^{rn} over S^n, represented by the parametrization (10.3) then multiplying (10.43) by $\partial \xi^p / \partial s^i$ yields

$$g_{\xi r}^{jl}\left(\frac{\partial^2 s^i}{\partial \xi^j \partial \xi^l} + \frac{\partial^2 \boldsymbol{f}[\boldsymbol{s}(\boldsymbol{\xi})]}{\partial \xi^j \partial \xi^l}\cdot\frac{\partial \boldsymbol{f}(\boldsymbol{s})}{\partial s^i}\right) = 0, \quad i,j,l = 1,\dots,n, \tag{10.49}$$

or introducing an operator $B_n^{r\xi}$, similar to (10.46),

$$B_n^{r\xi}[y] = g^{r\xi} g_{\xi r}^{ij}\frac{\partial^2 y}{\partial \xi^i \partial \xi^j}, \quad i,j = 1,\dots,n, \tag{10.50}$$

the system (10.49) is as:

$$B_n^{r\xi}[s^i] + B_n^{r\xi}[\boldsymbol{f}]\cdot\frac{\partial \boldsymbol{f}}{\partial s^i} = 0, \quad i = 1,\dots,n. \tag{10.51}$$

10.5 Formulation of Comprehensive Grid Generator

Sections 10.2 and 10.4 give a schematic description of a mapping approach in which an adaptive grid in the physical geometry S^{xn} represented by (10.1) is generated as a smooth mapping subject to the inverted Beltrami equation with respect to the metric of a monitor hypersurface S^{rn} over S^{xn}. This metric provides the generation of grids with node clustering in the zones of large variations of a specified function $\boldsymbol{f}(\boldsymbol{x})$. However, the application of such metric does not provide the generation of grids with other very important properties, in particular, grids with node clustering according to given function values, grid alignment with given vector fields, and combinations thereof. So it is necessary to generalize the metrics (10.8) and (10.9) of the monitor hypersurface S^{rn} in order to generate adaptive grids for a more wide variety of applied problems.

Thus, for the purpose of providing more efficient control of grid generation in the hypersurface S^{xn} we introduce the notion of a monitor manifold over S^{xn}. The points of the manifold are the points of S^{xn} while its metric may differ from the metric of the given hypersurface. We shall refer to this more general metric as the control metric.

This section reviews mathematical models for robust grid generators based on energy and diffusion functionals and corresponding differential Beltrami and diffusion equations, with respect to the control metrics.

10.5.1 Energy and Diffusion Functionals

Let the covariant (contravariant) elements of the control metric in the coordinates $v^1, \dots, v^n$ of S^{xn} be designated as $g^{\boldsymbol{v}}_{ij}(g^{ij}_{\boldsymbol{v}})$.

Similar to (10.29) the energy functional I_E with respect to the contravariant elements $g^{ij}_{\boldsymbol{s}}$ of the control metric $g^{\boldsymbol{s}}_{ij}$ in the parametric coordinates $s^1, \dots,$ s^n is as follows:

$$I_E[\boldsymbol{\xi}] = \frac{1}{2}\int_{S^n} \sqrt{g^{\boldsymbol{s}}}\, g^{ml}_{\boldsymbol{s}} \frac{\partial \xi^i}{\partial s^m}\frac{\partial \xi^i}{\partial s^l} \mathrm{d}\boldsymbol{s}, \quad i, l, m = 1, \dots, n, \tag{10.52}$$

where $g^{\boldsymbol{s}} = \det(g^{\boldsymbol{s}}_{ij})$.

Assuming in (10.52) $\sqrt{g^{\boldsymbol{s}}} = w(\boldsymbol{s})$, where $w(\boldsymbol{s}) > 0$ is a weight function, produces the functional of diffusion I_D with respect to the control metric $g^{\boldsymbol{s}}_{ij}$:

$$I_D[\boldsymbol{\xi}] = \frac{1}{2}\int_{S^n} w(\boldsymbol{s}) g^{ml}_{\boldsymbol{s}} \frac{\partial \xi^i}{\partial s^m}\frac{\partial \xi^i}{\partial s^l} \mathrm{d}\boldsymbol{s}, \quad i, l, m = 1, \dots, n. \tag{10.53}$$

The inverse of the function $\boldsymbol{\xi}(\boldsymbol{s}) : S^n \to \Xi^n$ which is a critical point of the functionals (10.52) or (10.53) determines the intermediate transformation $\boldsymbol{s}(\boldsymbol{\xi}) : \Xi^n \to S^n$ applied to specify the grid nodes in the physical geometry S^{xn}, represented by (10.1), by mapping the nodes of a reference grid in Ξ^n with the transformation

$$\boldsymbol{x}(\boldsymbol{s}(\boldsymbol{\xi})) : \Xi^n \to S^{xn} \subset R^{n+k}. \tag{10.54}$$

The novel feature of the functionals (10.52) and (10.53) is the recognition that the control metric $g^{\boldsymbol{s}}_{ij}$ can be freely chosen to give the grid in S^{xn} desired properties. The incorporation of the weight function w provides additional control over the redistribution of grid nodes in selected regions of S^{xn}. The job of constructing the grid then rests entirely on the choice of the control metric and the weight function.

The functionals (10.52) and (10.53) give some guess how to specify the control metric $g^{\boldsymbol{s}}_{ij}$ in order to obtain the grid with a desired property. Namely, the metric $g^{\boldsymbol{s}}_{ij}$ should be specified so that the quantity

$$m(\boldsymbol{s}) = w(\boldsymbol{s}) g^{jl}_{\boldsymbol{s}} \frac{\partial \xi^i}{\partial s^j} \frac{\partial \xi^i}{\partial s^l}, \quad i, j, l = 1, \dots, n, \tag{10.55}$$

reflects some measure of departure from the desired grid in S^{xn}. In this case minimization of the functionals with this metric gives the intermediate transformations $\boldsymbol{s}(\boldsymbol{\xi})$ which are able to produce grids with the desired property. In particular, in accordance with Sect. 10.3.2, for the metric $g^{\boldsymbol{s}}_{ij} = g^{rs}_{ij}$ of a monitor hypersurface S^{rn} over S^{xn}, the quantity (10.55) with $w(\boldsymbol{s}) = \sqrt{g^{\boldsymbol{s}}}$ is a measure of departure from a uniform grid in S^{rn} and consequently from an adaptive grid in S^{xn}.

Notice that the dimensionless functionals considered in Sect. 10.3.3 are readily formulated for the hypersurface S^{xn} with the introduced control metric $g^{\boldsymbol{s}}_{ij}$ through the invariants of the control metric tensor $(g^{\boldsymbol{\xi}}_{ij})$ in the grid coordinates $\xi^1, \dots, \xi^n$. However, these functionals are the same for the metrics $g^{\boldsymbol{s}}_{ij}$ and $v(\boldsymbol{s}) g^{\boldsymbol{s}}_{ij}$, therefore, the functionals should also include weight functions, similar to how it was done for formulating the diffusion functional (10.53).

10.5.2 Relation to Harmonic Functions

This subsection discusses another interpretation of the energy functional (10.52), which is related to the harmonic-functions approach for generating adaptive grids. This interpretation corresponds to a similar one discussed in Sect. 9.5.2 for ordinary two-dimensional surfaces lying in R^3.

According to the terminology adopted in multidimensional differential geometry, the integral (10.52) is the total energy associated with the mapping $\boldsymbol{\xi}(\boldsymbol{s}) : S^n \to \Xi^n$ representing a transformation between S^n, with its control metric tensor $(g^{\boldsymbol{s}}_{ij})$ in the coordinates s^i, and the computational domain Ξ^n, with the Cartesian coordinates ξ^i. A function which is a critical point of the energy functional is called a harmonic function. It follows from the theory of harmonic functions on manifolds that if there is a diffeomorphism $\boldsymbol{f}(\boldsymbol{s}) : S^n \to \Xi^n$ and the boundary of the manifold Ξ^n is convex, and its curvature is nonpositive, then a harmonic function coinciding with $\boldsymbol{f}$ on the boundary of the manifold S^n exists and is isotopic to $\boldsymbol{f}$. In the case under consideration, the coordinates of the manifold Ξ^n are Cartesian, and therefore its curvature is nonpositive. As in the case of ordinary two-dimensional surfaces, if S^2 is diffeomorphic to Ξ^2 and Ξ^2 is convex then the mapping $\boldsymbol{\xi}(\boldsymbol{s})$ that minimizes the functional of energy (10.52) is a one-to-one transformation, and the grid obtained by the proposed variational method is therefore nondegenerate in this case. However, this property, in general, is not valid for any dimension $n > 2$, in particular, for $n = 3$, which is important for the generation of three-dimensional adaptive grids in three-dimensional domains, in particular, by projecting quasiuniform grids from monitor hypersurfaces S^{r3}.

10.5.3 Beltrami and Diffusion Equations

Similar to (10.30), the intermediate transformation $\mathbf{s}(\boldsymbol{\xi})$ for generating grids in S^{xn} is determined as the inverse of the transformation

$$\boldsymbol{\xi}(\mathrm{s}) : S^n \to \Xi^n, \quad \boldsymbol{\xi}(\mathrm{s}) = \left[\xi^1(\mathrm{s}), \ldots, \xi^n(\mathrm{s})\right]$$

which is a solution of the following Dirichlet boundary value problem for the Euler-Lagrange equations derived from the functional of energy (10.52)

$$\frac{\partial}{\partial s^j}\left(\sqrt{g^{\mathbf{s}}} g_{\mathbf{s}}^{jk} \frac{\partial \boldsymbol{\xi}}{\partial s^k}\right) = 0, \quad j, k = 1, \ldots, n,$$
$$\boldsymbol{\xi}(\boldsymbol{s})|_{\partial S^n} = \boldsymbol{\varphi}(\mathrm{s}) : \partial S^n \to \partial \Xi^n, \quad \boldsymbol{\varphi}(\mathrm{s}) = \left[\varphi^1(\mathrm{s}), \ldots, \varphi^n(\mathrm{s})\right],$$

where $g_{\mathbf{s}}^{jk}$ are the contravariant components of the control metric in the coordinates $s^1, \ldots, s^n$, ∂S^n and $\partial \Xi^n$ are the boundaries of S^n and Ξ^n, respectively, while $\boldsymbol{\varphi}(\mathrm{s})$ is a one-to-one continuous transformation between the boundaries of S^n and Ξ^n. This boundary value problem has the following form for the components $\xi^i(\mathrm{s})$

$$\begin{aligned}\frac{\partial}{\partial s^j}\left(\sqrt{g^{\mathbf{s}}} g_{\mathbf{s}}^{jk} \frac{\partial \xi^i}{\partial s^k}\right) &= 0, \quad i, j, k = 1, \ldots, n, \\ \xi^i(\boldsymbol{s})|_{\partial S^n} &= \varphi^i(\mathrm{s}), \quad i = 1, \ldots, n.\end{aligned} \tag{10.56}$$

In the theory of Riemannian manifolds, the equations in (10.56) are called generalized Laplace equations or Beltrami equations of the second order. We shall hereafter call them Beltrami equations.

The functions $\xi^1, \ldots, \xi^n$ satisfying (10.56) form a curvilinear coordinate system in S^n and S^{xn}. These curvilinear coordinates are further referred to as the grid coordinates.

It is easily shown that an arbitrary one-to-one twice differentiable transformation $\boldsymbol{\psi}(\boldsymbol{s}) : S^n \to \Xi^n$, $\boldsymbol{\psi}(\boldsymbol{s}) = (\psi^1(\boldsymbol{s}), \ldots, \psi^n(\boldsymbol{s}))$, is a solution of the Dirichlet problem (10.56) with respect to the control metric $g_{ij}^{\boldsymbol{s}}$ specified in the parametric coordinates $s^1, \ldots, s^n$, by

$$g_{ij}^{\boldsymbol{s}} = \frac{\partial \boldsymbol{\psi}}{\partial s^i} \cdot \frac{\partial \boldsymbol{\psi}}{\partial s^i}, \quad i, j = 1, \ldots, n, \tag{10.57}$$

and with the following boundary conditions

$$\boldsymbol{\xi}(\boldsymbol{s})|_{\partial S^n} = \boldsymbol{\psi}(\boldsymbol{s}), \quad i = 1, \dots, n.$$

Indeed, it is readily obtained from (10.57) that

$$\sqrt{g^{\boldsymbol{s}}} = \det\left(\frac{\partial \psi^l}{\partial s^k}\right), \qquad g_{\boldsymbol{s}}^{jk} = \frac{\partial s^j}{\partial \psi^m} \frac{\partial s^k}{\partial \psi^m}, \quad j, k, l, m = 1, \dots, n,$$

so the ith equation in (10.56) with $\partial \xi^i / \partial s^k = \partial \psi^i / \partial s^k$ is in fact the identity of the form (2.47)

$$\frac{\partial}{\partial s^j}\left(\det\left(\frac{\partial \psi^l}{\partial s^k}\right) \frac{\partial s^j}{\partial \psi^i}\right) \equiv 0, \quad i, j = 1, \dots, n,$$

i.e., the functions $\psi^i(\boldsymbol{s})$, $i = 1, \dots, n$, are the solutions of the Dirichlet problem (10.56) for the Beltrami equations with respect to the metric (10.57).

The substitution of $w(\mathbf{s})$ for $\sqrt{g^{\mathbf{s}}}$ in the system of equations in (10.56) yields the Dirichlet problem for more general equations

$$\begin{aligned} &\frac{\partial}{\partial s^j}\left(w(\mathbf{s}) g_{\mathbf{s}}^{jk} \frac{\partial \xi^i}{\partial s^k}\right) = 0, \quad i, j, k = 1, \dots, n, \\ &\xi^i(\boldsymbol{s})|_{\partial S^n} = \varphi^i(\mathbf{s}), \quad i = 1, \dots, n. \end{aligned} \tag{10.58}$$

Here $w(\mathbf{s}) > 0$ is a weight function aimed at increasing or decreasing the effect of the control metric in the necessary zones of S^{xn}. The equations in (10.58) are in fact the Euler–Lagrange equations for the diffusion functional (10.53) so they will be referred to as the diffusion equations.

The diffusion equations in (10.58) are equivalent to the Beltrami equations if $w(\mathbf{s}) = \sqrt{g^{\mathbf{s}}}$, $g^{\mathbf{s}} = \det(g^{\mathbf{s}}_{ij})$. Moreover, for $n \neq 2$ they are always equivalent to the Beltrami equations, with respect to the metric

$$g_{ij} = (g^{\mathbf{s}})^{\frac{1}{2-n}} [w(\mathbf{s})]^{\frac{2}{n-2}} g^{\mathbf{s}}_{ij}, \quad i, j = 1, \dots, n, \tag{10.59}$$

regardless of the weight function $w(\mathbf{s}) > 0$. Indeed, for this metric,

$$\begin{aligned} g &= \det(g_{ij}) = (g^{\mathbf{s}})^{\frac{n}{2-n}} [w(\mathbf{s})]^{\frac{2n}{n-2}} g^{\mathbf{s}} = (g^{\mathbf{s}})^{\frac{2}{2-n}} [w(\mathbf{s})]^{\frac{2n}{n-2}}, \\ g^{ij} &= (g^{\mathbf{s}})^{\frac{1}{n-2}} [w(\mathbf{s})]^{\frac{2}{2-n}} g^{ij}_{\mathbf{s}}, \quad i, j = 1, \dots, n, \end{aligned}$$

so

$$\sqrt{g} g^{ij} = w(\mathbf{s}) g^{ij}_{\mathbf{s}}, \quad i, j = 1, \dots, n,$$

i.e., the system in (10.58) is the system of Beltrami equations with respect to the metric (10.59).

Though the Beltrami equations in (10.56) are comprehensive, i.e., an arbitrary nondegenerate twice differentiable intermediate transformation (10.2) can be computed as the inverse of the solution of these equations, the form of the diffusion equations in (10.58) appears to be simpler for formulating, especially for $n = 2$, in order to realize the necessary requirements for the grid properties in different zones of S^{xn}. However, one is to remember that for another coordinate system $v^1, \dots, v^n$ the equations in (10.58) become equivalent to

$$\frac{\partial}{\partial v^j}\left(w[\mathbf{s}(\mathbf{v})]Jg_{\mathbf{v}}^{jk}\frac{\partial \xi^i}{\partial v^k}\right) = 0, \quad i,j,k = 1,\dots,n,$$

where $J = \det(\partial s^i/\partial v^j)$. This very system in the parametric coordinates $v^1, \dots, v^n$ should be solved in order to obtain the same grid computed by the solution of the boundary value (10.58) in the parametric coordinates $s^1, \dots, s^n$.

10.5.4 Inverted Beltrami and Diffusion Equations

Note that, in order to find the grid nodes through the intermediate transformation $\mathbf{s}(\boldsymbol{\xi})$ the inverse of which is a solution of the problem (10.56) or (10.58) there is no necessity to compute the transformation $\mathbf{s}(\boldsymbol{\xi})$ at all points $\boldsymbol{\xi} \in \Xi^n$. It is sufficient to solve numerically the inverted boundary value problem obtained by interchanging in (10.56) or (10.58) dependent and independent variables, i.e., considering the variables $\xi^1, \dots, \xi^n$ as independent while the variables $s^1, \dots, s^n$ of the parametric domain S^n as dependent ones. The inverted problem with respect to the components $s^i(\boldsymbol{\xi})$, $i = 1, \dots, n$, of the intermediate transformation $\mathbf{s}(\boldsymbol{\xi})$ should be solved on the reference grid in Ξ^n. The values of this same numerical solution

$$\mathbf{s}(\boldsymbol{\xi}) = \left[s^1(\boldsymbol{\xi}), \dots, s^n(\boldsymbol{\xi})\right]$$

at the points of the reference grid determine grid nodes in S^n and consequently in S^{xn} by mapping them through $\mathbf{x}(\mathbf{s})$.

Similar to (10.47) it is readily shown from the equations in (10.56), that the general inverted Beltrami equations with respect to the control metric $g_{ij}^{\mathbf{s}}$ are as follows:

$$B_n^{\boldsymbol{\xi}}[s^l] = g^{\boldsymbol{\xi}}\Delta_B[s^l], \quad l = 1,\dots,n, \tag{10.60}$$

where

$$B_n^{\boldsymbol{\xi}}[s^l] = g^{\boldsymbol{\xi}}g_{\boldsymbol{\xi}}^{ij}\frac{\partial^2 s^l}{\partial \xi^i \partial \xi^j}, \quad i,j,l = 1,\dots,n,$$

$$g^{\boldsymbol{\xi}} = \det(g_{ij}^{\boldsymbol{\xi}}) = g^{\mathbf{s}}(J)^2 = 1/\det(g_{\boldsymbol{\xi}}^{ij}),$$

$$\Delta_B[s^l] = \frac{1}{\sqrt{g^{\mathbf{s}}}}\frac{\partial}{\partial s^j}\left(\sqrt{g^{\mathbf{s}}}g_{\mathbf{s}}^{lj}\right), \quad i,j,l = 1,\dots,n,$$

$J = \det(\partial s^i/\partial \xi^j)$, $g_{ij}^{\boldsymbol{\xi}}(g_{\boldsymbol{\xi}}^{ij})$ are the covariant (contravariant) elements of the control metric in the grid coordinates $\xi^1, \dots, \xi^n$, i.e.

$$g_{ij}^{\boldsymbol{\xi}} = g_{kl}^{\boldsymbol{s}} \frac{\partial s^k}{\partial \xi^i} \frac{\partial s^l}{\partial \xi^j}, \qquad g_{\boldsymbol{\xi}}^{ij} = g_{\boldsymbol{s}}^{kl} \frac{\partial \xi^i}{\partial s^k} \frac{\partial \xi^j}{\partial s^l}, \quad i,j,k,l = 1,\dots,n. \tag{10.61}$$

Notice that the operator $B_n^{\boldsymbol{\xi}}$ is in fact the operator $B_n^{x\xi}$ (see (10.46)) if $g_{ij}^{\boldsymbol{\xi}} = g_{ij}^{x\xi}$, while it is the operator $B_n^{r\xi}$ (see (10.50)) if $g_{ij}^{\boldsymbol{\xi}} = g_{ij}^{r\xi}$, and the operator B_n in the formula (6.16) if $g_{ij}^{\boldsymbol{\xi}} = g_{ij}$.

Consequently, the boundary value problem (10.56) is transformed into the following inverted problem with respect to the components $s^i(\boldsymbol{s})$ of the intermediate function $\boldsymbol{s}(\boldsymbol{\xi}) : \Xi^n \to S^n$

$$\begin{aligned} B_n^{\boldsymbol{\xi}}[s^i] &= (J)^2 \sqrt{g^{\boldsymbol{s}}} \frac{\partial}{\partial \xi^k} \left(\sqrt{g^{\boldsymbol{s}}} g_{\boldsymbol{s}}^{ij}\right) \frac{\partial \xi^k}{\partial s^j}, \quad i,j,k = 1,\dots,n, \\ s^i(\boldsymbol{\xi})|_{\partial \Xi^n} &= \psi^i(\boldsymbol{\xi}), \quad i = 1,\dots,n, \end{aligned} \tag{10.62}$$

where $\boldsymbol{\psi}(\boldsymbol{\xi}) : \partial \Xi^n \to \partial S^n$, $\boldsymbol{\psi}(\boldsymbol{\xi}) = [\psi^1(\boldsymbol{\xi}), \dots, \psi^n(\boldsymbol{\xi})]$, is the inverse of $\boldsymbol{\varphi}(\mathrm{s}) : \partial S^n \to \partial \Xi^n$, $J = \det(\partial s^i / \partial \xi^j)$.

Similar to (10.62) there is written out the Dirichlet boundary value problem obtained from (10.58) for the inverted diffusion equations

$$\begin{aligned} B_n^{\boldsymbol{\xi}}[s^i] &= \frac{g^{\boldsymbol{s}}(J)^2}{w(\boldsymbol{s})} \frac{\partial}{\partial \xi^k} \left(w(\boldsymbol{s}) g_{\boldsymbol{s}}^{ij}\right) \frac{\partial \xi^k}{\partial s^j}, \quad i,j,k = 1,\dots,n, \\ s^i(\boldsymbol{\xi})|_{\partial \Xi^n} &= \psi^i(\boldsymbol{\xi}), \quad i = 1,\dots,n. \end{aligned} \tag{10.63}$$

The inverted diffusion equations in (10.63) are also transformed into the following divergent form

$$\frac{\partial}{\partial \xi^j} \{ J w[\mathbf{s}(\boldsymbol{\xi})] g_{\boldsymbol{\xi}}^{ij} \} = 0, \quad i,j = 1,\dots,n, \tag{10.64}$$

obtained by applying the identity (2.47) to the equations in (10.58). Substituting in (10.64) $\sqrt{g^s}$ for $w(\boldsymbol{s})$ gives a divergent form for the inverted Beltrami equations.

Another formulation of a grid model through the Beltrami equations with respect to the intermediate transformation $\mathbf{s}(\boldsymbol{\xi})$:

$$\begin{aligned} \frac{\partial}{\partial \xi^j} \left(\sqrt{g} g^{jk} \frac{\partial s^i}{\partial \xi^k} \right) &= 0, \quad i,j,k = 1,\dots,n, \\ s^i(\boldsymbol{\xi})|_{\Xi^n} &= \psi^i(\boldsymbol{\xi}), \end{aligned} \tag{10.65}$$

where g^{jk} are the contravariant metric components of a monitor manifold over Ξ^n, $g = \det(g_{jk}) = 1/\det(g^{jk})$, was proposed by Godunov and Prokopov (1967), and Ryskin and Leal (1983). The equations in (10.65) seem to be more natural than the nonlinear inverted Beltrami equations in (10.62) obtained from the Beltrami equations in (10.56) for the implementation into numerical codes, since they are linear and of divergent form with respect to

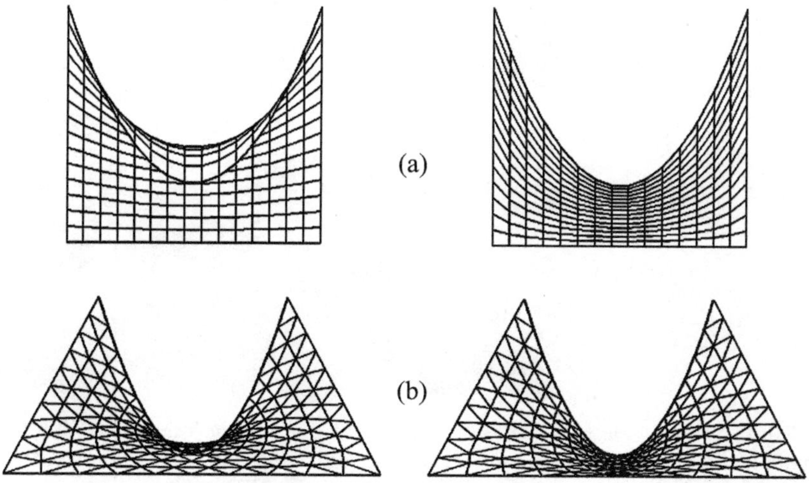

Fig. 10.2. Quadrilateral (**a**) and triangular (**b**) grids in concave domains generated by the solution of equations (10.65) (*left*) and by the solution of equations in (10.62) (*right*); both with respect to the Euclidean metric

the intermediate transformation $\mathbf{s}(\boldsymbol{\xi})$. However, such divergent model, owing to the maximum principle, does not guarantee that all grid points will be inside of the physical geometry S^{xn} when the parametric domain S^n is not convex (see Fig. 10.2, left), moreover the grid cells may be folded. Providing grid nondegeneracy through the solution of equations in (10.65) depends on devising a suitable metric in Ξ^n, which hasn't been done so far in a general form.

The considerations mentioned are such that it is more reasonable to make the formulation of grid systems through the Beltrami equations with respect to the function $\boldsymbol{\xi}(\mathbf{s}) : S^n \to \Xi^n$ the inverse of which yields the intermediate transformation $\mathbf{s}(\boldsymbol{\xi}) : \Xi^n \to S^n$.

10.6 Numerical Algorithms

The boundary value problems (10.62) or (10.63) allow one to generate grids on surfaces or in domains of an arbitrary dimension $n > 0$, and hence these problems can be applied to obtain grids in n-dimensional blocks by means of the successive generation of grids on curvilinear edges, faces, parallelepipeds, etc., using the solution at a step $i < n$ as the Dirichlet boundary condition for the following step $i + 1 \leq n$ (see Fig. 10.3 for $n = 3$). Thus, both the interior and the boundary grid points of a domain or surface can be calculated by the similar elliptic solvers.

Note that the basic system implemented in the present codes to generate adaptive and fixed structured grids is obtained by inverting the Poisson system

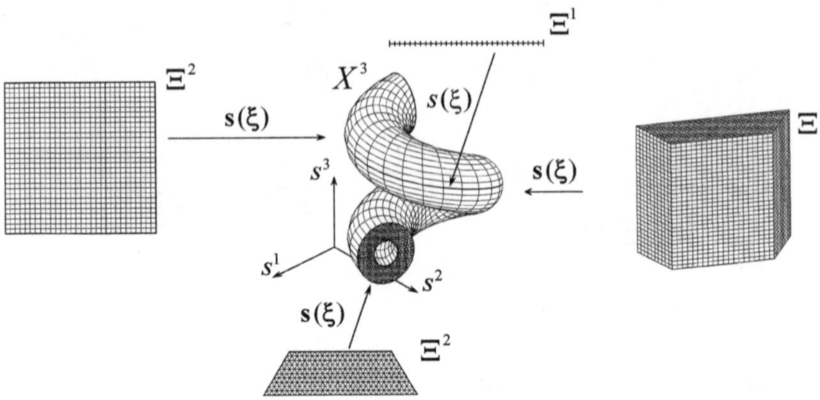

Fig. 10.3. Scheme for successive generation of one-dimensional, two-dimensional triangular and quadrilateral boundary grids, and three-dimensional prismatic interior grids

(10.31) and has the form (6.16):

$$B_n^{x\xi}[x^i] + gP^j \frac{\partial x^i}{\partial \xi^j} = 0, \quad i, j = 1, \ldots, n, \tag{10.66}$$

where $B_n^{x\xi}$ is the operator (10.46). Formally, the systems in (10.62) and (10.63) have some sort of similarity to the system (10.66). Therefore, the numerical algorithms developed in the current codes for the numerical solution of the inverted Poisson equations (10.66) can be applied with some modification to the solution of the boundary value problems (10.62) or (10.63).

In this section, we describe some such numerical algorithms for solving the Dirichlet boundary value problem for the inverted Beltrami and diffusion equations.

10.6.1 Finite-Difference Algorithm

Solutions of the transformed non-linear boundary-value problem (10.62) are found in the following way. First, this problem is replaced by a nonstationary boundary-value problem with respect to the components $s^i(\boldsymbol{\xi}, t)$, $i = 1, \ldots, n$, of the vector function $\boldsymbol{s}(\boldsymbol{\xi}, t) : \Xi^n \times [0, T] \to S^n$:

$$\begin{aligned} &\frac{\partial s^i}{\partial t} = (J)^p \left\{ B_n^{\boldsymbol{\xi}}[s^i] - (J)^2 \sqrt{g^{\boldsymbol{s}}} \frac{\partial}{\partial \xi^j} (\sqrt{g^{\boldsymbol{s}}} g_{\boldsymbol{s}}^{im}) \frac{\partial \xi^j}{\partial s^m} \right\}, \quad i, j, m = 1, \ldots, n, \\ &s^i(\boldsymbol{\xi}, t) = \psi^i(\boldsymbol{\xi}), \quad \boldsymbol{\xi} \in \partial \Xi^n, \ t \geq 0, \\ &s^i(\boldsymbol{\xi}, 0) = s_0^i(\boldsymbol{\xi}), \quad \boldsymbol{\xi} \in \Xi^n, \end{aligned} \tag{10.67}$$

where $J = \det(\partial s^i / \partial \xi^j)$, $p \geq 0$, $s_0^i(\boldsymbol{\xi})$ is the i-th component of the initial transformation

$$\mathbf{s}_0(\boldsymbol{\xi}) : \Xi^n \to S^n, \quad \mathbf{s}_0(\boldsymbol{\xi}) = \left[s_0^1(\boldsymbol{\xi}), \dots, s_0^n(\boldsymbol{\xi})\right],$$

specified by the user. The factor $(J)^p$, $p \geq 0$ in (10.67) is introduced to rule out the Jacobian J being a denominator after replacing in the right-hand part of the equations the derivatives $\partial\xi^i/\partial s^j$ with the derivatives $\partial s^k/\partial\xi^m$, in particular in the case of the metric of a monitor hypersurface. This incorporation of $(J)^p$ allows one to produce the final nondegenerate grid even if the initial and intermediate grids may be singular ones.

The problem (10.67) is written out in the following vector form

$$\begin{aligned} &\frac{\partial \boldsymbol{s}}{\partial t} = (J)^p\{B_n^{\boldsymbol{\xi}}[\boldsymbol{s}] - \boldsymbol{R}[\boldsymbol{s}]\}, \\ &\boldsymbol{s}(\boldsymbol{\xi}, t)|_{\partial\Xi^n} = \boldsymbol{\psi}(\boldsymbol{\xi}), \quad t \geq 0, \\ &\boldsymbol{s}(\boldsymbol{\xi}, 0) = \boldsymbol{s}_0(\boldsymbol{\xi}), \quad \boldsymbol{\xi} \in \Xi^n, \end{aligned} \tag{10.68}$$

where

$$\begin{aligned} \boldsymbol{R}[\boldsymbol{s}] &= \left(R^1[\boldsymbol{s}], \dots, R^n[\boldsymbol{s}]\right), \\ R^i[\boldsymbol{s}] &= (J)^2\sqrt{g^{\boldsymbol{s}}}\frac{\partial}{\partial\xi^j}\left(\sqrt{g^{\boldsymbol{s}}}g_{\boldsymbol{s}}^{im}\right)\frac{\partial\xi^j}{\partial s^m}, \quad i, j, m = 1, \dots, n. \end{aligned}$$

Notice that, for the metric (10.8) of the monitor hypersurface S^{rn} over S^n, i.e. $g_{ij}^{\boldsymbol{s}} = g_{ij}^{rs}$, $B_n^{\boldsymbol{\xi}} = B_n^{r\xi}$ (see (10.50)) and in accordance with (10.51),

$$R^i[\boldsymbol{s}] = -B_n^{r\xi}[\boldsymbol{f}] \cdot \frac{\partial \boldsymbol{f}[\boldsymbol{s}(\boldsymbol{\xi})]}{\partial\xi^j}\frac{\partial\xi^j}{\partial s^i}, \quad i, j = 1, \dots, n, \tag{10.69}$$

thus, for this case, $p = 1$ in (10.67) and (10.68).

When $B_n^{\boldsymbol{\xi}}$ is an elliptic operator, then the solution of the problem (10.68) relaxes to the solution of (10.62) as $t \to \infty$.

The boundary value problem (10.68) is usually solved by alternating direction implicit (ADI) methods. We discuss here a method of fractional steps which is one of such ADI methods.

10.6.1.1 One-Dimensional Equation

For $n = 1$ we have

$$B_1^{\xi}[s] = g_{11}^{\xi}\frac{\partial^2 s}{\partial\xi^1\partial\xi^1}, \qquad R^1[s] = -\frac{ds}{d\xi^1}\sqrt{g_{11}^s}\frac{d}{d\xi^1}\left[\frac{1}{\sqrt{g_{11}^s}}\right]$$

and a numerical solution of the boundary value problem (10.68) for $n = 1$ is found through the following scheme

$$\frac{s_i^{k+1} - s_i^k}{\tau} = (J)^p[s^k]\{g_{11}^{\xi}[s^k](s_{i+1}^{k+1} - 2s_i^{k+1} + s_{i-1}^{k+1}) + R^1[s^k]\},$$

where $s^k = s(\xi, k\tau)$, $k = 0, 1, 2, \dots$.

10.6.1.2 Two-Dimensional Equations

For $n = 2$ we have

$$B_2^{\boldsymbol{\xi}}[\boldsymbol{s}] = g_{22}^{\boldsymbol{\xi}} \frac{\partial^2 \boldsymbol{s}}{\partial \xi^1 \partial \xi^2} - 2g_{12}^{\boldsymbol{\xi}} \frac{\partial^2 \boldsymbol{s}}{\partial \xi^1 \partial \xi^2} + g_{11}^{\boldsymbol{\xi}} \frac{\partial^2 \boldsymbol{s}}{\partial \xi^2 \partial \xi^2},$$

$$R^i[\boldsymbol{s}] = (-1)^{j+m} J \sqrt{g^{\boldsymbol{s}}} \frac{\partial}{\partial \xi^j} \left(\sqrt{g^{\boldsymbol{s}}} g_{\boldsymbol{s}}^{im}\right) \frac{\partial s^{3-m}}{\partial \xi^{3-j}}, \quad i, j, m = 1, 2.$$

In particular, for the metric (10.8) of the monitor hypersurface S^{r2} over S^2, i.e. $g_{ij}^{\boldsymbol{s}} = g_{ij}^{rs}$, $B_2^{\boldsymbol{\xi}} = B_2^{r\xi}$ (see (10.50)) and in accordance with (10.69) for $n = 2$

$$\begin{aligned}
B_2^{\boldsymbol{\xi}}[y] \equiv g^{r\xi} g_{\xi r}^{ij} \frac{\partial^2 y}{\partial \xi^i \partial \xi^j} &= g_{22}^{r\xi} \frac{\partial^2 y}{\partial \xi^1 \partial \xi^1} - 2g_{12}^{r\xi} \frac{\partial^2 y}{\partial \xi^1 \partial \xi^2} + g_{11}^{r\xi} \frac{\partial^2 y}{\partial \xi^2 \partial \xi^2} \\
&= B_2^{x\xi}[y] + \frac{\partial \boldsymbol{f}(\boldsymbol{s}(\boldsymbol{\xi}))}{\partial \xi^2} \cdot \frac{\partial \boldsymbol{f}(\boldsymbol{s}(\boldsymbol{\xi}))}{\partial \xi^2} \frac{\partial^2 y}{\partial \xi^1 \partial \xi^1} \\
&\quad - 2 \frac{\partial \boldsymbol{f}(\boldsymbol{s}(\boldsymbol{\xi}))}{\partial \xi^1} \cdot \frac{\partial \boldsymbol{f}(\boldsymbol{s}(\boldsymbol{\xi}))}{\partial \xi^2} \frac{\partial^2 y}{\partial \xi^1 \partial \xi^2} \\
&\quad + \frac{\partial \boldsymbol{f}(\boldsymbol{s}(\boldsymbol{\xi}))}{\partial \xi^1} \cdot \frac{\partial \boldsymbol{f}(\boldsymbol{s}(\boldsymbol{\xi}))}{\partial \xi^1} \frac{\partial^2 y}{\partial \xi^2 \partial \xi^2}, \quad i, j = 1, 2, \\
R^i[\boldsymbol{s}] &= -(-1)^{j+i} \frac{1}{J} \left(B_2^{r\xi}[\boldsymbol{f}] \cdot \frac{\partial \boldsymbol{f}}{\partial \xi^j} \right) \frac{\partial s^{3-i}}{\partial \xi^{3-j}}, \quad i, j = 1, 2, \ i \text{ fixed}.
\end{aligned}$$

A numerical solution of (10.68) for $n = 2$ is found through the following scheme

$$\begin{aligned}
\frac{\mathbf{s}^{k+1/2} - \mathbf{s}^k}{\tau/2} &= (J)^p[\boldsymbol{s}^k] \{ g_{22}^{\boldsymbol{\xi}}[\boldsymbol{s}^k] L_{11}^h[\mathbf{s}^{k+1/2}] - 2g_{12}^{\boldsymbol{\xi}}[\boldsymbol{s}^k] L_{12}^h[\mathbf{s}^k] \\
&\quad + g_{11}^{\boldsymbol{\xi}}[\boldsymbol{s}^k] L_{22}^h[\mathbf{s}^k] + \boldsymbol{R}[\boldsymbol{s}^k] \}, \qquad (10.70) \\
\frac{\mathbf{s}^{k+1} - \mathbf{s}^{k+1/2}}{\tau/2} &= (J)^p[\boldsymbol{s}^k] g_{11}^{\boldsymbol{\xi}}[\boldsymbol{s}^k] \{ L_{22}^h[\mathbf{s}^{k+1}] - L_{22}[\mathbf{s}^k] \}.
\end{aligned}$$

Here $\boldsymbol{s}^{k+\alpha} = \boldsymbol{s}(\boldsymbol{\xi}, (k+\alpha)\tau)$, $k = 0, 1, 2, \ldots$, $\alpha = 0, 1/2, 1$, L_{ij}^h is a finite-difference operator approximating the operator $\partial^2/(\partial \xi^i \partial \xi^j)$, by the central differences. The first derivatives in the expressions $g_{ij}^{\boldsymbol{\xi}}[\boldsymbol{s}^k]$, $J[\boldsymbol{s}^k]$ and $R[\boldsymbol{s}^k]$ are also approximated by the central differences. The initial transformation $\mathbf{s}(\boldsymbol{\xi}, 0) = \mathbf{s}_0(\boldsymbol{\xi})$ is found through the formulas of transfinite interpolations described in Chap. 5. In two dimensions, the logical domain Ξ^2 may be taken as a unit square (Fig. 10.4 (*left*)) or, for example, as a symmetric trapezoid (Fig. 10.4 (*right*)), in particular, a triangle.

An analogous algorithm is written out for the boundary value problem (10.63) with $n = 2$.

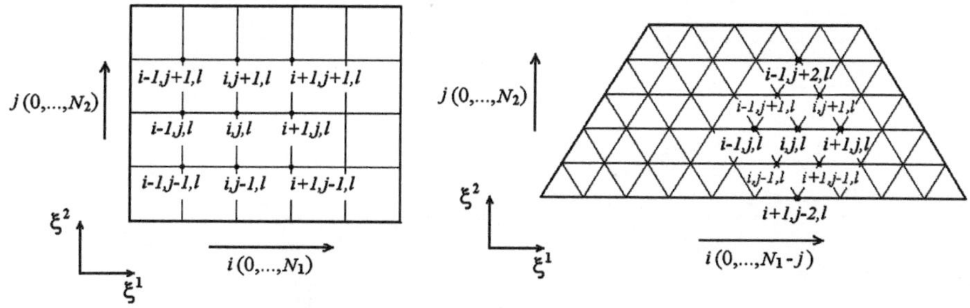

Fig. 10.4. Quadrilateral and triangular stencils for finite differences

10.6.1.3 Three-Dimensional Equations

For $n = 3$ in (10.68), we obtain

$$
\begin{aligned}
B_3^{\boldsymbol{\xi}}[\boldsymbol{s}] =& g^{\boldsymbol{\xi}} g_{\boldsymbol{\xi}}^{ij} \frac{\partial^2 \boldsymbol{s}}{\partial \xi^i \partial \xi^j} \\
=& [g_{22}^{\boldsymbol{\xi}} g_{33}^{\boldsymbol{\xi}} - (g_{23}^{\boldsymbol{\xi}})^2] \frac{\partial^2 \boldsymbol{s}}{\partial \xi^1 \partial \xi^1} + [g_{11}^{\boldsymbol{\xi}} g_{33}^{\boldsymbol{\xi}} - (g_{13}^{\boldsymbol{\xi}})^2] \frac{\partial^2 \boldsymbol{s}}{\partial \xi^2 \partial \xi^2} \\
&+ [g_{11}^{\boldsymbol{\xi}} g_{22}^{\boldsymbol{\xi}} - (g_{12}^{\boldsymbol{\xi}})^2] \frac{\partial^2 \boldsymbol{s}}{\partial \xi^3 \partial \xi^3} + 2[g_{23}^{\boldsymbol{\xi}} g_{13}^{\boldsymbol{\xi}} - g_{12}^{\boldsymbol{\xi}} g_{33}^{\boldsymbol{\xi}}] \frac{\partial^2 \boldsymbol{s}}{\partial \xi^1 \partial \xi^2} \\
&+ 2[g_{12}^{\boldsymbol{\xi}} g_{23}^{\boldsymbol{\xi}} - g_{22}^{\boldsymbol{\xi}} g_{13}^{\boldsymbol{\xi}}] \frac{\partial^2 \boldsymbol{s}}{\partial \xi^1 \partial \xi^3} + 2[g_{13}^{\boldsymbol{\xi}} g_{12}^{\boldsymbol{\xi}} - g_{23}^{\boldsymbol{\xi}} g_{11}^{\boldsymbol{\xi}}] \frac{\partial^2 \boldsymbol{s}}{\partial \xi^2 \partial \xi^3}, \\
& i, j = 1, 2, 3, \qquad (10.71)
\end{aligned}
$$

and applying formula (2.4)

$$
R^i[\boldsymbol{s}] = J\sqrt{g^{\boldsymbol{s}}} \frac{\partial}{\partial \xi^j} \left(\sqrt{g^{\boldsymbol{s}}} g_{\boldsymbol{s}}^{im}\right) \left(\frac{\partial s^{m+1}}{\partial \xi^{j+1}} \frac{\partial s^{m+2}}{\partial \xi^{j+2}} - \frac{\partial s^{m+1}}{\partial \xi^{j+2}} \frac{\partial s^{m+2}}{\partial \xi^{j+1}} \right),
$$
$$
i, j, m = 1, 2, 3.
$$

In particular, for the metric (10.8) of the monitor hypersurface S^{r3} over S^3 we have $B_3^{\boldsymbol{\xi}} = B_3^{r\xi}$, while

$$
R^i[\boldsymbol{s}] = -\frac{1}{J} B_3^{r\xi}[\boldsymbol{f}] \cdot \frac{\partial \boldsymbol{f}[\boldsymbol{s}(\boldsymbol{\xi})]}{\partial \xi^j} \left(\frac{\partial s^{i+1}}{\partial \xi^{j+1}} \frac{\partial s^{i+2}}{\partial \xi^{j+2}} - \frac{\partial s^{i+1}}{\partial \xi^{j+2}} \frac{\partial s^{i+2}}{\partial \xi^{j+1}} \right),
$$
$$
i, j = 1, 2, 3, \ i \text{ fixed}.
$$

where operator $B_3^{r\xi}$ is defined by the formula (10.50) for $n = 3$. If fact the operator $B_3^{r\xi}$ is described by (10.71) with the identification $g_{ij}^{\boldsymbol{\xi}} = g_{ij}^{r\xi}$.

Similar to (10.70) a numerical solution of (10.68) for $n = 3$ is found through the following scheme

$$\frac{\mathbf{s}^{k+1/3}-\mathbf{s}^k}{\tau/3} = (J)^p[\boldsymbol{s}^k]\{a^{11}[\mathbf{s}^k]L^h_{11}[\mathbf{s}^{k+1/3}] + a^{22}[\mathbf{s}^{\mathbf{k}}]L^h_{22}[\mathbf{s}^k]$$
$$+ a^{33}[\mathbf{s}^k]L^h_{33}[\mathbf{s}^k] + 2a^{12}[\mathbf{s}^k]L^h_{12}[\mathbf{s}^k] + 2a^{13}[\mathbf{s}^k]L^h_{13}[\mathbf{s}^k]$$
$$+ 2a^{23}[\mathbf{s}^k]L^h_{23}[\mathbf{s}^k] - \boldsymbol{R}[\boldsymbol{s}^k]\},$$
$$\frac{\mathbf{s}^{k+2/3}-\mathbf{s}^{k+1/3}}{\tau/3} = (J)^p[\boldsymbol{s}^k]a^{22}[\mathbf{s}^k]\{L^h_{22}[\mathbf{s}^{k+2/3}] - L^h_{22}[\mathbf{s}^k]\},$$
$$\frac{\mathbf{s}^{k+1}-\mathbf{s}^{k+2/3}}{\tau/3} = (J)^p[\boldsymbol{s}^k]a^{33}[\mathbf{s}^k]\{L^h_{33}[\mathbf{s}^{k+1}] - L^h_{33}[\mathbf{s}^k]\},$$

where $\boldsymbol{s}^{k+\alpha} = \boldsymbol{s}(\boldsymbol{\xi},(k+\alpha)\tau),\ k=0,1,2,\ldots,\ \alpha = 0,1/3,2/3,1,$

$$a^{ij} = g^{\boldsymbol{\xi}} g^{ij}_{\boldsymbol{\xi}} = (g^{\boldsymbol{\xi}}_{i+1j+1}g^{\boldsymbol{\xi}}_{i+2j+2} - g^{\boldsymbol{\xi}}_{i+1j+2}g^{\boldsymbol{\xi}}_{i+2j+1}), \quad i,j=1,2,3.$$

An analogous algorithm is written out for the boundary value problem (10.63) with $n = 3$.

10.6.2 Spectral Element Algorithm

The divergent inverted diffusion equations (10.64) as well as the divergent inverted Beltrami equations may be solved by a parallel code, using spectral elements for spatial discretization, Newton–Krylov methods for solution, and an adaptive time step.

Spatial discretization by high-order spectral elements is a method of exploiting the best features of both grid-based methods and global spectral representations. Grid-based methods, such as the finite difference approach described above, lead to nearest neighbor coupling and its resultant sparse matrix structure, and lend themselves to parallelization by domain decomposition and the kind of adaptive gridding. On the other hand, convergence of the spatial truncation error in these methods is relatively slow, typically a low power of the grid spacing h. Global spectral methods overcome the latter problem, offering exponential convergence with increasing numbers of basis functions, but lead to large, dense matrices and offer no obvious way to use adaptive gridding and parallelization by domain decomposition. With spectral elements, there is a relatively coarse grid, and within each grid cell there is a local expansion in basis functions based on orthogonal polynomials. The grid provides nearest-neighbor coupling while the spectral expansion provides exponential convergence.

All equations for spectral elements are to be expressed in flux-source form,

$$\frac{\partial u^k}{\partial t} + \nabla \cdot \mathbf{F}^k = S^k. \qquad (10.72)$$

This very form has the following system

$$\frac{\partial s^i}{\partial t} - \frac{\partial}{\partial \xi^j}(Jw(\mathbf{s})g^{ji}_{\boldsymbol{\xi}}) = 0, \quad i,j=1,\ldots,n, \qquad (10.73)$$

obtained from the inverted diffusion equations (10.64). The dependent variables u^k in (10.72) within each grid cell are expanded in a spectral basis $\alpha_j(\boldsymbol{\xi})$,

$$u^k(t,\boldsymbol{\xi}) \approx \sum_{j=0}^{n} u_j^k(t)\alpha_j(\boldsymbol{\xi}). \tag{10.74}$$

Spatially discretized equations are obtained by a Galerkin method, taking the scalar product of (10.72) with each basis function and integrating by parts to obtain

$$\ddot{M}\dot{\mathbf{u}} = \mathbf{r} \equiv \int_{\Xi^n} (S^k\alpha_i + \mathbf{F}^k \cdot \nabla\alpha_i)\mathrm{d}\boldsymbol{\xi} - \int_{\partial\Xi^n} \mathbf{F}_i^k \cdot \hat{\mathbf{n}}\mathrm{d}\boldsymbol{\xi} \tag{10.75}$$

with $\ddot{M}$ the mass matrix, $M_{i,j} \equiv (\alpha_i, \alpha_j)$, and the $\mathbf{u}$ the vector of mode amplitudes $u_j^k(t)$. Integrals are evaluated by Gaussian quadrature to an order appropriate to the degree of the Jacobi polynomials. Fluxes and sources may depend in an arbitrary nonlinear manner on t, $\boldsymbol{\xi}$, u^k, and ∇u^k. The code is structured in such a way that the details of discretization and the specification of physics equations are separated into different subroutines, making it as simple as possible to encode complex physics. The discretized flux-source form preserves conservation properties to high order. Elliptic equations are treated by zeroing the mass matrix.

Time discretization of (10.75) is fully implicit in order to treat multiple time scales efficiently and accurately,

$$\ddot{M}\left(\frac{\mathbf{u}^+ - \mathbf{u}^-}{h}\right) = \theta\mathbf{r}^+(\mathbf{u}^+) + (1-\theta)\mathbf{r}^-(\mathbf{u}^-) \tag{10.76}$$

with the time-centering parameter θ normally chosen as $1/2$ (Crank–Nicholson) for accuracy. Solution of (10.76) requires finding the roots of the nonlinear residual,

$$\mathbf{R}(u^+) \equiv \ddot{M}(\mathbf{u}^+ - \mathbf{u}^-) - h[\theta\mathbf{r}^+ + (1-\theta)\mathbf{r}^-] = \mathbf{0}, \tag{10.77}$$

solved by Newton's iteration,

$$\mathbf{R} + \ddot{J}\delta\mathbf{u}^+ = \mathbf{0}, \quad \delta\mathbf{u}^+ = -\ddot{J}^{-1}\mathbf{R}(\mathbf{u}^+), \quad \mathbf{u}^+ \to \mathbf{u}^+ + \delta\mathbf{u}^+ \tag{10.78}$$

with the Jacobian defined as $\ddot{J} \equiv \ddot{M} - h\theta\{\partial r_i^+/\partial u_j^+\}$.

Efficient solution of the large sparse linear system in (10.78) is greatly enhanced by the method of static condensation. Because of the C^0 nature of the spectral element representation, discussed above, higher-order elements in one grid cell couple to those in neighboring grid cells only through the shared linear finite elements which straddle cell boundaries.

10.7 Formulation of Control Metrics

The control metric serves for controlling the properties of grids in the physical geometry S^{xn}. It is evident that the metric should take into account the quantities requiring grid adaptation: geometric characteristics of S^{xn}, physical variables and their derivatives, specified vector fields, error of the numerical solution applied, etc.

For the purpose of better handling grid control, it is reasonable to restrict the whole set of the control metrics to a basic subset which, however, can be adequate for realizing the necessary grid properties. Also, these basic control metrics are to be described by simple formulas which allow one to establish readily the relations between them and the grid characteristics. Further, one of the natural ways to satisfy balanced grid properties, each of which is realized by an individual control metric, is to combine these metrics linearly. Therefore, the basic metric tensors formulated should be subject to the operation of summation in the sense that the sum of the two control metrics from the subset is also the metric (nonsingular tensor). Note, in general, the sum of two metrics may not be a metric, since the sum of two nonsingular matrices may be a singular matrix. In addition, a mathematical formulation of the control metric should be simple and comprehensive, so that a robust grid code could be developed for automatic generation of grids with required properties.

This section describes an approach for formulating such basic control metrics.

The most general and simple formulation of the imposed control metric on S^{xn} whose covariant elements in the parametric coordinates $s^1, \ldots, s^n$, are designated by $g^{\mathbf{s}}_{ij}$, is given through a set of covariant tensors of the first rank

$$\mathbf{F}^k(\mathbf{s}) = [F^k_1(\mathbf{s}), \ldots, F^k_n(\mathbf{s})], \quad k = 1, \ldots, l,$$

by the following formula

$$g^{\mathbf{s}}_{ij} = z(\mathbf{s}) g^{xs}_{ij} + F^k_i(\mathbf{s}) F^k_j(\mathbf{s}), \quad i, j = 1, \ldots, n, \ k = 1, \ldots, l, \tag{10.79}$$

where g^{xs}_{ij} is the metric of S^{xn}, $z(\mathbf{s}) \geq 0$ is a weight function specifying the contribution of this metric to the control metric $g^{\mathbf{s}}_{ij}$. In particular, the metric (10.79) with $z(s) = 1$, $F^k_i = \partial f^k / \partial s^i$ has the form (10.8) and (10.9). Also suggesting in (10.79) $z(s) = 0$, $F^k_i = \partial \psi^k / \partial s^i$, $i, k = 1, \ldots, n$, yields the metric (10.57).

Of course, it is assumed that the function $z(\mathbf{s})$ and the vectors $\mathbf{F}^k(\mathbf{s})$, $k = 1, \ldots, l$, in (10.79) are subject to the restriction

$$g^{\mathbf{s}} = \det(g^{\mathbf{s}}_{ij}) > 0,$$

at each point $\mathbf{s} \in S^n$.

Notice, if we introduce in R^{n+k+l} vectors $\mathbf{w}_i(\mathbf{s})$, $i = 1, \ldots, n$ by the formula

$$\mathbf{w}_i(\mathbf{s}) = \left[\sqrt{z(\mathbf{s})}\frac{\partial \mathbf{x}}{\partial s^i}, F_i^1, \dots, F_i^l\right], \quad i = 1, \dots, n,$$

where $\mathbf{x}(\mathbf{s})$ is the parametrization (10.1) of the physical geometry S^{xn}, then it is obvious that

$$g_{ij}^{\mathbf{s}} = \mathbf{w}_i \cdot \mathbf{w}_j, \quad i, j = 1, \dots, n.$$

So, for nonsingularity of the control metric tensor (10.79) the vectors $\mathbf{w}_i(\mathbf{s})$, $i = 1, \dots, n$, must be independent. In particular, the vectors $\mathbf{w}_i(\mathbf{s})$, $i = 1, \dots, n$, will be independent if $z(\mathbf{s}) > 0$ at each point $\mathbf{s} \in S^n$ since the tangent vectors $\partial \mathbf{x}/\partial s^i$, $i = 1, \dots, n$, of S^{xn} are independent.

It is evident that the linear combination of two metric tensors of the form (10.79) with corresponding nonnegative coefficients $\varepsilon_1(\mathbf{s})$ and $\varepsilon_2(\mathbf{s})$ is the matrix of the same form (10.79), and it is nonsingular (metric tensor) if $\varepsilon_1(\mathbf{s}) + \varepsilon_2(\mathbf{s}) > 0$ at each point of $\boldsymbol{s} \in S^n$.

Sometimes, instead of the covariant metric components $g_{ij}^{\mathbf{s}}$ it is convenient, in particular in order to define the measure of grid departure (10.55) from a required grid, to formulate the contravariant components $g_{\mathbf{s}}^{ij}$ of the control metric, for example, in the form (10.79), namely as

$$g_{\mathbf{s}}^{ij} = \epsilon(\mathbf{s})g_{sx}^{ij} + B_k^i B_k^j, \quad i, j = 1, \dots, n, \ k = 1, \dots, l, \tag{10.80}$$

where $\epsilon(\mathbf{s}) \geq 0$, g_{sx}^{ij} are the contravariant metric elements of S^{xn}, while $B_k^i, i = 1, \dots, n$, are the components of a contravariant vector $\mathbf{B}_k = (B_k^1, \dots, B_k^n)$, $k = 1, \dots, l$.

10.7.1 Specification of Individual Control Metrics

10.7.1.1 Control Metric for Generating Field-Aligned Grids

A contravariant metric tensor in the form (10.80) allows one to control the angle between the normal to a grid coordinate hypersurface in S^{xn} and a specified vector field, in particular, to generate vector field-aligned grids. As a tensor of the first rank in the formula (10.80) one may take either the same or a transformed vector field.

The need for a vector field-aligned coordinate system can be appreciated in the case $S^{xn} = S^n$ by considering the heat equation in a magnetized plasma, written in the grid coordinates ξ^i,

$$\frac{\partial T}{\partial t} = \nabla \cdot (\chi \cdot \nabla T) = \frac{1}{J}\frac{\partial}{\partial \xi^i}\left(J\chi : \nabla\xi^i \nabla\xi^j \frac{\partial T}{\partial \xi^j}\right), \quad i, j = 1, 2, 3. \tag{10.81}$$

By far the largest component of the anisotropic thermal conductivity tensor χ equals $\chi_{\parallel}\mathbf{bb}$, where $\chi_{\parallel}$ is parallel conductivity and $\mathbf{b}$ is the unit vector along the magnetic field $\mathbf{B} = (B^1, \dots, B^n)$. If $\mathbf{b} \cdot \nabla\xi^i$ is of order unity for all grid coordinates ξ^i, then the much smaller transverse terms in (10.81) involve the difference between large terms, resulting in a loss of numerical accuracy.

If it vanishes, or nearly so, for all but one coordinate, this inaccuracy can be avoided. Similar considerations hold for other manifestations of magnetic anisotropy.

For simple magnetic fields, such as those in the core region of the tokamak, it is possible to define a flux coordinate ψ labeling the magnetic surfaces, satisfying $\mathbf{B} \cdot \nabla\psi = 0$ exactly. In more complicated cases, such as nonaxisymmetric magnetic fields with multiple islands and regions of stochasticity, this is not possible. For the choice of the contravariant control metric tensor $(\mathbf{g}_{\mathbf{s}}^{ij}) = \mathbf{BB} = (B^i B^j)$, (10.52) or (10.53) may be interpreted as a variational principle for minimizing $\mathbf{B} \cdot \nabla\xi^i$ for one of the coordinates. This is too simple, however, because such a tensor is singular, $g_{\mathbf{s}} = \det(\mathbf{g}_{\mathbf{s}}^{ij}) = 0$.

The matrix $(g_{\mathbf{s}}^{ij})$ whose elements are specified in the form (10.80) is nondegenerate for an arbitrary $\epsilon(\mathbf{s}) > 0$; in addition this matrix is close to the matrix $(B^i B^j)$ when both $\epsilon(\mathbf{s})$ and $\mathbf{B}_k$, $k = 2, \dots, l$ are small and $\mathbf{B}_1 = \mathbf{B}$. Assuming this matrix as a contravariant metric tensor of a monitor manifold over S^n yields, in accordance with (10.55), the following measure of grid nonalignment with the vector field $\boldsymbol{B}$

$$
\begin{aligned}
m(\boldsymbol{s}) &= w(\boldsymbol{s})\big[\epsilon(\boldsymbol{s})\delta_k^j + B^j B^k + B_m^j B_m^k\big]\frac{\partial \xi^i}{\partial s^j}\frac{\partial \xi^i}{\partial s^k},\\
&i, j, k = 1, \dots, n, \ \ m = 2, \dots, l,
\end{aligned}
$$

and the contravariant elements

$$
\begin{aligned}
g_{\boldsymbol{s}}^{ij} &= \epsilon(\boldsymbol{s})\delta_j^i + B^i B^j + B_m^i B_m^j,\\
&i, j, m = 1, \dots, n, \ \ m = 2, \dots, l,
\end{aligned}
\tag{10.82}
$$

of the control metric for generating vector field-aligned grids.

While this approach to grid alignment with the magnetic field was motivated by the work of Brackbill (1993), there is a substantial difference. Whereas Brackbill has a variational principle similar to (10.52) or (10.53), his integrand is designed to make a coordinate gradient *parallel* to a specified vector field; the approach described above is designed to make a coordinate gradient *perpendicular* to a specified vector field.

10.7.1.2 Control Metric for Generating Grids Adapting to the Values of a Function

For generating a numerical grid in a physical geometry S^{xn} with node clustering in the zones of large values of a function $|v(\mathbf{s})|$ the measure of departure from the necessary grid in S^{xn} can be expressed in the form

$$
m(\mathbf{s}) = Z[v](\mathbf{s})g_{sx}^{km}\frac{\partial \xi^i}{\partial s^k}\frac{\partial \xi^i}{\partial s^m}, \quad i, j, k, m = 1, \dots, n,
\tag{10.83}
$$

where $Z[v] > 0$ is a positive operator such that $Z[v](\mathbf{s})$ is large (small) where $|v(\mathbf{s})|$ is small (large), in particular,

$$Z[v](\boldsymbol{s}) = \frac{1}{c_1|v(\boldsymbol{s})| + c_2}, \quad c_1 > 0,\ c_2 > 0. \tag{10.84}$$

So the contravariant elements of the control metric are as follows:

$$g_{\mathbf{s}}^{ij} = Z[v](\mathbf{s}) g_{sx}^{ij}, \quad i,j = 1,\dots,n. \tag{10.85}$$

Consequently, the covariant elements of the control metric are written out as

$$g_{ij}^{\mathbf{s}} = \Psi[v](\mathbf{s}) g_{ij}^{xs}, \quad i,j = 1,\dots,n, \tag{10.86}$$

where $\Psi[v](\mathrm{s}) = 1/Z[v](\mathrm{s})$, therefore $g^{s} = g^{xs}(z[v])^{-n/2}$. So the Beltrami equations in (10.56) with respect to the metric (10.86) are expressed in the form

$$\frac{\partial}{\partial s^j}\left((Z[v])^{(2-n)/2}\sqrt{g^{xs}}\, g_{sx}^{jk}\frac{\partial \xi^i}{\partial s^k}\right) = 0, \quad i,j = 1,\dots,n, \tag{10.87}$$

while the inverted Beltrami equations with respect to the metric (10.86) are as follows:

$$B_n^{x\xi}[s^k] = J^2\sqrt{g^{xs}}(Z[v])^{(n-2)/2}\frac{\partial}{\partial s^j}\left((Z[v])^{(2-n)/2}\sqrt{g^{xs}}\, g_{xs}^{jk}\right),$$
$$j,k = 1,\dots,n, \tag{10.88}$$

where

$$B_n^{x\xi}[y] = g^{x\xi} g_{\xi x}^{ij}\frac{\partial^2 y}{\partial \xi^i \partial \xi^j}, \quad i,j = 1,\dots,n.$$

In particular, if S^{xn} coincides with S^n, i.e. $g_{ij}^{xs} = \delta_j^i$ and consequently $g^{xs} = 1$, $g_{sx}^{ij} = \delta_j^i$, then the Beltrami equations (10.87) are as:

$$\frac{\partial}{\partial s^j}\left((Z[v])^{(2-n)/2}\frac{\partial \xi^i}{\partial s^j}\right) = 0, \quad i,j = 1,\dots,n.$$

This system of equations is equivalent to the system of Poisson equations (6.13)

$$\frac{\partial^2 \xi^i}{\partial s^j \partial s^j} = P^i, \quad i = 1,\dots,n,$$

where

$$P^i = \frac{n-2}{2}\frac{1}{Z[v]}\frac{\partial Z[v](\boldsymbol{s})}{\partial s^j}\frac{\partial \xi^i}{\partial s^j} = \frac{n-2}{2}\frac{1}{Z[v]}\frac{\partial Z[v](\boldsymbol{s}(\boldsymbol{\xi}))}{\partial \xi^m} g_{\xi x}^{im},$$
$$i,j,m = 1,\dots,n.$$

Consequently, the inverted equations are as follows:

$$B_n^{x\xi}[s^k] = \frac{2-n}{2}J^2\frac{1}{Z[v]}\frac{\partial}{\partial s^k}Z[v](\boldsymbol{s}), \quad k = 1,\dots,n. \tag{10.89}$$

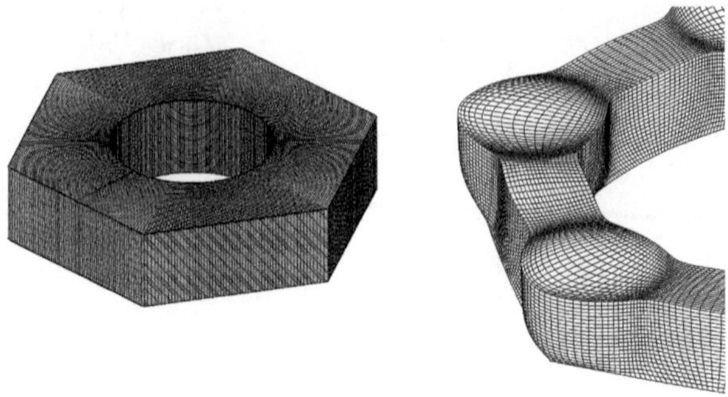

Fig. 10.5. Three-dimensional domain (*left*) and a fragment of an adaptive grid (*right*) in the interior of the domain

Figure 10.5 illustrates an adaptive three-dimensional grid obtained by the numerical solution of equations (10.89) through the algorithm described in Sect. 10.6.1. The grid nodes cluster in the zones of large values of a function $v(\boldsymbol{s})$.

The diffusion equations in (10.58) with respect to the metric (10.86) are expressed as:

$$\frac{\partial}{\partial s^j}\left(w(\boldsymbol{s})Z[v]g^{jk}_{sx}\frac{\partial \xi^i}{\partial s^k}\right) = 0, \quad i,j = 1,\dots,n. \tag{10.90}$$

So, the inverted diffusion equations in the metric (10.85) necessary to generate the adaptive grids are:

$$B^{x\xi}_n[s^k] = \frac{J^2 g^{xs}}{w(\boldsymbol{s})Z[v]}\frac{\partial}{\partial s^j}\left(w(\boldsymbol{s})Z[v]g^{jk}_{xs}\right), \quad j,k = 1,\dots,n. \tag{10.91}$$

These equations for $n = 2$ are more efficient for adaptation than the inverted Beltrami equations (10.88) and (10.89), which for $n = 2$ do not depend on the operator $Z[v]$.

10.7.1.3 Control Metrics for Generating Grids Adapting to the Gradient of a Function

The contravariant metric tensor (10.85) can also be used for providing node clustering in the zones of large variations of a function $\mathbf{f}(\mathbf{s}) = (f^1(\boldsymbol{s}),\dots, f^l(\boldsymbol{s}))$, introducing for this purpose an operator $Z[\mathrm{grad}\,\mathbf{f}]$ such that $Z[\mathrm{grad}\,\mathbf{f}](\mathbf{s})$ is large where $|\mathrm{grad}\,\mathbf{f}|(\mathbf{s})$ is small and vice versa. Figure 10.6 illustrates an adaptive prismatic grid with such node clustering.

One more control metric for generating grids with node clustering in the zones of the large variation of the function $\boldsymbol{f}(\boldsymbol{s})$ is of the form

$$g^{\boldsymbol{s}}_{ij} = \varepsilon(\boldsymbol{s})g^{xs}_{ij} + \varepsilon_k(\boldsymbol{s})\frac{\partial f^k}{\partial s^i}\frac{\partial f^k}{\partial s^j}, \quad i,j = 1,\dots,n, \; k = 1,\dots,l, \tag{10.92}$$

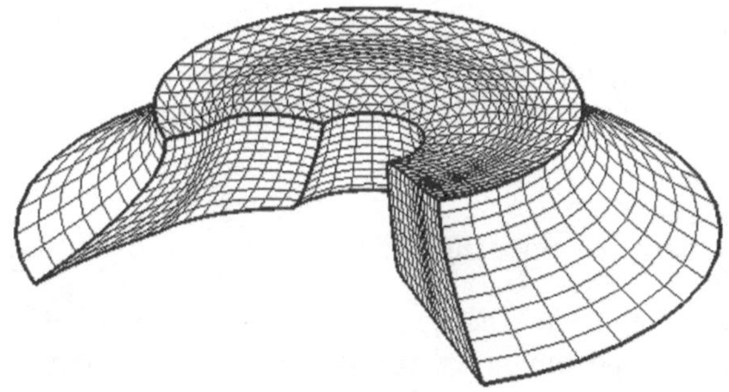

Fig. 10.6. Three-dimensional prismatic adaptive grid

where $\varepsilon(\boldsymbol{s}) \geq 0$, $\varepsilon_k(\boldsymbol{s}) \geq 0$, $k = 1, \ldots, l$. This control metric is a generalization of the metric of a monitor hypersurface (see formulas (10.8) and (10.9)).

10.7.2 Control Metrics for Generating Grids with Balanced Properties

For computing numerical grids that are field-aligned and adaptive to the values of one function and/or to the variation of another function, a natural way for defining a control metric consists in combining the corresponding metrics; i.e., the covariant elements of the balanced control metric are to have a form

$$g^{\mathbf{s}}_{ij}(\mathbf{s}) = \varepsilon_1(\mathbf{s})g^{al}_{ij} + \varepsilon_2(\mathbf{s})g^{adg}_{ij} + \varepsilon_3(\mathbf{s})g^{adv}_{ij}, \quad i,j = 1, \ldots, n, \tag{10.93}$$

where $\varepsilon_k(\mathbf{s}) \geq 0$, $k = 1, 2, 3$ are the weight functions specifying the contribution of the covariant elements g^{al}_{ij}, g^{adg}_{ij}, and g^{adv}_{ij}. The marks *al*, *adg* and *adv* in this formula mean that the corresponding metric elements are chosen to grid alignment, adaptation to gradients of a function, and adaptation to the values of the same of another function, respectively. It is evident that (10.93) will be a covariant metric tensor if

$$\varepsilon_1(\boldsymbol{s}) + \varepsilon_2(\boldsymbol{s}) + \varepsilon_3(\boldsymbol{s}) > 0, \quad \boldsymbol{s} \in S^n.$$

Analogously, there is written out a formula for the contravariant components $g^{ij}_{\boldsymbol{s}}$ of the balanced control metric

$$g^{ij}_{\mathbf{s}}(\mathbf{s}) = w_1(\mathbf{s})g^{ij}_{al} + w_2(\mathbf{s})g^{ij}_{adg} + w_3(\mathbf{s})g^{ij}_{adv}, \quad i,j = 1, \ldots, n. \tag{10.94}$$

There may be other effective ways for combining the corresponding tensor components; in particular, for generating grids that are field-aligned and adaptive to the values of a function $v(\boldsymbol{s})$, adequate results are demonstrated with the help of the following formula for the contravariant control metric elements

$$g^{ij}_{\mathbf{s}}(\mathbf{s}) = Z[v](\mathbf{s})g^{ij}_{al}, \quad i,j = 1, \ldots, n. \tag{10.95}$$

This metric provides a compromise between alignment and adaptation. An example of the grid generated by the spectral-element method through the metric (10.95) with $Z[v](\mathbf{s})$ specified by (10.84) was demonstrated in Glasser et al. (2006).

10.7.3 Application to Solution of Singularly-Perturbed Equations

As an example we consider the application of the adaptive grid technology described above to the numerical solution of the following boundary value problem for a singularly- perturbed equation

$$\begin{aligned} &-\varepsilon\Delta u + \mathbf{a}(\mathbf{s})\cdot\nabla u + f(\mathbf{s}, u) = 0, \quad \mathbf{s}\in S^n, \\ &u(\mathbf{s}) = u_0(\mathbf{s}), \quad \mathbf{s}\in\partial S^n, \end{aligned} \tag{10.96}$$

where

$$\Delta u = \sum_{i=1}^{n}\frac{\partial^2 u}{\partial s^i\partial s^i}, \quad i = 1,\ldots,n, \qquad \nabla u = \left(\frac{\partial u}{\partial s^1},\ldots,\frac{\partial u}{\partial s^n}\right),$$

$S^n \subset R^n$ ia a bounded domain, $\mathbf{a}(\mathbf{s}) = (a^1(s),\ldots,a^n(s))$ is the convection vector, $f(\mathbf{s}, u)$ is a specified function, while $0 < \varepsilon \ll 1$ is the coefficient of diffusion.

Depending on the function $\boldsymbol{a}(\boldsymbol{s})$, the solution of (10.96) has boundary and/or interior layers when the coefficient ε is small.

The boundary value problem (10.96) is written out in the grid coordinates ξ^i as:

$$\begin{aligned} &-\varepsilon B_n^{s\xi}[u] + F[u] = 0, \\ &u(\boldsymbol{\xi}) = \varphi(\boldsymbol{\xi}), \quad \boldsymbol{\xi}\in\partial\Xi^n, \end{aligned} \tag{10.97}$$

where $\varphi(\boldsymbol{\xi}) = u_0[\boldsymbol{s}(\boldsymbol{\xi})]$,

$$F[u] = \{(J)^2 a^i[\boldsymbol{s}(\boldsymbol{\xi})] + \varepsilon B_n^{s\xi}[s^i]\}\frac{\partial u}{\partial\xi^p}\frac{\partial\xi^p}{\partial s^i} + (J)^2 f(\boldsymbol{s}(\boldsymbol{\xi}), u), \quad i, p = 1,\ldots,n,$$

$$B_n^{s\xi}[u] = g^{s\xi}g^{ij}_{\xi s}\frac{\partial^2 u}{\partial\xi^i\partial\xi^j},$$

$$g^{ij}_{\xi s} = \frac{\partial\xi^i}{\partial s^k}\frac{\partial\xi^j}{\partial s^k}, \qquad g^{s\xi} = (J)^2, \qquad J = \det\left(\frac{\partial s^i}{\partial\xi^j}\right), \quad i, j = 1,\ldots,n.$$

Similar to (10.67), the problem (10.97) is replaced by the following nonstationary problem with respect to the function $u(\boldsymbol{\xi}, t)$

$$\begin{aligned} &\frac{\partial u}{\partial t} = J\{\varepsilon B_n^{s\xi}[u] - F[u]\}, \\ &u(\boldsymbol{\xi}, t) = \varphi_0(\boldsymbol{\xi}), \quad \boldsymbol{\xi}\in\partial\Xi^n, \\ &u(\boldsymbol{\xi}, 0) = u_0(\boldsymbol{\xi}), \quad \boldsymbol{\xi}\in\Xi^n. \end{aligned} \tag{10.98}$$

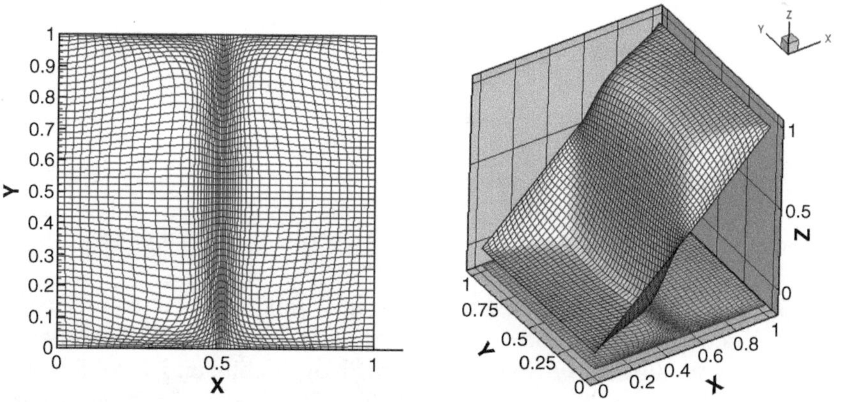

Fig. 10.7. Adaptive grid (*left*) and the solution with the interior layer (*right*)

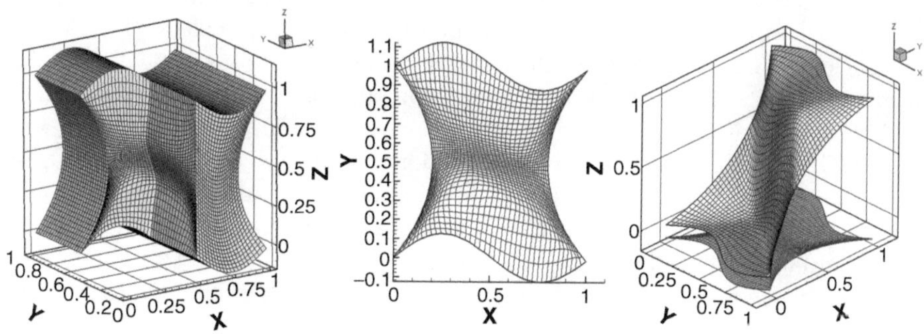

Fig. 10.8. Three-dimensional adaptive grid (*left*), its two-dimensional fragment (*center*), and the solution at the nodes of a coordinate surface (*right*)

Since the operator $B_n^{s\xi}$ is in fact the operator $B_n^{\boldsymbol{\xi}}$ in (10.67) with $g_{ij}^{\boldsymbol{\xi}} = g_{ij}^{s\xi}$, the problem (10.98) is solved with the problem (10.67) by the same method described in Sect. 10.6.1.

Figures 10.7 and 10.8 show the resulting two- and three-dimensional adaptive grids with node clustering in the zones of the interior layers of $u(\boldsymbol{s})$.

10.8 Comments

The method presented here, based on the utilization of the invariants of the metric tensor of the hypersurface to define the functional of smoothness, was formulated and justified by Liseikin (1991a, 1992, 1993a, 1999). All properties and interpretations of the functionals and the corresponding Beltrami and diffusion equations described here were studied and published by Liseikin 1991a, 1992, 1993a, 1996a, 1999, 2001a, 2001b, 2002a, 2002b, 2003, 2004, 2005, 2007).

Functionals of nonconformality to generate grids on hypersurfaces were formulated, in analogy to the functional (8.33) or (8.37), by Liseikin (1991a, 1999). Numerical implementations of the functional (10.23) with $n = 3$, $\alpha = 3/2$ for generating three-dimensional adaptive grids were performed by Branets and Garanzha (2002) and Azarenok (2007).

The formulas of the general control metrics were proposed by Liseikin (2004, 2007). An application of the geometry of manifolds and the control metrics to generating adaptive grids was performed by Liseikin (2004, 2005, 2007).

Implementations of the inverted Beltrami and diffusion equations with respect to the control metric in numerical codes to generate adaptive grids in domains and on surfaces were performed by Kupin and Liseikin (1994), Liseikin and Petrenko (1994), Liseikin (2004), Shokin et al. (2005), Glasser et al. (2006), Liseikin (2007), Liseikin, Likhanova, and Shokin (2007).

The measure (10.83) for generating adaptive grids in domains was introduced by Danaev, Liseikin, and Yanenko (1980) and Winslow (1981). The generation of grids through such a metric is helpful for numerically solving problems with strong anisotropy, and problems of magnetized plasmas, in particular.

The finite-difference algorithm of fractional steps described in this chapter was proposed by Yanenko (1971). Other versions of this algorithm that can be readily implemented for solving the resulting multidimensional grid equations, in particular the popular ADI (alternating direction implicit) method, were reviewed by Kovenya, Tarnavskii, and Chernyi (1990), Fletcher (1997), and Langtangen (2003).

Alternative finite-difference grid generation algorithms based on the minimization of the inverted energy and diffusion functionals in the Eucledian metric were developed by Ivanenko and Charakhch'yan (1988), while in the metric of a monitor hypersurface, proposed by Liseikin (1991a, 1991b), they were developed by Charakhch'yan and Ivanenko (1997), Garanzha (2000), and Azarenok and Ivanenko (2001).

The spectral element method for solution of plasma equations was developed by Glasser and Tang (2004). Applications of this method for computing plasmas and inverted diffusion grid equations were demonstrated by Glasser, Liseikin, and Kitaeva (2005), Glasser et al. (2005), and Liseikin (2007).

11

Unstructured Methods

11.1 Introduction

Unstructured mesh techniques occupy an important niche in grid generation. The major feature of unstructured grids consists, in contrast to structured grids, in a nearly absolute absence of any restrictions on grid cells, grid organization, or grid structure. Figuratively speaking, unstructured grids manifest the domination of anarchy while structured grids demonstrate adherence to order. The concept of unstructured grids allows one to place the grid nodes locally irrespective of any coordinate directions, so that curved boundaries can be handled with ease and local regions in which the solution variations are large can be resolved with a selective insertion of new points without unduly affecting the resolution in other parts of the physical domain.

Unstructured grid methods were originally developed in solid mechanics. Nowadays these methods influence many other fields of application beyond solid modeling, in particular, computational fluid dynamics where, they are becoming widespread.

Unstructured grids can, in principle, be composed of cells of arbitrary shapes built by connecting a given point to an arbitrary number of other points, but are generally formed from tetrahedra and hexahedra (triangles and quadrilaterals in two dimensions). The advantages of these grids lie in their ability to deal with complex geometries, while allowing one to provide natural grid adaptation by the insertion of new nodes.

At the present time the methods of unstructured grid generation have reached the stage where three-dimensional domains with complex geometry can be successfully meshed. The most spectacular theoretical and practical achievments have been connected with the techniques for generating tetrahedral (or triangular) grids. There are at least two basic approaches that have been used to generate these meshes: Delaunay, and advancing-front. This chapter presents a review of some popular techniques realizing these approaches.

V.D. Liseikin, *Grid Generation Methods*, Scientific Computation,
DOI 10.1007/978-90-481-2912-6_11, © Springer Science+Business Media B.V. 2010

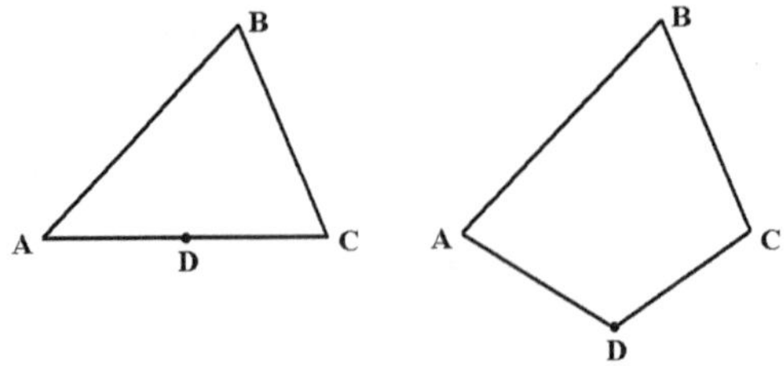

Fig. 11.1. Convex (*left*) and strongly convex (*right*) quadrilateral cells

Note that the chapter addresses only some general aspects of unstructured grid methods. The interested reader who wishes to learn more about the wider aspects of unstructured grids should study, for example, the monographs by Carey (1997), George and Borouchaki (1998), and Frey and George (2008).

11.2 Consistent Grids and Numerical Relations

This section presents some considerations associated with consistent unstructured grids. Such grids are composed of convex cells. The consistency property of unstructured grids is often verified through relations which connect the numbers of edges, faces, and cells of the grids. These relations are also discussed in this section.

11.2.1 Convex Cells

A convex n-dimensional cell S is the convex hull of some $n+k,\ k>1$, points $\boldsymbol{P}_1,\dots,\boldsymbol{P}_{n+k}$ from X^n which do not lie in any $(n-1)$-dimensional plane. Thus S is composed of all points $\boldsymbol{x}\in R^n$ which are defined through $\boldsymbol{P}_i$ by the equation

$$\boldsymbol{x}=\sum_{i=1}^{n+k}\alpha^i\boldsymbol{P}_i,$$

$$\sum_{i=1}^{n+k}\alpha^i=1,\quad 1\geq\alpha^i\geq 0.$$

We call all those points $\boldsymbol{P}_l$ of the set $\{\boldsymbol{P}_i,\ i=1,\dots,n+k\}$ which lie on the boundary of S vertices of the convex cell S.

A m-dimensional face of the convex n-dimensional cell S $(n>m)$ is a convex hull of $m+1$ vertices P_l, which does not contain any other vertices of S.

We call the cell S strongly convex if for all $m<n$ it does not have any two m-dimensional faces which lie in an m-dimensional plane. Fig. 11.1 demonstrates the difference between convex (Fig. 11.1, left) and strongly convex

(Fig. 11.1, right) cells. Evidently, if $\boldsymbol{P}$ is an interior point of the strongly convex cell S with the vertices $\boldsymbol{P}_1, \ldots, \boldsymbol{P}_{n+k}$ then there will be at least $n+1$ points in the expansion of $\boldsymbol{P}$

$$\boldsymbol{P} = \sum_{i=1}^{n+k} \alpha^i \boldsymbol{P}_i, \qquad \sum_{i=1}^{n+k} \alpha^i = 1, \quad \alpha^i \geq 0, \; i = 1, \ldots, n+k,$$

with nonzero values of their coefficients α^i.

11.2.1.1 Simplexes and Simplex Cells

The simplest n-dimensional element applied to the discretization of domains is an n-dimensional cell which is the hull of $n+1$ points $\boldsymbol{x}_1, \ldots, \boldsymbol{x}_{n+1}$ that do not lie in any $(n-1)$-dimensional plane. Such cells are called simplexes. Thus a simplex is formed by the points $\boldsymbol{x}$ from R^n satisfying the equation

$$\begin{aligned} \boldsymbol{x} &= \sum_{i=1}^{n+1} \alpha^i \boldsymbol{x}_i, \quad i = 1, \ldots, n+1, \\ &\sum_{i=1}^{n+1} \alpha^i = 1, \quad \alpha^i \geq 0. \end{aligned} \tag{11.1}$$

This simplex is a strongly convex cell whose vertices are the points $\boldsymbol{x}_1, \ldots, \boldsymbol{x}_{n+1}$. Each m-dimensional face of the simplex is an m-dimensional simplex defined through $m+1$ vertices. The point $\boldsymbol{x}$ is an interior point of the simplex if $\alpha^i > 0$ for all $i = 1, \ldots, n+1$. It is obvious that a three-dimensional simplex is a tetrahedron with the vertices $\boldsymbol{x}_1$, $\boldsymbol{x}_2$, $\boldsymbol{x}_3$, $\boldsymbol{x}_4$ while a two-dimensional simplex is a triangle and a one-dimensional simplex is an interval.

In practical discretizations of domains, convex cells whose boundary faces are simplexes are also applied. Such cells are referred to as simplex cells.

For each n-dimensional simplex cell S the following relation between the number of faces is valid:

$$\begin{aligned} &\sum_{i=k}^{n-1} (-1)^i \binom{i+1}{k+1} N_i = (-1)^{n-1} N_k, \\ &k = -1, \ldots, n-2, \quad N_{-1} = 1, \\ &\binom{l}{m} = \frac{l(l-1)\cdots(l-m+1)}{m!}, \quad m \geq 1, \binom{l}{0} = 1, \end{aligned} \tag{11.2}$$

where N_i, $i = 1, \ldots, n$, is the number of i-dimensional boundary simplexes of S, and N_0 is the number of vertices of S.

11.2.2 Consistent Grids

By a consistent grid or a consistent discretization we mean a set of points V from R^n and a collection of n-dimensional strongly convex cells T satisfying the following conditions:

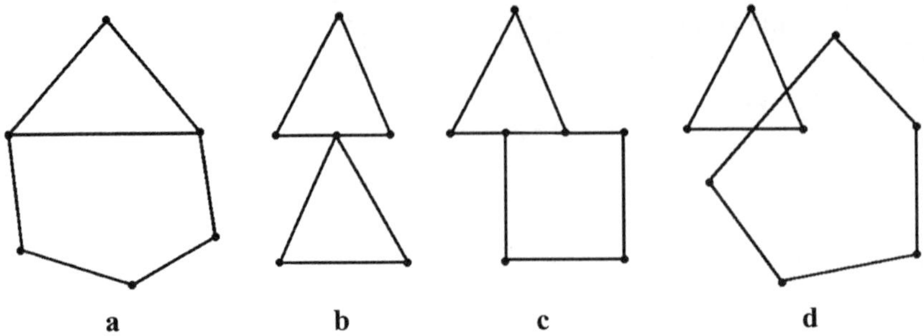

Fig. 11.2. Admitted (**a**) and nonadmitted (**b**, **c**, **d**) intersections of cells

(1) the set of the vertices of the cells of T coincides with V;
(2) if two different cells S_1 and S_2 intersect, then the region of the intersection is a common face for both cells.

This definition does not admit the fragments of discretizations depicted in Fig. 11.2(b), (c), (d).

If the union of the cells of the consistent discretization constitutes a simply connected n-dimensional domain, i.e. a domain which is homeomorphic to an n-dimensional cube, then, in accordance with the Euler theorem,

$$\sum_{i=0}^{n-1}(-1)^i N_i = 1 + (-1)^{n-1}, \tag{11.3}$$

where N_i, $i > 0$, is the number of i-dimensional boundary faces of the domain discretization, while N_0 is the number of boundary vertices. In particular, N_1 is the number of boundary edges. The relation (11.3) and the following ones can be used to verify the consistency of a generated grid.

11.2.2.1 Three-Dimensional Discretization

In three dimensions we have, for each consistent discretization, the following generalization of the Euler formula (11.3):

$$N_0 - N_1 + N_2 = 2(k - t), \tag{11.4}$$

where k is the number of simply connected subdomains and t is the number of holes.

For any convex three-dimensional cell, we also have the relations

$$\begin{aligned} &N_0 - N_1 + N_2 = 2, \\ &4 \le N_0 \le 2N_2 - 4, \\ &4 \le N_2 \le 2N_0 - 4. \end{aligned} \tag{11.5}$$

Let $N_0(k)$ be the number of those vertices each of which is common to k edges of a convex cell S. Analogously, by $N_2(k)$ we denote the number of faces each of which is formed by k edges. Obviously,

$$N_0 = \sum_{k\geq 3} N_0(k), \qquad N_2 = \sum_{k\geq 3} N_2(k).$$

Using (11.3) we obtain

$$N_0(3) + N_2(3) = 8 + \sum_{k\geq 4} (k-4)\big[N_0(k) + N_2(k)\big] \geq 8. \tag{11.6}$$

If the cell is a simplex cell, then we have from (11.2) the relations

$$\begin{aligned} N_0 - N_1 + N_2 &= 2, \\ -2N_1 + 3N_2 &= 0 \end{aligned} \tag{11.7}$$

or, in equivalent, form

$$\begin{aligned} N_1 &= 3N_0 - 6, \\ N_2 &= 2N_0 - 4. \end{aligned} \tag{11.8}$$

Also, the following nonlinear inequalities for a simplex cell are valid:

$$\begin{aligned} 2N_1 &< (N_0 - 1)N_0, \\ 3N_2 &< (N_0 - 2)N_1. \end{aligned} \tag{11.9}$$

11.2.2.2 Discretization by Triangulation

A consistent discretization by simplexes is called a triangulation. Let the number of edges be maximal for a given set of vertices of a two-dimensional triangulation. Let $C(P)$ be the boundary of the hull formed by all vertices of the triangulation and N_C be the number of vertices which lie in $C(P)$. Then the following relations are valid:

$$\begin{aligned} N_T &= 2(N_V - 1) - N_C, \\ N_E &= 3(N_V - 1) - N_C, \end{aligned} \tag{11.10}$$

where N_T is the number of triangles, N_E is the number of edges, and N_V is the number of vertices of the triangulation. Thus the maximal triangulation for the given vertices has a fixed number of triangles and edges.

11.3 Methods Based on the Delaunay Criterion

Much attention has been paid in the development of methods for unstructured discretizations to triangulations which are based upon the very simple geometrical constraint that the hypersphere of each n-dimensional simplex defined by $n + 1$ points is void of any other points of the triangulation. For example, in three dimensions the four vertices of a tetrahedron define a circumsphere which contains no other nodes of the tetrahedral mesh. This restriction is referred to as the Delaunay or incircle criterion, or the empty-circumcircle property. Triangulations obeying the Delaunay criterion are called Delaunay triangulations. They are very popular in practical applications owing to the following optimality properties valid in two dimensions:

(1) Delaunay triangles are nearly equilateral;
(2) the maximum angle is minimized;
(3) the minimum angle is maximized.

These properties give some grounds to expect that the grid cells of a Delaunay triangulation are not too deformed.

The Delaunay criterion does not give any indication as to how the grid points should be defined and connected. One more drawback of the Delaunay criterion is that it may not be possible to realize it over the whole region with a prespecified boundary triangulation. This disadvantage gives rise to two grid generation approaches of constrained triangulation which preserve the boundary connectivity and take into account the Delaunay criterion. In the first approach of constrained Delaunay triangulation the Delaunay property is overriden at points close to the boundaries and consequently the previously generated boundary grid remains intact. Alternatively, or in combination with this technique, points can be added in the form of a skeleton to ensure that breakthroughs of the boundary do not occur. Another approach, which observes the Delaunay criterion over the whole domain, is to postprocess the mesh by recovering the boundary simplexes which are missed during the generation of the Delaunay triangulation and by removing the simplexes lying outside the triangulated domain.

There are a number of algorithms to generate unstructured grids based on the Delaunay criterion in constrained or unconstrained forms.

Some methods for Delaunay triangulations are formulated for a preassigned distribution of points which are specified by means of some appropriate technique, in particular, by a structured grid method. These points are connected to obtain a triangulation satisfying certain specific geometrical properties which, to some extent, are equivalent to the Delaunay criterion.

Many Delaunay triangulations use an incremental Bowyer–Watson algorithm which can be readily applied to any number of dimensions. It starts with an initial triangulation of just a few points. The algorithm proceeds at each step by adding points one at a time into the current triangulation and locally reconstructing the triangulation. The process allows one to provide both solution-adaptive refinement and mesh quality improvement in the framework of the Delaunay criterion. The distinctive characteristic of this method is that point positions and connections are computed simultaneously.

One more type of algorithm is based on a sequential correction of a given triangulation, converting it into a Delaunay triangulation.

11.3.1 Dirichlet Tessellation

A very attractive means for generating a Delaunay triangulation of an assigned set of points is provided by a geometrical construction first introduced by Dirichlet (1850).

Consider an arbitrary set of points P_i, $i = 1, \ldots, N$, in the n-dimensional domain. For any point P_i we define a region $V(P_i)$ in R^n characterized by the property that it is constituted by the points from R^n which are nearer to P_i then to any other P_j, i.e.

$$V_i = \{x \in R^n | \mathrm{d}(x, P_i) \le \mathrm{d}(x, P_j),\ i \ne j,\ j = 1, \ldots, N\},$$

where $\mathrm{d}(a, b)$ denotes the distance between the points a and b. These areas V_i are called the Voronoi polyhedrons. Thus, the polyhedra are intersections of half-spaces and therefore they are convex, though not necessarily bounded. The set of Voronoi polyhedra corresponding, to the collection of points P_i is called the Voronoi diagramor Dirichlet tessellation. The common boundary of two facing Voronoi regions $V(P_i)$ and $V(P_j)$ is an $(n-1)$-dimensional polygon. A pair of points P_i and P_j whose Voronoi polyhedra have a face in common is called a configuration pair. By connecting only the contiguous points, a network is obtained. In this network, a set of $n+1$ points which are contiguous with one another forms an n-dimensional simplex. The circumcenter, i.e. the center of the hypersphere, of any simplex is a vertex of the Voronoi diagram. The hypersphere of the simplex is empty, that is, there is no point inside the hypersphere. Otherwise, this point would be nearer to the circumcenter than the points on the hypersphere. Thus the set of simplexes constructed in such a manner from the Dirichlet tessellation constitutes a new tessellation which satisfies the Delaunay criterion and is, therefore, a Delaunay triangulation. The boundary of the Delaunay triangulation built from the Voronoi diagram is the convex hull of the set of points P_i.

It should be noted that Delaunay triangulations and Dirichlet tessellations can be considered the geometrical duals of each other, in the sense that for every simplex S_i there exists a vextex P_i of the tessellation and, conversely, for every Voronoi region $V(P_j)$ there exists a vertex P_j of the triangulation. In addition, for every edge of the triangulation there exists a corresponding $(n-1)$-dimensional segment of the Dirichlet tessellation.

11.3.2 Incremental Techniques

The empty-hypercircle criterion of the Delaunay triangulation can be utilized to create incremental triangulation algorithms for arbitrary dimensions. Recall that by the Delaunay triangulation of a set V_N of N points in n-dimensional space we mean the triangulation of V_N by simplexes with the vertices taken from V_N such that no point lies inside the hypersphere of any n-dimensional simplex.

Here two incremental methods are presented. In the first method a new n-dimensional simplex is constructed during each stage of the triangulation, using for this purpose the given set of points. In the second technique each step produces several simplexes which are generated after inserting a new point.

11.3.2.1 A-Priori-Given Set of Points

Let a set of points V_N in a bounded n-dimensional domain X^n be given. We assume that these points do not lie in any $(n-1)$-dimensional hyperplane. The incremental technique starts by taking an $(n-1)$-dimensional face e (edge in two dimensions and triangle in three dimensions), commonly the one with the smallest size, and constructing hyperspheres through the vertices of e and any one of the remaining points of V_N. One of these hyperspheres formed by a point, say P_1, does not contain inside it any point of V_N. The $(n-1)$-dimensional simplex e and P_1 define a new n-dimensional simplex. In the next step the $(n-1)$-dimensional simplex e is taken out of consideration. The algorithm stops, and the triangulation is complete, when every boundary face corresponds to the side of one simplex and every internal $(n-1)$-dimensional simplex forms the common face of precisely two n-dimensional simplexes. It is clear that this algorithm is well suited to generate a Delaunay triangulation with respect to a prescribed boundary triangulation.

The set of points used to generate the triangulation can be built with a structured method or an octree approach, or by embedding the domain into a Cartesian grid. However, the most popular approach is to utilize the strategy of a sequential insertion of new points.

11.3.2.2 Modernized Bowyer–Watson Technique

Another incremental method, proposed by Baker (1989) and which is a generalization of the Bowyer–Watson technique, starts with some triangulation not necessarily a Delaunay one, of the set of N points $V_N = \{P_i | i = 1, \dots, N\}$ by an assambly of simplexes $T_N = \{S_j\}$. For any simplex $S \in T_N$, let R_S be the circumradius and Q_S the circumcenter of S. In the sequential-insertion technique, a new point P is introduced inside the convex hull of V_N. Let $B(P)$ be the set of the simplexes whose circumspheres contain the point P, i.e.

$$B(P) = \big\{S | S \in T_N, \ \mathrm{d}(P, Q_S) < R_S\big\},$$

where $\mathrm{d}(P, Q)$ is the distance between P and Q. All these simplexes from $B(P)$ form a region $\Gamma(P)$ surrounding the point P. This region is called the generalized cavity. The maximal simply connected area of $\Gamma(P)$ that contains the point P is called the principal component of $\Gamma(P)$ and denoted by Γ_P. The point P is checked to determine if it is visible from all boundary segments of the principal component or if it is obscured by some simplex. In the former case the algorithm generates new simplexes associated with P by joining all of the vertices of the principal component with the point P. In the latter case, either this point is rejected and a new one is introduced or the principal component Γ_P is reduced by excluding the redundant simplexes from $B(P)$ to obtain an area whose boundary is not obscured from P by any simplex. Then the new simplexes are formed as in the former case. The union of these simplexes and those which do not form the reduced region of the retriangulation defines a new triangulation of the set of $N+1$ points $V_{N+1} = V_N \cup \{P\}$. In this manner, the process proceeds by inserting new points, checking visibility, adjusting the principal component, and generating new simplexes. The new triangulation differs from the previous one only locally around the newly inserted point P.

In two dimensions we have that if the initial triangulation is the Delaunay triangulation then the region $\Gamma(P)$ is of star shape and consequently the boundary is visible from the point P and each step of the Bowyer–Watson algorithm produces a Delaunay triangulation. Thus in this case the Bowyer–Watson algorithm is essentially a "reconnection" method, since it computes how an existing Delaunay triangulation is to be modified because of the insertion of new points. In fact, the algorithm removes from the existing grid all the simplexes which violate the empty-hypersphere property because of the insertion of the new point. The modification is constructed in a purely sequential manner, and the process can be started from a very simple initial Delaunay triangulation enclosing all points to be triangulated (for example, that formed by one very large simplex or one obtained from a given set of boundary points) and adding one point after another until the necessary requirements for grid quality have been satisfied.

11.3.3 Approaches for Insertion of New Points

The sequential nature of the Bowyer–Watson algorithm gives rise to a problem of choosing the position where to insert the new point in the existing mesh, because a poor point distribution can eventually lead to an unsatisfactory triangulation. The new point should be chosen according to some suitable geometrical and physical criteria which depend on the existing triangulation and the behavior of the physical solution. The geometrical criteria commonly consist in the requirement for the grid to be smooth and for the cells to be of a standard uniform shape and of a necessary size. The physical criterion commonly requires the grid cells to be concentrated in the zones of large solution variations. With respect to the geometrical criterion of generating uniform cells, the vertices and segments of the Dirichlet tessellation are promising locations for placing a new point since they represent a geometrical locus which falls, by construction, midway between the triangulation points. Thus, in order to control the size and shape of the grid cells, there are commonly considered two different ways in which the new point is inserted. In the first, the new point is chosen at the vertex of the Voronoi polyhedron corresponding to the "worst" simplex. In the second way, the new point is inserted into a segment

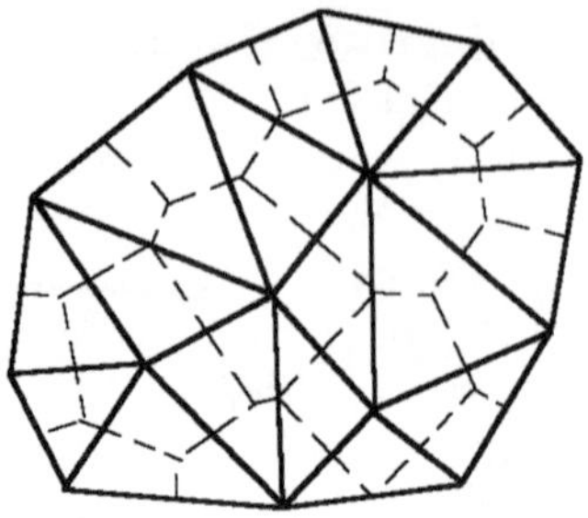

Fig. 11.3. Voronoi diagram and Delaunay triangulation

of the Dirichlet tessellation, in a position that guarantees the required size of the newly generated simplexes.

11.3.4 Two-Dimensional Approaches

This subsection discusses the major techniques delineated in Sects. 11.3.1–11.3.3 for generating planar triangulations based on the Delaunay criterion.

11.3.4.1 Voronoi Diagram

The Delaunay triangulation has a dual set of polygons referred to as the Voronoi diagram or the Dirichlet tessellation. The Voronoi diagram can be constructed for an arbitrary set of points in the domain. Each polygon of the diagram corresponds to the point that it encloses. The polygon for a given point is the region of the plane which is closer to that point then to any other points. These regions have polygonal shapes and the tessellation of a closed domain results in a set of nonoverlapping convex polygons covering the convex hull of the points. It is clear that the edge of a Voronoi polygon is equidistant from the two points which it separates and is thus a segment of the perpendicular bisector of the line joining these two points. The Delaunay triangulation of the given set of points is obtained by joining with straight lines all point pairs whose Voronoi regions have an edge in common. For each triangle formed in this way there is an associated vertex of the Voronoi diagram which is at the circumcentre of the three points which form the triangle. Thus each Delaunay triangle contains a unique vertex of the Voronoi diagram, and no other vertex within the Voronoi structure lies within the circle centered at this vertex. Figure 11.3 depicts the Voronoi polygons and the associated Delaunay triangulation.

It is apparent from the definition of a Voronoi polygon that degeneracy problems can arise in the triangulation procedure when

(1) three points of a potential triangle lie on a straight line;
(2) four or more points are cyclic.

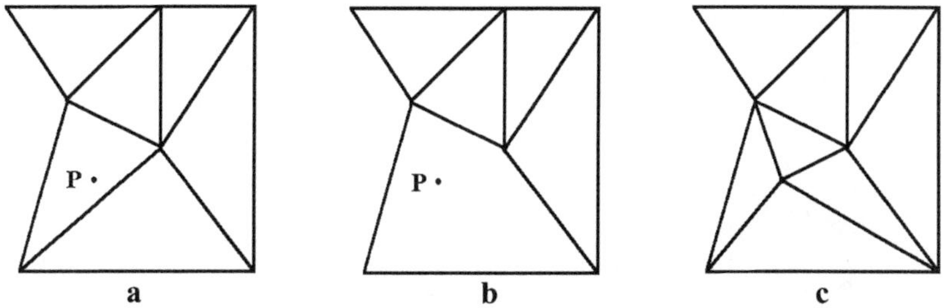

Fig. 11.4. Stages of the planar incremental algorithm

These cases are readily eliminated by rejecting or slightly moving the point which causes the degeneracy from its original position.

11.3.4.2 Incremental Bowyer–Watson Algorithm

The two-dimensional incremental technique, introduced independently by Bowyer (1981) and Watson (1981), triangulates a set of points in accordance with the requirement that the circumcircle through the three vertices of a triangle does not contain any other point. The accomplishment of this technique starts from some Delaunay triangulation which is considered as an initial triangulation. The initial triangulation commonly consists of a square divided into two triangles which contain the given points. With this starting Delaunay triangulation, a new grid node is chosen from a given set of points or is found in accordance with some user-specified rule to supply new vertices. In order to define the grid cells which contain this point as a vertex, all the cells whose circumcircles enclose the inserted point are identified and removed. The union of the removed cells forms the region which is referred to as the Delaunay or inserting cavity. A new triangulation is then formed by joining the new point to all boundary vertices of the inserting cavity created by the removal of the identified triangles. Fig. 11.4 represents the stages of the planar incremental algorithm.

The distinctive feature of the two-dimensional Delaunay triangulations is that all edges of the Delaunay cavity are visible from this inserted point, i.e. each point of the edges can be joined to it by a straight line which lies in the cavity.

Properties of the Planar Delaunay Cavity

In order to prove the fact that all boundary edges of the Voronoi cavity are visible from the introduced point, we consider an edge AB lying on the boundary of the cavity. Let ABC be the triangle with the vertices A, B, and C, which lies in the Delaunay cavity formed by the insertion of the point, denoted by P (Fig. 11.5). It is obvious that all edges of triangle ABC are visible if P lies inside the triangle. Let P lie outside the triangle. As this triangle lies

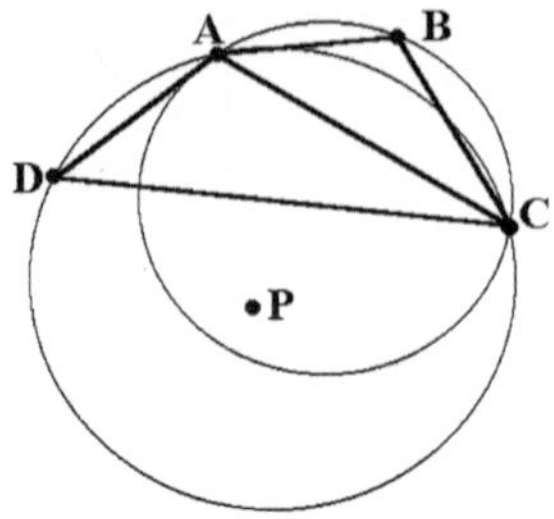

Fig. 11.5. Illustration of the inserted point P and the *triangles* of the Delaunay cavity

in the Delaunay cavity, it follows that P lies inside circle ABC. In this case the quadrilateral whose vertices are the points $ABCP$ is convex. Thus P has to be visible from edge AB unless we have a situation like the one depicted in Fig. 11.5, in which some triangle ACD separates the edge from P. As triangle ACD belongs to the initial Delaunay triangulation, the vertex D lies outside circle ABC. However, since a chord of a circle subtends equal angles at its circumference, we readily find that P belongs to circle ACD, i.e. the triangle lies inside the Delaunay cavity formed by P. Thus triangle ACD does not prevent those edges of ABC which are the boundary edges of the cavity from being visible from P. Repeating the argument with the other triangles, the number of which is finite, we come to the conclusion that there are no triangles between the boundary of the Delaunay cavity and P which do not lie in the cavity. Also, we find that the Delaunay cavity is simply connected. We emphasize that these facts are valid if the original triangulation satisfies the Delaunay criterion.

Thus, in accordance with the incremental algorithm, the Delaunay cavity is triangulated by simply connecting the inserted point with each of the nodes of the initial grid that lie on the boundary of the cavity. The union of these triangles with those which lie outside of the cavity (Fig. 11.4(c)) completes one loop of the incremental grid construction. The subsequent steps are accomplished in the same fashion.

It is apparent that in two dimensions the creation of these new cells results in a Delaunay triangulation, i.e. the Delaunay criterion is valid for all new triangles. Here we present a schematic proof of this fact.

Let AB be an edge of the Delaunay cavity formed by the insertion of point P. Suppose that the new triangle ABP does not satisfy the Delaunay criterion. Then there exists some point D on the same side of AB as P and which lies inside circle ABP (Fig. 11.6). Consider the original triangle that had AB as an edge. There are two possibilities: either ABD is this original triangle or there is another point, say E, on the cavity boundary lying outside circle ABP. In the former case P lies outside circle ABD, i.e. triangle ABD does not lie in the Delaunay cavity and consequently edge AB is not the edge of the cavity, contrary to our assumption. In the latter case arc ABP lies

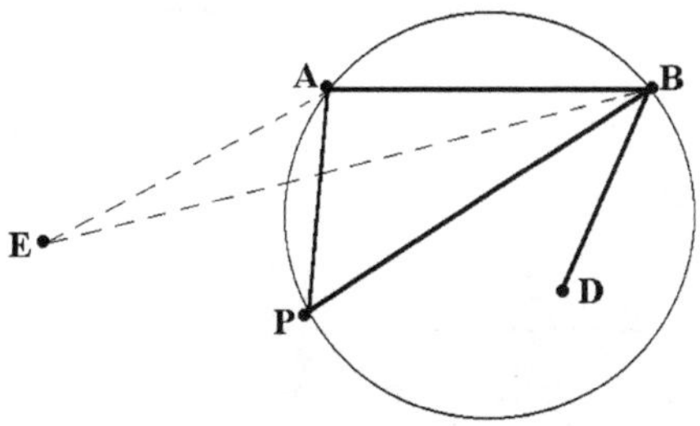

Fig. 11.6. Illustration of the proof that the Delaunay criterion is satisfied by all new triangles created by the incremental algorithm

inside circle ABE. However, this contradicts the assumption that the original triangulation was of Delaunay type. Therefore circle ABP does not contain other points, i.e. the triangle ABP satisfies the Delaunay criterion.

Thus, we find that the planar Bowyer–Watson algorithm is a valid procedure for generating Delaunay triangulations. One more issue that has received attention is that the point placement selected to generate Delaunay triangulations can be used to generate meshes with a good aspect ratio.

Initial Triangulation

Because the mesh points are introduced in a sequential manner, in the initial stages of this construction an extremely coarse grid containing a small subset of the total number of mesh points and consisting of a small number of very large triangles can be chosen. For example, for generating grids in general two-dimensional domains, an initial triangulation can be formed by dividing a square lying in the domain or containing it into two triangles. Then interior and boundary points are successively added to build successive triangulations until the necessary requirements of domain approximation are observed.

It is desirable to make the initial triangulation boundary-conforming, i.e. all boundary edges are included in the triangulation. One natural way is to triangulate initially only the prescribed boundary nodes, by means of the Bowyer–Watson algorithm. Since the Delaunay triangulation of a given set of points is a unique construction, there is no guarantee that the triangulation built through the boundary points will be boundary-conforming. However, by repeated insertion of new mesh points at the midpoints of the missing boundary edges, a boundary-conforming triangulation may be obtained. Another way to maintain boundary integrity is obtained by rejecting any point that would result in breaking boundary connectivity.

11.3.4.3 Diagonal-Swapping Algorithm

The diagonal-swapping algorithm makes use of the equiangular property of a Delaunay-type triangulation, which states that the minimum angle of each triangle in the mesh in maximized.

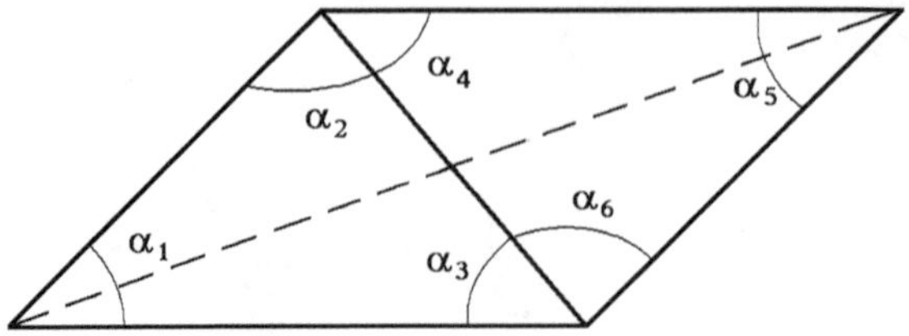

Fig. 11.7. The triangulation which maximizes the minimum angle. The *dashed line* indicates a possible original triangulation

Assuming we have some triangulation of a given set of points, the swapping algorithm transforms it into a Delaunay triangulation by repeatedly swapping the positions of the edges in the mesh in accordance with the equiangular property. For this purpose, each pair of triangles which constitutes a convex quadrilateral is considered. This quadrilateral produces two of the required triangles when one takes the diagonal which maximizes the minimum of the six interior angles of the quadrilaterals, as shown in Fig. 11.7. Each time an edge swap is performed, the triangulation becomes more equiangular. The end of the process results in the most equiangular (the Delaunay) triangulation.

This technique based on the Delaunay criterion retriangulates a given triangulation in a unique way, such that the minimum angle of each triangle in the mesh is maximized. This has the advantage that the resulting meshes are optimal for the given point distribution, in that they do not usually contain many extremely skewed cells.

11.3.5 Constrained Form of Delaunay Triangulation

One way to ensure that the boundary triangulation remains intact in the process of retriangulation by inserting new points is to use a constrained version of the Delaunay triangulation algorithm of Sect. 11.3.2 that does not violate the point connections made near the boundary.

11.3.5.1 Principal Component

For the purpose of generating a constrained two-dimensional triangulation we consider the modernized Bowyer–Watson algorithm for an arbitrary triangulation T that may not satisfy the Delaunay criterion. Let P be a new, introduced point. The Delaunay cavity is the area constituted by all triangles whose circumcircles contain P. Let this be denoted by $\Gamma(P)$.

An important fact is that the Delaunay cavity created by the introduction of the point P contains no points other than P in its interior. In order to show this we consider a point A in the triangulation T that is a vertex of at least one triangle in $\Gamma(P)$. If there is a triangle $S \notin \Gamma(P)$ that has A as a vertex then the point A is not an interior point of $\Gamma(P)$. Thus we need to show that

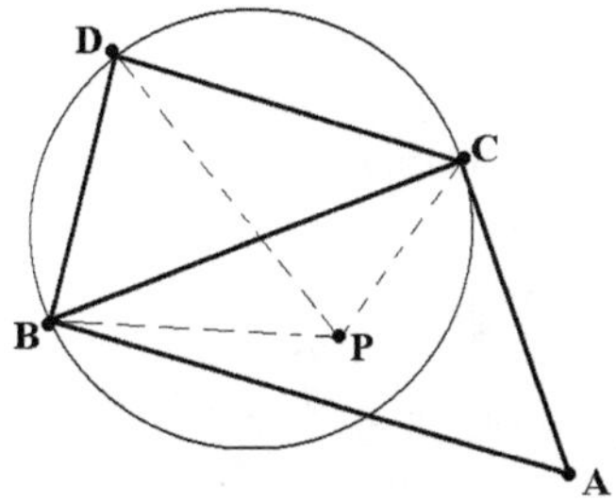

Fig. 11.8. Illustration of the principal component

there exists such a triangle. Let $\{S_i\}$ be the set of all triangles that have A as a vertex, and let C_i be the circumcircle associated with triangle S_i. Now $S_i \in \Gamma(P)$ if and only if the new point P lies inside C_i. Thus, for vertex A to be an interior point of $\Gamma(P)$, point P must lie inside $\cap C_i$. However, if the point A is an interior point of $\Gamma(P)$ then the interior of $\cap C_i$ is empty, since the vertex A is the only point that lies on all the circles of $\{S_i\}$. Thus at least one triangle of $\{S_i\}$ does not lie in $\Gamma(P)$ and hence the vertex A is not an interior point of $\Gamma(P)$.

In the case of a general triangulation, the cavity $\Gamma(P)$ need no longer be simply connected. For the purpose of retriangulation, we consider the maximal simply connected region of the cavity that contains the new point P. This region is called the principal component of the Delaunay cavity and is designated by Γ_P.

It is apparent that the principal component possesses the property that all its boundary edges are visible from P. To prove this, we first note that Γ_P is not empty, since it includes the triangle containing P. Let this be the triangle ABC (Fig. 11.8). Now consider all neighboring triangles sharing a common edge with the triangle ABC. In particular, let triangle BCD lie in Γ_P. Point P must, therefore, lie inside circle BCD. As points P, B, C, and D define a convex quadrilateral, all edges of this quadrilateral are visible from P. Continuing this process by means of a tree search through all triangles in Γ_P, we clearly see that all edges of Γ_P are visible from P.

11.3.5.2 Formulation of the Constrained Triangulation

Now we can formulate the generation of a constrained planar Delaunay triangulation developed by Baker (1989).

We assume that certain triangles of a triangulation T are fixed, in particular those adjacent to the boundary. Let this subset of T be denoted by $\overline{T}$. The triangles from $\overline{T}$ do not participate in the building of any Delaunay cavity, i.e. if the cavity created by the introduction of a new point contains one or more of the fixed triangles, we restrict the reconnections to the part of the cavity that does not contain any fixed triangle. Let $\Upsilon(P)$ be this part of the cavity, i.e. $\Upsilon(P) = \Gamma(P) - \overline{T}$. By Υ_P we denote the maximal simply connected region of $\Upsilon(P)$ that contains P. In analogy with Γ_P, we call the region Υ_P the principal component of $\Upsilon(P)$. It is clear that the principal component Υ_P exists only if P does not lie inside any of the triangles belonging to the collection $\overline{T}$ of the fixed triangles.

It is apparent that the boundary edges of the principal component Υ_P are visible from P. As the analogous fact has been proved for Γ_P, we can restrict our consideration to the case $\Gamma_P \cap \overline{T} \neq \emptyset$. Let the edges of the principal component Γ_P be given by $\overline{A_1, A_2}$, $\overline{A_2, A_3}, \dots, \overline{A_{n-1}, A_n}$, $\overline{A_n, A_1}$, where $\{A_i\}i = 1, n$ are the vertices on the boundary of Γ_P. These edges, and consequently the vertices A_i, are visible from P. The subcavity obtained by removing one of the triangles from Γ_P contains at most three new edges. These internal edges lie wholly inside the cavity Γ_P and divide Γ_P into disjoint polygonal regions. The principal component Υ_P is the polygonal region that contains the point P, and this polygon is made up of one these internal cavity edges and other edges which come from the cavity boundary. The vertices of the polygon containing P must, therefore, remain visible from P. Hence all edges of the polygon are visible from P. By repeating this argument with other triangles removed from Γ_P, we conclude that the boundary edges of Υ_P are visible from P.

Now, the vertices of Υ_P can be connected with P, thus building the constrained retriangulation. This retriangulation keeps the fixed triangles of $\overline{T}$ intact.

11.3.5.3 Boundary-Conforming Triangulation

A key requirement of a mesh generation procedure is to ensure that the mesh is boundary-conforming, i.e. the edges of the assembly of triangles conform to the boundary curve. The procedure of constrained triangulation allows one to keep a subset of the boundary triangles, built from the edges forming the boundary, intact. These boundary triangles can be generated by any one of the suitable procedures. Thus the resulting triangulation will be boundary-conforming and its interior triangles obey the Delaunay criterion.

Another approach developed by Weatherill and Hassan (1994) to applying the Delaunay criterion to generate boundary-conforming grids consists in recovering the boundary edges which are missing during the process of Delaunay triangulation and then deleting all triangles that lie outside the domain.

11.3.6 Point Insertion Strategies

The Bowyer–Watson algorithm proceeds by sequentially inserting a point inside the domain at selected sites and reconstructing the triangulation so as to include new points. This subsection presents two approaches to sequential point insertion which provide a refinement of planar Delaunay triangulations. In both cases, bounds on some measures of grid quality such as the minimum angle, the ratio of maximum to minimum edge length, and the ratio of circumradius to inradius are estimated.

11.3.6.1 Point Placement at the Circumcenter of the Maximum Triangle

One simple but effective approach consists in placing a new point at the circumcenter of the cell with the largest circumradius and iterating this process until the maximum circumradius is less than some prescribed threshold. In this way, by eliminating bad triangles, the quality of the grid is improved at every new point insertion, terminating with a grid formed only by suitable triangles. In this subsection it will be shown that the Bowyer–Watson incremental algorithm together with point insertion at the circumcenters of maximal triangles will lead to a triangulation with a guaranteed level of triangle quality.

Unconstrained Triangulation

Let $\{T_n\}$, $n = 0, 1, \ldots$, be a sequence of Delaunay triangulations built by the repeated application of the Bowyer–Watson algorithm with point insertion at the circumcenter of the maximal triangle. By the maximal triangle of a triangulation we mean the triangle with the maximum value of its circumradius. We assume that the initial Delaunay triangulation T_0 conforms to a prescribed set of boundary edges. Now let l_n, L_n, $n = 0, 1, \ldots$, be the minimum and maximum edge lengths, respectively, of T_n, and let R_n be the radius of the maximal triangle of T_n. Furthermore, for any triangle S we denote its circumradius by R_S and its inradius by r_S. Thus $R_n = \max\{R_S,\ S \in T_n\}$. We have the following relations:

(1) $R_{i+1} \leq R_i$;
(2) when $R_{n-1} \geq l_0$, then $l_n = l_0$, and when $R_{n-1} < l_0$, then $l_n = R_{n-1}$;
(3) when $R_n \leq l_0$, then $L_n/l_n \leq 2$, $\theta \geq 30°$ for all angles of the triangulation T_n, and $\min R_S/r_S \leq 2 + 4\sqrt{3}$ for all triangles S of T_n.

To prove the first relation we consider an edge e_n of the Delaunay cavity of the triangulation T_n formed by an inserted point P. There exist triangles S_1 and S_2 in T_n which share the common edge e_n, such that S_1 lies inside while S_2 lies outside the Delaunay cavity. Let S_1 be defined by the points A, B, and C and S_2 be defined by the points A, B, and D. Then edge e_n is the line segment $\overline{AB}$. Since P lies outside circle ABD, P lies on the same side of e_n as C. If the center of circle ABP lies on the same side of e_n as D then angle APB is obtuse and, consequently, the circumradius of triangle ABP is smaller than the circumradius of triangle ABD. We denote these circumradii by R_{ABP} and R_{ABD}, respectively.

If the center of circle ABP lies on the same side of e_n as C then the angle θ_1 subtended by chord AB at C is less than the angle θ_2 subtended at P. Since the centers of circles ABP and ABC lie on the same side of $\overline{AB}$ as points C and P, it follows that $\theta_1 < \pi/2$ and $\theta_2 < \pi/2$. If the length of chord $\overline{AB}$ is l then

$$R_{ABP} = \frac{l}{2\sin\theta_2} < \frac{l}{2\sin\theta_1} = R_{ABC},$$

where R_{ABC} is the circumradius of ABC. Thus we obtain

$$R_{ABP} < R_{ABD} \quad \text{and} \quad R_{ABP} < R_{ABC}.$$

Since this is true for all edges of the Delaunay cavity, we obtain the proof of the first relation, that the maximum circumradius R_n decreases, i.e. $R_{n+1} \leq R_n$, with strict inequality if there is only one triangle with the maximum radius R_n. As there can be only a limited number of maximal triangles in T_n, after several applications of the procedure we obtain $R_{n+k} < R_n$.

It follows that the maximum radius can be reduced to any required size after a sufficiently large number of iterations. When R_n falls below the value of l_0, so that $l_{n+1} = R_n$, we obtain the following obvious inequality:

$$L_{n+1} \leq 2R_{n+1} \leq 2R_n = 2l_{n+1}. \tag{11.11}$$

It is evident that repeated point insertion at the circumcenter reduces the value $\lambda = L_n/l_n$ to a value no greater than 2. The upper bound of 2 for λ is achieved when $R_n \leq l_0$. Let $\theta_{\min}$ be the minimum angle. We have

$$\sin\theta_{\min} \geq \frac{l_{n+1}}{2R_{n+1}}, \tag{11.12}$$

with equality if the minimum edge length of any maximal triangle is equal to l_{n+1}, the minimum edge length for the triangulation T_{n+1}. From the inequalities (11.11) and (11.12), we obtain

$$\sin\theta_{\min} \geq \frac{l_{n+1}}{2R_{n+1}} = \frac{R_n}{2R_{n+1}} \geq \frac{1}{2}, \tag{11.13}$$

so that

$$\theta_{\min} \geq \pi/6.$$

For each triangle the quantity $\mu = R/r$, where R is the circumradius and r is the inradius, is a characteristic of cell deformity. The maximum value of μ occurs for an isosceles triangle with an angle between sides of $\theta_{\min}$ and assumes the value

$$\mu_{\max} = \frac{1}{2\cos\theta_{\min}(1 - \cos\theta_{\min})}.$$

From (11.13), we obtain

$$\mu \leq 2 + 4/\sqrt{3}$$

after a sufficient number of retriangulations with the insertion of new points at the circumcenters of maximal triangles.

These considerations prove the properties (2) and (3) stated above.

Generalized Choice of the Insertion Triangles

In the approach considered, a new point is inserted at the circumcentre of the largest triangle. The choice of the insertion triangle, namely the triangle where the point is inserted, can be formulated in accordance with more general principles.

One simple formulation is based on the specification of a function $f(\boldsymbol{x})$ which prescribes a measure of grid size or quality, say the radius of the circumscribed circle, at the point $\boldsymbol{x}$. The actual expression for $f(\boldsymbol{x})$ can be obtained by interpolating prescribed nodal values over a convenient background mesh. The function $f(\boldsymbol{x})$ defines a quantity $\alpha(S)$ for each triangle S:

$$\alpha(S) = \frac{R_S}{f(Q_S)},$$

where Q_S is the position of the centre of the circle circumscribed around the triangle S. The largest value of $\alpha(S)$ determines the choice of the insertion triangle S. By repeatedly inserting the new point at the circumcenters of such triangles it is possible to reach eventually a mesh in which $\max_S \alpha(S) < 1$.

11.3.6.2 Voronoi-Segment Point Insertion

The second approach proposed by Rebay (1993) to placing a new point consists in inserting the point along a segment of the Dirichlet tessellation. In contrast to the first approach, where the position of the inserted point is predetermined, and the required cell size is reached after a number of iterations, this technique provides an opportunity to generate one or possibly several new triangles having, from the very beginning, the size prescribed for the final grid. This is achieved by choosing a suitable position for point placement in the Dirichlet tessellation, between a triangle whose circumradius falls below the required value and a neighboring triangle whose circumradius is still too large. This point insertion results in almost equilateral triangles over most of the interior of the domain.

Formulation of the Algorithm

At each stage of the process of generating the triangulations T_n, $n = 1, 2, \dots$, the triangles of T_n are divided into two groups, which are referred to as the groups of accepted (small enough) and nonaccepted (too large) triangles, respectively. In most cases the accepted triangles are the boundary triangles and those whose circumradii are below 3/2 times the prescribed threshold. The remaining triangles constitute the group of nonaccepted triangles.

The algorithm proceeds by always considering a maximal nonaccepted triangle which borders one of the accepted triangles (Fig. 11.9). Let ABC be the accepted triangle and ADB the nonaccepted triangle. The Voronoi segment connecting the circumcenters of these triangles is the interval EF

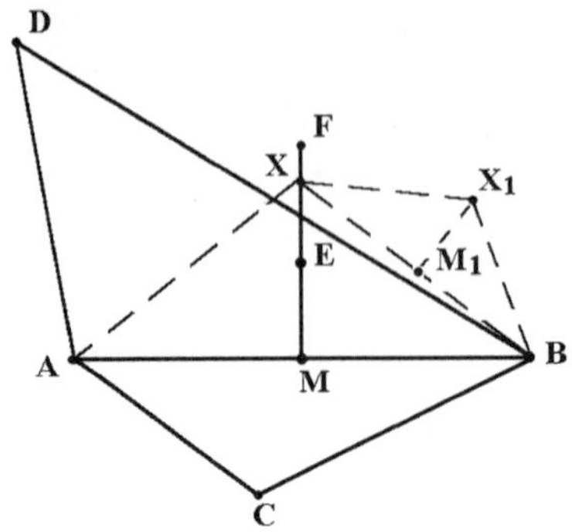

Fig. 11.9. Voronoi-segment point insertion

which is perpendicular to the common edge AB and divides it into two equal parts. In the algorithm, a new point X is inserted on the Voronoi segment edge EF in a position chosen so that the triangle formed by connecting X with A and B has the prescribed size. This point is inserted in the interval between the midpoint M of the common edge and the circumcenter F of the nonaccepted triangle ADB.

Let p be one half the length of edge AB, and q the length of FM. As point F is the circumcenter of the triangle ADB we find that $q \geq p$. Let f_M be the prescribed threshold value for the circumradius at the point M. It may seem that we can locate the new point X on segment EF at the intersection of EF with the circle that passes through points A and B and has a radius equal to f_M. However, it might happen that this exact value f_M for the circumradius is not appropriate, since any circle through A and B has a radius $\rho \geq p/2$. Furthermore, a real intersection point X exists only for circles having a radius ρ smaller than that of the circle passing through AB and F, i.e. $\rho \leq (p^2 + q^2)/2q$. For these reasons the circumradius for the triangle AXB is defined by the equation

$$R_{AXB} = \min\left[\max(f_M, p), \frac{p^2 + q^2}{2q}\right]. \tag{11.14}$$

Since

$$\frac{p^2 + q^2}{2q} = \frac{(p - q)^2 + 2pq}{2q} \geq p,$$

we find that $R_{AXB} \geq p$. In accordance with the algorithm, the new point X will lie on the interval EF between M and F at a distance

$$d = R_{AXB} + \sqrt{(R_{AXB})^2 - p^2} \tag{11.15}$$

from the point M.

Properties of the Triangulation

The condition

$$R_{AXP} \leq \frac{p^2 + q^2}{2q}$$

and (11.15) ensure that $d \leq q$. We also have, from (11.15), that $d \geq p$. Angle AXB is a right angle when $d = p$ and it decreases as d increases.

If the accepted triangle ABC is equilateral then angle AFB must be no greater than $2\pi/3$, since otherwise the Delaunay triangulation would have given rise to an edge connecting C to F.

At the first stage we expect $p \ll q$. Recall that the threshold of f_M is such that $f_M < p < 3f_M/2$. It follows that $f_M < p \leq (p^2 + q^2)/2q$ and hence $d = p$ and $R_{AXB} = p$. Thus triangle AXB has a right angle at vertex X. Since $2 < 3f_M/2$, triangle AXB will be tagged as accepted and each segment AX and XB will be a candidate for the next accepted triangle, built in the same way as AXB. Now, we denote the quantity p, equal to one half the length of the accepted edge of the ith iteration, by p_i, and thus $p_1 = p_0/\sqrt{2}$. Analogously, we use d_i, R_i, and M_i at the ith iteration of the procedure. It turns out that on repeating the procedure, p_i and d_i show the following behavior:

$$p_i \to \sqrt{3} f_{M_i}/2, \qquad d_i \to 3 f_{M_i}/2,$$

i.e. the generated triangles tend to become equilateral, with circumradius f_{M_i}. To show this, let

$$f_{M_n} = \left(\frac{2}{\sqrt{3}} + \epsilon_n\right) p_n = \left(\frac{2}{\sqrt{3}} + \epsilon_{n+1}\right) p_{n+1}. \tag{11.16}$$

If $|\epsilon_n|$ is sufficiently small, we obtain $p_n < f_{M_n}$ so that $R_n = f_{M_n}$ and, from (11.15),

$$d_n = f_{M_n} + \sqrt{f_{M_n}^2 - p_n^2}.$$

Further, we have

$$4p_{n+1}^2 = p_n^2 + d_n^2 = 2\left(f_{M_n}^2 + f_{M_n}\sqrt{f_{M_n}^2 - p_n^2}\right).$$

Thus

$$\frac{p_{n+1}^2}{p_n^2} = \frac{1}{2}\frac{f_{M_n}^2}{p_n^2} + \frac{1}{2}\frac{f_{M_n}}{p_n}\sqrt{\frac{f_{M_n}^2}{p_n^2} - 1}.$$

Using (11.16), we obtain

$$\frac{p_{n+1}^2}{p_n^2} = \frac{1}{2}\left(\frac{2}{\sqrt{3}} + \epsilon_n\right)^2 + \frac{1}{2}\left(\frac{2}{\sqrt{3}} + \epsilon_n\right)\sqrt{\left(\frac{2}{\sqrt{3}} + \epsilon_n\right)^2 - 1},$$

which results in

$$\frac{p_{n+1}}{p_n} = 1 + \frac{3}{4}\sqrt{3} + \epsilon_n + O(\epsilon_n^2).$$

From (11.16), we also have

$$\frac{p_{n+1}}{p_n} = \frac{2/\sqrt{3} + \epsilon_n}{2/\sqrt{3} + \epsilon_{n+1}}.$$

Comparing the last two equations and neglecting terms $O(\epsilon_n^2)$, we find that

$$\epsilon_{n+1} \simeq -\epsilon_{n/2}.$$

Thus, for $|\epsilon_n|$ sufficiently small, the algorithm ensures that $\epsilon_n \to 0$ and

$$p_n \to \sqrt{3} f_{M_n}/2.$$

Therefore it can be expected that a large number of the interior triangles will be nearly equilateral. Close to the boundary there may be isosceles right-angled triangles, and in regions where the boundary has large curvature there may be some obtuse triangles. A maximum angle of 120° and minimum angle of 30° may be realized by an obtuse triangle formed when the vertex D of a nonaccepted triangle is sufficiently close to an active edge.

In analogy with the first approach to inserting new points the choice of the triangle into which the new popint is inserted can be modified by introducing a quality measure function $f(\boldsymbol{x})$ and a corresponding control quantity $\alpha(S)$.

11.3.7 Surface Delaunay Triangulation

A surface Delaunay triangulation is defined by analogy with the planar Delaunay triangulation.

Let P_i be the vertices of the surface triangulation T. A triangle S from T satisfies the Delaunay criterion if the interior of the circumsphere through the vertices of S and centered on the plane formed by S does not contain any points. If all triangles satisfy the Delaunay criterion then the triangulation T is called a surface Delaunay triangulation.

In practice, all methods for planar Delaunay triangulations are readily reformulated for a surface Delaunay triangulation. However, a study of the properties of the triangulations generated by these methods has not been performed.

11.3.8 Three-Dimensional Delaunay Triangulation

In three dimensions the network of the Delaunay triangulation is obtained by joining the vertices of the Voronoi polyhedrons that have a common face. Each vertex of a Voronoi polyhedron is the circumcenter of a sphere that passes through four points which form a tetrahedron and no other point in the construction can lie within the sphere.

11.3.8.1 Unconstrained Technique

The most popular three-dimensional algorithm providing a Delaunay structure is the one based on the Bowyer–Watson sequential process: each point of the grid is introduced into an existing Delaunay triangulation, which is broken and then reconnected to form a new Delaunay triangulation.

In general the algorithm follows the same steps as in the two-dimensional construction described above. It starts with an initial Delaunay triangulation formed by a supertetrahedron or supercube, partitioned into five tetrahedrons which contain all other points. The remaining points which comprise the mesh to be triangulated are introduced one at a time, and the Bowyer–Watson algorithm is applied to create the Delaunay cavity and the corresponding retriangulation after each point insertion.

An important feature of a mesh generation procedure is its ability to produce a boundary-conforming mesh, i.e. the triangular faces of the assembly of tetrahedrons conform to the boundary surface. Unfortunately, the unconstrained technique does not guarantee that the boundary faces will be contained within such a triangulation. Thus an modified procedure must be introduced to ensure that the resulting triangulation is boundary-conforming.

11.3.8.2 Constrained Triangulation

The purpose of the constrained Delaunay triangulation is to generate a triangulation which preserves the connections imposed on the boundary points. The three-dimensional constrained triangulation is carried out in the same way as for two-dimensional triangulations.

In the first approach the tetrahedrons whose faces constitute the boundary surface are fixed during the process of retriangulation. These boundary tetrahedrons are generated in the first step of triangulation. The next steps include the insertion of a point, the definition of a star-shaped cavity containing the point, and retriangulation of the cavity. The resulting grid is boundary-conforming and its interior subtriangulation is a Delaunay triangulation.

The second approach to the constrained triangulation of a domain developed by Weatherill and Hassan (1994) starts with inputting the boundary points and boundary point connectivities of the faces of the boundary triangulation. After performing a Delaunay triangulation of the boundary points, a new Delaunay triangulation is built by inserting interior points and applying the Bowyer–Watson algorithm. After this, the tetrahedrons intersecting the boundary surface are transformed to recover the boundary triangulation. If a boundary face is not present in the new Delaunay triangulation, this is due to the fact that edges and faces of the tetrahedrons of the Delaunay triangulation intersect this face. Since the face is formed from three edges, it is necessary to recover first the face edges and then the face. This is achieved by first finding the tetrahedrons which are intersected by the face edges. There is a fixed combination of possible standard intersections of each tetrahedron by any mixed boundary edge, which allows one to perform direct transformations to recover the edge. Having established the intersection types, these tetrahedrons are then locally transformed into new tetrahedrons so that the required edges are present. A similar procedure then follows to recover the boundary faces.

11.4 Advancing-Front Methods

Advancing-front techniques extend the grid into the region in the form of marching layers, starting from the boundary and proceeding until the whole region has been covered with grid cells. Such a procedure allows an initial unstructured mesh to be automatically generated from a surface representation of the geometry. Thus advancing-front techniques need some initial triangulation of the boundaries of the geometry, and this triangulation forms the initial front. The marching process includes the construction of a new simplex, which is built by connecting either some appropriate points on the front, or some inserted new point with the vertices of a suitable face on the front.

The advancing-front approaches offer the advantages of high-quality point placement and integrity of the boundary. The efficiency of the grid-marching process largely depends on the arrangement of grid points in the front, especially at sharp corners. A new grid point is placed at a position which is determined so as to result in a simplex with prescribed optimal quality features. In some approaches, the grid points are positioned along a set of predetermined vectors. To ensure a good grid quality and to facilitate the advancing process, these vectors are commonly determined once at each layer mesh point by simply averaging the normal vectors of the faces sharing the point and then smoothing the vectors. Other approaches to selecting new points for moving the front use the insertion techniques applied in the Delaunay triangulations described above.

The fronts continue to advance until either

(1) opposite fronts approach to within a local cell size; or
(2) certain grid quality criteria are locally satisfied.

Grid quality measures which are to be observed in the process of grid generation include the cell spacing and sizes of angles. The desired mesh spacings and other gridding preferences in the region are commonly specified by calculations on a background grid.

11.4.1 Procedure of Advancing-Front Method

In order to generate cells with acceptable angles and lengths of edges by a marching process, the advancing-front concept inherently requires a preliminary specification of local grid spacing and directionality at every point of the computational mesh. The spacing is prescribed by defining three (two in two dimensions) orthogonal directions together with some length scale for each direction. The directions and length scales are commonly determined from background information, in particular, by carrying out computations on a coarse grid and interpolating the data.

The advancing-front procedure proceeds by first listing all faces which constitute the front and then selecting an appropriate face (edge in two dimensions) on the front. The operation of the selection is very important since the quality of the final grid may by affected by the choice. According to a common rule, the face is selected where the grid spacing is required to be the smallest. A collection of vertices on the front which are appropriate for connection to the vertices of the selected face to form a tetrahedron (triangle in two dimensions) is searched. The collection may be formed by the vertices which lie inside a sphere centered at the barycenter of the face, with an appropriate radius based upon the height of a unit equilateral tetrahedron. A new point is also created which is consistent with the ideal position determined from the background information about grid spacing and directionality. The selected vertices and the new point are ordered according to their distance from the barycenter of the selected face. Each sequential tetrahedron formed by the face and the ordered points is then checked to find out whether it intersects any face in the front. The first point which satisfies the test and gives a tetrahedron of good quality is chosen as the fourth vertex for the new tetrahedron. The current triangle is then removed from the list of front faces, since it is now obscured by the new tetrahedron. This process continues until there are no more faces in the list of front faces.

In many cases, the use of the background mesh to define the local grid spacing can be replaced by sources in the form of points, lines, and surfaces.

One of the advantages of such a procedure is that all operations are performed locally, on neighboring faces only. Additionally, boundary integrity is observed, since the boundary triangulation constitutes the initial front.

The disadvantages of the advancing-front approach relate mainly to the phase in which a local direction and length scale are determined and to the checking phase for ensuring the acceptability of a new tetrahedron.

11.4.2 Strategies to Select Out-of-Front Vertices

One of the critical items of advancing-front methods is the placement of new points. Upon generating a new simplex, a point is placed at a position which is determined so as to result in the required shape and size of the new simplex. The parameters which define the desirable cell at each domain position are specified by a function which is determined a priori or found in the process of computation.

In one approach, the new point is placed along a line which is orthogonal to a chosen face on the front and passes through its circumcenter. This placement is aimed at the creation of a new simplex whose boundary contains the chosen face.

If the simplex generated with the new point results in a crossover with the front it is discarded. Alternately, if the new point is located very close to a vertex on the front it is replaced by this vertex in order to avoid the appearance of a cell with a very small edge at some later stage.

Another approach, generally applied in two dimensions, takes into account a vertex on the front and the angle at which the edges cross at this point. The point is created with the aim of making the angles in the new triangles as near to 60° as possible. In particular, a very large angle between the edges is bisected or even trisected. On the other hand, if the vertex has a small interior angle the two adjacent vertices on the front are connected. This approach can be extended to three dimensions by analyzing a dihedral angle at the front.

11.4.3 Grid Adaptation

The frontal approach is well suited to generating adaptive grids near the boundary segments, where the grid cells are commonly required to be highly stretched.

Highly stretched grid cells begin forming individually from the boundary and march into the domain. However, unlike the conventional procedure in which cells are added in no systematic sequence, the construction of a stretched grid needs to be performed by advancing one layer of cells at a time, with the minimum congestion of the front and a uniform distribution of stretched cells. The new points are positioned along a set of predetermined vectors in accordance with the value of a stretching function. The criterion by which the points are evaluated has a significant impact on the grid quality and the marching process. Because of the requirement high aspect ratio of cells in the boundary layer, the conventional criteria based on the cell angles are not appropriate for building highly stretched cells.

In a criterion based on a spring analogy, the points forming a new layer are assumed to be connected to the end points of the face by tension springs. Among these points, the one with the smallest spring force is considered the most suitable to form the new cell, and consequently to change the front boundary. The spring concept allows one to indicate when an opposing front is very close to the new location, namely, when an existing point on the front has the smallest spring force. The adaptive advancing process terminates on a front face when the local grid characteristics on the front, influenced by the stretching function, no longer match those determined by the background grid in that location. When the proximity and/or grid quality criteria are satisfied on all faces of the front, the process switches from an advancing-layers method to the conventional advancing-front method to form regular isotropic cells in the rest of the domain.

11.4.4 Advancing-Front Delaunay Triangulation

A combination of the advancing-front approach and the Delaunay concept gives rise to the advancing-front Delaunay methods.

If the boundary of a domain is triangulated and in the interior of the domain a set of points to be triangulated is given, then the advancing-front Delaunay triangulation is carried out by forming the cells adjoining the front in accordance with the empty-circumcircle property.

The procedure for the triangulation can be outlined as follows. A face on the front is chosen, and a new simplex is tentatively built by joining the vertices of the face to an arbitrary point on the front, in the interior of the domain with regard to the front. If this simplex contains any points within its circumcircle, it is not added to the triangulation. By checking all points, the appropriate vertex which produces a simplex containing no points interior to its circumcircle is eventually found. The simplex formed through this vertex is accepted and the front is advanced.

Another algorithm is based on the strategy of placing new points ahead of the front and triangulating them according to the Delaunay criterion.

11.4.5 Three-Dimensional Prismatic Grid Generation

The use of prismatic cells is justified by the fact that the requirement of high aspect ratio can be achieved without reducing the values of the angles between the cell edges.

The procedure for generating a prismatic grid begins by triangulating the boundary surface of a domain. The next stage in the procedure computes a quasinormal direction at each node of the surface triangulation. Then the initial surface is shifted along these quasinormal directions by a specified distance d. This gives the first layer of prismatic cells. This shifting process is repeated a number of times using suitable values of d at each stage, and either the same or newly computed normal directions. The value of the quantity d can be chosen in the form of any of the stretching functions described in Chap. 4.

The efficiency of the algorithm is essentially dependent on the choice of quasinormal directions. The generation of the quasinormals is carried out in three stages, depending on a position of the vertices.

(1) Normals are first computed at the vertices which lie on the corners of the boundary. These are calculated as the angle-weighted average of the adjacent surface normals. The angle used is the one between the two edges adjacent to the boundary surface and meeting at the corner.
(2) Normals at grid points on the geometrical edges of the boundary surface are computed. These normals are the average of the two adjacent surface normals.
(3) Finally, the normals at grid nodes on the boundary surfaces are calculated.

11.5 Comments

Unstructured grid methods were originally developed in solid mechanics. The paper by Field (1995) reviews some early techniques for unstructured mesh generation that rely on solid modelling. An informal survey that illustrates the wide range of unstructured mesh generation was conducted by Owen (1998) and described in the handbook of grid generation edited by Thompson, Soni, and Weatherill (1999).

Though unstructured technology deals chiefly with tetrahedral (triangular in two dimensions) elements, some approaches rely on hexahedrons (or quadrilaterals) for the decomposition of arbitrary domains. Recent results have been presented by Tam and Armstrong (1991) and Blacker and Stephenson (1991).

Properties of n-dimensional triangulations were reviewed by Lawson (1986). The relations between the numbers of faces were proved in the monograph by Henle (1979) and in the papers by Steinitz (1922), Klee (1964), and Lee (1976).

The Delaunay triangulation and Voronoi diagram were originally formulated in the papers of Delaunay (1934, 1947) and Voronoi (1908), respectively. Algorithms for computing Voronoi diagrams have been developed by Green and Sibson (1978), Brostow, Dussault, and Fox (1978), Finney (1979), Bowyer (1981), Watson (1981), Tanemura, Ogawa, and Ogita (1983), Sloan and Houlsby (1984), Fortune (1985), and Zhou et al. (1990). Results of studies of geometrical aspects of Delaunay triangulations and their dual Voronoi diagrams were presented in the monographs by Edelsbrunner (1987), Du and Hwang (1992), Okabe, Boots, and Sugihara (1992), and Preparata and Shamos (1985). Proofs of the properties of planar Delaunay triangulations were given by Guibas and Stolfi (1985) and by Baker (1987, 1989).

A technique for creating the Delaunay triangulation of an a priori given set of points was proposed by Tanemura, Ogawa, and Ogita (1983). The incremental two-dimensional Delaunay triangulation which starts with an initial triangulation was developed by Bowyer (1981) and Watson (1981). Watson has also shown the visibility of the edges of the cavity associated with the inserted point. Having demonstrated that the Delaunay criterion is equivalent to the equiangular property, Sibson (1978) devised and later Lee and Schachter (1980) investigated a diagonal-swapping algorithm for generating a Delaunay triangulation by using the equiangular property.

A novel approach, based on the aspect ratio and cell area of the current triangles, to the generation of points as the Delaunay triangulation proceeds was developed by Holmes and Snyder (1988). In their approach a new point is introduced in the existing triangulation at the Voronoi vertex corresponding to the worst triangle. Ruppert (1992) and Chew (1993) have shown that in the planar case the procedure leads to a Delaunay triangulation with a minimum-angle bound of 30 degrees. An alternative procedure of inserting the new point on a Voronoi segment was proposed by Rebay (1993). A modification of the Rebay technique was made by Baker (1994). Haman, Chen, and Hong (1994) inserted points into a starting Delaunay grid in accordance with the boundary curvature and distance from the boundary, while Anderson (1994) added nodes while taking into account cell aspect ratio and proximity to boundary surfaces.

Approaches to the generation of boundary-conforming triangulations based upon the Delaunay criterion have been proposed by Lee (1978), Lee and Lin (1986), Baker (1989), Chew (1989), Cline and Renka (1990), George, Hecht, and Saltel (1990), Weatherill (1990), George and Hermeline (1992), Field and Nehl (1992), Hazlewood (1993), and Weatherill and Hassan (1994). All techniques and methods considered in the present chapter for proving the results associated with the constrained Delaunay triangulation were described on the basis of papers by Weatherill (1988), Baker (1989, 1994), Mavriplis (1990), Rebay (1993), and Weatherill and Hassan (1994).

Further development of unstructured grid techniques based on the Delaunay criterion and aimed at the solution of three-dimensional problems has been performed by Cavendish, Field, and Frey (1985), Shenton and Cendes 1985), Perronet (1988), Baker (1987, 1989), Jameson, Baker, and Weatherill (1986), and Weatherill (1988). The application of the Delaunay triangulation for the purpose of surface interpolation was discussed by DeFloriani (1987).

The octree approach originated from the pioneering work of Yerry and Shephard (1985). The octree data structure has been adapted by Lohner (1988b) to produce efficient search procedures for the generation of unstructured grids by the moving front technique. Octree-generated cells were used by Shephard et al. (1988b) and Yerry and Shephard (1990) to cover the domain and the surrounding space and then to derive a tetrahedral grid by cutting the cubes. The generation of hexahedral unstructured grids was developed by Schneiders and Bunten (1995).

The moving-front technique has been successfully developed in three dimensions by Peraire et al. (1987), Lohner (1988a) and Formaggia (1991). Some methods using Delaunay connectivity in the frontal approach have been created by Merriam (1991), Mavriplis (1991, 1993), Rebay (1993), Muller, Roe, and Deconinck (1993), and Marcum and Weatherill (1995).

Advancing-front grids with layers of prismatic and tetrahedral cells were formulated by Lohner (1993). A more sophisticated procedure, basically using bands of prismatic cells and a spring analogy to stop the advancement of approaching layers, was described by Pirzadeh (1992). The application of adaptive prismatic meshes to the numerical solution of viscous flows was demonstrated by Parthasarathy and Kallinderis (1995).

Some procedures for surface triangulations have been developed by Peraire et al. (1988), Lohner and Parikh (1988), and Weatherill et al. (1993).

A survey of adaptive mesh refinement techniques was published by Powell, Roe, and Quirk (1992). The combination of the Delaunay triangulation with adaptation was performed by Holmes and Lamson (1986), Mavriplis (1990), and Muller (1994). The implementation of solution adaptation into the advancing-front method with directional refinement and regeneration of the original mesh was studied by Peraire et al. (1987). Approaches based on the use of sources to specify the local point spacing have been developed by Pirzadeh (1993, 1994), and Weatherill et al. (1993).

The prospects and trends for unstructured grid generation in its application to computational fluid dynamics were discussed by Baker (1995) and Venkatakrishan (1996). The first application of the Delaunay triangulation in computational fluid dynamics was carried out by Bowyer (1981) and Baker (1987). The advancing-front technique was introduced, in computational fluid dynamics, primarily by Peraire et al. (1987), Lohner (1988a), and Lohner and Parikh (1988). The techniques of George (1971), Wordenweber (1981, 1983), Lo (1985), and Peraire (1986) foreshadowed the more recent advancing-front methods. Muller (1994) and Marchant and Weatherill (1994) applied a combination of frontal and Delaunay approaches to treat problems with boundary layers. Muller (1994) generated triangular grids in the boundary layer by a frontal technique, with high-aspect-ratio triangles, and filled the remainder of the domain with triangles built by the Delaunay approach. Another way to treat a boundary layer with the advancing-front approach was applied by Hassan et al. (1994). In the first step the boundary layer is covered by a single layer of tetrahedral cells. Then the newly generated nodes are moved along the cell edges towards the boundary by a specified distance. These steps, in the original layer, are repeated until a required resolution has been reached. After this the advancing front proceeds to fill up the remainder of the domain.

An algorithm for the generation of a high-quality well-graded quadrilateral element mesh from a triangular element mesh was presented by Lee and Lo (1994). Very important applications to parallel unstructured mesh generation were discussed by Chrisochoides (2006) and Ivanov (2008).

References

Ablow, C.M. (1982): Equidistant mesh for gas dynamics calculations. Appl. Math. Comput. **10**(11), 859–863

Ablow, C.M., Schechter, S. (1978): Campylotropic coordinates. J. Comput. Phys. **27**, 351–363

Acharya, S., Moukalled, F.H. (1990): An adaptive grid solution procedure for convection–diffusion problems. J. Comput. Phys. **91**, 32–54

Adjerid, S., Flaherty, J.F. (1986): A moving finite element method with error estimation and refinement for one-dimensional time-dependent partial differential equations. SIAM J. Numer. Anal. **23**, 778–796

Ahuja, D.V., Coons, S.A. (1968): Geometry for construction and display. IBM Syst. J. **7**, 188–205

Alalykin, G.B., Godunov, S.K., Kireyeva, L.L., Pliner, L.A. (1970): *On Solution of One-Dimensional Problems of Gas Dynamics in Moving Grids.* Nauka, Moscow (Russian)

Albone, C.M. (1992): Embedded meshes of controllable quality synthesised from elementary geometric features. AIAA Paper 92-0633

Albone, C.M., Joyce, M.G. (1990): Feature-associated mesh embedding for complex configurations. In *AGARD Conference Proceedings 464.13*

Allwright, S. (1989): Multiblock topology specification and grid generation for complete aircraft configurations. In Schmidt, W. (ed.), *AGARD Conference Proceedings 464, Applications of Mesh Generation to Complex 3-D Configurations.* Loen, Norway. Advisory Group for Aerospace Research and Development, NATO

Amsden, A.A., Hirt, C.W. (1973): A simple scheme for generating general curvilinear grids. J. Comput. Phys. **11**, 348–359

Anderson, D.A. (1983): Adaptive grid methods for partial differential equations. In Ghia, K.N., Ghia, U. (eds.), *Advances in Grid Generation*, pp. 1–15. ASME, Houston

Anderson, D.A. (1987): Equidistribution schemes, Poisson generators, and adaptive grids. Appl. Math. Comput. **24**, 211–227

V.D. Liseikin, *Grid Generation Methods*, Scientific Computation,
DOI 10.1007/978-90-481-2912-6, © Springer Science+Business Media B.V. 2010

Anderson, W.K. (1994): A grid generation and flow solution method for the Euler equations in unstructured grids. J. Comput. Phys. **110**, 23–38

Anderson, D.A., Rajendran, N. (1984): Two approaches toward generating orthogonal adaptive grids. AIAA Paper 84-1610

Anderson, D.A., Steinbrenner, J. (1986): Generating adaptive grids with a conventional grid scheme. AIAA Paper 86-0427

Andrew, B., Whrite, J.R. (1979): On selection of equidistributing meshes for two-point boundary-value problems. SIAM J. Numer. Anal. **16**(3), 472–502

Andrews, A.E. (1988): Progress and challenges in the application of artificial intelligence to computational fluid dynamics. AIAA J. **26**, 40–46

Arina, R., Casella, M. (1991): A harmonic grid generation technique for surfaces and three-dimensional regions. In Arcilla, A.S., Hauser, J., Eiseman, P.R., Thompson, J.F. (eds.), *Numerical Grid Generation in Computational Fluid Dynamics and Related Fields*, pp. 935–946. North-Holland, Amsterdam

Atta, E.H., Vadyak, J. (1982): A grid interfacing zonal algorithm for three-dimensional transonic flows about aircraft configurations. AIAA Paper 82-1017

Atta, E.H., Birchelbaw, L., Hall, K.A. (1987): A zonal grid generation method for complex configurations. AIAA Paper 87-0276

Azarenok, B.N. (2007): A method of constructing adaptive hexahedral moving grids. J. Comput. Phys. **226**(10), 1102–1121

Azarenok, B.N., Ivanenko, S.A. (2001): Application of moving adaptive grids for simulation of supersonic gas flow. Comput. Fluid Dyn. J. Jpn. **10**(3), 400–404

Babuŝka, I., Aziz, A.K. (1976): On the angle condition in the finite element method. SIAM J. Numer. Anal. **13**(2), 214–226

Babuŝka, I., Rheinboldt, W.C. (1978): A-posteriori error estimates for the finite element method. Int. J. Numer. Methods Eng. **12**, 1597–1615

Bahvalov, N.S. (1969): On optimization of the methods of the numerical solution of boundary-value problems with boundary layers. J. Comput. Math. Math. Phys. **9**(4), 842–859 (Russian). [English transl.: USSR Comput. Math. Math. Phys. **9** (1969)]

Baker, T.J. (1987): Three-dimensional mesh generation by triangulation of arbitrary points sets. AIAA Paper 87-1124-CP

Baker, T.J. (1989): Automatic mesh generation for complex three-dimensional region using a constrained Delaunay triangulation. Eng. Comput. **5**, 161–175

Baker, T.J. (1994): Triangulations, mesh generation and point placement strategies. In Caughey, D. (ed.), *Computing the Future*, pp. 1–15. Wiley, New York

Baker, T.J. (1995): Prospects and expectations for unstructured methods. In *Proceedings of the Surface Modeling, Grid Generation and Related Issues in Computational Fluid Dynamics Workshop*. NASA Conference Publication **3291**, pp. 273–287. NASA Lewis Research Center, Cleveland, OH

Baker, T.J. (1997): Mesh adaptation strategies for problems in fluid dynamics. Finite Elem. Anal. Des. **25**, 243–273

Bank, R.E., Xu, J. (1996): An algorithm for coarsening unstructured meshes. Numer. Math. **73**, 1–36

Bank, R.E., Smith, R.K. (1997): Mesh smoothing using a posteriori error estimates. SIAM J. Numer. Anal. **34**, 979–997

Barfield, W.D. (1970): An optimal mesh generator for Lagrangian hydrodynamic calculations in two space dimensions. J. Comput. Phys. **6**, 417–429

Bayliss, A., Garbey, M. (1995): Adaptive pseudo-spectral domain decomposition and the approximation of multiple layers. J. Comput. Phys. **119**(1), 132–141

Belinsky, P.P., Godunov, S.K., Ivanov, Yu.V., Yanenko, I.K. (1975): The use of one class of quasiconformal mappings to generate numerical grids in regions with curvilinear boundaries. Z. Vycisl. Mat. Mat. Fiz. **15**, 1499–1511 (Russian)

Belk, D.M., Whitfield, D.L. (1987): Three-dimensional Euler solutions on blocked grids using an implicit two-pass algorithms. AIAA Paper 87-0450

Bell, J.B., Shubin, G.R. (1983): An adaptive grid finite-difference method for conservation laws. J. Comput. Phys. **52**, 569–591

Bell, J.B., Shubin, G.R., Stephens, A.B. (1982): A segmentation approach to grid generation using biharmonics. J. Comput. Phys. **47**(3), 463–472

Benek, J.A., Buning, P.G., Steger, J.L. (1985): A 3-d chimera grid embedding technique. AIAA Paper 85-1523

Benek, J.A., Steger, J.L., Dougherty, F.C. (1983): A flexible grid embedding technique with application to the Euler equations. AIAA Paper 83-1944

Berger, M.J., Oliger, J. (1983): Adaptive mesh refinement for hyperbolic partial differential equations. Manuscript NA-83-02, Stanford University, March

Berger, A.E., Han, H., Kellog, R.B. (1984): A priori estimates of a numerical method for a turning point problem. Math. Comput. **42**(166), 465–492

Blacker, T.D., Stephenson, M.B. (1991): Paving a new approach to automated quadrilateral mesh generation. Int. J. Numer. Methods Eng. **32**, 811–847

Blom, J.G., Verwer, J.G. (1989): On the use of arclength and curvature in a moving-grid method which is based on the methods of lines. Report NM-N8402, CWI, Amsterdam

Boor, C. (1974): *Good Approximation by Splines with Variable Knots.* Lect. Notes Math. **363**, pp. 12–20. Springer, Berlin

Bowyer, A. (1981): Computing Dirichlet tessellations. Comput. J. **24**(2), 162–166

Brackbill, J.U. (1993): An adaptive grid with directional control. J. Comput. Phys. **108**, 38–50

Brackbill, J.U., Saltzman, J. (1982): Adaptive zoning for singular problems in two directions. J. Comput. Phys. **46**, 342–368

Branets, L., Carey, G.F. (2005): A local cell quality metric and variational grid smoothing algorithm. Eng. Comput. **21**, 19–28

Branets, L.V., Garanzha, V.A. (2002): Distortion measure of trilinear mapping. Application 3-D grid generation. Number. Linear Algebra Appl. **302**, 1–16

Brish, N.J. (1954): On boundary-value problems for the equation $\epsilon y'' = f(x, y, y')$ with small ϵ. Dokl. Akad. Nauk USSR **XCV**(3), 429–432 (Russian)

Brostow, W., Dussault, J.P., Fox, B.L. (1978): Construction of Voronoi polyhedra. J. Comput. Phys. **29**, 81–92

Buscaglia, G.C., Dari, E.A. (1997): Anisotropic mesh optimization and its application in adaptivity. Int. J. Numer. Methods Eng. **40**, 4119–4136

Carey, G.F. (1979): Adaptive refinement and nonlinear fluid problems. Comput. Methods Appl. Mech. Eng. **17/18**, 541–560

Carey, G.F. (1997): *Computational Grids. Generation, Adaptation, and Solution Strategies.* Taylor and Francis, London

Castilio, J.E. (1991): The discrete grid generation method on curves and surfaces. In Arcilla, A.S., Hauser, J., Eiseman, P.R., Thompson, J.F. (eds.), *Numerical Grid Generation in Computational Fluid Dynamics and Related Fields*, pp. 915–924. North-Holland, Amsterdam

Catheral, D. (1991): The adaption of structured grids to numerical solutions for transonic flow. Int. J. Numer. Methods Eng. **32**, 921–937

Cavendish, J.C., Field, D.A., Frey, W.H. (1985): An approach to automatic three-dimensional finite element mesh generation. Int. J. Numer. Methods Eng. **21**, 329–347

Chan, W.M., Buning, P.G. (1995): Surface grid generation methods for overset grids. Comput. Fluids **24**(5), 509–522

Chan, W.M., Steger, L.G. (1992): Enhancement of a three-dimensional hyperbolic grid generation scheme. Appl. Math. Comput. **51**(1), 181–205

Charakhch'yan, A.A., Ivanenko, S.A. (1997): A variational form of the Winslow grid generator. J. Comput. Phys. **136**, 385–398

Chen, K. (1994): Error equidistribution and mesh adaptation. SIAM J. Sci. Comput. **15**(4), 798–818

Chen, K., Baines, M.J., Sweby, P.K. (1993): On an adaptative time stepping strategy for solving nonlinear diffusion equations. J. Comput. Phys. **105**, 324–332

Chen, Z., Tristano, J.R., Kwok, W. (2003): Combined Laplacian and optimization-based smoothing for quadratic mixed surface meshes. In *Proceedings of the 12th International Meshing Roundtable*, pp. 361–370

Chew, L.P. (1989): Constrained Delaunay triangulations. Algorithmica **4**, 97–108

Chew, P. (1993): Mesh generation, curved surfaces and guaranteed quality triangles. Technical Report, IMA, Workshop on Modeling, Mesh Generation and Adaptive Numerical Methods for Partial Differential Equations, University of Minnesota, Minneapolis

Chiba, N., Nishigaki, I., Yamashita, Y., Takizawa, C., Fujishiro, K. (1998): A flexible automatic hexahedral mesh generation by boundary-fit method. Comput. Methods Appl. Mech. Eng. **161**, 145–154

Chrisochoides, N. (2006): Parallel mesh generation. In *Numerical Solution of PDEs on Parallel Computers*. Lecture Notes in Computational Science and Engineering **51**, pp. 237–264. Springer, Berlin. Chap. 7

Chu, W.H. (1971): Development of a general finite difference approximation for a general domain. J. Comput. Phys. **8**, 392–408

Clement, P., Hagmeijer, R., Sweers, G. (1996): On the invertibility of mapping arising in 2D grid generation problems. Numer. Math. **73**, 37–51

Cline, A.K., Renka, R.L. (1990): A constrained two-dimensional triangulation and the solution of closest node problems in the presence of barriers. SIAM J. Numer. Anal. **27**, 1305–1321

Coons, S.A. (1967): Surfaces for computer aided design of space forms. Project MAC, Design Division Department of Mechanical Engineering, MII, 1964; Revised to MAC-TR-41, June, 1967

Cordova, J.Q., Barth, T.J. (1988): Grid generation for general 2-D regions using hyperbolic equations, AIAA Paper 88-0520

Cougny, H.L., Shephard, M.S., Georges, M.K. (1990): Explicit node point smoothing within the octree mesh generator. Report 10-1990, SCOREC, RPI, Troy, NY

Coyle, J.M., Flaherty, J.E., Ludwig, R. (1986): On the stability of mesh equidistribution strategies for time-dependent partial differential equations. J. Comput. Phys. **62**, 26–39

Crowley, W.P. (1962): An equipotential zoner on a quadrilateral mesh. Memo, Lawrence Livermore National Lab., 5 July 1962

Danaev, N.T., Liseikin, V.D., Yanenko, N.N. (1978): A method of nonstationary coordinates in gas dynamics. Foreign Technology Division, Wright-Patterson Air Force Base, Ohio, FTD-ID(RS)-1673-78, October

Danaev, N.T., Liseikin, V.D., Yanenko, N.N. (1980): Numerical solution on a moving curvilinear grid of viscous heat-conducting flow about a body of revolution. Chisl. Metody Mekhan. Sploshnoi Sredy **11**(1), 51–61 (Russian)

Dannenhoffer, J.F. (1990): A comparison of adaptive-grid redistribution and embedding for steady transonic flows. AIAA Paper 90-1965

Dannenhoffer, J.F. (1995): Automatic blocking for complex three-dimensional configurations. In *Proceedings of the Surface Modeling, Grid Generation, and Related Issues in Computational Fluid Dynamics Workshop*, p. 123. NASA Lewis Research Center, Cleveland OH. May

Dannelongue, H.H., Tanguy, P.A. (1991): Three-dimensional adaptive finite element computations and applications to non-Newtonian flows. Int. J. Numer. Methods Fluids **13**, 145–165

Dar'in, N., Mazhukin, V.I. (1989): Mathematical modelling of non-stationary two-dimensional boundary-value problems on dynamic adaptive grids. Mat. Model. **1**(3), 29–30 (Russian)

Dar'in, N., Mazhukin, V.I., Samarskii, A.A. (1988): Finite-difference solution of gas dynamics equations using adaptive grids which are dynamically associated with the solution. Z. Vycisl. Mat. Mat. Fiz. **28**, 1210–1225 (Russian)

Darmaev, T.G., Liseikin, V.D. (1987): A method for generating multidimensional adaptive grids. Model. V Meh. **1**(1), 49–57 (Russian)

Davis, S.F., Flaherty, J.E. (1982): An adaptive finite element method for initial boundary-value problems for partial differential equations. SIAM J. Sci. Stat. Comput. **3**, 6–27

DeFloriani, L. (1987): Surface representations on triangular grids. Vis. Comput. **3**, 27–50

Delaunay, B. (1934): Sur la sphere vide. Bull. Acad. Sci. USSR VII: Class Sci. Mat. Nat. **6**, 793–800

Delaunay, B. (1947): *Petersburg School of Number Theory.* Akad. Sci. USSR, Moscow (Russian)

Denny, V.E., Landis, R.B. (1972): A new method for solving two-point boundary-value problems using optimal node distribution. J. Comput. Phys. **9**(1), 120–137

Desbois, F., Jacquotte, O.-P. (1991): Surface mesh generation and optimization. In Arcilla, A.S., Hauser, J., Eiseman, P.R., Thompson, J.F. (eds.), *Numerical Grid Generation in Computational Fluid Dynamics and Related Fields*, pp. 131–142. North-Holland, Amsterdam

Dirichlet, G.L. (1850): Uber die Reduction der positiven quadratischen Formen mit drei underbestimmten ganzen Zahlen. J. Reine Angew. Math. **40**(3), 209–227

Dompierre, J., Vallet, M.-G., Labbe, P., Guibault, F. (2005): An analysis of simplex shape measures for anisotropic meshes. Comput. Methods Appl. Mech. Eng. **194**, 4895–4914

Dorfi, E.A., Drury, L. O'C. (1987): Simple adaptive grids for 1-D initial value problems. J. Comput. Phys. **69**, 175–195

Du, D.-Z., Hwang, F. (eds.) (1992): *Computing in Euclidean Geometry.* World Scientific, Singapore

Dvinsky, A.S. (1991): Adaptive grid generation from harmonic maps on Riemannian manifolds. J. Comput. Phys. **95**, 450–476

Dwyer, H.A., Onyejekwe, O.O. (1985): Generation of fully adaptive and/or orthogonal grids. In *Proc. 9th Int. Cong. Numerical Methods in Fluid Dynamics*, Saclay, pp. 422–426. Springer, Berlin

Dwyer, H.A., Kee, R.J., Sanders, B.R. (1980): Adaptive grid method for problems in fluid mechanics and heat transfer. AIAA J. **18**(10), 1205–1212

Edelsbrunner, H. (1987): *Algorithms in Combinatorial Geometry.* Springer, Berlin

Edwards, T.A. (1985): Noniterative three-dimensional grid generation using parabolic partial differential equations. AIAA Paper 85-0485

Eells, J., Lenaire, L. (1988): Another report on harmonic maps. Bull. Lond. Math. Soc. **20**(5), 385–524

Eiseman, P.R. (1980): Geometric methods in computational fluid dynamics. ICASE Report 80-11 and Von Karman Institute for Fluid Dynamics Lecture Series Notes

Eiseman, P.R. (1985): Grid generation for fluid mechanics computations. Annu. Rev. Fluid Mech. **17**, 487–522

Eiseman, P.R. (1987): Adaptive grid generation. Comput. Methods Appl. Mech. Eng. **64**, 321–376

Eriksson, L.E. (1982): Generation of boundary-conforming grids around wing-body configurations using transfinite interpolation. AIAA J. **20**, 1313–1320

Eriksson, L.E. (1983): Practical three-dimensional mesh generation using transfinite interpolation. Lecture Series Notes 1983-04, von Karman Institute for Fluid Dynamics, Brussels

Field, D.A. (1986): Implementing Watson's algorithm in three dimensions. In *Proc. 2nd Ann. ACM Symp. on Comput. Geometry*, pp. 246–259

Field, D.A. (1995): The legacy of automatic mesh generation from solid modeling. Comput. Aided Geom. Des. **12**, 651–673

Field, D.A. (2000): Qualitative measures for initial meshes. Int. J. Numer. Methods Eng. **47**, 887–906

Field, D.A., Nehl, T.W. (1992): Stitching together tetrahedral meshes. In Field, D., Komkov, V. (eds.), *Geometric Aspects of Industrial Design*, pp. 25–38. SIAM, Philadelphia. Chap. 3

Finney, J.L. (1979): A procedure for the construction of Voronoi polyhedra. J. Comput. Phys. **32**, 137–143

Fletcher, C.A.J. (1997): *Computational Techniques for Fluid Dynamics 1: Fundamental and General Techniques*. Springer, Berlin

Formaggia, L. (1991): An unstructured mesh generation algorithm for three-dimensional aeronautical configurations. In Arcilla A.S., Hauser J., Eiseman P.R., Thompson J.F. (eds.), *Numerical Grid Generation in Computational Fluid Dynamics and Related Fields*, pp. 249–260. North-Holland, Amsterdam

Fortune, S. (1985): A sweepline algorithm for Voronoi diagrams. AT&T Bell Laboratory Report, Murray, Hill, NJ

Frey, P.J., George, P.-L. (2008): *Mesh Generation Application to Finite Elements*. Wiley, New York

Furzland, R.M., Verwer, J.G., Zegeling, P.A. (1990): A numerical study of three moving grid methods for 1-D PDEs which are based on the method of lines. J. Comput. Phys. **89**, 349–388

Garanzha, V.A. (2000): Barrier variational generation of quasi-isimetric grids. Comput. Math. Math. Phys. **40**, 1617–1637

Garon, A., Camarero, R. (1983): Generation of surface-fitted coordinate grids. In Ghia, K.N., Ghia, U. (eds.), *Advances in Grid Generation*, pp. 117–122. ASME, Houston

Gelfand, I.M., Fomin, S.V. (1963): *Calculus of Variations*. Prentice-Hall, Englewood Cliffs

Georgala, J.M., Shaw, J.A. (1989): A discussion on issues relating to multiblock grid generation. In Schmidt, W. (ed.), *AGARD Conference Proceedings 464, Applications of Mesh Generation to Complex 3-D Configurations.* Loen, Norway. Advisory Group for Aerospace Research and Development, NATO

George, A.J. (1971): Computer implementation of the finite element method. Stanford University, Department of Computer Science, STAN-CS-71-208

George, P.L., Borouchaki, H. (1998): *Delaunay Triangulation and Meshing.* Hermes, Paris

George, P.L., Hermeline, F. (1992): Delaunay's mesh of convex polyhedron in dimension *d*: application for arbitrary polyhedra. Int. J. Numer. Methods Eng. **33**, 975–995

George, P.L., Hecht, F., Saltel, E. (1990): Automatic 3d mesh generation with prescribed meshed boundaries. IEEE Trans. Magn. **26**(2), 771–774

Ghia, K.N., Ghia, U., Shin, C.T. (1983): Adaptive grid generation for flows with local high gradient regions. In Ghia, K.N., Ghia, U. (eds.), *Advances in Grid Generation*, pp. 35–47. ASME, Houston

Giannakopoulos, A.E., Engel, A.J. (1988): Directional control in grid generation. J. Comput. Phys. **74**, 422–439

Glasser, A.H., Tang, X.Z. (2004): The SEL macroscopic modeling code. Comput. Phys. Commun. **164**, 237–243

Glasser, A.H., Liseikin, V.D., Kitaeva, I.A. (2005): Control of grid properties with the help of monitor metrics. Comput. Math. Math. Phys. **45**(8), 1416–1432

Glasser, A.H., Kitaeva, I.A., Likhanova, Yu.V., Liseikin, V.D., Lukin, V.S. (2005): Specification of monitor metrics for generating balanced numerical grids. J. Bull. NCC IIS, Numer. Anal. **13**, 1–13

Glasser, A.H., Liseikin, V.D., Shokin, Ju.I., Vaseva, I.A., Likhanova, Ju.V. (2006): *Grid Generation with the Use of Beltrami and Diffusion Equations.* Nauka, Novosibirsk (Russian)

Gnoffo, P.A. (1983): A finite-volume, adaptive grid algorithm applied to planetary entry flowfields. AIAA J. **21**(9), 1249–1254

Godunov, S.K., Prokopov, G.P. (1967): Calculation of conformal mappings in the construction of numerical grids. J. Comput. Math. Math. Phys. **7**, 1031–1059 (Russian)

Godunov, S.K., Prokopov, G.P. (1972): On utilization of moving grids in gasdynamics computations. J. Vycisl. Mat. Mat. Phys. **12**, 429–440 (Russian). [English transl.: USSR Comput. Math. Math. Phys. **12**, 182–195 (1972)]

Gordon, W.J. (1969): Distributive lattices and the approximation of multivatiate functions. In Shwenberg, I.J. (ed.), *Symposium on Approximation with Special Emphasis on Spline Functions*, pp. 223–277. Academic Press, San Diego

Gordon, W.J. (1971): Blending-function methods of bivariate and multivariate interpolation and approximation. SIAM J. Numer. Anal. **8**, 158–177

Gordon, W.J., Hall, C.A. (1973): Construction of curvilinear coordinate systems and applications to mesh generation. Int. J. Numer. Methods Eng. **7**, 461–477

Gordon, W.J., Thiel, L.C. (1982): Transfinite mappings and their application to grid generation. In Thompson J.F. (ed.), *Numerical Grid Generation*, pp. 171–192. North-Holland, Amsterdam

Green, P.J., Sibson, R. (1978): Computing Dirichlet tessellations in the plane. Comput. J. **21**(2), 168–173

Greenberg, J.B. (1985): A new self-adaptive grid method. AIAA J. **23**, 317–320

Guibas, L., Stolfi, J. (1985): Primitives for the manipulation of general subdivisions and the computation of Voronoi diagrams. ACM Trans. Graph. **4**, 74–123

Gurtin, M.E. (1981): *An Introduction to Continuum Mechanics.* Academic Press, San Diego

Hagmeijer, R. (1994): Grid adaption based on modified anisotropic diffusion equations formulated in the parametric domain. J. Comput. Phys. **115**(1), 169–183

Hall, D.J., Zingg, D.W. (1995): Viscous airfoil computations adaptive grid redistribution. AIAA J. **33**(7), 1205–1210

Haman, B., Chen, J.-L., Hong, G. (1994): Automatic generation of unstructured volume grids inside or outside closed surfaces. In Weatherill, N.P., Eiseman, P.R., Hauser, J., Thompson, J.F. (eds.), *Numerical Grid Generation in Computational Field Simulation and Related Fields*, p. 187. Pineridge, Swansea

Harten, A., Hyman, J.M. (1983): A self-adjusting grid for the computation of weak solutions of hyperbolic conservation laws. J. Comput. Phys. **50**, 235–269

Hassan, O., Probert, E.J., Morgan, K., Peraire, J. (1994): Unstructured mesh generation for viscous high speed flows. In Weatherill, N.P., Eiseman, P.R., Hauser, J., Thompson, J.F. (eds.), *Numerical Grid Generation in Computational Field Simulation and Related Fields*, p. 779. Pineridge, Swansea

Haussling, H.J., Coleman, R.M. (1981): A method for generation of orthogonal and nearly orthogonal boundary-fitted coordinate systems. J. Comput. Phys. **43**, 373–381

Hawken, D.F., Gottlieb, J.J., Hansen, J.S. (1991): Review of some adaptive node-movement techniques in finite-element and finite-difference solutions of partial differential equations. J. Comput. Phys. **95**, 254–302

Hazlewood, C. (1993): Approximating constrained tetrahedrizations. Comput. Aided Geom. Des. **10**, 67–87

Hedstrom, G.W., Rodrigue, C.M. (1982): *Adaptive-Grid Methods for Time-Dependent Partial Differential Equations.* Lect. Notes Math. **960**, pp. 474–484. Springer, Berlin

Henle, M. (1979): *A Combinatorial Introduction to Topology.* Freeman, New York

Hindman, R.G., Kutler, P., Anderson, D. (1981): Two-dimensional unsteady Euler equation solver for arbitrary shaped flow regions. AIAA J. **19**(4), 424–431

Ho-Le, K. (1988): Finite element mesh generation methods: a review and classification. Comput. Aided Des. **20**, 27–38

Hodge, J.K., Leone, S.A., McCarry, R.L. (1987): Non-iterative parabolic grid generation for parabolized equations. AIAA J. **25**(4), 542–549

Holcomb, J.F. (1987): Development of a grid generator to support 3-dimensional multizone Navier–Stokes analysis. AIAA Paper 87-0203

Holmes, D.G., Lamson, S.H. (1986): Adaptive triangular meshes for compressible flow solutions. In Hauser, J., Taylor, C. (eds.), *Numerical Grid Generation in Computational Fluid Dynamics*, p. 413. Pineridge, Swansea

Holmes, D.G., Snyder, D.D. (1988): The generation of unstructured triangular meshes using Delaunay triangulation. In Sengupta, S., Hauser, J., Eiseman, P.R., Thompson, J.F. (eds.), *Numerical Grid Generation in Computational Fluid Dynamics*, pp. 643–652. Pineridge, Swansea

Huang, W., Sloan, D.M. (1994): A simple adaptive grid method in two dimensions. SIAM J. Sci. Comput. **15**(4), 776–797

Huang, H., Prosperetti, A. (1994): Effect of grid orthogonality on the solution accuracy of the two-dimensional convection diffusion equation. Numer. Heat Transfer, Part B **26**, 1–20

Huang, W., Ren, Y., Russel, R.D. (1994): Moving mesh PDEs based on the equidistribution principle. SIAM J. Numer. Anal. **31**, 709–730

Jacquotte, O.-P. (1987): A mechanical model for a new grid generation method in computational fluid dynamics. Comput. Methods Appl. Mech. Eng. **66**, 323–338

Jameson, A., Baker, T.J., Weatherill, N.P. (1986): Calculation of inviscid transonic flow over a complete aircraft. AIAA Paper 86-0103, January

Jeng, Y.N., Shu, Y.-L. (1995): Grid combination method for hyperbolic grid solver in regions with enclosed boundaries. AIAA J. **33**(6), 1152–1154

Ivanenko, S.A., Charakhch'yan, A.A. (1988): An algorithm for constructing curvilinear grids consisting of convex quadrangles. Sov. Math. Dokl. **36**(1), 51

Ivanov, E. (2008): *Parallel Tetrahedral Meshing Based on A-priori Domain Decomposition: From Scratch to Results by Utilizing Off-the-Shelf Sequential Software*. VDM Verlag Dr. Muller, Saarbrucken

Kerlic, C.D., Klopfer, G.H. (1982): Assessing the quality of curvilinear meshes by decomposing the Jacobian matrix. Appl. Math. Comput. **10/11**, 787

Khamayseh, A., Mastin, C.W. (1996): Computational conformal mapping for surface grid generation. J. Comput. Phys. **123**, 394–401

Kim, B., Eberhardt, S.D. (1995): Automatic multiblock grid generation for high-lift configuration wings. In *Proceedings of the Surface Modeling, Grid Generation, and Related Issues in Computational Fluid Dynamics Workshop*, p. 671. NASA, Lewis Research Center, Cleveland OH, May

Kim, H.J., Thompson, J.F. (1990): Three-dimensional adaptive grid generation on a composite block grid. AIAA J. **28**, 470–477

Klee, V. (1964): The number of vertices of a convex polytope. Can. Math. **16**, 37

Klopfer, G.H., McRae, D.D. (1981): The nonlinear modified equation approach to analysing finite difference schemes. AIAA Paper 81-1029

Knupp, P.M. (1991): The direct variational grid generation method extended to curves. Appl. Math. Comput. **43**, 65–78

Knupp, P.M. (1992): A robust elliptic grid generator. J. Comput. Phys. **100**, 409–418

Knupp, P.M. (1995): Mesh generation using vector-fields. J. Comput. Phys. **119**, 142–148

Knupp, P.M. (1996): Jacobian-weighted elliptic grid generation. SIAM J. Sci. Comput. **17**(6), 1475–1490

Knupp, P.M. (2001): Algebraic mesh quality metrics. SIAM J. Sci. Comput. **23**, 193–218

Knupp, P.M. (2007): Remarks on mesh quality. AIAA, Reno, NV

Knupp, P.M., Steinberg, S. (1993): *Fundamentals of Grid Generation*. CRC Press, Boca Raton

Knupp, P., Margolin, L., Shashkov, M. (2002): Reference Jacobian optimization-based rezone strategies for arbitrary Lagrangian–Eulerian methods. J. Comput. Phys. **176**, 93–128

Kochin, N.E. (1951): *Vector Calculus and Principles of Tensor Calculus*. Nauka, Moscow (Russian)

Kovenya, V.M., Tarnavskii, G.A., Chernyi, S.G. (1990): *Application of a Splitting Method to Fluid Problems*. Nauka, Novosibirsk (in Russian)

Krugljakova, L.V., Neledova, A.V., Tishkin, V.F., Filatov, A.Yu. (1998): Unstructured adaptive grids for problems of mathematical physics (survey). Math. Model. **10**(3), 93–116 (Russian)

Kupin, E.P., Liseikin, V.D. (1994): A method of projection for generating multidimensional adaptive grids. J. Comput. Technol. **3**(8), 189–198 (Russian)

Ladygenskaya, O.A., Uraltseva, N.N. (1973): *Linear and Quasilinear Equations of Elliptic Type*. Nauka, Moscow (Russian)

Langtangen (2003): *Computational Partial Differential Equations. Numerical Methods and Diffpack Programming*. Springer, Berlin

Lawson, C.L. (1986): Properties of n-dimensional triangulations. Comput. Aided Geom. Des. **3**, 231–246

Lee, D.T. (1978): Proximity and reachibility in the plane. Technical Report R-831, University of Illinois, Urbana, IL

Lee, D.T., Lin, A.K. (1986): Generalized Delaunay triangulation for planar graphs. Discrete Comput. Geom. **1**, 201–217

Lee, D.T., Schachter, B.J. (1980): Two algorithms for constructing a Delaunay triangulation. Int. J. Comput. Inf. Sci. **9**(3), 219–241

Lee, D.T., Tsuei, Y. (1992): A formula for estimation of truncation errors of convection terms in a curvilinear coordinate system. J. Comput. Phys. **98**, 90–100

Lee, C.K., Lo, S.H. (1994): A new scheme for the generation of a graded quadrilateral mesh. Comput. Struct. **52**, 847–857

Lee, K.D. (1976): On finding k-nearest neighbours in the plane. Technical Report 76–2216, University of Illinois, Urbana, IL

Lee, K.D., Loellbach, J.M. (1989): Geometry-adaptive surface grid generation using a parametric projection. J. Aircr. **2**, 162–167

Lee, K.D., Huang, M., Yu, N.J., Rubbert, P.E. (1980): Grid generation for general three-dimensional configurations. In Smith, R.E. (ed.), *Proc. NASA Langley Workshop on Numerical Grid Generation Techniques*, p. 355, October

Li, S., Petzold, L. (1997): Moving mesh methods with upwinding schemes for time-dependent PDEs. J. Comput. Phys. **131**(2), 368–377

Liao, G. (1991): On harmonic maps. In Castilio, J.E. (ed.), *Mathematical Aspects of Numerical Grid Generation.* Frontiers in Applied Mathematics **8**, pp. 123–130. SIAM, Philadelphia

Liao, G., Anderson, D. (1992): A new approach to grid generation. Appl. Anal. **44**, 285–298

Liao, G.G., Cai, X., Fleitas, D., Luo, X., Wang, J., Xue, J. (2008): Optimal control approach to data aligment. Appl. Math. Lett. **21**, 898–905

Lin, K.L., Shaw, H.J. (1991): Two-dimensional orthogoal grid generation techniques. Comput. Struct. **41**(4), 569–585

Liseikin, V.D. (1984): On the numerical solution of singularly perturbed equations with a turning point. J. Comput. Math. Math. Phys. **24**(12), 1812–1818 (Russian). [English transl.: USSR Comput. Math. Math. Phys. **24** (1984)]

Liseikin, V.D. (1986): On the numerical solution of equations with a power boundary layer. J. Comput. Math. Math. Phys. **26**(12), 1813–1820 (Russian)

Liseikin, V.D. (1991a): On generation of regular grids on n-dimensional surfaces. J. Comput. Math. Math. Phys. **31**, 1670–1689 (Russian). [English transl.: USSR Comput. Math. Math. Phys. **31**, 47–57 (1991)]

Liseikin, V.D. (1991b): Techniques for generating three-dimensional grids in aerodynamics (review). Probl. At. Sci. Tech. Ser. Math. Model. Phys. Process. **3**, 31–45 (Russian)

Liseikin, V.D. (1992): On a variational method of generating adaptive grids on n-dimensional surfaces. Sov. Math. Dokl. **44**(1), 149–152

Liseikin, V.D. (1993a): On some interpretations of a smoothness functional used in constructing regular and adaptive grids. Russ. J. Numer. Anal. Math. Model. **8**(6), 507–518

Liseikin, V.D. (1993b): Estimates for derivatives of solutions to differential equations with boundary and interior layers. Sib. Math. J. **July**, 1039–1051

Liseikin, V.D. (1996a): Adaptive grid generation on the basis of smoothness functional. In Soni, B.K., Thompson, J.F., Hauser, J., Eiseman, P.R.

(eds.), *Numerical Grid Generation in CFD*. Mississippi State University **2**, pp. 1131–1140

Liseikin, V.D. (1996b): Construction of structured adaptive grids—a review. Comput. Math. Math. Phys. **36**(1), 1–32

Liseikin, V.D. (1999): *Grid Generation Methods*. Springer, Berlin

Liseikin, V.D. (2001a): *Layer Resolving Grids and Transformations for Singular Perturbation Problems*. VSP, Utrecht

Liseikin, V.D. (2001b): Application of notions and relations of differential geometry to grid generation. Russ. J. Numer. Anal. Math. Model. **16**(1), 57–75

Liseikin, V.D. (2002a): Analysis of grids derived by a comprehensive grid generator. Russ. J. Numer. Anal. Math. Model. **17**(2), 183–202

Liseikin, V.D. (2002b): On geometric analysis of grid properties. Russ. Dokl. Akad. Nauk **383**(2), 167–170

Liseikin, V.D. (2003): On analysis of clustering of numerical grids produced by elliptic models. Russ. J. Numer. Anal. Math. Model. **18**(2), 159–183

Liseikin, V.D. (2004): *A Computational Differential Geometry Approach to Grid Generation*. Springer, Berlin

Liseikin, V.D. (2005): On a universal monitor metric for numerical grid generation. Dokl. Math. **71**(1), 15–19

Liseikin, V.D. (2007): *A Computational Differential Geometry Approach to Grid Generation*, 2nd edn. Springer, Berlin

Liseikin, V.D., Petrenko, V.E. (1989): *Adaptive Invariant Method for the Numerical Solution of Problems with Boundary and Interior Layers*. Computer Center SD AS USSR, Novosibirsk (Russian)

Liseikin, V.D., Petrenko, V.E. (1994): On analytical and numerical investigations of a projection method for the generation of adaptive grids. J. Comput. Technol. **3**(9), 108–120 (Russian)

Liseikin, V.D., Yanenko, N.N. (1977): Selection of optimal numerical grids. Chisl. Metody Mekhan. Sploshnoi Sredy **8**(7), 100–104 (Russian)

Liseikin, V.D., Likhanova, Ju.V., Shokin, Ju.I. (2007): *Numerical Grids and Coordinate Transformations for Solution of Singular Perturbation Problems*. Nauka, Novosibirsk (Russian)

Liu, A., Joe, B. (1994a): On the shape of tetrahedra from bisection. Math. Comput. **63**, 141–154

Liu, A., Joe, B. (1994b): Relationship between tetrahedron shape measures. BIT **34**, 268–287

Lo, S.H. (1985): A new mesh generation scheme for arbitrary planar domains. Int. J. Numer. Methods Eng. **21**, 1403–1426

Lohner, R. (1988a): Generation of three-dimensional unstructured grids by the advancing-front method. AIAA Paper 88-0515, January

Lohner, R. (1988b): Some useful data structures for the generation of unstructured grids. Commun. Appl. Numer. Methods **4**, 123–135

Lohner, R. (1993): Matching semi-structured and unstructured grids for Navier–Stokes calculations. AIAA Paper 933348-CP

Lohner, R., Parikh, P. (1988): 3-dimensional grid generation by the advancing front method. Int. J. Numer. Methods Fluids **8**, 1135–1149

Lomov, S.A. (1964): Power boundary layer in problems with a small parameter. Dokl. Akad. Nauk USSR **184**(3), 516–519 (Russian)

Lorenz, J. (1982): *Nonlinear Boundary-Value Problems with Turning Points and Properties of Difference Schemes.* Lect. Notes Math. **942**, pp. 150–169. Springer, Berlin

Lorenz, J. (1984): Analysis of difference schemes for a stationary shock problems. SIAM J. Numer. Anal. **21**(6), 1038–1053

Marchant, M.J., Weatherill, N.P. (1994): Unstructured grid generation for viscous flow simulations. In Weatherill, N.P., Eiseman, P.R., Hauser, J., Thompson, J.F. (eds.), *Numerical Grid Generation in Computational Field Simulations and Related Fields*, p. 151. Pineridge, Swansea

Marcum, D.L., Weatherill, N.P. (1995): Unstructured grid generation using iterative point insertion and local reconnection. AIAA J. **33**(9), 1619–1625

Mastin, C.W. (1982): Error induced by coordinate systems. In Thompson, J.F. (ed.), *Numerical Grid Generation*, pp. 31–40. North-Holland, Amsterdam

Mastin, C.W. (1992): Linear variational methods and adaptive grids. Comput. Math. Appl. **24**(5/6), 51–56

Mavriplis, D.J. (1990): Adaptive mesh generation for viscous flows using Delaunay triangulation. J. Comput. Phys. **90**, 271–291

Mavriplis, D.J. (1991): Unstructured and adaptive mesh generation for high Reynolds number viscous flows. In Arcilla, A.S., Hauser, J., Eiseman, P.R., Thompson, J.F. (eds.), *Numerical Grid Generation in Computational Fluid Dynamics and Related Fields*, pp. 79–92. North-Holland, Amsterdam

Mavriplis, D.J. (1993): An advancing front Delaunay triangulation algorithm designed for robustness. AIAA Paper 93-0671

McNally, D. (1972): FORTRAN program for generating a two-dimensional orthogonal mesh between two arbitrary boundaries. NASA, TN D-6766, May

Merriam, M. (1991): An efficient advancing front algorithm for Delaunay triangulation. Technical Report, AIAA Paper 91-0792

Miki, K., Takagi, T. (1984): A domain decomposition and overlapping method for the generation of three-dimensional boundary-fitted coordinate systems. J. Comput. Phys. **53**, 319–330

Miller, K. (1981): Moving finite elements II. SIAM J. Numer. Anal. **18**, 1033–1057

Miller, K. (1983): Alternate codes to control the nodes in the moving finite element method. In *Adaptive Computing Methods for Partial Differential Equations*, pp. 165–182. SIAM, Philadelphia

Miller, K., Miller, R.N. (1981): Moving finite elements I. SIAM J. Numer. Anal. **18**, 1019–1032

Morrison, D. (1962): Optimal mesh size in the numerical integration of an ordinary differential equation. J. Assoc. Comput. Mach. **9**, 98–103

Moser, J. (1965): The volume elements on a manifold. Trans. Am. Math. Soc. **120**, 286

Muller, J.D. (1994): Quality estimates and stretched meshes based on Delaunay triangulation. AIAA J. **32**, 2372–2379

Muller, J.D., Roe, P.L., Deconinck, H. (1993): A frontal approach for internal node generation in Delaunay triangulations. Int. J. Numer. Methods Fluids **17**(3), 241–256

Nagumo, M. (1937): Uber die Differentialgleichung $y'' = f(x, y, y')$. Proc. Phys. Math. Soc. Jpn. **19**, 861–866

Nakahashi, K., Deiwert, G.S. (1985): A three-dimensional adaptive grid method. AIAA Paper 85-0486

Nakamura, S. (1982): Marching grid generation using parabolic partial differential equations. Appl. Math. Comput. **10**(11), 775–786

Nakamura, S. (1983): Adaptive grid relocation algorithm for transonic full potential calculations using one-dimensional or two-dimensional diffusion equations. In Ghia, K.N., Ghia, U. (eds.), *Advances in Grid Generation*, pp. 49–58. ASME, Houston

Nakamura, S., Suzuki, M. (1987): Noniterative three-dimensional grid generation using a parabolic-hyperbolic hybrid scheme. AIAA Paper 87-0277

Noack, R.W. (1985): Inviscid flow field analysis of maneuvering hypersonic vehicles using the SCM formulation and parabolic grid generation. AIAA Paper 85-1682

Noack, R.W., Anderson, D.A. (1990): Solution adaptive grid generation using parabolic partial differential equations. AIAA J. **28**(6), 1016–1023

Ogawa, S., Ishiguto, T. (1987): A method for computing flow fields around moving bodies. J. Comput. Phys. **69**, 49–68

Okabe, A., Boots, B., Sugihara, K. (1992): *Spatial Tessellations Concepts and Applications of Voronoi Diagrams*. Wiley, New York

Owen, S. (1998): A survey of unstructured mesh generation technology. In *7th International Roundtable*

Parthasarathy, V., Kallinderis, Y. (1995): Directional viscous multigrid using adaptive prismatic meshes. AIAA J. **33**(1), 69–78

Parthasarathy, V.N., Graichen, C., Hathaway, A.F. (1993): A comparison of tetrahedron quality measures. Finite Elem. Anal. Des. **15**, 255–261

Pathasarathy, T. (1983): *On Global Univalence Theorems*. Lecture Notes in Mathematics **977**. Springer, New York

Pearce, D. (1990): Optimized grid generation with geometry definition encoupled. AIAA Paper 90-0332

Peraire, J. (1986): A finite element method for convection dominated flows. Ph.D. Dissertation, University of Wales

Peraire, J., Peiro, J., Formaggia, L., Morgan, K., Zienkiewicz, O.C. (1988): Finite element Euler computations in three dimensions. AIAA Paper 88-0032

Peraire, J., Vahdati, M., Morgan, H., Zienkiewicz, O.C. (1987): Adaptive remeshing for compressible flow computations. J. Comput. Phys. **72**, 449–466

Pereyra, V., Sewell, E.G. (1975): Mesh selection for discrete solution of boundary-value problems in ordinary differential equations. Numer. Math. **23**, 261–268

Perronet, A. (1988): A generator of tetrahedral finite elements for multimaterial objects or fluids. In Sengupta, S., Hauser, J., Eiseman, P.R., Thompson, J.F. (eds.), *Numerical Grid Generation in Computational Fluid Mechanics*, pp. 719–728. Pineridge, Swansea

Petzold, L.R. (1987): Observations on an adaptive moving grid method for one-dimensional systems of partial differential equations. Appl. Numer. Math. **3**, 347–360

Pierson, B.L., Kutler, P. (1980): Optimal nodal point distribution for improved accuracy in computational fluid dynamics. AIAA J. **18**, 49–53

Pirzadeh, S. (1992): Recent progress in unstructured grid generation. AIAA Paper 92-0445

Pirzadeh, S. (1993): Structured background grids for generation of unstructured grids by advancing front method. AIAA J. **31**(2), 257–265

Pirzadeh, S. (1994): Viscous unstructured three-dimensional grids by the advancing-layers method. AIAA Paper 94-0417

Potter, D.E., Tuttle, G.H. (1973): The construction of discrete orthogonal coordinates. J. Comput. Phys. **13**, 483–501

Powell, K.G., Roe, P.L., Quirk, J.J. (1992): Adaptive-mesh algorithms for computational fluid dynamics. In Hussaini, M.Y., Kumar, A., Salas, M.D. (eds.), *Algorithmic Trends in Computational Fluid Dynamics*, pp. 301–337. Springer, New York

Preparata, F.P., Shamos, M.I. (1985): *Computational Geometry: An Introduction*. Springer, New York

Prokopov, G.P. (1989): Systematic comparison of algorithms and programs for constructing regular two-dimensional grids. Probl. At. Sci. Tech. Ser. Math. Model. Phys. Process. **3**, 98–107 (Russian)

Rai, M.M., Anderson, D.A. (1981): Grid evolution in time asymptotic problems. J. Comput. Phys. **43**, 327–344

Rai, M.M., Anderson, D.A. (1982): Application of adaptive grids to fluid flow problems with asymptotic solution. AIAA J. **20**, 496–502

Rebay, S. (1993): Efficient unstructured mesh generation by means of Delaunay triangulation and Bowyer–Watson algorithm. J. Comput. Phys. **106**, 125–138

Reed, C.W., Hsu, C.C., Shiau, N.H. (1988): An adaptive grid generation technique for viscous transonic flow problems. AIAA Paper 88-0313

Rheinboldt, W.C. (1981): Adaptive mesh refinement process for finite element solutions. Int. J. Numer. Methods Eng. **17**, 649–662

Rizk, Y.M., Ben-Shmuel, S. (1985): Computation of the viscous flow around the shuttle orbiter at low supersonic speeds. AIAA Paper 85-0168

Roberts, G.O. (1971): *Proceedings of the Second International Conference on Numerical Methods in Fluid Dynamics*. Springer, Berlin

Ronzheimer, A., Brodersen, O., Rudnik, R., Findling, A. (1994): A new interactive tool for the management of grid generation processes around arbitrary configurations. In Weatherill, N.P., Eiseman, P.R., Hauser, J., Thompson, J.F. (eds.), *Numerical Grid Generation in Computational Field Simulation and Related Fields*, p. 441. Pineridge, Swansea

Rubbert, P.E., Lee, K.D. (1982): Patched coordinate systems. In Thompson, J.F. (ed.), *Numerical Grid Generation*, p. 235. North-Holland, Amsterdam

Ruppert, J. (1992): Results on triangulation and high quality mesh generation. Ph.D. Thesis, University of California, Berkeley

Russell, R.D., Christiansen, J. (1978): Adaptive mesh selection strategies for solving boundary-value problems. SIAM J. Numer. Anal. **15**, 59–80

Ryskin, G., Leal, L.G. (1983): Orthogonal mapping. J. Comput. Phys. **50**, 71–100

Saltzman, J.S. (1986): Variational methods for generating meshes on surfaces in three dimensions. J. Comput. Phys. **63**, 1–19

Samareh-Abolhassani, J., Stewart, J.E. (1994): Surface grid generation in a parameter space. J. Comput. Phys. **113**, 112–121

Schneiders, R., Bunten, R. (1995): Automatic generation of hexahedral finite element meshes. Comput. Aided Geom. Design. **12**(7), 693

Schonfeld, T., Weinerfelt, P., Jenssen, C.B. (1995): Algorithms for the automatic generation of 2d structured multiblock grids. In *Proceedings of the Surface Modeling, Grid Generation, and Related Issues in Computational Fluid Dynamics Workshop*, p. 561. NASA, Lewis Research Center, Cleveland OH, May

Schwarz, W. (1986): Elliptic grid generation system for three-dimensional configurations using Poisson's equation. In Hauser, J., Taylor, C. (eds.), *Numerical Grid Generation in Computational Fluid Dynamics*, pp. 341–351. Pineridge, Swansea

Semper, B., Liao, G. (1995): A moving grid finite-element method using grid deformation. Numer. Methods Partial Differ. Equ. **11**, 603–615

Serezhnikova, T.I., Sidorov, A.F., Ushakova, O.V. (1989): On one method of construction of optimal curvilinear grids and its applications. Sov. J. Numer. Anal. Math. Model. **4**(1), 137–155

Shaw, J.A., Weatherill, N.P. (1992): Automatic topology generation for multiblock grids. Appl. Math. Comput. **52**, 355–388

Shenton, D.N., Cendes, Z.J. (1985): Three-dimensional finite element mesh generation using Delaunay tessellation. IEEE Trans. Magn. **MAG-21**, 2535–2538

Shephard, M.S., Grice, K.R., Lot, J.A., Schroeder, W.J. (1988a): Trends in automatic three-dimensional mesh generation. Comput. Strict. **30**(1/2), 421–429

Shephard, M.S., Guerinoni, F., Flaherty, J.E., Ludwig, R.A., Bachmann, P.L. (1988b): Finite octree mesh generation for automated adaptive three-dimensional flow analysis. In Sengupta, S., Hauser, J., Eiseman, P.R., Thompson, J.F. (eds.), *Numerical Grid Generation in Computational Fluid Mechanics*, pp. 709–718. Pineridge, Swansea

Shewchuck, J.R. (2002): What is a good linear finite element? Interpolation, conditioning, anisotropy, and quality measures. Preprint, University of California at Berkelly, CA 94720

Shokin, Ju.I., Liseikin, V.D., Lebedev, A.S., Danaev, N.T., Kitaeva, I.A. (2005): *Method of Riemannian Geometry for Generating Numerical Grids*. Nauka, Novosibirsk (Russian)

Sibson, R. (1978): Locally equiangular triangulations. Comput. J. **21**(3), 243–245

Sidorov, A.F. (1966): An algorithm for generating optimal numerical grids. Trudy MIAN USSR **24**, 147–151 (Russian)

Slater, J.W., Liou, M.S., Hindman, R.G. (1995): Approach for dynamic grids. AIAA J. **33**(1), 63–68

Sloan, S.W., Houlsby, G.T. (1984): An implementation of Watson's algorithm for computing 2D Delaunay triangulations. Adv. Eng. Softw. **6**(4), 192–197

Smith, R.E. (1981): Two-boundary grid generation for the solution of the three-dimensional Navier–Stokes equations. NASA TM-83123

Smith, R.E. (1982): Algebraic grid generation. In Thompson, J.F. (ed.), *Numerical Grid Generation*, pp. 137–170. North-Holland, Amsterdam

Smith, R.E., Eriksson, L.E. (1987): Algebraic grid generation. Comput. Methods Appl. Mech. Eng. **64**, 285–300

Sokolnikoff, I.S. (1964): *Tensor Analysis*. Wiley, New York

Soni, B.K. (1985): Two and three dimensional grid generation for internal flow applications of computational fluid dynamics. AIAA Paper 85-1526

Soni, B.K. (1991): Grid optimization: a mixed approach. In Arcilla, A.S., Hauser, J., Eiseman, P.R., Thompson, J.F. (eds.), *Numerical Grid Generation in Computational Fluid Dynamics and Related Fields*, p. 617. North-Holland, Amsterdam

Soni, B.K., Yang, J.-C. (1992): General purpose adaptive grid generation system. AIAA Paper 92-0664

Soni, B.K., Huddleston, D.K., Arabshahi, A., Yu, B. (1993): A study of CFD algorithms applied to complete aircraft configurations. AIAA Paper 93-0784

Sorenson, R.L. (1986): Three-dimensional elliptic grid generation about fighter aircraft for zonal finite-difference computations. AIAA Paper 86-0429

Sparis, P.D. (1985): A method for generating boundary-orthogonal curvilinear coordinate systems using the biharmonic equation. J. Comput. Phys. **61**(3), 445–462

Sparis, P.N., Karkanis, A. (1992): Boundary-orthogonal biharmonic grids via preconditioned gradient methods. AIAA J. **30**(3), 671–678

Spekreijse, S.P. (1995): Elliptic grid generation based on Laplace equations and algebraic transformations. J. Comput. Phys. **118**, 38–61

Spekreijse, S.P., Nijhuis, G.H., Boerstoel, J.W. (1995): Elliptic surface grid generation on minimal and parametrized surfaces. In *Proceedings of the Surface Modeling, Grid Generation and Related Issues in Computational Fluid Dynamics Workshop*, p. 617. NASA Conference Publication **3291**, NASA Lewis Research Center, Cleveland, OH

Starius, G. (1977): Constructing orthogonal curvilinear meshes by solving initial value problems. Numer. Math. **28**, 25–48

Steger, J.L. (1991): Grid generation with hyperbolic partial differential equations for application to complex configurations. In Arcilla, A.S., Hauser, J., Eiseman, P.R., Thompson, J.F. (eds.), *Numerical Grid Generation in Computational Fluid Dynamics and Related Fields*, pp. 871–886. North-Holland, Amsterdam

Steger, J.L., Chaussee, D.S. (1980): Generation of body fitted coordinates using hyperbolic differential equations. SIAM. J. Sci. Stat. Comput. **1**(4), 431–437

Steger, J.L., Rizk, Y.M. (1985): Generation of three-dimensional body-fitted coordinates using hyperbolic partial differential equations. NASA, TM 86753, June

Steger, J.L., Sorenson, R.L. (1979): Automatic mesh-point clustering near a boundary in grid generation with elliptic partial differential equations. J. Comput. Phys. **33**, 405–410

Steinberg, S., Roache, P.J. (1986): Variational grid generation. Numer. Methods Partial. Differ. Equ. **2**, 71–96

Steinberg, S., Roache, P.J. (1990): Anomalies in grid generation on curves. J. Comput. Phys. **91**, 255–277

Steinberg, S., Roache, P.J. (1992): Variational curve and surface grid generation. J. Comput. Phys. **100**, 163–178

Steinbrenner, J.P., Chawner, J.R., Fouts, C.L. (1990): Multiple block grid generation in the interactive environment. AIAA Paper 90-1602

Steinitz, E. (1922): Polyeder and Raumeintailungen. Enzykl. Mathematischen Wiss. **3**, 163

Stewart, M.E.M. (1992): Domain decomposition algorithm applied to multi-element airfoil grids. AIAA J. **30**(6), 1457

Sweby, P.K., Yee, H.C. (1990): On the dynamics of some grid adaptation schemes. In Weatherill, N.P., Eiseman, P.R., Hauser, J., Thompson, J.F. (eds.), *Numerical Grid Generation in Computational Field Simulation and Related Fields*, p. 467. Pineridge, Swansea

Tai, C.H., Chiang, D.C., Su, Y.P. (1996): Three-dimensional hyperbolic grid generation with inherent dissipation and Laplacian smoothing. AIAA J. **34**(9), 1801–1806

Takagi, T., Miki, K., Chen, B.C.J., Sha, W.T. (1985): Numerical generation of boundary-fitted curvilinear coordinate systems for arbitrarily curved surfaces. J. Comput. Phys. **58**, 69–79

Takahashi, H., Shimizu, H. (1991): A general purpose automatic mesh generation using shape recognition technique. Comput. Eng. **1**, 519–526

Tam, T.K.H., Armstrong, C.G. (1991): 2D finite element mesh generation by medial axis subdivision. Adv. Eng. Softw. Workst. **13**, 313–344

Tamamidis, P., Assanis, D.N. (1991): Generation of orthogonal grids with control of spacing. J. Comput. Phys. **94**, 437–453

Tanemura, M., Ogawa, T., Ogita, N. (1983): A new algorithm for three-dimensional Voronoi tessellation. J. Comput. Phys. **51**, 191–207

Thacker, W.C. (1980): A brief review of techniques for generating irregular computational grids. Int. J. Numer. Methods Eng. **15**(9), 1335–1341

Theodoropoulos, T., Bergeles, G.C. (1989): A Laplacian equation method for numerical generation of boundary-fitted 3d orthogonal grids. J. Comput. Phys. **82**, 269–288

Thomas, P.D. (1982): Composite three-dimensional grids generated by elliptic systems. AIAA J. **20**(9), 1195–1202

Thomas, P.D., Middlecoff, J.F. (1980): Direct control of the grid point distribution in meshes generated by elliptic equations. AIAA J. **18**(6), 652–656

Thomas, M.E., Bache, G.E., Blumenthal, R.F. (1990): Structured grid generation with PATRAN. AIAA Paper 90-2244

Thomas, P.D., Vinokur, M., Bastianon, R.A., Conti, R.J. (1972): Numerical solution for three-dimensional inviscid supersonic flow. AIAA J. **10**(7), 887–894

Thompson, J.F. (1984a): Grid generation techniques in computational fluid dynamics. AIAA J. **22**(11), 1505–1523

Thompson, J.F. (1984b): A survey of dynamically-adaptive grids in the numerical solution of partial differential equations. AIAA Paper 84-1606

Thompson, J.F. (1985): A survey of dynamically-adaptive grids in the numerical solution of partial differential equations. Appl. Numer. Math. **1**, 3–27

Thompson, J.F. (1987a): A composite grid generation code for general 3-d regions. AIAA Paper 87-0275

Thompson, J.F. (1987b): A general three-dimensional elliptic grid generation system on a composite block structure. Comput. Methods Appl. Mech. Eng. **64**, 377–411

Thompson, J.F. (1996): A reflection on grid generation in the 90s: trends, needs influences. In Soni, B.K., Thompson, J.F., Hauser, J., Eiseman, P.R. (eds.), *Numerical Grid Generation in CFD*. Mississippi State University **1**, pp. 1029–1110

Thompson, J.F., Weatherill, N.P. (1993): Aspects of numerical grid generation: current science and art. AIAA Paper 93-3539

Thompson, J.F., Thames, F.C., Mastin, C.W. (1974): Automatic numerical generation of body-fitted curvilinear coordinate system for field containing any number of arbitrary two-dimensional bodies. J. Comput. Phys. **15**, 299–319

Thompson, J.F., Warsi, Z.U.A., Mastin, C.W. (1982): Boundary-fitted coordinate systems for numerical solution of partial differential equations—a review. J. Comput. Phys. **47**, 1–108

Thompson, J.F., Warsi, Z.U.A., Mastin, C.W. (1985): *Numerical Grid Generation. Foundations and Applications*. North-Holland, Amsterdam

Thompson, J.F., Soni, B.K., Weatherill, N.P. (eds.) (1999): *Handbook of Grid Generation*. CRC Press, Boca Raton

Tolstykh, A.I. (1973): A method for the numerical solution of Navier–Stokes equations of a compressible gas flow for a wide range of Reynolds number. Dokl. Akad. Nauk SSSR **210**, 48–51 (Russian)

Tolstykh, A.I. (1978): Concentration of grid points in the process of solving and using high-accuracy schemes for the numerical investigation of viscous gas flows. Comput. Math. Math. Phys. **18**, 139–153 (Russian). [English transl.: USSR Comput. Math. Math. Phys. **18**, 134–138 (1979)]

Tu, Y., Thompson, J.F. (1991): Three-dimensional solution-adaptive grid generation on composite configurations. AIAA J. **29**, 2025–2026

Venkatakrishan, V. (1996): Perspective on unstructured grid flow solvers. AIAA J. **34**(3), 533–547

Vinokur, M. (1974): Conservation equation of gasdynamics in curvilinear coordinate systems. J. Comput. Phys. **14**(2), 105–125

Vinokur, M. (1983): On one-dimensional stretching functions for finite-difference calculations. J. Comput. Phys. **50**, 215–234

Visbal, M., Knight, D. (1982): Generation of orthogonal and nearly orthogonal coordinates with grid control near boundaries. AIAA J. **20**(3), 305–306

Viviand, H. (1974): Conservative forms of gas dynamic equations. Rech. Aeros. **1**, 65–68

Vogel, A.A. (1990): Automated domain decomposition for computational fluid dynamics. Comput. Fluids **18**(4), 329–346

Voronoi, G.F. (1908): Nouvelles applications des parameters countinus a la theorie des formes quadratiques. Deuxieme Memoire: Recherches sur la parallelloedres primitifs. J. Reine Angew. Math. **134**, 198–287

Vulanovic, R. (1984): Mesh construction for numerical solution of a type of singular perturbation problems. In *Numer. Methods Approx. Theory*, pp. 137–142.

Wan, D., Turek, S. (2006a): Direct numerical simulation of particulate flow via multigrid FEM techniques and the fictitious boundary method. Int. J. Numer. Methods Fluids **51**(5), 531–566. doi:10.1002/fld.1129

Wan, D., Turek, S. (2006b): Fictitious boundary and moving mesh methods for the numerical simulation of rigid particulate flows. J. Comput. Phys. **22**, 28–56. http://dx.doi.org/10.1016/j.jcp.2006.06.002

Wan, D., Turek, S. (2007): An efficient multigrid-FEM method for the simulation of solid–liquid two phase flows. J. Comput. Appl. Math. **203**(2), 561–580

Warsi, Z.U.A. (1981): Tensors and differential geometry applied to analytic and numerical coordinate generation. MSSU-EIRS-81-1, Aerospace Engineering, Mississippi State University

Warsi, Z.U.A. (1982): Basic differential models for coordinate generation. In Thompson, J.F. (ed.), *Numerical Grid Generation*, pp. 41–78. North-Holland, Amsterdam

Warsi, Z.U.A. (1986): Numerical grid generation in arbitrary surfaces through a second-order differential-geometric model. J. Comput. Phys. **64**, 82–96

Warsi, Z.U.A. (1990): Theoretical foundation of the equations for the generation of surface coordinates. AIAA J. **28**(6), 1140–1142

Warsi, Z.U.A., Thompson, J.F. (1990): Application of variational methods in the fixed and adaptive grid generation. Comput. Math. Appl. **19**(8/9), 31–41

Wathen, A.J. (1990): Optimal moving grids for time-dependent partial differential equations. J. Comput. Phys. **101**, 51–54

Watson, D. (1981): Computing the n-dimensional Delaunay tessellation with application to Voronoi polytopes. Comput. J. **24**(2), 167–172

Weatherill, N.P. (1988): A method for generating irregular computational grids in multiply connected planar domains. Int. J. Numer. Methods Fluids **8**, 181–197

Weatherill, N.P. (1990): The integrity of geometrical boundaries in the two-dimensional Delaunay triangulation. Commun. Appl. Numer. Methods **6**, 101–109

Weatherill, N.P., Forsey, C.R. (1984): Grid generation and flow calculations for complex aircraft geometries using a multi-block scheme. AIAA Paper 84-1665

Weatherill, N.P., Hassan, O. (1994): Efficient three-dimensional Delaunay triangulation with automatic point creation and imposed boundary constraints. Int. J. Numer. Methods Eng. **37**, 2005–2039

Weatherill, N.P., Soni, B.K. (1991): Grid adaptation and refinement in structured and unstructured algorithms. In Arcilla, A.S., Hauser, J., Eiseman, P.R., Thompson, J.F. (eds.), *Numerical Grid Generation in Computational Fluid Dynamics and Related Fields*, pp. 143–158. North-Holland, Amsterdam

Weatherill, N.P., Marchant, M.F., Hassan, O., Marcum, D.L. (1993): Adaptive inviscid flow solutions for aerospace geometries on efficiently generated unstructured tetrahedral meshes. AIAA Paper 93-3390

Weilmuenster, K.J., Smith, R.E., Everton, E.L. (1991): Gridding strategies and associated results for winged entry vechicles. In Arcilla, A.S., Hauser, J., Eiseman, P.R., Thompson, J.F. (eds.), *Numerical Grid Generation in Computational Fluid Dynamics and Related Fields*, pp. 217–228. North-Holland, Amsterdam

White, A.B. (1979): On selecting of equidistributing meshes for two-point boundary-value problems. SIAM J. Numer. Anal. **16**(3), 472–502

White, A.B. (1982): On the numerical solution of initial boundary-value problems in one space dimension. SIAM J. Numer. Anal. **19**, 683–697

White, A.B. (1990): Elliptic grid generation with orthogonality and spacing control on an arbitrary number of boundaries. AIAA Paper 90-1568

Widhopf, G.D., Boyd, C.N., Shiba, J.K., Than, P.T., Oliphant, P.H., Huang, S.-C., Swedberg, G.D., Visich, M. (1990): RAMBO-4G: An interactive general multiblock grid generation and graphics package for complex multibody CFD applications. AIAA Paper 90-0328

Winkler, K.H., Mihalas, D., Norman, M.L. (1985): Adaptive grid methods with asymmetric time-filtering. Comput. Phys. Commun. **36**, 121–140

Winslow, A.M. (1967): Equipotential zoning of two-dimensional meshes. J. Comput. Phys. **1**, 149–172

Winslow, A.M. (1981): Adaptive mesh zoning by the equipotential method. UCID-19062, Lawrence Livermore National Laboratories

Wordenweber, B. (1981): Automatic mesh generation of 2- and 3-dimensional curvilinear manifolds. Ph.D. Dissertation (available as Computer Laboratory Report N18), University of Cambridge

Wordenweber, B. (1983): Finite-element analysis from geometric models. Int. J. Comput. Math. Elect. Eng. **2**, 23–33

Wulf, A., Akrag, V. (1995): Tuned grid generation with ICEM CFD. In *Proceedings of the Surface Modeling, Grid Generation, and Related Issues in Computational Fluid Dynamics Workshop*, p. 477. NASA, Lewis Research Center, Cleveland OH, May

Yanenko, N.N. (1971): *The Method of Fractional Steps.* The Solutions of Problems of Mathematical Physics in Several Variables. Springer, Berlin

Yanenko, N.N., Danaev, N.T., Liseikin, V.D. (1977): A variational method for grid generation. Chisl. Metody Mekhan. Sploshnoi Sredy **8**(4), 157–163 (Russian)

Yerry, M.A., Shephard, M.S. (1985): Automatic three-dimensional mesh generation for three-dimensional solids. Comput. Struct. **20**, 31–39

Yerry, M.A., Shephard, M.S. (1990): Automatic three-dimensional mesh generation by the modified-octree technique. Int. J. Numer. Methods Eng. **20**, 1965–1990

Zegeling, P.A. (1993): Moving-grid methods for time-dependent partial differential equations. CWI Tract 94, Centrum voor Wiskund en Informatica, Amsterdam

Zhou, J.M., Shao, K.-R., Zhou, K.-D., Zhan, Q.-H. (1990): Computing constrained triangulation and Delaunay triangulation: a new algorithm. IEEE Trans. Magn. **26**(2), 694–697

Index

V.D. Liseikin, *Grid Generation Methods*, Scientific Computation,
DOI 10.1007/978-90-481-2912-6,

FSC
www.fsc.org
MIX
Papier | Fördert
gute Waldnutzung
FSC® C083411

Kolibri 360 GmbH
Ferdinand-Jühlke-Straße 7
99095 Erfurt, Deutschland
produktsicherheit@kolibri360.de